CONTAMINATION CONTROL AND DEFECT REDUCTION IN SEMI-CONDUCTOR MANUFACTURING I

Edited by

Dennis N. Schmidt
U.S. Filter—Membralox
Warrendale, Pennsylvania

Assistant Editors

David Reedy
Alex Schwarz

ELECTRONICS AND DIELECTRIC SCIENCE AND TECHNOLOGY DIVISIONS

Proceedings Volume 92-21

THE ELECTROCHEMICAL SOCIETY, INC., 10 South Main St., Pennington, NJ 08534-2896

PREFACE

This Electrochemical Society Softbound Symposium Volume contains papers presented at the Contamination Control and Defect Reduction in Semiconductor Manufacturing I Symposium held October 12-14, 1992 in Toronto, Ontario, Canada. This symposium was co-sponsored by the Electronics and the Dielectrics Science and Technology Divisions of The Electrochemical Society, Inc.

I would especially like to thank the invited speakers of the Symposium for their participation. Invited speakers are recognized experts in the field whose work has earned them international respect through publication in archival journals. The participation of these authorities adds credibility to any scientific forum, and their invited talks lay a strong foundation upon which subsequent contributed papers can build during a session. I deeply thank the fourteen invited speakers of the Contamination Control and Defect Reduction Symposium for lending their considerable prestige to the meeting and to these proceedings.

Semiconductor manufacturing has evolved into a mature industry, and, in so doing, has changed from being primarily science driven, with new process technologies and new chemistries being key, to technology driven, in which perfecting established technologies has become critical to success. At this time (mid 1992) a semiconductor organization's ability to survive economically is less dependent upon its ability to come up with new scientific discoveries (the scientific foundation of semiconductor manufacturing being well established during the 1970's and 1980's) and more dependent on its manufacturing prowess. These bottom line realities have made working in Contamination Control and Defect Reduction, heretofore often considered to be a discipline in semiconductor manufacturing having little prestige, a respected profession having equal status to that of process engineer or device designer.

My primary purpose in producing this symposium and volume was to provide a legitimate scientific forum for the Contamination Control Community free of commercial considerations or connections. Certainly, the goal was not to produce a contamination control book to be read only by contamination control people. My two objectives were:

1) To bring contamination control topics into the scope of The Electrochemical Society and therefore broaden contamination control technology's exposure to the scientific and academic community that The Electrochemical Society traditionally serves. I hope that this will further establish contamination control technology as a legitimate scientific discipline.

2) Teach process engineers and other semiconductor professionals the basics of
   contamination control so that they can use this information within their
   organizations to drive yields and therefore profitability. If they succeed in
   doing this they will insure their company's survival in the difficult global
   semiconductor marketplace.

It is my hope that the papers contained in this volume will accomplish these
goals.

The production of a volume such as this requires an enormous amount of
work, and certainly, no one individual could accomplish such a formidable task
without considerable help. Therefore, I would like to sincerely thank my two
symposium co-chairman, David Reedy and Alex Schwarz and the headquarters
staff of The Electrochemical Society, Inc., for the invaluable assistance which
they provided me.

Dennis N. Schmidt, Editor
June 1992

TABLE OF CONTENTS

* Denotes an Invited Paper

* Denotes an Invited Paper

FACTS ABOUT THE ELECTROCHEMICAL SOCIETY, INC.

The Electrochemical Society, Inc., is a nonprofit, scientific, educational, international organization founded for the advancement of the theory and practice of electrochemistry, electrothermics, electronics, and allied subjects. The Society was founded in Philadelphia in 1902 and incorporated in 1930. There are currently over 5000 scientists and engineers from more than 40 countries who hold individual membership; the Society is also supported by more than 100 corporations through Patron and Sustaining Memberships.

The technical activities of the Society are carried on by Divisions and Groups. Local Sections of the Society have been organized in a number of cities and regions.

Major international meetings of the Society are held in the Spring and Fall of each year. At these meetings, the Divisions and Groups hold general sessions and sponsor symposia on specialized subjects.

The Society has an active publications program which includes the following.

JOURNAL OF THE ELECTROCHEMICAL SOCIETY - The JOURNAL is a monthly publication containing technical papers covering basic research and technology of interest in the areas of concern to the Society. Papers submitted for publication are subjected to careful evaluation and review by authorities in the field before acceptance, and high standards are maintained for the technical content of the JOURNAL.

EXTENDED ABSTRACTS - Extended abstracts of the technical papers presented at the Spring and Fall Meetings of the Society are published in serialized softbound volumes.

PROCEEDINGS VOLUMES - Papers presented in symposia at Society and Topical Meetings are published from time to time as serialized softbound Proceedings Volumes. These provide up-to-date views of specialized topics and frequently offer comprehensive treatment of rapidly developing areas.

MONOGRAPH VOLUMES - The Society has, for a number of years, sponsored the publication of hardbound Monograph Volumes, which provide authoritative accounts of specific topics in electrochemistry, solid state science and related disciplines.

CONDENSATION INDUCED PARTICLE FORMATION DURING VACUUM PUMP DOWN

Yan Ye, Benjamin Y. H. Liu and David Y.H. Pui
Particle Technology Laboratory
Mechanical Engineering Department
University of Minnesota
Minneapolis, MN 55455

ABSTRACT

When a vacuum system containing particle free, cleanroom air is pumped down from atmospheric pressure to produce a vacuum, the gas in the chamber expands and cools to cause water vapor condensation and droplet formation. Experiments have been performed to show that a stable aerosol of submicrometer residue particles on the order of 0.2 μm are formed when the water droplets have subsequently re-evaporated. The residue particles are spherical in shape and appear to contain mainly sulfuric acid. The theoretical and experimental studies conducted strongly suggest that the main steps involved in the residue particle formation process are: (1) simultaneous absorption of SO_2 and H_2O_2 in the ppb range from the air into the water droplets during pump down, (2) concentration of SO_2 and H_2O_2 in the liquid when the droplets re-evaporate, and (3) oxidation of SO_2 by H_2O_2 in the concentrated liquid solution during the final period of evaporation to produce stable residue particles of H_2SO_4. The mathematical model developed gives results that are consistent with the experimental observations.

INTRODUCTION

Particulate contamination in semiconductor processing equipment has received a great deal of attention in recent years because of their effect on product yield. Particles formed in such equipment can deposit on the wafer to cause failure of the product. A very important source of particulate contamination is the particles formed when the processing chamber is evacuated. Although the phenomenon has been studied by several investigators, the hypotheses put forth are not entirely consistent and the observed phenomenon are not fully explainable by the available theories. In order to better understand the mechanisms of particle formation, and the importance of such particles on product degradation and yield, further theoretical and experimental studies are needed.

PREVIOUS STUDIES

Particle generation during pump down and venting has been studied by different investigators. One of the first such reported studies is that by Hoh (1984). He placed witness wafers in a 75 liter chamber, and measured particles deposited on the wafers after pump down and venting. He found that the number of particles deposited on a wafer strongly depended on the pumping and venting speeds. If the chamber was first evacuated and then vented at high speeds, the number of particles on the wafer was too large to be counted with his instrument. If pumping and venting speeds were low, very few particles were detected. He attributed the observed particle deposition to the traditional concept of particle reentrainment from surfaces.

Chen at el. (1989) studied particle generation in a vacuum load-lock during pump down by measuring the time-dependent particle counts with a High Yield Technology PM-100 in-situ particle counter. The particle counter was installed between a chamber and the inlet to a roughing pump, and the chamber was filled with class 10-100 cleanroom air. They found that particle counts during pump down were much higher than those observed in the chamber before pump down began--generally higher the pumping speed, higher the particle count. The particle counts observed were attributed to vapor condensation on particles smaller than 0.005µm induced by turbulence. They defined an instant Reynolds number with gas velocity and tube diameter in the exhaust line, and used the Re to predict the turbulence, and thus condensation. A critical Re was proposed to predict the occurrence of large number of particles. In their followed-on work (Chen, et al. 1990), they indicated the possibility of homogeneous nucleation of water vapor triggered by turbulence. However, no explanation was given as to how turbulence could have triggered nucleation and condensation.

Zhao (1990) in his Ph.D. thesis investigated particle formation during pump down by measuring the time-dependent gas temperature during pump down and the total number of particles generated. Using a thermocouple of 0.001" diameter, he found that the gas temperature in the chamber decreased due to adiabatic expansion, and then increased due to heat transfer from the chamber walls. A numerical model was developed to calculate the change in temperature, pressure, and water vapor supersaturation to determine the onset of condensation and particle formation. He also measured the number of residue particles generated in a 260 cm^3 chamber that had been filled with particle free air at various relative humidities and then pumped down at different pumping speeds. He obtained empirical relationships between the number concentration of particles formed, the initial relative humidity and pumping speed. His study demonstrated the formation of residue particles as a result of water vapor condensation due to homogeneous nucleation induced by gas expansion. A mathematical model was then developed to predict the on-set of condensation for chamber of different sizes, different initial relative humidities and pumping speeds.

The particle formation mechanism during pump down was also studied by Wu et al. (1990). They developed a numerical model to calculate water vapor concentration, and the number and mass concentration of condensation nuclei and condensed droplets. They showed that a large number of particles can be formed and grow to sizes larger than 10 µm if the evacuation process can be considered as an adiabatic expansion process. The temperature and number of particles generated were also measured with a thermocouple and an in-situ particle counter in their work. Their experimental results on droplet formation were similar to those given by Zhao (1990) and Chen et al. (1989).

Although many of these studies have shown the formation of water droplets during vacuum pump down, it is not clear why the subsequent evaporation of the water droplets would lead to the formation of more permanent and stable residue particles in the vacuum chamber. As reported initially by Szymanski et al. (1988) and studied in more detail by Zhao (1990), when the water droplets formed during pump down were given enough time to evaporate, small residue particles are found to remain in the system. These residue particles have been found to be quite stable in their physical characteristics and do not completely evaporate even upon heating to temperatures as high as 180 ºC.

The purpose of this study is to perform theoretical and experimental studies to elucidate the mechanisms and processes leading to the formation of such small residue particles in a vacuum system following water vapor condensation and re-evaporation during vacuum pump down. First, experiments were performed to firmly establish that the residue particles were indeed produced as a result of nucleation and condensation of water vapor. Then, the physical and chemical characteristics of these residue particles were measured to provide data for the formulation of a more comprehensive theory on residue particle formation during pump down. The theory was then further developed to provide quantitative results for comparison with the experimental results.

NUCLEATION AND DROPLET GROWTH

In order to identify the particle formation mechanisms, experiments must be performed under well controlled conditions. A special vacuum system was developed to minimize the effect due to pre-existing airborne particles and particles reentrainment from the chamber walls. Figure 1

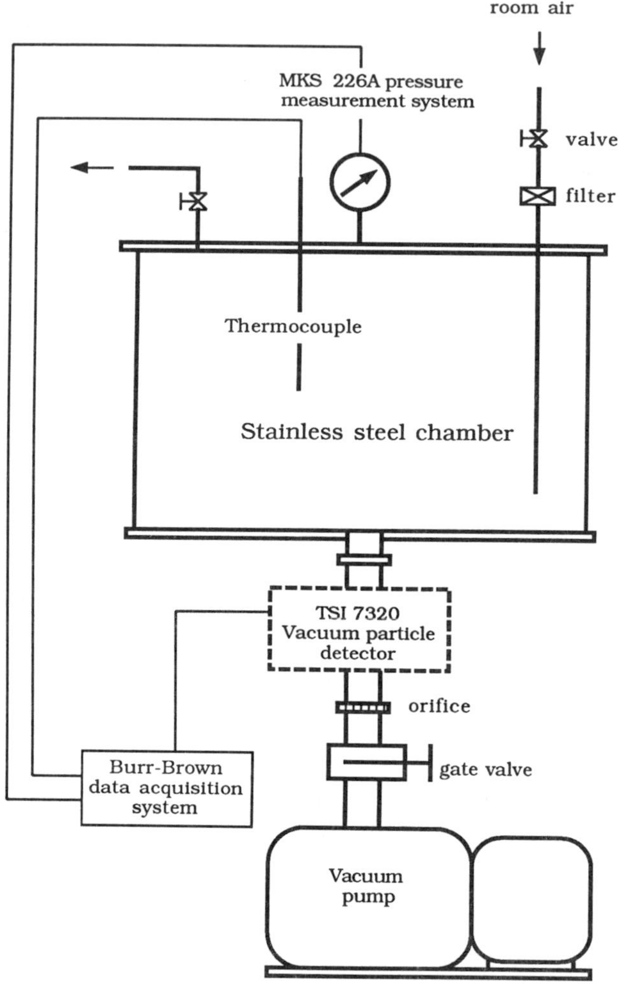

Figure 1   Schematic diagram of vacuum system for particle formation study

shows the schematic diagram of the system used. The system consists of a 50 liter stainless steel chamber of a tight construction, a rotary vane pump, and instruments for measuring the temperature and pressure of the gas in the chamber. The particle measurement instruments are also connected to the system based on the measuring requirements. In the system shown in Figure 1, an in-situ vacuum particle counter is also included. The chamber is separated by a gate valve from the pump and the pumping speed is controlled by the size of the orifice located at the chamber exit. Before pump down, gas is filled into the chamber through a Wafergard filter. Either room air or dry nitrogen can be introduced into the chamber through the filter. Since the Wafergard filter is known to have high particle removal efficiency, the initial particle count in the chamber is very low.

Particle formation during pump down was observed with the in-situ particle counter installed at the exit of chamber. The in-situ particle counter is connected to a Burr-Brown data acquisition system, and the sampling time interval of the particle counter was set to 0.05 sec. Figure 2 shows the measurement of time dependent particle counts and pressure during evacuation of the chamber at a pumping speed of 290 lpm. When the chamber was filled with dry nitrogen, particle counts measured by the in-situ particle counter was very low as shown in figure 2 (a). When the chamber was filled with nitrogen with 12% relative humidity at 25°C, the variation of particle counts with time is shown in figure 2 (b). Particle counts was very low in the first several seconds of the pump down. A large number of particles were formed and counted by the particle counter, when the chamber was evacuated to a pressure of about 400 torr. As the evacuation proceeded, the particles counts reached a maximum value of about 150 particle in 0.05 sec, and then decreased with further reduction in pressure. Figure 2(c) shows the particle generation when the chamber was filled with nitrogen at 18% relative humidity and 25°C. A large number of particles began to appear at pressure close to that shown in Figure 2 (b). However, the number of particles generated were so high that the particle counter became saturated in the pressure range between 400 to 100 torr; after which, the particle count resumed normal operation showing a general decrease in count as the pressure decreased. It should be noted that the nominal lower detection limit of the in-situ particle counter is 0.5 μm. Therefore, presence of particle count only indicated the presence of particles larger than 0.5 μm.

Since very few particles were observed during evacuation of a chamber with dry nitrogen, it was evident that particle generation in the low vacuum range (>20 Torr) did not result from particle re-entrainment from the chamber surfaces. The effect of water vapor concentration on the particle counts indicated that particle formation was related to nucleation and condensation of water vapor. The occurrence of water vapor nucleation during pump down can be further confirmed with saturation ratio calculations.

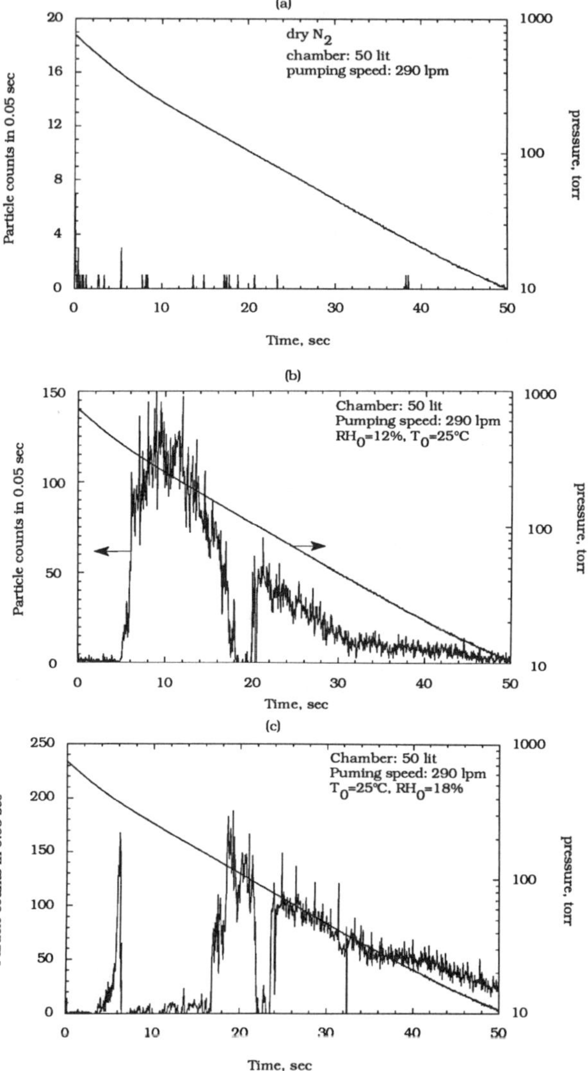

Figure 2. Change of particle counts with time during pump down with various relative humidity

Since gas in the chamber undergoes expansion during pump down, the gas temperature will decrease as the pressure decreases. Because the stainless steel chamber walls have a high heat capacity, the chamber wall would remain essentially at a constant temperature as the gas expands and its temperature decreases. The temperature difference between the chamber wall and the gas will cause heat to be transferred from the wall to the gas in the chamber, leading to the reheating of the gas back to its initial temperature after a short time. Figure 3 (a) shows some typical temperature and pressure vs. time curves during pump down for a chamber filled with dry nitrogen and for a pumping speed of 430 lpm. The temperature was measured with a 0.001" type-E thermocouple, and the pressure was monitored with a MKS variable capacitance manometer.

The water vapor saturation ratio, defined as the ratio of the partial pressure of water vapor and the saturation pressure, varies with gas temperature and pressure during pump down. Figure 3 (b) shows the variation of saturation ratio with time. The calculation is made for the case of the temperature and pressure given in Figure 3(a), and does not take into account depletion of water vapor in the gas phase as the water vapor condenses. The saturation ratio is found to increase in the first several seconds, since the decrease in saturation pressure resulting from the decrease in gas temperature is much more rapid than the decrease in the partial pressure caused by expansion. For the same pumping speed, the saturation ratio depends on the initial water vapor concentration (or relative humidity at the initial temperature). The higher the initial water vapor concentration the higher is the saturation ratio achieved. For the pumping conditions shown in figure 4, saturation ratio as high as 14 has been calculated for an initial relative humidity of 20 %. For an initial relative humidity of 50% the saturation ratio is found to be as high as 30 if no nucleation and condensation are assumed to take place.

During the actual pump down process, homogeneous nucleation will increase rapidly once a certain critical saturation ratio is achieved. Several alternative theoretical expressions can be used to estimate the homogeneous nucleation rate (Springer, 1978). The following equation is generally used to predict the nucleation rate.

$$I = C^* Z\, n_e\,(g) \tag{1}$$

where C^* is the rate of arrival of monomer at the critical cluster surface, Z is the Zeldovich factor accounting for the difference between the steady state concentration and the equilibrium concentration of the clusters, and $n_e\,(g^*)$ is the equilibrium concentration of cluster of the critical size g^* (Girshick et al, 1990),

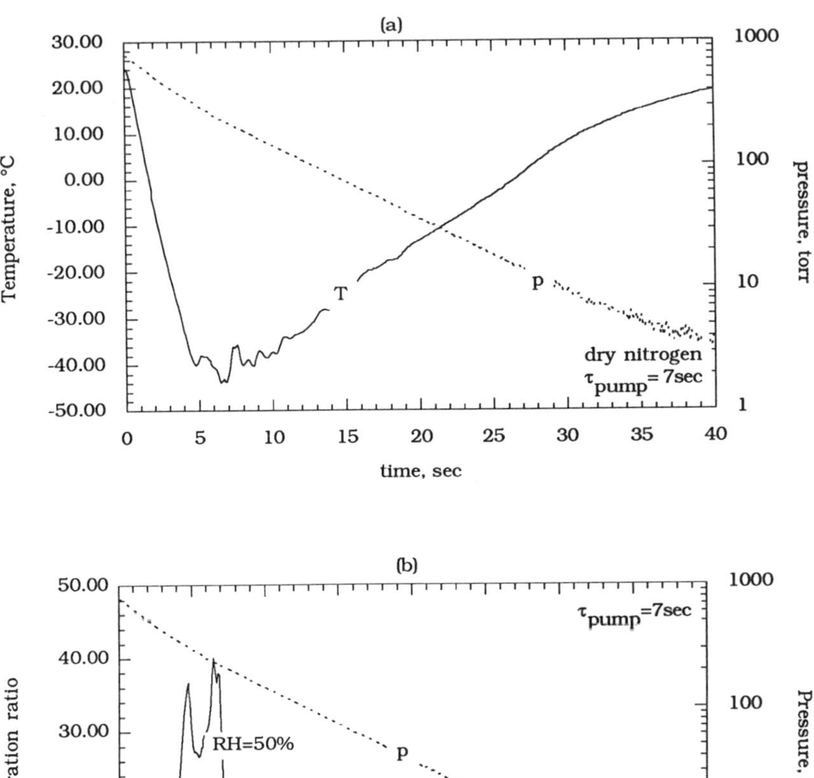

Figure 3. Change of (a) temperature and (b) saturation ration during pump down

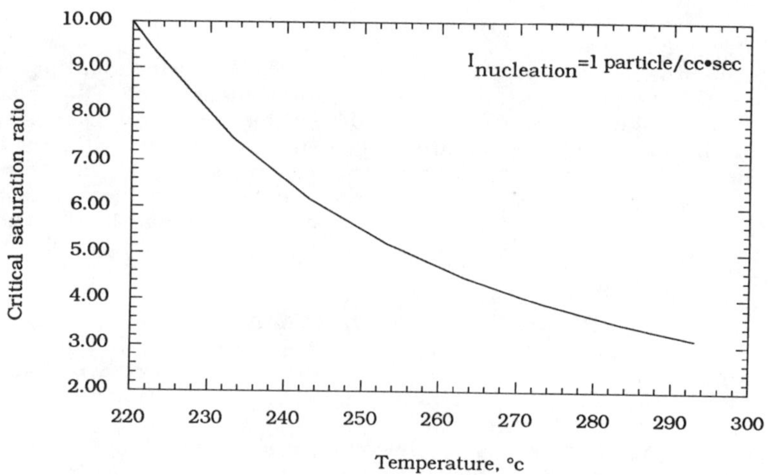

Figure 4. Critical saturation ratio for homogeneous nucleation

The nucleation rate increases as the water vapor saturation ratio is increased. If the onset of droplet formation is defined as the point at which the nucleation rate is equal to 1 per cm^3 per sec, the critical saturation ratio for droplet formation can be calculated by solving Eq. (1). Figure 4 shows the critical saturation ratio for droplet formation at various temperatures. For temperature between -20 to -40 °C, the critical saturation ratio range from 5 to 8. As shown in figure 3 (b), possible saturation ratios during pressure reduction can be much higher then the critical saturation ratio. It is evident that water vapor will condense to form droplets by the process of homogeneous nucleation during typical conditions encountered in vacuum pump down.

After the droplets are formed by homogeneous nucleation, the water droplets will continue to grow by water vapor condensation. The detailed numerical calculation (Ye, 1992) shows that water vapor will be consumed once a large number of droplets are formed, and the droplets will grow to a large size. This reduces the water vapor concentration in the gas, thereby reducing the maximum saturation ratio that can be achieved to a level only slightly higher than the critical saturation ratio, instead of the values shown in Figure 3(b) based on the assumption of no water vapor depletion.

RESIDUE PARTICLE FORMATION

Droplet formation by nucleation of water vapor has been observed since C. R. T. Wilson did his famous cloud chamber experiments near the end of the last century (Friedlander, 1979). The work described in the last section confirmed that the same phenomenon as observed by Wilson also took place in a vacuum chamber during pump down. Although the phenomenon of homogeneous nucleation is well known for droplet formation, very little is known about the mechanisms and steps involved in the formation of non-volatile trace chemical species in the droplets that later provided the material for the residue particles. It is important to understand the physics and chemistry of the processes involved so that the phenomenon of residue particle formation can be better understood.

Several experiments were performed to study the physical and chemical properties of the residue particles formed during vacuum pump down. The same system as that shown in Figure 1 was used in these experiments. The chamber was first filled with particle-free air from the ambient and then pumped down to a final pressure of 200 to 300 Torr. The chamber was then vented and re-pressurized with particle-free air to atmospheric pressure. After a short time delay (on the order of a minute) to allow the water droplets to re-evaporate, the residue particles in the chamber were then measured with a TSI 3932 Differential Mobility Particle Sizer (DMPS) or collected on a substrate with an impactor for examination in an electron microscope.

If the only species involved in the condensation and droplet formation process is water, then the droplets would completely re-evaporate upon reheating of the gas. However, it was found that this is not the case. Figure 5 shows a typical measurement result for the size distribution of the residue particles formed following nucleation, condensation and evaporation in a typical pump down process. The mean size of the residue particles was found to be about 0.19 μm and the size distribution of the residue particles was close to log-normal, with a geometric standard deviation of about 1.2. The figure also shows the particle size distribution measured at different times, i.e. immediately after pump down, 1 hour after pump down, and 24 hours after pump down. It is seen that over these time periods, the measured particle size distribution did not change significantly. The reduction in particle concentration can be attributed to particle loss on the chamber walls by diffusion and dilution resulting from samples being withdrawn from the vacuum chamber by the size distribution measuring instruments. The residue particles, therefore, must be non-volatile so that they can exist in the chamber over such an extended period of time.

The nature of the residue particles was further investigated by examining the particle deposit collected by the inertial impactor. Figure 6 shows a photograph of the collected particles as seen through a Zeiss Axioskope microscope at 400x magnification. The particles are seen to be spheres with very smooth surfaces. The shape suggests that they are small liquid droplets, some of which have coalesced to form larger droplets during the collection process. It was also found that by exposing the collected liquid droplets to high concentration of NH_3 the droplets would be transformed to solid particles.

The particles were also collected on a beryllium substrate by the single nozzle impactor, and analyzed with an Energy Dispersive X-Ray (EDX) spectrometer installed in a high resolution electron microscope. Figure 7 shows a typical EDX spectrum showing the existence of sulfur, oxygen, and carbon in the droplets. The existence of the sulfur peak in the collected droplets, and their reaction with ammonia strongly suggest that a major component of the droplets is sulfuric acid. Since sulfuric acid has a very low vapor pressure at room temperatures, it can exist as stable droplets even in the submicrometer size range, as we have observed.

In order to further confirm the existence of sulfuric acid in the residue particle, the bulk density, and hygroscopic property were also measured. The bulk density and hygroscopic properties of the residue particles are very close to those for sulfuric acid. More detailed information about these measurements is given by Ye (1992).

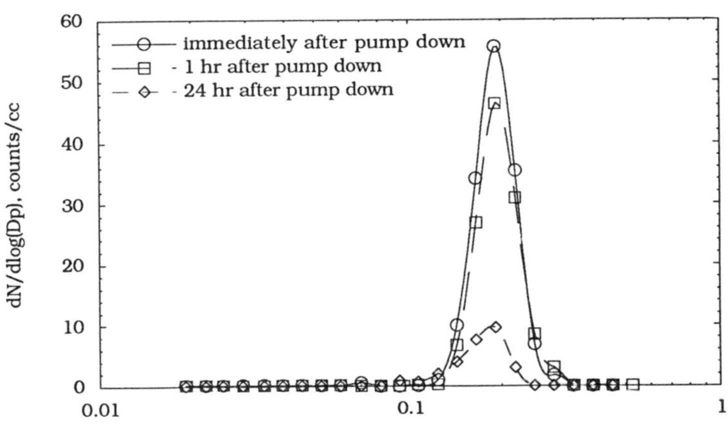

Figure 5.   Residue particle size distribution measured at different time

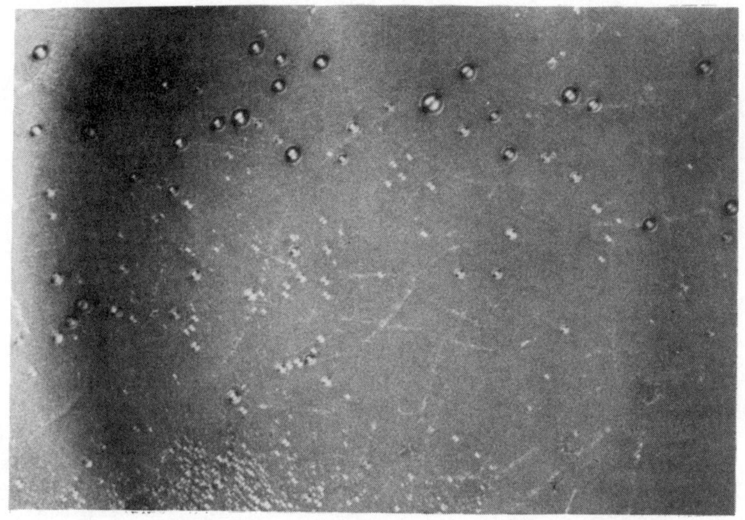

Figure 6. Rsidue particles collected on a Lexan graphic film substrate

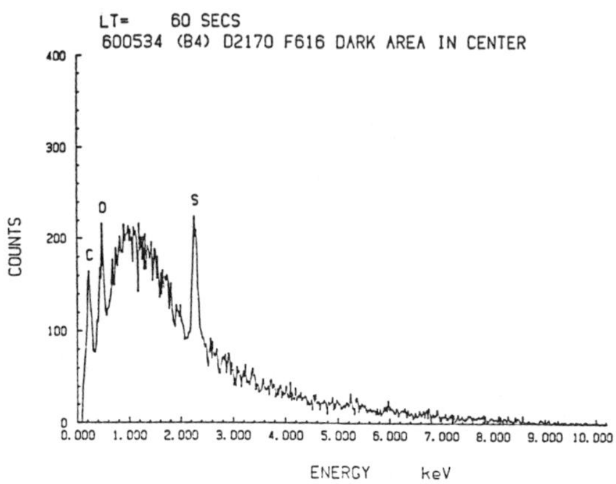

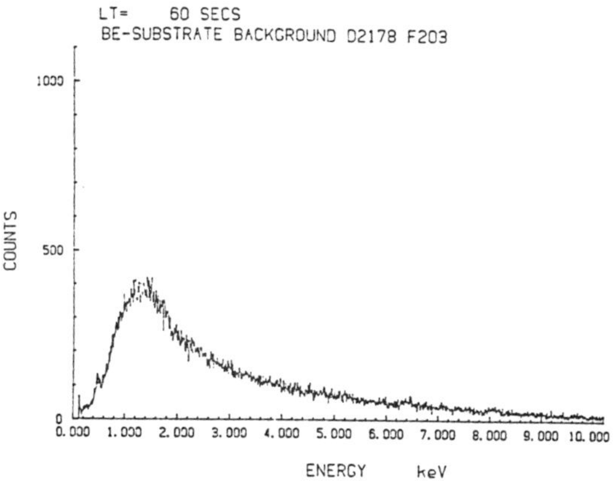

Figure 7.   Chemical components in the residue particles detected by EDX

MECHANISMS OF RESIDUE PARTICLE FORMATION

The physical and chemical characteristics of the residue particles as described above suggests that homogeneous nucleation and condensation of water vapor during pressure reduction are not the only processes involved and that trace chemical species present in the gas phase can react to form chemical species needed for residue particle formation. The steps involved can only be postulated. For the case of residue particles of sulfuric acid, we believe the following steps may be involved.

Since filtered laboratory air was used in many of the experiments described above, they would contain trace quantities of SO_2, O_3, H_2O_2, gas phase H_2SO_4, and other normal air pollutants present in the urban atmosphere. Table 1 shows some typical concentrations of trace gases in clean tropospheric air as well as SO_2 and O_3 measured in the room air where these experiments were conducted. A Meloy Labs Model 285 Sulfur Analyzer and a Dasibi Environmental Model 1003-AH Ozone Analyzer were used respectively. The concentration of species that were not measured can be estimated with data for clean troposphere in the table.

Table 1   Typical Concentrations of Trace Gases

Species	Concentration, ppb		
	Clean Troposphere	Polluted Air	Room Air[b]
SO_2	1-10[a]	20-200[a]	~10
O_3	20-80[a]	50-250[a]	10-50
NH_3	1[a]	3-50[a]	----
H_2O_2	0.1-10		----
H_2SO_4	~0.001		----

[a] Seinfeld, 1986

[b] Measured in this work.

Since sulfuric acid vapor in normal ambient air has a very low concentration, it cannot be be the major source of sulfuric acid found in the residue particles. In addition, residue particles were found to form only when the relative humidity was sufficiently high to cause water vapor condensation, the steps may involve absorption of SO_2 into the liquid droplets and oxidation of the absorbed SO_2 by liquid phase oxidization by such species as H_2O_2 that are simultaneously absorbed into the water droplets during condensation.

The hypothesis underlying the theory of residue particle formation is depicted in Figure 8. As shown in the figure, before pump down, water and other trace chemical species exist in the gas phase. During evacuation, gas temperature decreases due to gas expansion, leading to an increase in saturation ratio of water vapor. Once the saturation ratio exceeds the critical value for homogeneous nucleation, a large number of water droplet are formed, converting water vapor to water droplets. During the droplet formation process, trace gases in the air such as SO_2, H_2O_2, and O_3 would diffuse to the droplets and be simultaneously absorbed into the droplets. When the droplets re-evaporate, the concentration of the trace chemical species in the liquid will greatly increase. Calculation--described below-- shows that during the final period of the droplet evaporation phase, the concentration of H_2O_2 in the liquid becomes so high that it reacts quickly with the dissolved SO_2 to form sulfuric acid. This sulfuric acid, along with the equilibrium concentration of water then becomes the residue particles detected in the pump down experiments.

In order to establish the plausibility of the above described hypothesis of sulfuric acid residue particle formation, a numerical model was developed to simulate the nucleation, condensation, SO_2 oxidation, and droplet re-evaporation process involved in residue particle formation (Ye, 1992).   The following time dependent parameters are calculated simultaneously in the numerical model:

- gas temperature and pressure resulting from evacuation and heat transfer from chamber surface.

- water vapor concentration due to evacuation, nucleation, and condensation.

- saturation ratio and nucleation rate.

- number concentration, mean size, and geometric standard deviation of formed particles

- variation of the concentration of trace gases in the gas phase due to evacuation and gas absorption.

- concentration of trace gases in the droplets

- concentration of reaction products in the droplets

16

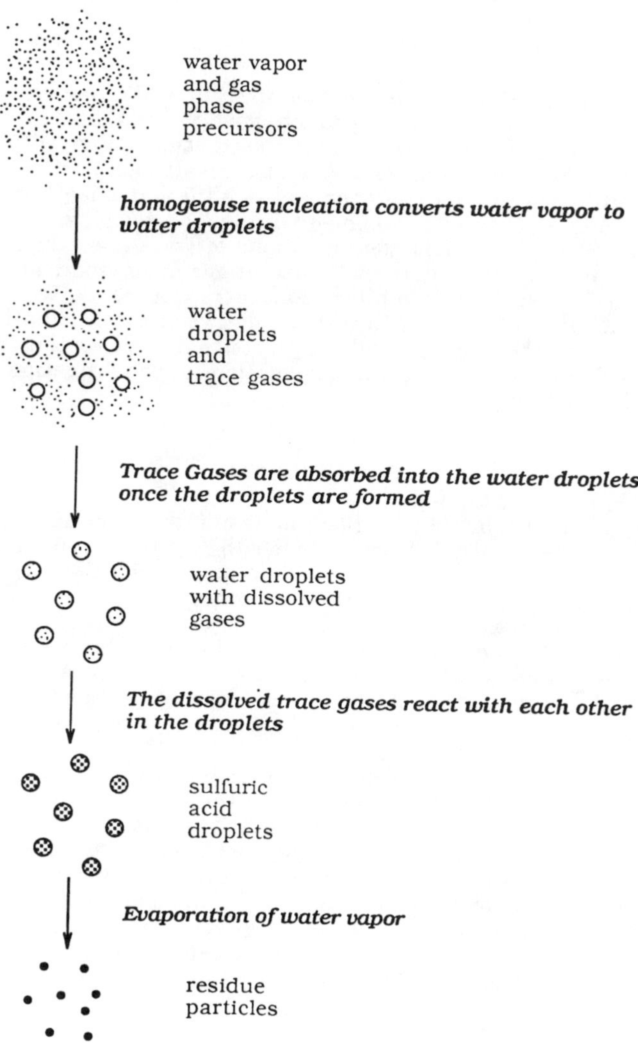

Figure 8. Hypothesis of Residue particle formation process

- nucleation rate of ice from supercooled water

- fraction of ice particles

The numerical model developed in this work has been used to calculate particle formation during pump down under the same evacuation conditions as those used for the experimental study. Figure 9 shows the comparison of numerical and experimental results for particle formation during pump down for the chamber and at different pumping speeds. In the experiments, the actual concentrations of trace gases were not measured. When the typical concentrations of trace gases in air listed in Table 1 are used in the numerical calculation, the numerical results for residue particle size and number concentration agree well with the experimental data. The good agreement between the calculation results and the experimental data strongly suggests that the hypothesis put forth in this paper provides a plausible explanation for the formation of residue particles during pump down.

CONCLUSION

The phenomenon of residue particle formation during pressure reduction has been studied experimentally. It was found that the formation of stable sub-micron particles is induced by the nucleation and condensation of water vapor produced by pressure reduction. The residue particles were found to be non-volatile and very stable with mean diameters on the order of 0.2 μm. The physical and chemical characteristics of the residue particles shows that they contain sulfur, oxygen and carbon. The data strongly suggest that sulfuric acid is a major component in the residue particles.

The mechanisms of sulfuric acid and residue particle formation have been hypothesized and a quantitative theory developed to predict the rate of sulfuric acid particle formation and the concentration and size of the particles formed under different pumping conditions. The theory envisions the nucleation and condensation of water vapor as the primary step during which trace amounts of SO_2 and H_2O_2 in the ambient are absorbed into the droplets by molecular diffusion. These trace gases then react in the liquid phase during evaporation to form stable sulfuric acid. The numerical simulation results based on the hypothesized mechanisms have been found to agree well with the experimental observation.

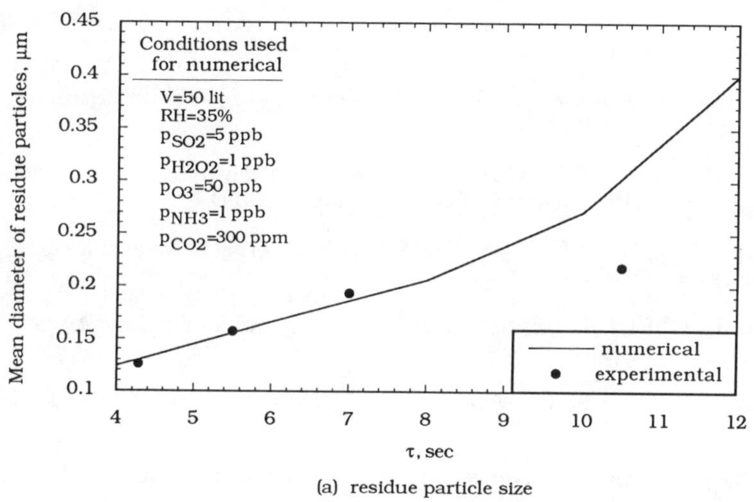

(a) residue particle size

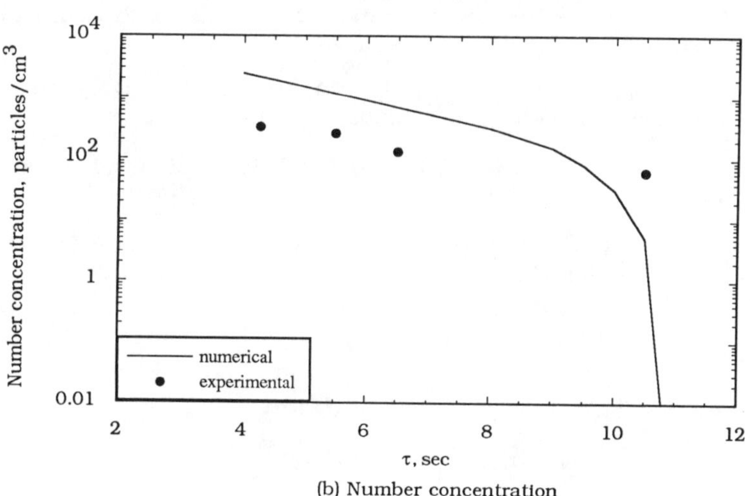

(b) Number concentration

Figure 9.   Comparison of the numerical and experimental results.
Effect of pumping speed on residue particle formation

REFERENCES

Chen, D., Seidel, T., Belinski, S. and Hackwood, S. (1989). Journal of Vacuum Science and Technology A 7(5): 3105-3111.

Chen, D. and Hackwood, S. (1990). Journal of Vacuum Science and Technology A 8(2): 933-940.

Friedlander (1983). "Dynamics of aerosol formation by chemical reaction." Annals. New. York. Academy of Sciences 404: 354-364.

Girshick, S. L., Chiu, C. P. and McMurry, P. H. (1990). Aerosol Science and Technology 13(4): 465-477.

Hoh, P. D. (1984). Journal of Vacuum Science and Technology A 2(2): 198-199.

Seinfeld, J. H. (1986). "Atmospheric Chemistry and Physics of Air Pollution". John Wiley & Sons, New York.

Springer, G. S. (1978). Advances in heat transfer, 14:281-346

Szymanski, W. W., Liu, B. Y. H. and Xu, Z. (1988). Presentation in 6th Review Meeting of Particle Contamination Control Research Consortium, Minneapolis, MN,

Wu, J. J., Cooper, D. W. and Miller, R. J. (1990). Journal of Vacuum Science and Technology A 8(3): 1961-1968.

Ye, Y. (1992). Ph. D. Thesis, "A fundamental Study of Residue Particle Formation During Pressure Reduction". University of Minnesota.

Zhao, Z. (1990). Ph. D. Theses, "Thermodynamics and Particle Formation During Vacuum Pump-Down". University of Minnesota.

MECHANISMS OF PARTICLE TRANSPORT IN PROCESS EQUIPMENT

R.P. Donovan, T. Yamamoto, R. Periasamy, and A.C. Clayton
Research Triangle Institute
P.O. Box 12194
Research Triangle Park, NC 27709-2194

ABSTRACT

Mechanisms affecting particle transport in wafer processing environments include: sedimentation, convective diffusion, thermophoresis, electrophoresis, and photophoresis. The first two mechanisms are universal and beyond the control of the process engineer. The latter three mechanisms, however, depend on variables that can be controlled and used to minimize particle deposition on product wafers. This paper reviews theoretical models of all five mechanisms and describes procedures and hardware configurations that use one or more of the three controllable mechanisms to protect wafers from particulate contamination.

INTRODUCTION

Mechanisms affecting particle transport in wafer processing environments include:

- sedimentation,
- convective diffusion,
- thermophoresis,
- electrophoresis, and
- photophoresis.

The first two mechanisms are universal and beyond the control of the process engineer. The latter three mechanisms, however, depend on variables that can be controlled and, when understood and used appropriately, can be made to reduce or minimize particle deposition on product wafers. The purpose of this paper is to review theoretical models of these mechanisms and describe practical process equipment configurations that take advantage of these particle transport properties to shield wafers from particulate contamination.

An underlying assumption of this work is that contamination free manufacturing will require control of particle transport as well as control of particle-generating sources. This emphasis on particle transport reflects the growing awareness of the importance of particles generated by the processes themselves. Such particle sources are now forecast to be the dominant sources of particles in future wafer processing environments. Where the process itself generates most of the particles, processing environments that are particle-free become a contradiction and are no longer a complete option for achieving contamination free manufacturing. The only choice remaining is to include control of particle transport and deposition as part of the particle-control strategy.

PARTICLE DEPOSITION VELOCITY

Particle deposition velocity is the variable commonly used to relate the accumulation of particles on a surface to the concentration of particles in the environment adjacent to the surface. It is defined as follows:

$$V_M = \frac{J}{C_o} = \frac{\Delta n}{tAC_o} \tag{1}$$

where

$$
\begin{aligned}
V_M \quad &= \text{ the experimentally measured particle deposition velocity} \\
J \quad &= \text{ particle flux to the wafer surface} \\
C_o \quad &= \text{ particle concentration in the wafer environment} \\
\Delta n \quad &= \text{ increase in total particles per wafer of surface area, A, brought about} \\
&\quad\;\; \text{by exposure to a particle environment of } C_o \text{ particles/unit volume for} \\
&\quad\;\; \text{time, t.}
\end{aligned}
$$

Particle deposition velocity varies with mechanism of particle transport and deposition. It is the variable used in this paper to compare the relative importance of the various deposition mechanisms in wafer processing environments.

Sedimentation

Sedimentation describes the gravitational settling of particles that occur in all wafer processing environments.

For a wafer placed horizontally in a verticle laminar flow field (Figure 1), the calculated particle deposition velocity attributable to sedimentation, V_s, obtained by equating the gravitational force on the particle to the Stokes drag force on the particle, is [1-5]:

$$V_s = \frac{\rho_p C_c d_p^2 g}{18\eta} \tag{2}$$

where

$$
\begin{aligned}
\rho_p \quad &= \text{ the particle density} \\
d_p \quad &= \text{ the particle diameter}
\end{aligned}
$$

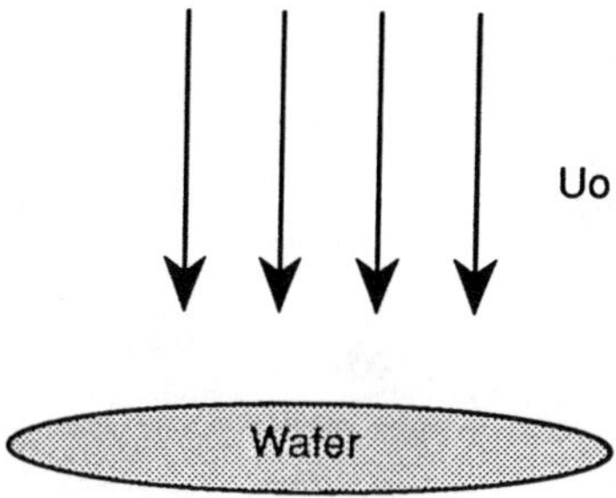

Figure 1. The geometrical configuration: A horizontal wafer in a vertical laminar flow.

g   = the gravitational constant
η = the viscosity of the medium in which the particle is settling
C_c = the Cunningham correction factor.

This mechanism usually dominates the deposition of large particles (> 1-10 µm) in the atmosphere and of even smaller particles in vacuum equipment. It is less important in liquid media.

The Cunningham correction factor, C_c, appearing in Eq. (2) is an empirical correction to the Stokes drag equation that allows the use of Stokes law under conditions not quite fulfilling the Stokes law assumption of zero gas velocity at the particle surface. This assumption is invalid when the particle diameter is of the same order of magnitude as the mean free path of gas molecules (0.066 µm at atmospheric pressure).

An empirical equation for expressing this correction factor is given by Hinds [5]:

$$C_c = 1 + \frac{2}{Pd_p} \left[6.32 + 2.01 \ \exp \ (-0.1095Pd_p) \right] \tag{3}$$

where

P   = absolute pressure in cm Hg
d_p = particle diameter in µm.

Figure 2 is a plot of Eq. (3) as a function of particle diameter with pressure as a parameter. As is clear in Eq. (3), C_c depends on the Pd_p product; reducing either one increases C_c.

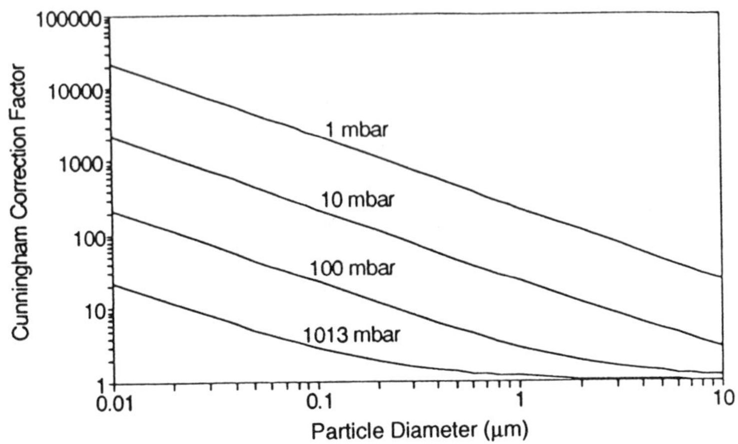

**Figure 2. The pressure dependent Cunningham correction
factor (C_c)as a function of particle size.**

The primary pressure-dependent factor in Eq. (2) is C_c as given in Eq. (3) and
Figure 2. The pressure dependence of gas viscosity is not expected to be large [6] and
is assumed to be negligible in this discussion. Under this assumption the pressure
dependence of V_s derives totally from C_c and is plotted in Figure 3 for particles of unit
density in air.

Convective Diffusion

Convective diffusion describes particle transport by a combination of convection
remote from the wafer and diffusion across a concentration boundary layer adjacent to the
wafer surface. The expression used here to predict the particle deposition velocity
attributable to convective diffusion, V_{dif}, is taken from Liu and Ahn [1] who derived the
following mass transfer relationship by analogy with heat transfer problems:

$$V_{dif} = \frac{ShD}{D_w} \tag{4}$$

where

Sh = the Sherwood Number

$$= 1.08 R_e^{1/2} S_c^{1/3} = 1.08 \left(\frac{U_o D_w \rho_f}{\eta} \right)^{1/2} \left(\frac{\eta}{\rho_f D} \right)^{1/3}$$

D = the particle diffusion coefficient

$$= \frac{kTC_c}{3\pi\eta d_p}$$

24

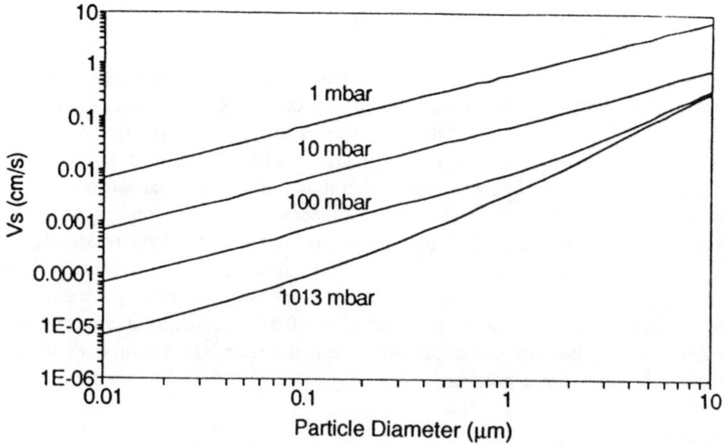

Figure 3. The pressure dependent particle deposition velocity due to sedimentation (V_s) as a function of particle size.

$$D_w \quad = \text{ the wafer diameter}$$
$$k \quad\ \ = \text{ the Boltzmann constant}$$
$$T \quad\ \ = \text{ absolute temperature}$$
$$U_o \quad = \text{ the free stream velocity}$$
$$\rho_f \quad\ = \text{ the fluid density}$$

In addition to C_c, the fluid density ρ_f is a primary pressure dependent term in determining V_{dif}. Assuming an ideal gas, ρ_f depends directly on pressure:

$$\rho_f = \frac{PM}{RT} \tag{5}$$

where

$$P \quad = \text{ pressure}$$
$$M \quad = \text{ molecular weight}$$
$$R \quad = \text{ the gas constant}$$

Adding the convective diffusion expression for deposition velocity to the sedimentation expression yields a total deposition velocity, V_t:

$$V_t = V_s + V_{dif} \tag{6}$$

Figure 4 is a plot of V_t, using the same values for V_s that appeared in Figure 3 and the values of V_{dif} calculated for a wafer diameter of 12.5 cm, a free stream velocity of 50.8 cm/s and 20 °C temperature. This plot shows that at atmospheric pressure V_{dif} dominates V_t for particle diameters less than 0.1 µm, while V_s dominates for particle diameters greater than 0.5 µm. Between these two diameters is a transition region where both mechanisms are important. As pressure decreases, this transition region shifts to smaller diameters, meaning that sedimentation is predicted to become relatively more important as pressure decreases. Note, however, that both the diffusion-dominated branch and the sedimentation-dominated branch of V_t increase significantly with decreasing pressure. Wafers exposed to the same concentration of particles in their environment at both atmospheric and subatmospheric pressures are predicted to become dirtier after exposure to the subatmospheric pressure.

Thermophoresis

Thermophoresis is a particle motion induced by a thermal gradient in the medium in which the particle is suspended. The side of the particle exposed to the higher temperature is bombarded by higher energy molecules than the side of the particle facing the cooler region of the medium. This difference in adjacent medium temperatures creates a force on the particle in the direction of the cooler region.

The expression used in this paper to calculate the thermophoretic contribution to particle deposition velocity, V_{th}, comes from Batchelor and Shen [7]:

$$V_{th} = - \frac{H\eta \nabla T}{T\rho_f} \tag{7}$$

where

H = the thermophoresis coefficient

$$= 2.294 \, C_c \, \frac{(k_a/k_p + 2.2 \, Kn)}{(1 + 3.438 \, Kn)(1 + 2(k_a/k_p) + 4.4 \, Kn)}$$

k_a, k_p = the thermal conductivity of the air and particle, respectively

Kn = the Knudsen number

$$= \frac{2\lambda}{d_\rho}$$

λ = the mean free path of the fluid molecules

$$= \frac{\eta}{\rho} \sqrt{\frac{\pi RT}{2M}}$$

∇T = the temperature gradient perpendicular to the wafer surface

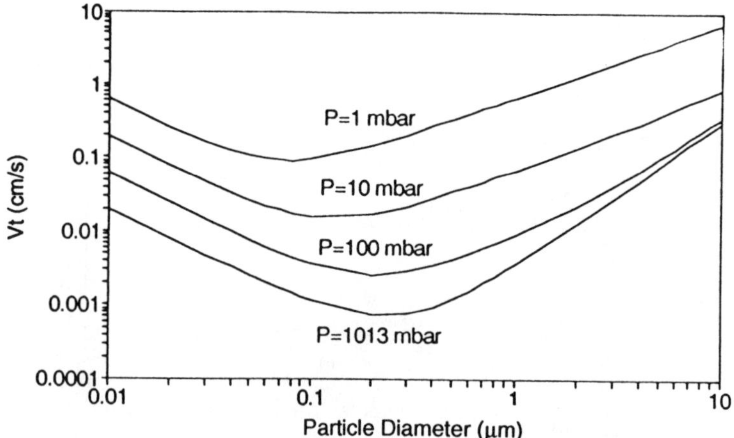

Figure 4. The pressure dependent net deposition velocity (V_t) due to sedimentation and diffusion, as a function of particle size in a free stream velocity of 50.8 cm/s.

The primary pressure dependence of V_{th} comes from C_c and Kn. These two terms essentially compensate each other's dependence on particle diameter so that V_{th} is predicted to have only a minor dependence on particle diameter (Figure 5). Figure 5 plots V_{th} (Eq. [7]) for a temperature gradient of 10 K/cm at a T of 293 K, the molecular weight of air (28.9 g/mol) and a k_a/k_p of 0.1. Decreasing k_a/k_p by a factor of 10 ($k_a/k_p = 0.01$) makes a noticeable difference only for particle diameters greater than 1 μm at atmospheric pressure. At lower pressures and smaller particle diameters the k_a/k_p contribution is negligible.

Figure 5 shows that Eq. (7) predicts an inverse dependence on pressure with the values of V_{th} at a ∇T of 10 K/cm being of the same order of magnitude as those previously predicted for V_t. The negative sign in Eq. (7) means that when the wafer temperature is greater than that of the surrounding medium, the thermophoretic effect repels particles from the wafer surface. Adding this negative contribution from thermophoresis (Figure 5) to the V_t values plotted in Figure 4 produces the values of V_{tt} shown in Figure 6.

$$V_{tt} = V_t + V_{th}$$

$$(8)$$

The negative sign of V_{th} makes V_{tt} negative over a range of particle diameters, meaning that thermophoresis under the values of variables assumed is predicted to shield horizontally placed wafers from particle deposition in this size range that otherwise would occur because of sedimentation or convective diffusion.

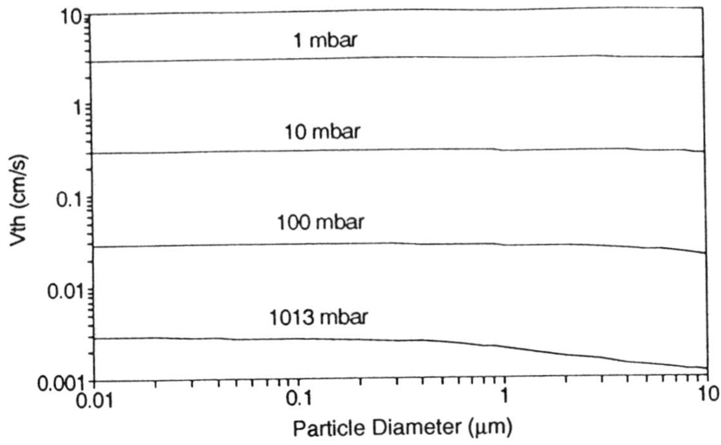

Figure 5. The pressure dependent particle deposition velocity due to thermophoretic force alone (V_{th}) when the temperature gradient, ∇T, is 10 K/cm.

Electrophoresis

Electrophoresis is the drift of charged particles in an electric field, the consequence of a classical Coulomb interaction. Polarization forces, such as dielectrophoretic or image forces, are also present but are usually smaller in magnitude than the Coulomb force and will not be included in the modeling of electrical forces presented here.

Hinds [5] gives the following expression for particle deposition velocity, V_c, attributable to electrophoresis by equating the Coulomb force on a particle to the Stokes drag force:

$$V_c = \frac{n_e e C_c E}{3\pi\eta d_p} = \frac{C_c q E}{3\pi\eta d_p} = \frac{C_c q_{rms}}{3\pi\eta d_p}\,\psi E \qquad (9)$$

where

n_e	= the net number of elementary charges on the particle
e	= the elementary electronic charge
E	= the electric field perpendicular to the wafer surface
q	= the net charge on the particle
	= $n_e e$
q_{rms}	= the rms charge on a particle in Boltzmann equilibrium
ψ	= normalized particle charge
	= q/q_{rms}.

The q_{rms} value of particle charge represents a minimum value to be expected. It varies with particle diameter and, for the calculations performed here, a relationship given by Hinds was used [5]:

$$q_{rms} = 2.37\sqrt{d_p \ (\mu m)} \quad . \tag{10}$$

The ΨE parameter in Eq. (9) embodies the electrophoretic variables. With E in volts/cm, ΨE values on the order of 10 represent very low electrophoretic forces; ΨE of 100, low; ΨE of 1,000, moderate; ΨE of 10,000, high; and $\Psi E >10,000$, very high. Even low values of ΨE can make significant contributions to the total deposition velocity, as illustrated in Figure 7 where

$$V_{tc} = V_t + V_c \ . \tag{11}$$

Figure 7 compares V_t (solid curves) at atmospheric pressure (1013 mbars) and 10 mbars with V_{tc} (dashed curves) at those same pressures calculated for ΨE values of 100 volts/cm. The addition of this modest electrical force produces a major increase in net deposition velocity at 1013 mbars but an even larger increase at 10 mbars, indicating that at both pressures electrophoretic effects are larger than those attributable to sedimentation or convective diffusion for particle diameters in the submicrometer range. Above 1 μm particle diameter sedimentation begins to dominate even as it does without the electrophoretic contribution. Note that the transition region between sedimentation and electrophoretic dominance shifts to larger particle diameters at lower pressures.

The Figure 7 calculations were carried out under the assumption that the electrophoretic force between the particle and the wafer is one of attraction. Coulomb forces, unlike the dielectrophoretic or image forces which are always attractive, can also be repulsive. All that is required is a change of field direction (or a change of the charge sign on the particle). This switch makes the V_c contribution in Eq. (11) subtractive rather than additive. Under these conditions the plots of V_{tc} in Figure 7 are altered to those of Figure 8. All values used in the calculation of V_{tc} are the same for the two figures, including a ΨE value of 100 v/cm, except that the sign of V_c has been changed to negative.

Figure 8 implies that modest electrophoretic forces, like the thermophoretic forces depicted in Figure 6, should be able to shield horizontal wafers from submicrometer settling or diffusing particles over the entire range of pressures considered. The electrophoretic forces modeled here dominate even the sub 0.1 μm particle regime, a dominance not achieved by the thermophoretic conditions modeled in Figure 6.

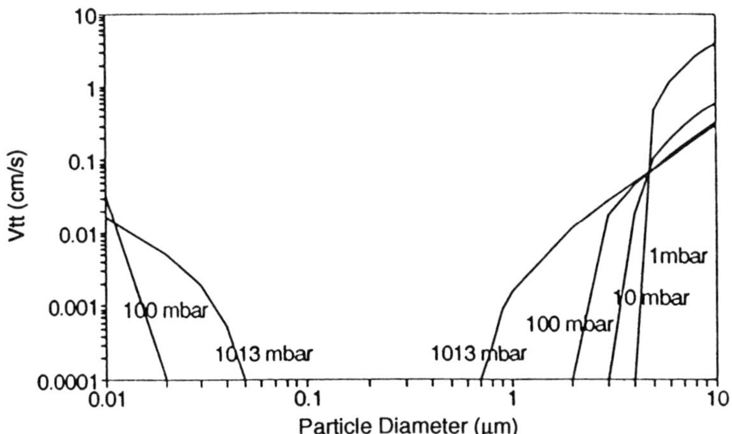

Figure 6. The pressure dependence of (V_{tt}) when the temperature gradient is 10 K/cm, $k_a/k_p = 0.1$.

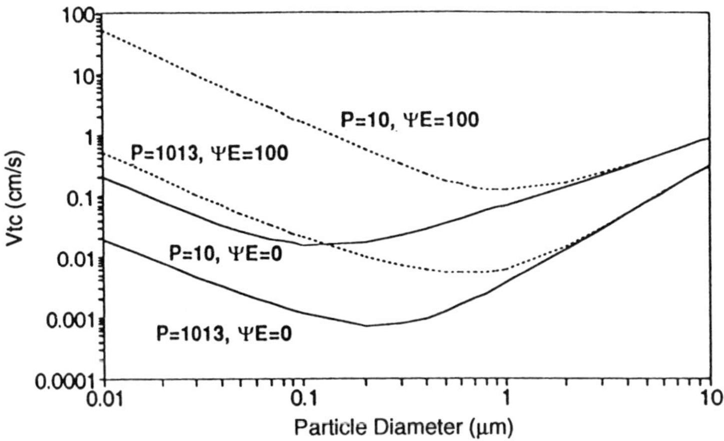

Figure 7. The effects of electrostatics and pressure on V_{tc}.

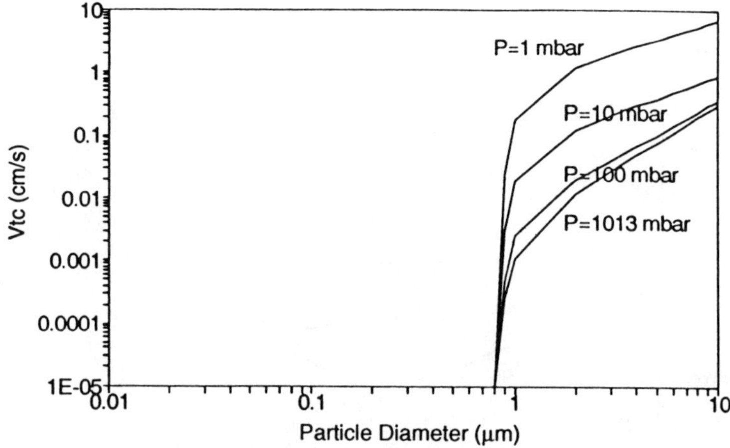

Figure 8. V_{tc} when V_c has a negative sign.

The relative contributions of thermophoresis and electrophoresis to net particle deposition velocity can also be assessed by adding both contributions simultaneously to the basic V_t expression:

$$V_{ttc} = V_t + V_{th} + V_c .$$

(12)

Since the thermophoretic force at the wafer surface is most often directed away from the wafer (product wafers are more often at higher rather than lower temperatures than their surroundings), the sign of V_{th} in Eq. (12) will be negative. For comparison purposes V_c should then be made positive (electrophoresis creates an attractive interaction for the purposes of this comparison). V_t is always positive for a horizontally positioned surface. Under these assumptions and using the same parameters as before, Figures 9 and 10 plot V_{ttc} at atmospheric pressure and at 1 mbar, respectively. At atmospheric pressure (Figure 9) the electrophoretic attraction dominates over virtually the entire range of particle diameters. At 1 mbar, however, a significant band of diameters is predicted to still be thermophoretically shielded in spite of the attractive electrophoretic forces. These figures reflect the relative independence of particle diameter on thermophoresis (Figure 5) which could be of importance in subatmospheric processing.

Photophoresis

Photophoresis refers to particle motion induced by light. Two different interactions are part of this mechanism: (1) radiometric effects and (2) radiation pressure. Each will be modeled and discussed separately.

Radiometry--Radiometry is somewhat analogous to thermophoresis in that the illuminating light on the particle creates a temperature gradient in the particle which in

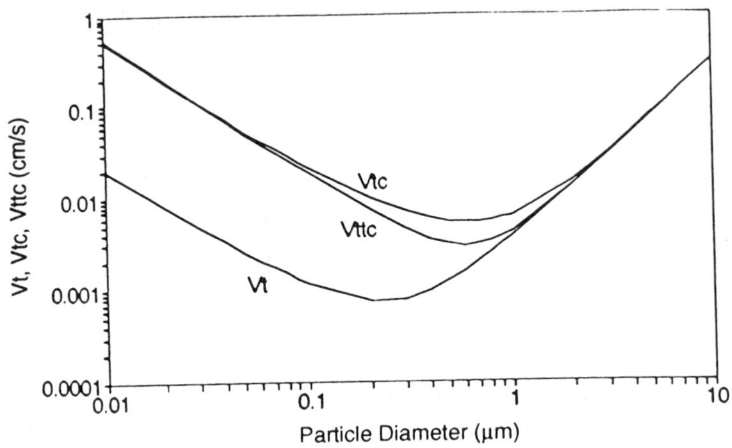

Figure 9. Comparison of the total deposition velocities, V_t, V_{tc} (for $\Psi E=100$), and $V_{ttc}(\nabla T=10$ K/cm and $\Psi E=100)$ when P=1013 mbar.

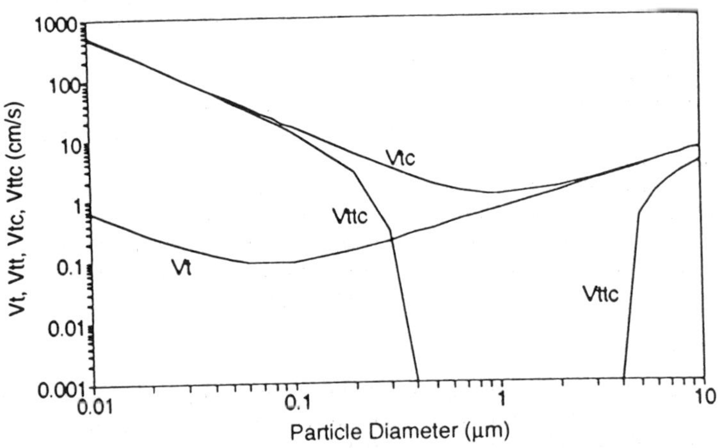

Figure 10. Comparison of the total deposition velocities, V_t, V_{tc} (for $\Psi E=100$), and V_{ttc} ($\nabla T=10$ K/cm and $\Psi E=100$) with P=1 mbar.

turn creates a temperature gradient in the surrounding media. This photo-induced temperature gradient in the media produces a thermophoretic force on the particle. The distinction between a radiometric force and a thermophoretic force is that in the former the temperature gradient originates from within the particle while in the latter it originates in the surrounding fluid.

Arnold and Lewittes give the following expression for the radiometric force on a particle [8] (force is the property modeled since measurements to date have just been of photophoretic force rather than the particle deposition velocity attributed to radiometry):

$$F_{rm} = - \frac{2\pi d_p \eta^2 JKI}{\rho_f T_p K_i} \tag{13}$$

where the newly introduced variables are

J	=	a measure of the asymmetry of the internal heating of the particle
K	=	the coefficient of thermal slip
I	=	the incident light intensity
T_p	=	the surface temperature of the illuminated particle
K_i	=	the interior thermal conductivity of the particle.

The factor J determines how, within the particle, the absorbed light is distributed. A positive J means that the dark side of the particle is absorbing more of the incident light than the side nearer the light source. Thus, the radiometric force will be directed toward the light source, a so-called "negative" photophoretic effect. In the more typical case, the illuminated side absorbs more light, J is negative and a positive radiometric force accelerates the particle away from the light source. All strongly absorbing particles fall in this second category.

Figure 11 is a plot of Eq. (13) for the values J = -0.4; K = 0.9; I = 22.6 watts/cm^2; T_p = 573 K and K_i = 0.062 watts/m·K. The pressure dependence reflected in Figure 11 arises solely from the fluid density term, ρ_f. At atmospheric pressure the magnitude of the predicted radiometric forces exceed the gravitational force on all particles smaller than about 10 µm (ρ_p = 2.1 g/cm). At pressures less than 10 mbar, the predicted radiometric forces are greater than the gravitational force for all particle diameters less than 100 µm and orders of magnitude greater for submicrometer particles (Figure 12).

While not shown in Figures 11 or 12, the radiometric force is expected to decrease as pressure decreases below the range shown. Maximum radiometric forces are reported to be measured when Knudsen numbers are in the 0.1 to 1 range [9]. These values of K_n are typical of submicrometer particles at the chamber pressures of many deposition and etching processes.

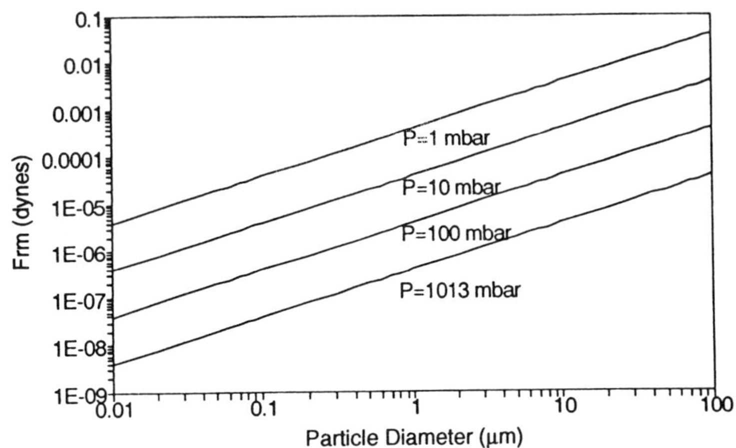

Figure 11. Predicted radiometric force on an absorbing particle illuminated with light intensity of 22.6 watts/cm$_2$ (0.3 watts in a 1.3 mm diameter spot).

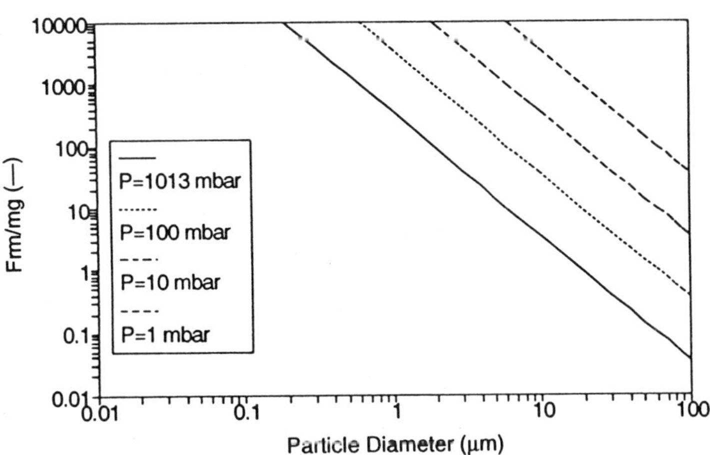

Figure 12. Predicted radiometric forces from Figure 12 normalized in terms of the gravitational force, mg.

Radiation Pressure--The second photophoretic interaction between a particle and an illuminating light beam is called radiation pressure and refers to the momentum transfer to the particle that accompanies light reflection, refraction, or absorption. Any change in the momentum of the incoming photons must be balanced by an equal and opposite particle momentum. Reflecting particles cause the largest changes in photon momentum. Refracting particles in gaussian light beams experience forces perpendicular to the direction of the incident beam in addition to forces in the direction of the incident beam (forces caused by radiation pressure are always with positive direction—away from the light source).

Hinds [5] and Ashkin [10] give the following equation for the radiation pressure force on a particle:

$$F_{rp} = \frac{I \pi d_p^2 Q_{rp}}{4c} \tag{14}$$

where

Q_{rp} = the fraction of light geometrically incident on the particle that is effective in transferring momentum to the particle in the direction away from the light source

c = the velocity of light.

Q_{rp} is a complicated function of particle size, wavelength, index of refraction, and absorption coefficient. Because values of Q_{rp} are not generally available, Q_{rp} was assumed to be equal to 0.5 in all calculations carried out here.

The force attributable to radiation pressure has no pressure dependence. It is the same at atmospheric pressures as in hard vacuum. Its magnitude, as plotted in Figure 13, is expected to be smaller than the radiometric force. Only with highly focused, high power lasers will the light intensity be great enough to produce forces comparable to the gravitational force.

Controlling Particulate Contamination of Product Wafers

The first two of the five particle transport mechanisms modeled—sedimentation and convective diffusion—are universal and operative in all environments. While steps can be taken to minimize particle deposition attributable to these mechanisms (for example, mounting wafers vertically or upside down in processing equipment), the assumption taken in the rest of this paper is that the V_t values calculated and plotted in Figure 4 represent fundamental minimum values of particle deposition velocity to be encountered by horizontally placed wafers in processing equipment unless preventive measures are taken.

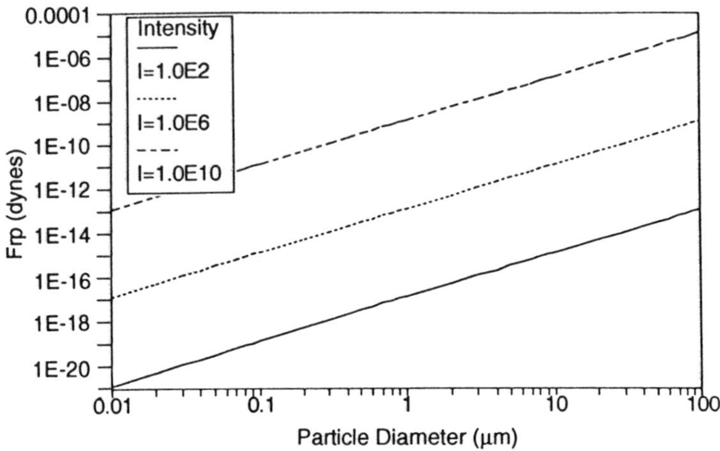

**Figure 13. Predicted radiation pressure forces as a
function of light intensity.**

The preceding section has shown that each of the three particle transport mechanisms discussed after sedimentation and convective diffusion, namely, thermophoresis, electrophoresis, and photophoresis, are predicted to be capable of making contributions that oppose this fundamental V_t and are of greater magnitude (see Figures 6, 8, and 12). Thus, any one of these three mechanisms can be the basis of hardware or procedures that reduce the net particle deposition velocity below the V_t values of Figure 4. This section presents data that support these predictions.

Thermophoretic Protection

One of the simplest procedures to implement is the inclusion of a heating stage for horizontally placed wafers in a subatmospheric process chamber. Data presented in Table 1 compare particle deposition on wafers subjected to a pumpdown and venting cycle only in an electron beam evaporator apparatus. No processing was carried out. The chamber was simply pumped down to a typical working pressure and then vented back to atmospheric pressure with dry nitrogen gas from the house supply.

The data in Table 1 reflect an order-of-magnitude decrease in particle deposition on those wafers that were heated just 30 °C above the ambient temperature. No control over particle properties was possible in these measurements; the particles collected were those present in this fab line metal coater.

Table 1. Wafer Protection Attributed to Thermophoresis

Wafer exposure	Particles added to horizontal wafer (cm^{-2})	
	Room temperature (20 °C)	Heated (50 °C)
Cabinet top, photolithography bay, 24 h	1.44 ± 0.32	0.46 ± 0.24
Electron beam evaporator, during vacuum pumpdown and vent cycle only (no processing)	5.66 ± 2.73	0.61 ± 0.42

Similar data, collected in the ambient atmosphere of a clean room, are also given in Table 1. These data, while not as dramatic as the evaporator, show a thermophoretically induced reduction in particulate contamination of monitor wafers exposed to the ambient of a lithography bay in a clean room.

The conclusion is that, all else being equal, modestly heated wafers are both predicted and shown to collect fewer particles from a given process or ambient environment than unheated wafers. And so, while quantitative evaluation of the model predictions has not yet been completed, ample evidence exists to support the qualitative effect.

Electrophoretic Protection

Electrophoretic effects were shown to either increase (Figure 7) or decrease (Figure 8) particle deposition velocity, depending on whether the Coulomb force attracted or repelled electrically charged particles. Figure 14 presents data that show both these effects in the same electron beam evaporator used to collect the data of Table 1. The dominant charge sign on the particles in this chamber is clearly negative. By electrically biasing the wafers during pumpdown and venting (again, no actual processing was performed while these bias measurements were being made), the particle deposition velocity could be decreased by about an order of magnitude with a negative bias of 5 kV on the wafers. A positive bias of 5 kV on the wafers increased deposition velocity over unbiased wafers by more than an order of magnitude.

When the electrical charge of the particles in the processing environment is both known and unipolar, it is simple to implement electrophoretic wafer protection by intro-

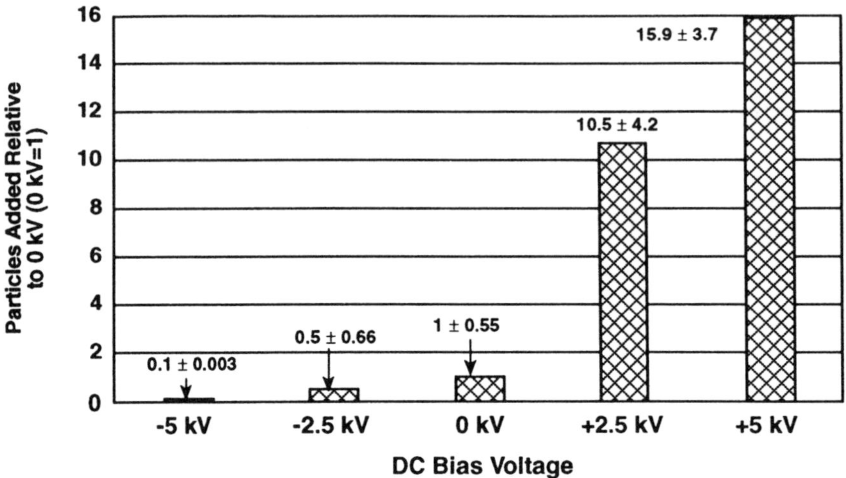

Figure 14. The influence of wafer bias upon particle deposition during pumpdown and venting of an electron beam evaporator.

ducing a dc field perpendicular to the wafer surface. Biasing the wafer holder with respect to ground is all that is needed if the wafers are already mounted on an electrically isolated wafer holder. If the wafers are electrically grounded by their fixture, adding an electrode in the wafer vicinity can create the needed electric field.

When the electrical charge of the particles in the processing chamber is bipolar, variable, or unknown, a dual electrode deflection plate scheme, such as illustrated in Figure 15, can be used. This electrode arrangement deflects particles of either sign from the wafer vicinity.

Figure 16 compares deposition data collected in an experimental chamber into which particles of known composition, size, and average electrical charge were injected. These particles were 1.3 μm dioctyl phthalate spheres of approximately 300 unit electrical charges each. They were injected at a rate of 200,000/s into a chamber held at about 400 mbars pressure. Activating the electric field between the plates dramatically reduced the particle deposition, as indicated by the data in Figure 16.

Thus, although electrical forces are usually recognized as undesirable because of their well known added contributions to particle deposition, with suitable control electrodes, electrical effects can be made beneficial. Frequently, the added fixturing requirements are minimal. Simple dc electric fields have proven effective.

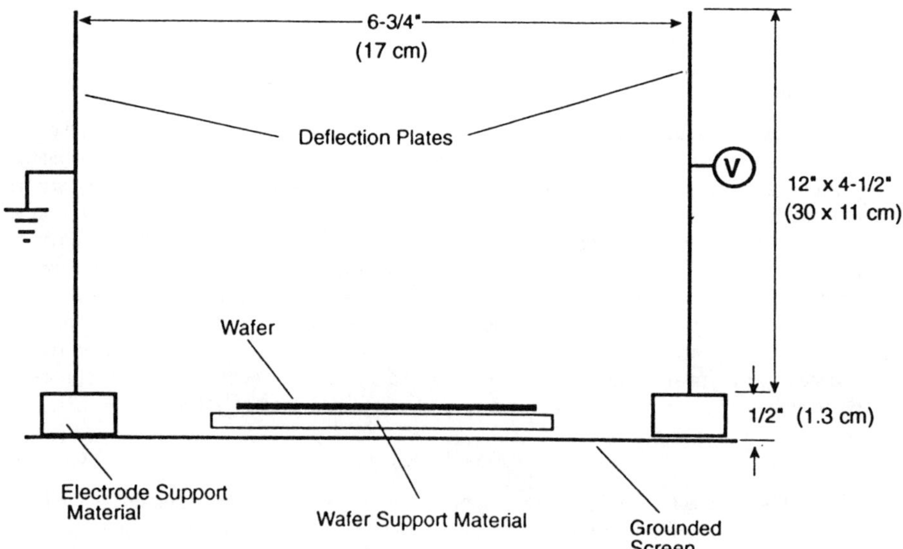

Figure 15. Deflection plate configuration for electrophoretic protection against bipolar-charged particles.

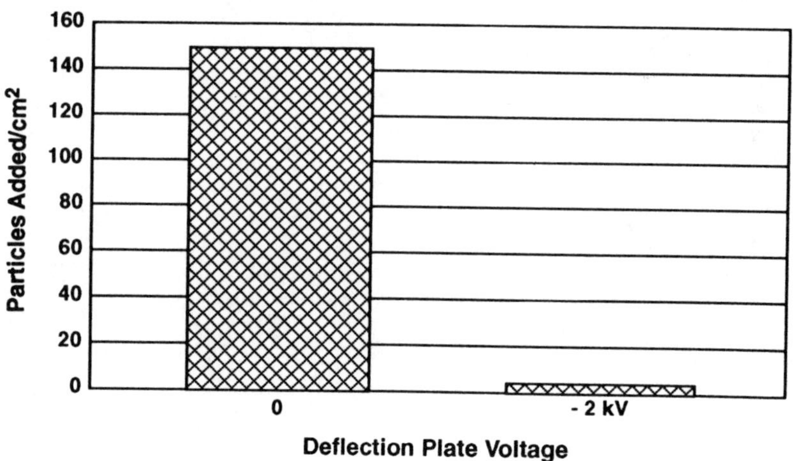

Figure 16. Reduction of particle deposition by deflection plate configuration.

Photophoretic Protection

Photophoretic effects are less well known and documented at present. What has been measured is the magnitude of the photophoretic forces on various illuminated particles such as plotted in Figure 17. To collect these data a 17.5 µm asphalt particle (a particle readily available at RTI and one with optical properties similar to those of silicon) was levitated in a electrodynamic balance. The force plotted is that required to restore the particle to its original position after being illuminated by laser beams of various intensities. The photophoretic force observed here increased with decreasing pressure as predicted by the radiometric type of interaction (Eq. [13], Figures 11 and 12). Even with an unfocused laser beam of over 1 mm spot size, photophoretic forces exceeding the gravitational force were observed. Thus, the potential of photophoretic protection seems real, even though wafer protection data are not yet available.

CONCLUSIONS

Five mechanisms of particle transport in wafer processing environments have been theoretically modeled. Sedimentation and convective diffusion were portrayed as universal, occurring in all wafer environments. Thermophoresis, electrophoresis, and photophoresis were shown to be particle transport mechanisms theoretically capable of reducing particle deposition on wafers and data were presented to indicate that the wafer protection potential predicted for each of these mechanisms is at least qualitatively observed. Quantitative verification of the theoretical models remains to be demonstrated.

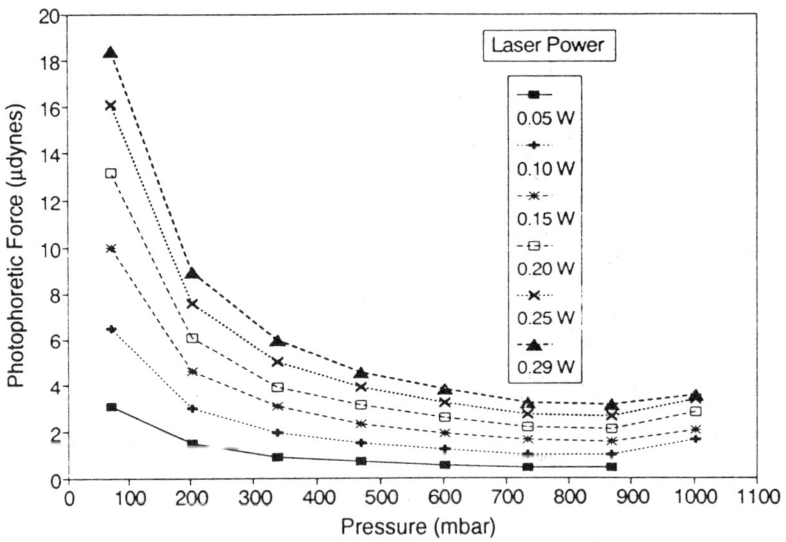

**Figure 17. Photophoretic force on an absorbing particle
(particle mg = 5.8 µ dynes).**

ACKNOWLEDGMENTS

This work was sponsored by the Semiconductor Research Corporation, Manufacturing Process Sciences, under Contract 91-MJ-199.

REFERENCES

[1] B.Y.H. Liu and K. Ahn, "Particle Deposition on Semiconductor Wafers," *J. Aerosol Sci. and Tech.* 6:215-44 (1987).

[2] T.W. Peterson, K.M. Scanners, F. Stratmann, and H. Fissan, "Particle Deposition on Wafers: A Comparison Between Two Modeling Approaches," *J. Aerosol Sci.* 19(7):1409-12 (1988).

[3] J.R. Turner, D.K. Liguras, and H.J. Fissan, "Clean Room Applications of Particle Deposition from Stagnation Flow: Electrostatic Effects," *J. Aerosol Sci.* 20(4):403-17 (1989).

[4] D.W. Cooper, R.J. Miller, J.J. Wu, and M.H. Peters, "Deposition of Submicron Aerosol Particles During Integrated Circuit Manufacturing: Theory," *Particulate Sci. and Tech* 8:209-24 (1990).

[5] W.C. Hinds, *Aerosol Technology: Properties, Behavior, and Measurement of Airborne Particles*, a Wiley Interscience Publication, John Wiley & Sons, New York, 1982.

[6] R.B. Bird, W.E. Stewart, and E.N. Lightfoot, *Transport Phenomena*, John Wiley & Sons, Inc., New York, 1960.

[7] G.K. Batchelor and C. Shen, "Thermophoretic Deposition in Gas Flowing over Cold Surface," *J. Colloid Interface Sci.* 107:21-37 (1985).

[8] S. Arnold and M. Lewittes, "Size Dependence of the Photophoretic Force," *J. Appl. Phys.* 53(7):5314-5319 (1982).

[9] A.I. Bogolepov, "Experimental Study of Photophoresis," *J. Aerosol Sci.* 22 Suppl. 1, 5439-5442 (1991).

[10] A. Ashkin, "Acceleration and Trapping of Particles by Radiation Pressure," *Phys. Rev. Ltrs.* 24(4):156-59 (1970).

PARTICULATE CONTROL DURING ECR-CVD DEPOSITION OF SiO_2

D. R. Denison and A. K. Yasuda
Lam Research Corp, Fremont, CA 94538

Particulates, a major problem in chemical vapor deposition systems, are a primary cause of yield loss in semiconductor device manufacture. A new, production level electron cyclotron resonance CVD tool, developed in partnership with Sematech, incorporates particle control hardware that has demonstrated a reduction from hundreds of particles/cm^2 to less than 0.1/cm^2 for particles 0.3 micrometers and larger. It is seen that two factors dominate particle generation: film adhesion and film stress of the material deposited on the system liners. These factors can be controlled by the selection of liner material and temperature, geometry, and surface conditioning.

Submicron particulate contamination is a major cause of yield loss in high density IC manufacturing. Ultraclean rooms have greatly reduced the environmental particle contribution and increased use of robotics has reduced contamination from the tool so that the major contamination source now occurs during processing. High rate, single wafer processing increases the magnitude and frequency of the thermal cycling of shields and wafer handling surfaces which makes particle control more difficult. In chemical vapor deposition (CVD) systems, these problems are exacerbated by the deposition of stressed films on these surfaces. The thermal mismatch between deposit and substrate material can result in spalling and creation of particle "showers" that become incorporated in the films deposited on the wafers.

Electron cyclotron resonance (ECR) CVD, described elsewhere [1,2,3], has been developed as a technique for depositing high quality SiO_2 for intermetal dielectric where gap fill in high aspect ratio spaces between metal lines is required. This high rate deposition technique, occurring at a low pressure of 1 to 2 mTorr, avoids the gas phase nucleation problems reported for higher pressure CVD systems [4]. The work presented here reports the results of efforts to reduce particulate contamination in ECR-CVD SiO2 films. The system, developed at Lam Research as a Joint Development Project with the Sematech of Austin, Texas, is shown schematically in Fig. 1.

Oxygen is introduced into the plasma chamber where the ECR conditions exist to produce a high density plasma. The divergent magnetic field causes the electrons and ions to stream toward the substrate holder. Silane is injected into the reaction chamber and the oxygen ions react with the silane precursor adsorbed on the wafer surface to produce SiO2.

The wafer is introduced into the system by a mechanism which places the wafer on the substrate holder which is then raised to the reaction position. In this manner, the transport system parts are completely shielded from deposition. The particulate control problem then resolves into four areas relating to the shield surfaces surrounding the wafer during deposition.

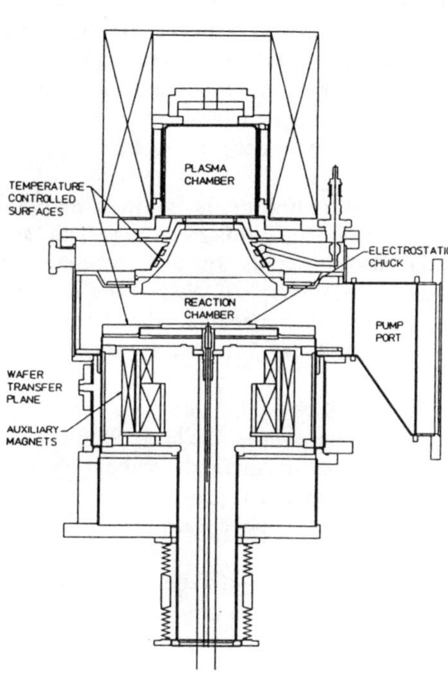

Fig. 1. The ECR-CVD system used in the particle reduction experiments.

These are:

1. Geometry to minimize the area surrounding the wafer and still not impede the gas flow.
2. Choice of surface materials to promote film adhesion.
3. Surface conditioning and preparation.

4. Temperature control to minimize film stress from thermal expansion mismatch.

<u>Geometry</u>: The geometry of the system surfaces is primarily determined by the shape of the ECR plasma stream as it exits the plasma chamber along the divergent magnetic field lines. A conical liner surrounds the plasma aperture and this horn shaped liner results from the need to minimize film growth on the liner surface by the plasma stream impinging on it. There is evidence [2] that the growth of the SiO_2 film results from the surface reaction of an adsorbed silane precursor and the stream of energetic ions.

The surface surrounding the wafer holder is simply a planar extension to provide shielding for the auxiliary magnets used to steer the ion stream in the vicinity of the wafer. This surface, of necessity, intercepts the ion stream and a heavy build up of SiO_x occurs. This surface is the primary source of particulates because the film grows at a high rate and, with a stress of 200 to 500 MPa, spalling can shower particles onto the wafers.

There are also shields on the chamber walls. These shields have two functions. The first is to provide a surface whose material promotes adhesion and the second is simply to provide a removable surface when the inevitable, manual clean is necessary.

<u>Materials</u>: The material chosen for these surfaces is aluminum because SiO_2 adheres strongly to a clean aluminum surface even when the deposition takes place at room temperature [6]. The emphasis is on the word "clean". The deposited oxide is first removed using a grit blast. This is followed by an HF/HNO_3 acid etch to release any grit imbedded in the surface and etch any residual oxide. The final step is an electropolish to reduce the overall surface area and roughness. The cleaned parts are sealed in plastic bags flushed with dry nitrogen.

Once installed and the system evacuated, an oxygen plasma is used to eliminate any organic surface residue and to desorb water vapor. This plasma treatment extends only for about 15 minutes to avoid building up a heavy surface oxide and during this time, the aluminum is maintained at 20° C by a circulating water system.

Excessive oxygen treatment results in homopolar surface arcing which has been shown to produce particles [5].

<u>Temperature Control</u>: The final consideration is the thermal stability of the system to prevent temperature cycling and the attendant stressing of the film which results in spalling. All of the surfaces seen by the wafer are water cooled and designed with sufficient crossection so that thermal cycling is less than 5° C.   Room temperature water, normally 20° C, is used for temperature control and has demonstrated good adhesion. System serviceability and temperature control are greatly simplified without hot surfaces.

Fig. 2 shows the chronological progress made by implementing the principles stated here.   At first, with no shields in place, the particle level was in the hundreds per cm^2. Introduction of the conical "horn" and the wafer holder cover, made of water cooled aluminum, reduced the level to about 10 per cm^2. Finally, an optimum length to the "horn" was developed which gave good particle control without sacrificing gas conductance. Aluminum wall shields were installed and the level further reduced to about 0.3 per cm^2.

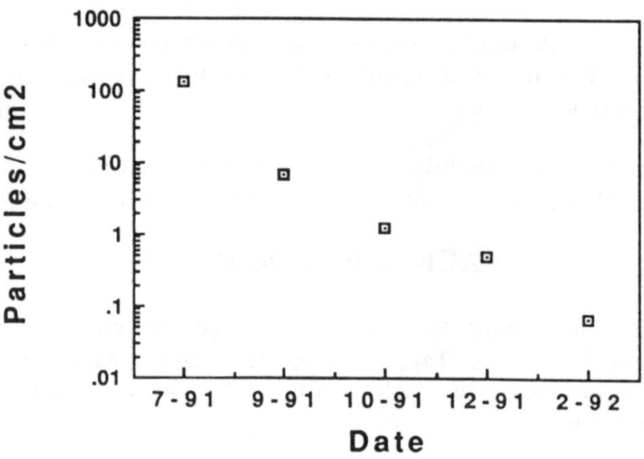

Fig. 2.   Particulate counts in particles per cm2, 0.3 micrometers and larger, spaced along an events line which relates to the chronological hardware development.

Fig. 3 presents the particle performance of an advanced ECR-CVD tool, operating with shields as described here, and designed to deposit interlayer dielectric on 200 mm wafers. The 200 mm measurement wafers were deposited with 10 kÅ of SiO2 and particles greater than 0.3 micrometer size counted using a Tencor 5500 particle counter. All other wafers received 30 kÅ deposit. It is seen that particulate contamination is now near 0.05 per cm2.

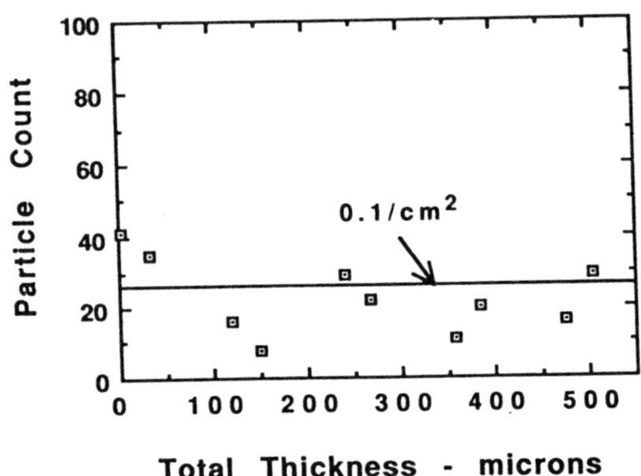

Fig. 3.   Particle count on 200 mm wafers as a function of total deposition. Counts were made on wafers that had received 10 kÅ of SiO2 deposition.

In summary, when attention is paid to the conditions set forth here, SiO2 can be deposited with a low level of particulate contamination.

ACKNOWLEDGMENTS

The authors would like to acknowledge the contributions made to this project by James Tappan, Stephen Lassig, Mike Walsh and the many other people who offered their suggestions and encouragement.

REFERENCES

1. D.R. Denison, Chien Chiang, and David B. Fraser, Electrochem. Soc. Ext. Abs. 89-1, 180 (1989).

2. D.R. Denison, SPIE Proceedings <u>1185</u>, 142 (1989(.

3. Seitaro Matsuo and Kenichi Ono, Jap. J. Appl. Phys. <u>22</u>, L210 (1983).

4. J.J. Wu and R.J. Miller, Am. Assoc. of Aerosol Res., Annual Meeting, 1988.

5. H. Oya, T. Ichihashi, and N. Wada, Jap. J. Appl. Phys. <u>21</u> , 554 (1982).

6. A. Volshchenkov, private communication.

# PARTICLE TRANSPORT AT SUBATMOSPHERIC PRESSURES:
MODELS AND VERIFICATION

R. Periasamy, T. Yamamoto, R. P. Donovan and A. C. Clayton
Research Triangle Institute
Research Triangle Park, North Carolina 27709

This paper reports on a program comparing calculated and experimental values of particle deposition velocity at pressures down to 100 pascals (1 millibar). The calculated values were obtained by incorporating the pressure dependence of gas density and the Cunningham slip correction factor into previously published particle deposition models. While the deposition mechanisms modeled included sedimentation, diffusion, thermophoresis, electrophoresis, and photophoresis, experimental measurements have been made so far to verify only the sedimentation model. The experimental results presented confirm the predicted increased importance of gravitational settling at subatmospheric pressures.

INTRODUCTION

Removal or elimination of all particle-generating sources in a vacuum chamber is one approach to reducing particulate contamination on wafer surfaces. However, that the primary sources of particulate contamination in both current and future wafer processing are forecast to be the process themselves and the equipment in which the processes are carried out implies that (1) the processing environment is unlikely to ever be completely particle-free (the process themselves generate particles), and (2) strategies to control particulate contamination during processing must therefore include control of the particle transport and deposition within the processing equipment. This paper reports on studies to understand and control particle deposition at subatmospheric pressures.

PARTICLE DEPOSITION MODEL

By extending well-known models of particle deposition velocity to subatmospheric pressures, predictions of the relative importance of the various particle deposition mechanisms at low pressures can be made. Two factors--the Cunningham correction factor and the gas density--account for virtually all the pressure dependence of the variables in existing deposition models [1]. This work presents calculated values of particle deposition velocity at pressures down to 10^2 pascals (1 millibar), obtained by incorporating the pressure dependencies of air density and the Cunningham correction factor into these previously published models. While this approach is in effect attempting to apply these models beyond the range over which they were intially derived, the limited number of experimental measurements made so far seems to verify that the extended models provide reasonably good predictions over the pressure range investigated (10^5 to 10^2 pascals). However, all measurements carried out so far have been limited to particle diameters of 0.3 μm and above.

The predicted pressure dependence of the particle deposition velocity attributable to gravitational settling, V_s, is plotted in Figure 1. Here, the deposition velocity V_s calculated for different pressures is presented as a function of particle diameter for deposition on a 10.2 cm diameter wafer, assuming a free stream velocity of 50.8 cm/s. The increase in the Cunningham correction factor with decreasing pressure accounts for all the increases predicted for V_s [1]. When the effects of gravity and convective diffusion are combined to determine the total deposition velocity, V_t ($=V_s + V_d$), the deposition velocity curves as shown in Figure 2 are obtained indicating that for particle diameters < 0.1-0.3 μm, diffusional deposition becomes important [1].

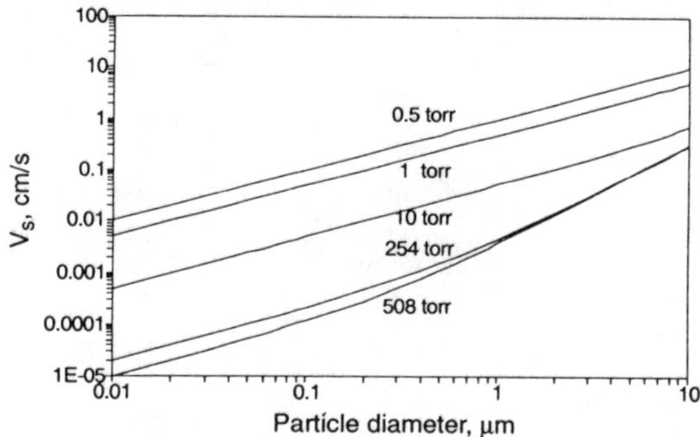

Figure 1. Pressure and particle diameter dependence of V_s.

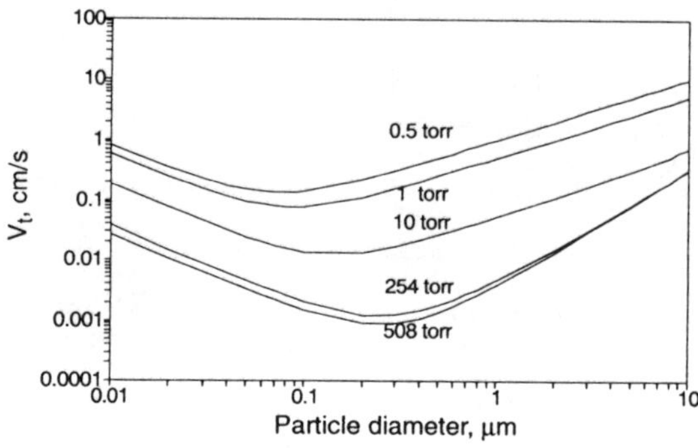

Figure 2. Pressure and particle diameter dependence of V_t.

PARTICLE DEPOSITION EXPERIMENTS

The predictions of Figures 1 and 2 were experimentally verified by measurements carried out in a custom-built vacuum chamber dedicated to particle research (Figure 3). This chamber incorporates a particle injection system consisting of a modified vibrating orifice aerosol generator (VOAG) which can generate particles of known and controlled composition, size, concentration and average electrical charge [2]. Thus, this chamber can be used to verify model predictions of subatmospheric pressure particle deposition velocity attributable to sedimentation, thermophoresis and electrophoresis. Diffusion dominated deposition occurs at particle diameters below those detectable by the laser surface scanner used in these experiments to count particles on the surface of the wafer (WIS 150). Photophoretic effects were investigated in another chamber and will be reported later.

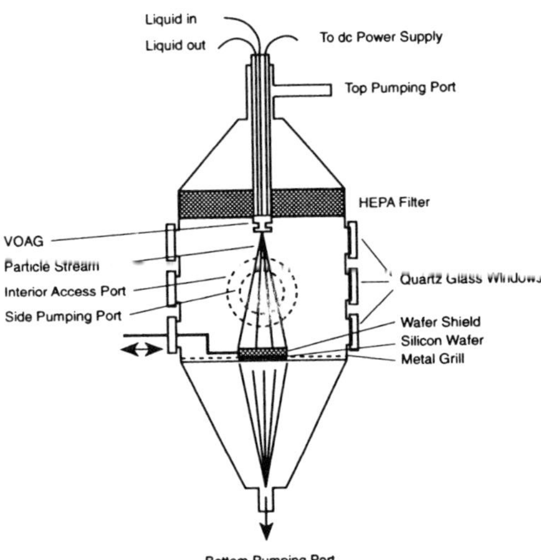

Figure 3. A cross-sectional view of the subatmospheric pressure chamber.

Sedimentation Experiments

To show that gravity is the dominant deposition mechanism for particles larger than 1 μm, particle deposition experiments were conducted at subatmospheric pressures in the chamber for 1.3 and 2.2 μm diameter dioctyl phthalate (DOP) droplets generated using the modified VOAG. For 1.3 μm DOP droplets, experiments were conducted at chamber pressures of 1, 10, 254 and 508 torr. For 2.2 μm DOP droplets, experiments were performed for 0.5 torr chamber pressure.

50

The particle deposition velocity on 4" silicon wafers placed in the vacuum chamber was measured at these pressures for both charged and uncharged particles produced by the modified VOAG. The procedure was to load the test wafer in a shielded position, seal the chamber and begin the pump down. Particle injection began shortly after the pump down started. After a steady state pressure was reached at the desired level, the silicon wafer was exposed to the particle stream for a known exposure time. After venting back to atmospheric pressure with the wafer shielded, the increase in the number of the particles on the wafer surface was measured using an Aeronca WIS 150 laser surface scanner.

To investigate the deposition of submicron particles at low pressures, experiments were conducted for PSL spheres 0.3 and 0.43 μm in size at a chamber pressure of 0.5 torr. To minimize the swelling of PSL in ethanol, methanol was used as the working liquid to generate methanol droplets containing PSL spheres. Particle deposition data were collected for both charged and uncharged PSL spheres in the low pressure chamber.

Thermophoretic Experiments

The influence of themophoresis on particle deposition as predicted by the deposition model developed earlier [1], was verified by conducting experiments in the low pressure chamber where the silicon wafers were heated to 50°, 85° and 120° C at a chamber pressure of 0.5 torr while maintaining the generation of the test particles at room temperature (approximately 11° C). The reduction in the number of particles deposited on the wafer due to the heating of the substrate for temperature differences of 40°, 75° and 110° C were measured in these experiments. PSL spheres, 0.3 and 0.43 μm in size and DOP droplets, 0.4 and 1.3 μm, in diameter were used for these experimental investigations.

RESULTS AND DISCUSSION
Particle Deposition Velocity Calculation

For the experiments reported here, the particle deposition velocity measured in the chamber was determined by:

$$V_m = \frac{\Delta n}{A\, t\, C_0} \tag{1}$$

where

V_m = measured particle deposition velocity, cm/s,
Δn = increase in total number of surface particles,
A = wafer surface area, cm^2,
t = time of exposure, s, and
C_0 = particle concentration adjacent to the wafer during deposition, $\#/cm^3$.

51

The particle concentration adjacent to the wafer surface depends upon both the particle injection rate, S, and the pumping rates of the vacuum system, Q_x, according to the following rate equation:

$$V \frac{dC_0}{dt} = S - Q_x C_0 \qquad (2)$$

where

S = number of particles injected per unit time, s^{-1},
Q_x = volume pumping rate of the system, cm^3/s,
V = volume over which the injected particles are distributed above the wafer surface, cm^3, and

$\dfrac{dC_0}{dt}$ = change of particle concentration with time, $\#/cm^3\text{-}s$

For steady state operation, equation (2) reduces to,

$$C_0 = \frac{S}{Q_x} \qquad (3)$$

S is known from the particle injection rate of the VOAG. Q_x varies with chamber pressure and throttle setting which is a primary control for pumping rate and the establishment of a constant pressure in the vacuum chamber. In the particle deposition experiments, it was measured after reaching a steady state.

Sedimentation Experiments

A comparison between the predicted pressure dependence of V_s (Figure 1) and that measured for DOP droplets in the vacuum chamber is illustrated in Figure 4. As evident in Figure 4, the measured V_m data agree well with the predicted magnitudes of V_s and show the predicted trends. However, these data verify only the sedimentation branch of the deposition curve without the inclusion of thermophoretic, electrophoretic or photophoretic forces. A match between the predicted V_t and the V_m measured for PSL spheres deposition in the low pressure chamber is also depicted in Figure 4. Here again gravity dominates at this pressure and particle size (1 torr, 0.3 μm), although the transition to diffusion-dominated deposition is beginning. Subatmospheric pressure particle deposition velocities measured for DOP and PSL spheres are listed in Table 1.

Thermophoretic Experiments

Repeatable thermophoretic particle deposition data could not be obtained with DOP droplets because heating the substrate tended to spread and coalesce the droplets deposited on the wafer. This provided

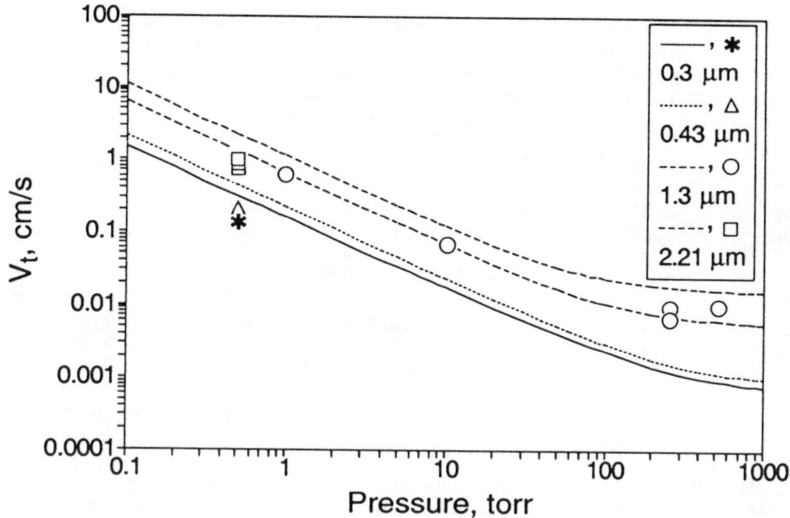

Figure 4. Comparison of calculated and experimental particle deposition velocities V_s and V_t for dioctyl phthalate (DOP) and polystyrene latex (PSL) spheres.

inconclusive particle deposition data because all the flaws in the surface scanner become saturated when there is a film on the surface under investigation. Particle deposition data, presented in Table 2, for 0.4 µm DOP droplet obtained at 260 torr pressure indicates that indeed a small reduction in number of particles deposited on the wafer is achievable when the wafer is heated to 50° C but the reduction was too small to be statistically significant. For DOP spheres, the measurement problems evidently dominated the thermophoretic phenomenon.

On the other hand, heating the wafer reduced the number of particles deposited on the wafer surface for the PSL particle studies which is in qualitative agreement with the trend predicted by the model. For example, particle deposition data depicted in Figure 5 shows that a significant reduction in the number of particles deposited on the wafer can be obtained simply by heating the substrate to 50° C for both charged and uncharged PSL spheres.

It was realized that during the performance of the thermophoretic experiments at low pressures that even though the PSL particles were detected in the correct size channel in the surface scanner, their number

Table 1. Subatmospheric Pressure Particle Deposition Data

Particle Species	Particle Size, μm	Pressure, torr	Charge, electrons	Particles Added/cm^2	V_m, cm/s
DOP	1.29	1.0	1,623	79.5	0.62
DOP	1.29	10.0	3,746	82.5	0.07
DOP	1.29	254.0	6,238	52.5	0.007
DOP	1.29	254.0	9,986	70.5	0.009
DOP	1.29	508.0	11,232	78.8	0.01
DOP	2.21	0.5	6,196	50.9	0.99
DOP	2.21	0.5	0	49.9	0.97
PSL	0.3	0.5	0	65.2	0.17
PSL	0.43	0.5	18,633	14.9	0.19
PSL	0.43	0.5	3,111	43.4	0.11

Table 2. Effect of Thermophoresis on Particle Deposition at Low Pressures

Particle Species	Particle Size, μm	Pressure, torr	Charge, electrons	Wafer Temp., $^{\circ}$C	Particles Added/cm^2	V_m, cm/s
DOP	0.4	260.0	1,644	16	24.4	0.01
DOP	0.4	260.0	1,644	50	23.1	0.0095
DOP	1.29	1.0	0	12	5.1	0.013
DOP	1.29	1.0	0	50	0.7	0.002
PSL	0.3	1.0	5,916	11	108.8	0.9
PSL	0.3	1.0	6,573	85	94.5	0.78
PSL	0.3	0.5	3,779	11	99.8	0.21
PSL	0.3	0.5	1,972	120	56.3	0.12
PSL	0.3	0.5	0	11	89.6	0.19
PSL	0.3	0.5	4,600	119	2.7	0.006
PSL	0.3	0.5	1,314	12	84.4	0.17
PSL	0.3	0.5	1,316	11	69.3	0.14
PSL	0.3	0.5	4,110	120	50.1	0.10
PSL	0.43	0.5	496	12	50.3	0.13
PSL	0.43	0.5	775	50	30.6	0.08
PSL	0.43	0.5	0	12	35.5	0.09
PSL	0.43	0.5	0	50	20.2	0.05

changed as the wafers were aged. That is, as the wafer was allowed to sit for some time after the deposition experiment, the number of particles in all the size bins changed dramatically, indicating that the droplets generated were not dry. This was not the case with heated subtrates in that the particle counts did not change as a function of time. Anamolous behaviour observed in the thermophoresis experiments needs further investigation.

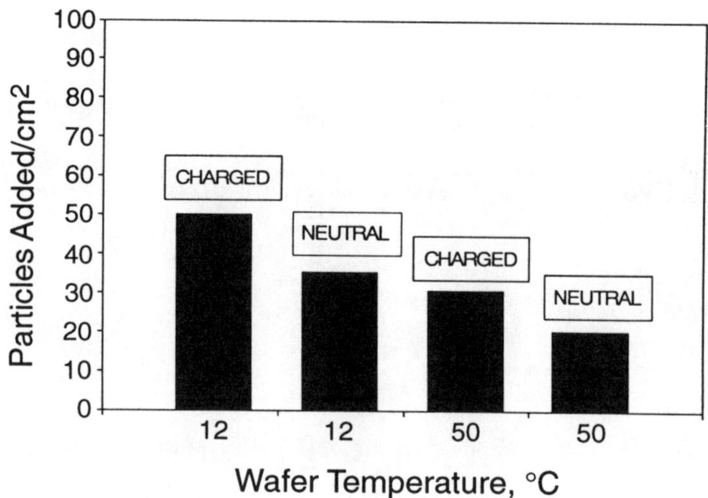

Figure 5.   Data from thermophoretic experiments for 0.43 μm polystyrene latex (PSL) spheres at 0.5 torr pressure.

Particle deposition velocities measured for the theromophoretic experiments performed at reduced pressures for PSL spheres and DOP Droplets are presented in Table 2. Future experiments will concentrate on obtaining repeatable thermophoretic particle deposition data that can be used to validate and improve the theoretical model developed.

CONCLUSIONS

Experimentally measured particle deposition velocities at reduced pressures for particles larger than 1 μm, where sedimentation was expected to dominate, agreed reasonably well with the predictions made by the deposition model. Similarly, for submicron particles, the model prediction qualitatively agrees with the experimental data obtained at subatmospheric pressures. Thermophoretic experiments conducted in the low pressure chamber indicates that the substrate can be heated to reduce particle contamination on its surface. Measurements to include electrophoresis at 100 pascals are under way.

REFERENCES

[1]   R. P. Donovan, T. Yamamoto, R. Periasamy and A. C. Clayton, "Mechanisms of Particle Transport in Process Equipment," Fall Meeting of the Electrochemical Society, Toronto, Canada, Oct. 1992.

[2]   R. Periasamy, A. C. Clayton, P. A. Lawless, R. P. Donovan and D. S. Ensor, "Generation of Uniformly Sized, Charged Particles in a Vacuum," Aerosol Sci. Tech. 15:256-265 (1991).

ACKNOWLEDGMENT

This research was supported by the Semiconductor Research Corporation, Manufacturing Process Sciences, under contract 91-MJ-199.

SIZING ACCURACY, COUNTING EFFICIENCY, LOWER DETECTION LIMIT AND REPEATABILITY OF A WAFER SURFACE SCANNER FOR IDEAL AND REAL-WORLD PARTICLES

Benjamin Y. H. Liu, Seung-Ki Chae
Particle Technology Laboratory
Department of Mechanical Engineering
University of Minnesota, Minneapolis, MN 55455

and

Gwi-Nam Bae
Thermal Fluid Engineering Laboratory
Korea Institute of Science and Technology, Seoul, Korea

ABSTRACT

The performance characteristics of a Tencor Surfscan 4000 wafer surface scanner have been evaluated using ideal, polystyrene latex (PSL) spheres and irregularly shaped, real-world particles of Si and SiO_2. The particles were uniform in size and were deposit on bare silicon wafers and used as standard calibration wafers to study the scanner response. Particles in the 0.1 to 1.0 µm diameter range were used. The sizing accuracy, counting efficiency, lower detection limit, and count repeatability of the scanner were studied systematically. The full Maxwell's electromagnetic equations have also been solved numerically on a supercomputer to obtain the light scattering cross-section of the particles on bare silicon wafers. The calculation compares favorably with the experimental results. Because of the high refractive index of silicon, the wafer surface scanner has been found to detect Si particles to a considerably smaller size than PSL. The data suggest that Si particles as small as 0.1 µm have been detected by the Surfscan 4000 with 90% counting efficiency even though the nominal lower limit of the instrument is 0.3 µm based on PSL calibration.

INTRODUCTION

Control of particulate contamination of wafer surfaces is very important in semiconductor manufacturing. To detect surface particulate contaminants, a wafer surface scanner is usually used. Such scanners are essential for many applications, such as inspection of incoming bare silicon wafers to determine their cleanliness level, measuring particles added by a semiconductor processing equipment during operation, evaluation of the efficiency of wafer cleaning systems, and determining the cleanliness of the wafer fabrication environment.

During the last decade, much effort has been directed to developing better and improved commercial wafer surface scanners, which generally make use of laser light scattering to detect both contaminant particles and defects on wafer surfaces. Measurement of the scattered laser light by a particle on a wafer surface by means of a photo-detector can provide information on the size of the particles and the ability of a specific scanner to detect submicron particles. Several wafer surface scanners are available commercially and they include those manufactured by Tencor, Estek, PMS, Censor and Hitachi. These are widely used by the semiconductor industry as measurement tools during the production of very large scale integrated circuit devices.

Detection of particles suspended in air or a liquid medium by laser light scattering is another well known technique of particle measurement and many air- or liquid-borne particle counters are available. For such counters, the measured intensity of scattered light depends on the refractive index, shape and diameter of particle, as well as the optics of the instrument, including the laser wavelength, polarization, and the collection angles used. These parameters are also important in surface particle detection by light scattering. In addition, the incident angle of laser light on the wafer surface and the optical properties of the wafer itself also become important. Due to the many parameters involved and the complex nature of the near field interaction between a particle and the substrate wafer surface, the scattered intensity field cannot be simply obtained by the existing light scattering theory, or by solving the Mie scattering equations.

Since each commercial wafer surface scanner has its own unique optical design, the response of these scanners is also expected to be different. For the Tencor Surfscan series scanners, and the ANS scanner of Censor, the incident laser beam is normal to the wafer surface. For the LS series scanners of Hitachi, WIS series scanner of Estek, and the SAS series of PMS, oblique incident angles are used. The incident angles are respectively 7°, 15° and 60° from the normal. The laser source used is generally He-Ne of $\lambda=0.633$ μm wavelength or Ar ion of $\lambda=0.488$ μm. A high power laser with short wavelength can increase the scattered ligtht signal, but it also increases the noise level. By suitably combining the laser source, and optimizing the incident and collection angle geometry, advanced wafer surface scanners have achieved a lower detection of 0.1 μm diameter for PSL spheres. Since the wafer is generally scanned by sweeping the beam while the wafer is transported through the system or rotated on an axis, some uncertainly in particle size or count can also result from the specific scanning system used. Some commonly used scanning techniques include arc raster scan, and telecentric and spiral scans.

The performance of a wafer surface scanner is generally given by the manufacturer on the basis of calibration using polystyrene latex (PSL) spheres on bare silicon wafers. Size calibration studies using PSL spheres have been reported by several authors and recently summarized by Chae et al [1]. The performance of the scanner for real-world particles of a non-spherical shape and a different refractive index than PSL is generally unknown. Further, the scattering cross-section of a PSL sphere for many scanners is not a monotonically increasing function of particle size. This gives rise to sizing uncertainties when the instrument is used to measure size of the contamiant particles on the wafer. In the PMS SAS scanner, a special optical design is used to obtain a monotonic, scattering cross-section vs. particle size curve. However, the actual response of this and other commercial scanners to irregular, non-spherical, real-world particles of a different refractive index remains largely unknown and large uncertainties remain in the interpretation of particle size data given by these scanners.

Besides the particle size response, other performance characteristics of wafer surface scanners are also important and they include the counting efficiency (sometimes referred to as the capture ratio), lower detection limit and count repeatability. The purpose of this paper is to describe some recent research on techniques to measure the important performance parameters of wafer surface scanners. Advances in particle technology has made it possible to generate real-world particles of a uniform and accurately known particle size, and these particles are used to study the performance of wafer surface scanners in the 0.1 to 1.0 μm diameter range. A state-of-the-art particle generation and deposition system has been developed in our Laboratory and used to deposit monodisperse Si, SiO_2 as well as ideal PSL spheres uniformly on bare silicon wafers. These are then used as standard test wafers for studying the performance characteristics of a Tencor Surfscan-4000 wafer surface scanner available in our Laboratory.

EXPERIMENTAL

The experimental methods used to study the response characteristics of a wafer surface scanner includes generating standard test particles, depositing them on bare silicon wafers and then using the wafers to determine the performance characteristics of the scanner experimentally. The procedure is described below.

Wafer Surface Scanner

The wafer surface scanner used in this study is the Tencor Surfscan-4000, shown schematically in Fig. 1. The actual instrument is located in a class 10 clean room at the University of Minnesota. The Surfscan 4000 is an older model instrument with a nominal 0.3 μm lower detection limit for PSL particles, as compared to the 0.1 μm lower detection limit for more modern instruments such as the Surfscan-6200. However, the optical systems of the Surfscan series instruments are quite similar. A He-Ne randomly polarized laser with λ=0.633 μm wavelength is used and the laser light beam is swept across the wafer surface by a rotating, scanning mirror, while the wafer is transported through the system. This is usually referred to as the arc raster scan technique. The laser beam is not always perfectly normal to the wafer in the scan direction. In the absence of particle or a surface defect, the laser beam is reflected through a slot in the cylindro-elliptical mirror. When particles or surface defects are present, light is scattered over a wide angle. The scattered light is then focused by the cylindro-elliptical mirror into the line opening in front of an integrating sphere, and directed to the photomultiplier.

Standard Particles and Test Wafers

Three different kinds of particles, including PSL spheres, and irregular Si and SiO_2 particles, are used in this study. The monodisperse PSL spheres are generated by atomizing a liquid suspension of the spheres and depositing them on bare silicon wafers using our laboratory particle deposition apparatus. Fifteen particle sizes in the 0.204 to 0.966 μm diameter range were used. Si and SiO_2 particles are chosen as real-world particles because they are frequently found contaminants in the semiconductor manufacturing environment. Since the particles are non-uniform in size, they are first classified into narrow size ranges and then deposited on the wafer. Nine discrete particle sizes in the range of 0.1 to 1.0 μm range for Si and eight sizes for SiO_2 are used. The refractive indices of Si and SiO_2 are respectively m=3.88 - i 0.02 and m=1.46 - i 0.0 at λ=0.633 μm at the He-Ne laser wavelength, while that for PSL is m=1.59 - i 0.0.

The standard test wafers described above are prepared in the same Class 10 cleanroom where the Tensor Surfscan 4000 is located. Since the monodisperse particles must be deposited in a contaminant-free environment, a new state-of-the-art particle deposition system has been constructed and used in these experiments to prepare the standard calibration wafers. The operating principle of the system is similar to that described by Pui et al [2] for particle deposition, and standard test wafers can be prepared for particles ranging from 0.05 μm to approximately 2.0 μm in diameter. For these experiments, bare silicon wafers showing < 40 particles are used, and usually the particles added range from 100 to 2000.

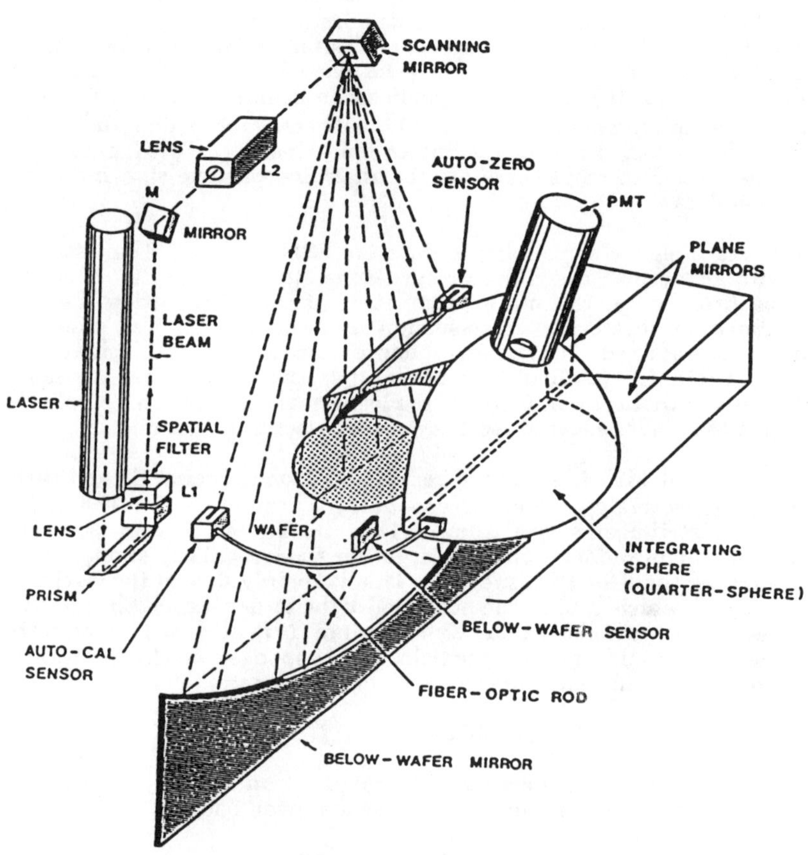

Fig. 1. Schematic optics diagram of Tencor Surfscan-4000 wafer surface scanner

DATA ANALYSIS

The Tencor Surfscan-4000 gives results in terms of scattering cross-sections to a lower limit of 0.06 μm^2. Besides a wafer map showing the location of the individually detected particles on the wafer, a histogram, and the total number of particles in each particle size channels, it also provides a table on the number of particles in different scattering cross-section ranges in increments of 0.01 μm^2 between 0.06 to 2.55 μm^2. The data from these tables are further analyzed using a DISTFIT statistical package developed in our Laboratory for analyzing particle size and other distribution data.

Figure 2 shows some sample data for PSL spheres. The data are fitted with lognormal distribution curves using the DISTFIT package. From the fitted data, the main parameters of the distribution, i.e. the number median scattering cross-section and the geometrical standard deviation, are obtained. These distribution parameters for PSL spheres are shown in Table 1. It should be noted that the geometrical standard deviation is a statistical parameter of the lognormal distribution. A value of 1.0 means that the measured scattering cross-sections are uniform.

Fig. 3. and Fig. 4. show the results for monodisperse Si and SiO_2 particles, respectively. The corresponding parameters for the fitted lognormal distributions are summarized in Tables 2 and 3. The geometrical standard deviations are larger for the irregularly shaped, real-world particles than for PSL spheres. This is mainly due to the irregular particle shape, which causes the scattered light to depend on the random orientation of the particles on the wafer surface. It is evident that both the lower detection limit and the particle size responses of the real-world particles are quite different from those for PSL spheres.

RESULTS

The principal experimental results on the performance characteristics of the wafer surface scanner are given below.

Particle Sizing Accuracy

The ability of a wafer surface scanner to measure accurately the size of a deposited particle on a wafer depends on the response characteristics of the scanner to particles of that specific material. Experiments were performed first with PSL spheres. Theoretical calculations were then made on the scanner response and compared with the experimental data.

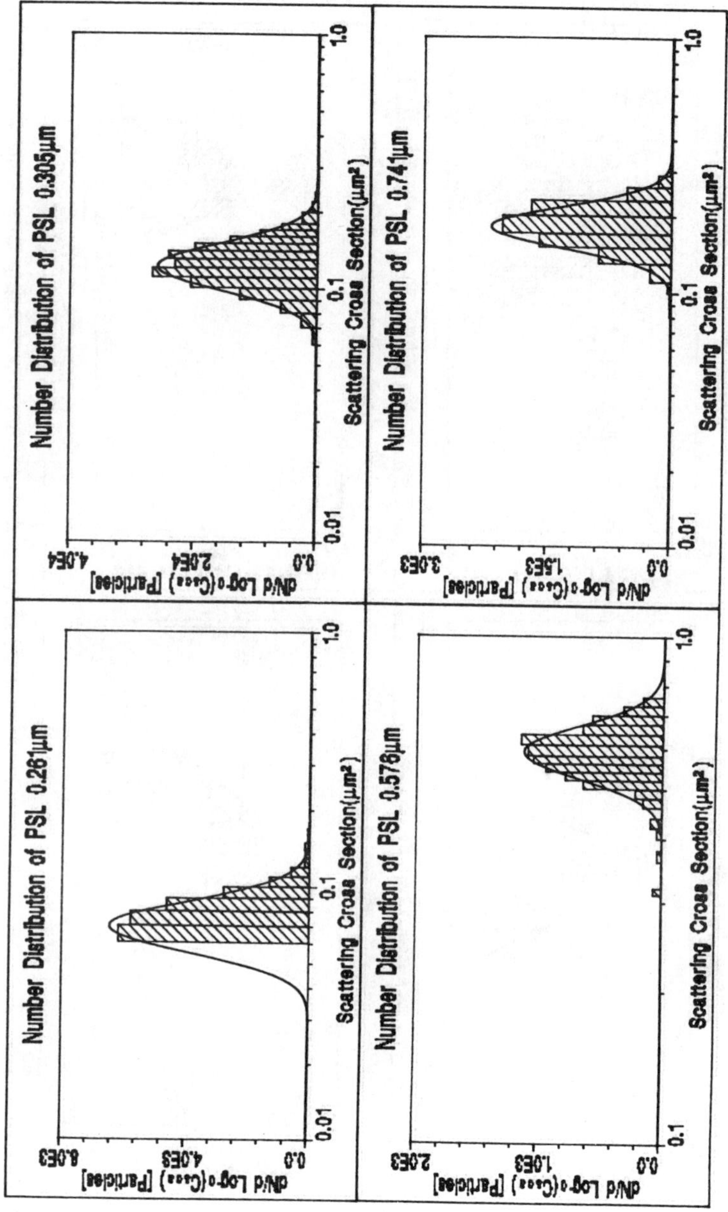

Fig. 2. Response of the Tencor Surfscan-4000 showing distribution of scattering cross section for monodisperse PSL spheres

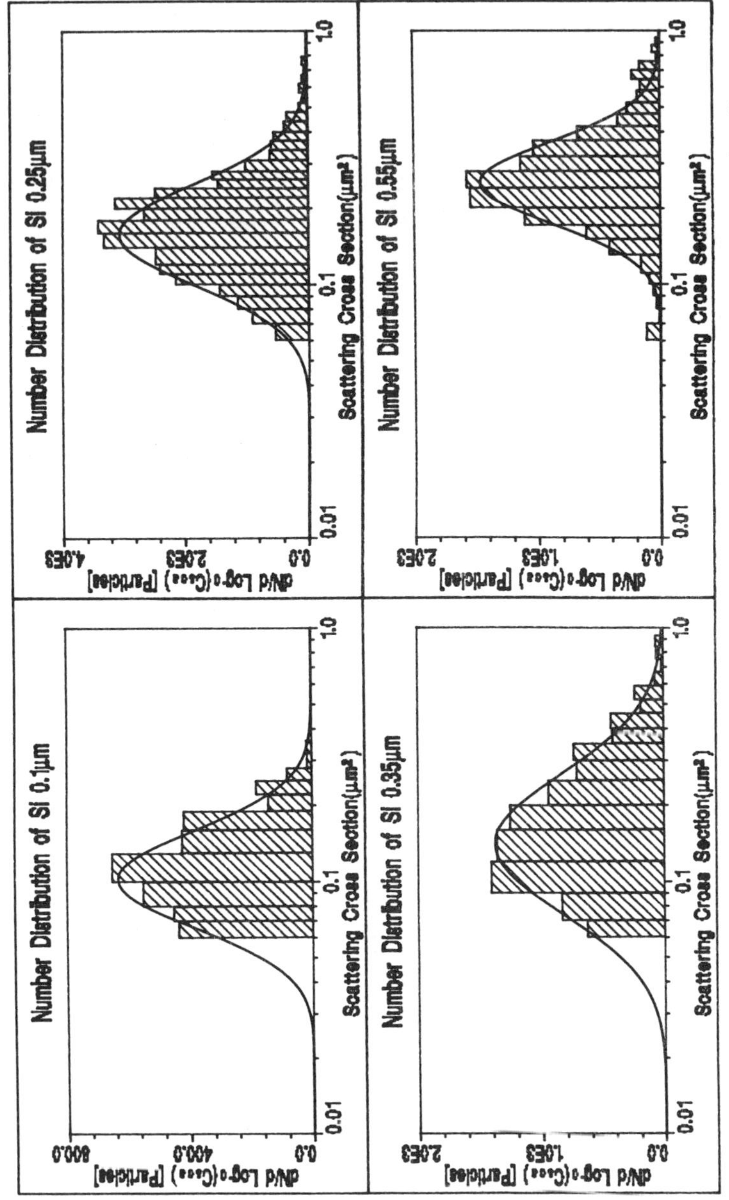

Fig. 3. Response of the Tencor Surfscan-4000 showing distribution of scattering cross section for monodisperse Si particles

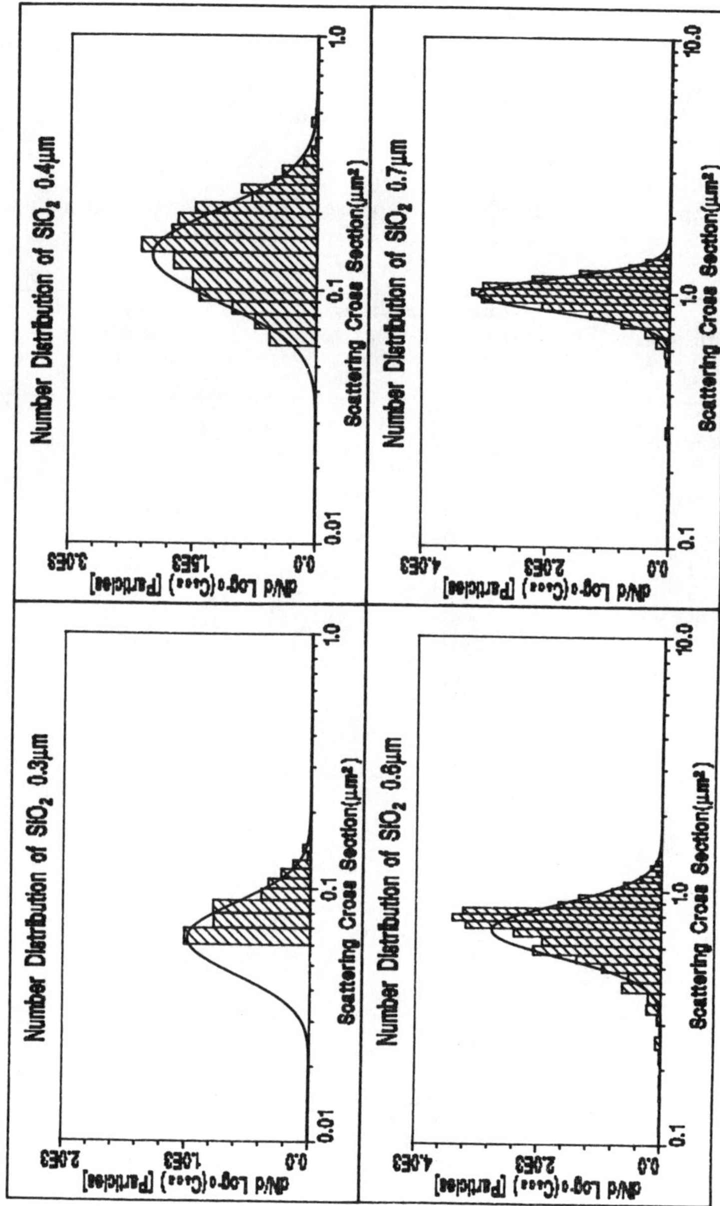

Fig. 4. Response of the Tencor Surfscan-4000 showing distribution of scattering cross section for monodisperse SiO₂ particles

Table 1. Experimental particle size response of the Tencor Surfscan-4000 for
PSL spheres

Particle diameter (μm)	Number median scattering cross section (μm^2)	Geometrical standard deviation
0.261	0.0694	1.2615
0.305	0.1215	1.2395
0.327	0.1276	1.2364
0.368	0.1947	1.1585
0.431	0.3245	1.1382
0.453	0.4422	1.1267
0.482	0.4674	1.1206
0.546	0.5861	1.1059
0.576	0.5941	1.1069
0.643	0.2736	1.1642
0.741	0.1792	1.1817
0.839	0.3175	1.1511
0.913	0.7194	1.1013
0.966	0.9258	1.1026

Table 2. Experimental size response of the Tencor Surfscan-4000 for
Si particles

Particle diameter (μm)	Number median scattering cross section (μm^2)	Geometrical standard deviation
0.1	0.1024	1.557
0.15	0.1393	1.5508
0.25	0.1574	1.5791
0.35	0.1426	1.8498
0.45	0.2122	1.5477
0.55	0.2511	1.4557
0.65	0.3540	1.2680
0.75	0.4507	1.1946
0.85	0.4964	1.2206

Table 3. Experimental size response of the Tencor Surfscan-4000 for
SiO_2 particles

Particle diameter (μm)	Number median scattering cross section (μm^2)	Geometrical standard deviation
0.3	0.0636	1.3802
0.4	0.1362	1.5144
0.5	0.3898	1.3746
0.6	0.6999	1.2689
0.7	0.9806	1.1631
0.8	0.9799	1.2957
0.9	0.8331	1.5645

Finally, the response of the scanner to real-world particles is experimentally studied using irregularly shaped particles of Si and SiO$_2$.

Fig. 5 shows the experimental result on the measured median scattering cross-sections for PSL spheres, taken from Table 1, plotted against the particle diameter. The theoretical results based on mathematical modeling and numerical calculation are also shown for comparison. The experimental data for PSL spheres >1.0 μm in diameter were taken from our previously reported study [3].

The above figure clearly illustrates the problem involved in size measurement based on scattering cross-section using calibration data for PSL. Since the scattering cross-section does not increase monotonically with increasing particle size, a large scattering cross-section does not necessary mean a large particle size. There are two peaks and valleys in the experimental calibration curve in the range between 0.26 μm and 2.003 μm. The first peak and valley are found to be located at D_p=0.576 μm (=0.91λ) and D_p=0.741 μm (=1.17λ), respectively, and the second, D_p=1.09 μm (=1.72λ) and D_p=1.25 μm (=1.98λ), respectively. The oscillatory nature of the response curve makes it impossible to precisely determine the particle size of PSL spheres from scattering cross-section measurement alone.

The origin of this oscillatory response can be understood by making theoretical light scattering calculations. The calculation was based on a modified version of the Maxwell's equations solver originally developed by Wojcik et al. [4]. In the program, which has been implements on a Cray supercomputer, the full Maxwell's electromagnetic equations were solved numerically using an axisymmetrical system geometry with a normal incident laser beam. The P- and S- polarized, angular intensity distribution functions, $I_p(\theta)$ and $I_s(\theta)$, were obtained for the PSL sphere. This angular intensities were then integrated within the collection angle of the wafer surface scanner to obtain the scattering cross-section using the following equation[5],

$$C_{sca}(\theta_1,\theta_2) = \int_0^{2\pi}\int_{\theta_1}^{\theta_2}(I_p(\theta)\cos^2\varphi + I_s(\theta)\sin^2\varphi)\sin\theta\, d\theta\, d\varphi \tag{1}$$

The results are then compared with the experiments. The numerical calculation is successful in predicting the oscillatory response of the wafer scanner. However, only six numerical points were obtained due to the large amount of supercomputer time need for each calculation.

Although the numerical calculation can provide an exact solution for this problem, other less computational intensive modeling approaches are needed for solving practical problems. We have developed two PC-based

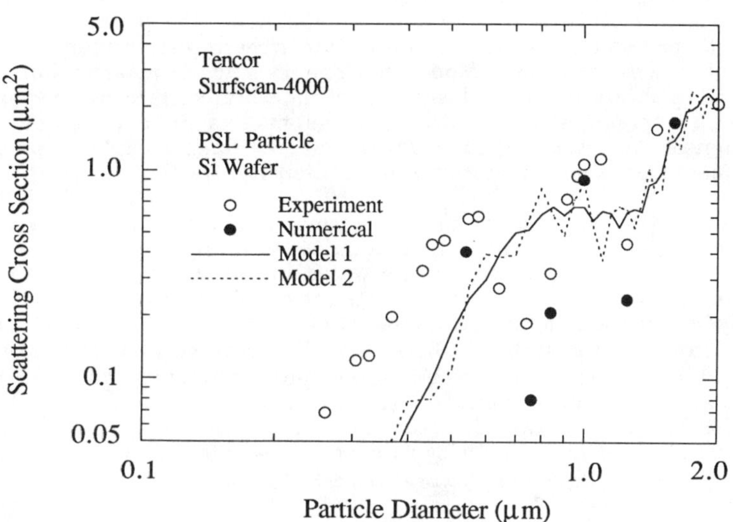

Fig. 5. Theoretical and experimental particle size response of Tencor Surfscan-4000 for PSL spheres

models for this purpose. These are referred to as Model 1 and Model 2, both of which incorporate the Mie light scattering theory and Fresnel surface boundary condition [5]. These models seek to obtain a distribution of angular scattered intensity by superimposing the scattered and reflected components calculated by Mie theory. Model 1 adds the two scattered components from the Mie theory calculation without considering their phase difference, the first component being the directly back-scattered component, and the second, forward-scattered component that has been reflected by the wafer surface. Only the first order interaction between a sphere and a surface is considered. In Model 2, the phase difference between these two components are taken into account, resulting in a higher degree of oscillation than Model 1. The P- and S- polarizations are integrated to obtain the overall response of the wafer surface scanner using equation (1) given above. Although the models do not follow the experimental data exactly, they do provide reasonable explanation and approximate trends for the scattered light intensity.

The response of a wafer surface scanner to real-world particles of an irregular shape and different refractive index is an important problem that need to be studied before data from a wafer surface scanner can be more clearly intepreted and understood. In our previously reported study [1], the size response of the Tencor Surfscan-4000 to transparent (SiO_2, Si_3N_4 and Al_2O_3) and light-absorbing (Si, Al and Zn) real-world particles were investigated. The concept of an equivalent scattering diameter was introduced. For more detailed size response comparisons and for evaluation of the true lower detection limit of the counter, a more detailed study using Si and SiO_2 particles in narrow size intervals is presented in this study and corresponding theoretical calculations made.

The theoretical scattering cross-sections for the Tencor 4000 calculated by means of Model 1 are shown in Figure 6 for Si, SiO_2 and PSL. Since the particles are assumed to be spheres, the results only shows the effect of particle refractive index, and not that due to particle shape. For SiO_2 whose refractive index is smaller than PSL, the scattering cross-section is seen to be smaller in the submicron size range. For Si, whose refractive index is much higher than PSL, the response is highly oscillatory. Further, the average slope of the response curve is much smaller compared to PSL as the lower detection limit of the instrument is approached.

Fig. 7 shows the experimental scattering cross-sections for Si, SiO_2 and PSL. For the real-world particles of Si, the oscillatory character of the response curve as predicted by theory has been largely damped out due to the irregular particle shape and the random orientation of the particles on the silicon wafer surface. In addition, the slow decrease in scattering cross-section as the lower detection limit is approach actually makes the real-

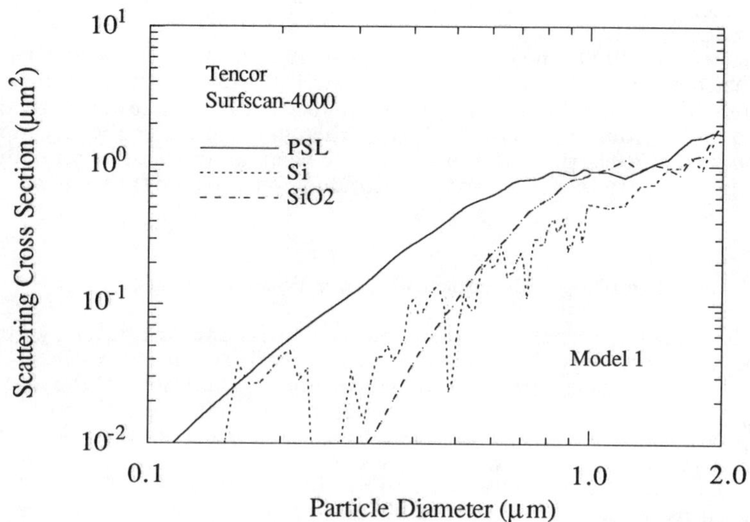

Fig. 6. Theoretical size response of Tencor Surfscan-4000 for PSL spheres and Si and SiO_2 particles

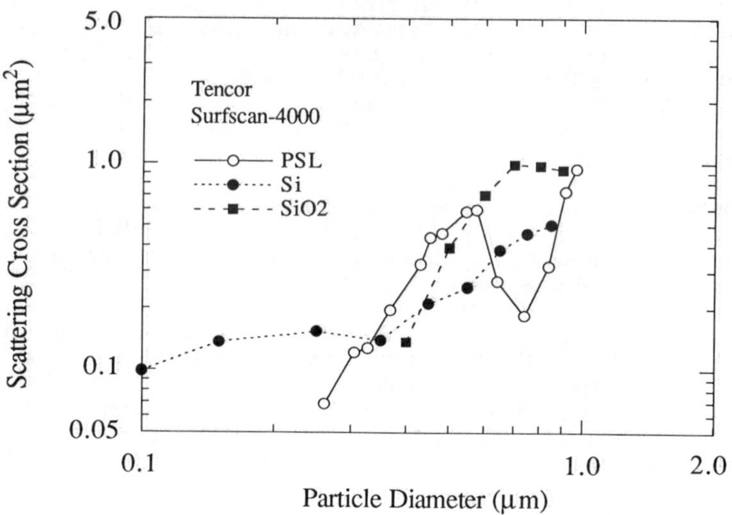

Fig. 7. Experimental size response of Tencor Surfscan-4000 for PSL spheres and Si and SiO_2 particles

world silicon particles to appear larger than PSL spheres of the same size. As a result, we have been able to see Si particles as small as 0.1 μm, using Tencor Surfscan 4000 whose nominal lower limit is 0.3 μm based on PSL sphere calibration. In the case of SiO_2 particles, whose refractive index is only slightly smaller than PSL, the response curve and lower detection limit are quite similar to those for PSL, although the peak and valley in the scattering vs. particle size curve have largely been damped out as a result of the irregular particle shape and the random particle orientation on the wafer surface.

Counting Efficiency and Lower Detection Limit

The counting efficiency (also referred to as capture ratio by some authors) of a wafer surface scanner can be defined as the ratio of the number of particles counted by a wafer surface scanner to the actual number of particles on the wafer surface. The counting efficiency is a function of particle refractive index, and generally drops as the lower limit of the instrument is approached. One can define the lower particle size limit of a scanner at the 50% counting efficiency level, or at some other levels such as 90%. The counting efficiency of wafer surface scanners has been estimated by repeatedly scanning a wafer and analyzing the data statistically [6]. A different approach is used in the present study.

The counting efficiency of the Tencor Surfscan 4000 is determined indirectly in this study by statistical curve fit through the distribution data for scattering cross-sections. As shown in Figures 2, 3, and 4, for particles of a uniform size on the wafer, the measured distribution of scattering cross-sections can be fitted by a lognormal curve. Such is the case for PSL spheres of 0.305, 0.578 and 0.74 μm in Figure 2. However, for PSL spheres of 0.26 μm, the distribution curve is truncated at the lower end by the finite lower limit of the system. By fitting a lognormal curve through such a truncated distribution, the fraction of counts missed by the system can be calculated as the ratio of area under the curve below the lower limit and the total area under the curve. By this means the counting efficiency of the Tencor Surfscan 4000 has been estimated for PSL spheres as well as Si and SiO_2 particles. The results are shown in Figure 8.

It is interesting to note in Figure 8 that although the counting efficiency of the Tencor Surfscan 4000 drops off sharply as the nominal lower limit of the instrument is reached for PSL and SiO_2 particles, the drop off is much more gradual for Si particles. This can be attributed to the high refractive index of Si and the strong scattering by the high refractive index material. The counting efficiency for Si remains as high as 90% at 0.1 μm. For PSL spheres, the counting efficiency is 100% at D_p=0.305 μm, 50% at D_p=0.261 μm, and 0% at D_p=0.204 μm. For particles of SiO_2 , 100% of the

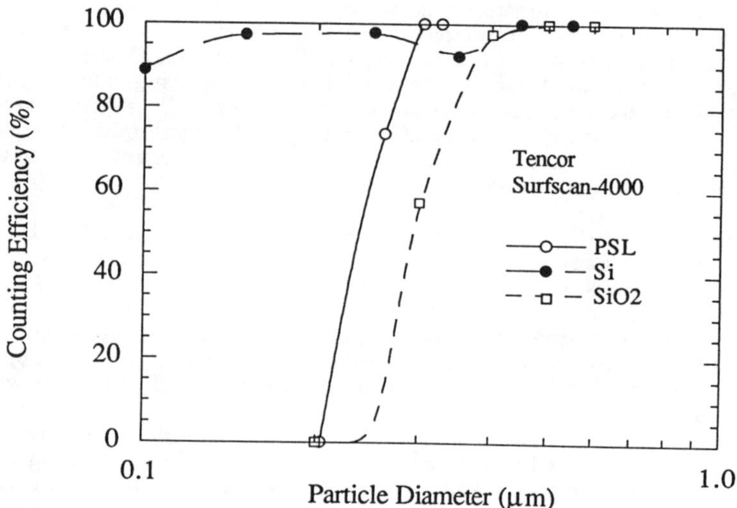

Fig. 8. Experimental counting efficiency of Tencor Surfscan-4000 for PSL spheres and Si and SiO_2 particles

particles are counted at D_p=0.4 µm, 50% at D_p=0.3 µm and 0% at D_p=0.2 µm.

Count Repeatability

The count repeatability is another important factor that must be taken into account when applying wafer surface scanner to actual particulate contaminant measuring problems. As most wafer surface scanners use a pixel counting algorithm, a scanner can under- or over-count the actual number of particles on a wafer surface. According to the manufacturer, a signal processing, Pulsed Position Correlator has been incorporated into the Surfscan to improve the counting accuracy [7]. However, there are other factors, such as uncertainty and variation caused by edge exclusion and wafer orientation, and possible contaminant particle addition during scan, that contribute to variability in count that can be established only by experiments.

Standard statistical treatment methods can be applied to the measurement of count repeatability of wafer surface scanners. For pure Poisson distributions, the standard deviation in count should be equal to the square root of the mean count. However, it has been found that the actual count repeatability is quasi-Poisson [8], with the standard deviation equal to one-half of the mean count. This behavior has been explained and applied to the prediction of detection probability, or counting efficiency, of wafer surface scanners [6]. The count repeatability of a wafer surface scanner is especially important when the scanner is used to determine the efficiency of wafer cleaning systems by comparing the wafer particle count before and after cleaning [9], where variability in count can lead to errors and serious data interpretation problems.

In the previously reported studies on count repeatability, standard calibration wafers carrying PSL spheres of a specific particle size were generally used. However, the influence of particle size, particle material and such factors as the number of particles on the wafer and the number of runs used has either been unknown, or only partially investigated.

In the present experiments, a series of standard 100 mm diameter calibration wafers carrying uniform sized PSL, Si and SiO_2 particles were prepared and the number of particles on the wafer was varied between 100 to 2000. The background particle count on the bare silicon wafer before particle deposition was kept below 40. A constant edge exclusion of 6 mm on a 100 mm wafer was used. Particle distribution on the wafer surface was found to be quite uniform. Each test wafer was scanned 10 times by passing each through the system by the automatic wafer transport system.

Fig. 9 shows the experimentally measured standard deviation in particle count, i.e. square root of variance, of the 10 repeated scans for each wafer plotted against the mean total count per wafer. The theoretical relationship between the standard deviation, σ, and the mean total count, X_m, given by $\sigma^2 = X_m$ for a pure Poisson distribution, is shown for comparison. The experimental standard deviation is seen to be generally below than the value given by the Poisson distribution, and no strong trends can be seen on the influence of particle size, material or refractive index on the standard deviation in count can be discerned.

To further analyze the data, we define a residual variance by the following equation

$$\sigma_{residual}^2 = X_m - \sigma^2 \tag{2}$$

The residual variance is the difference between the variance for a Poisson distribution, X_m, and the experimentally measured variance for the wafer surface scanner, σ^2. It provides a measure for the variance in count other than that caused by a pure Poisson distribution.

When the residual variance is plotted against the mean total count as in Fig. 10, a strong correlation is obtained between the residual standard deviation and the mean total count. The data can be correlated by the equation:

$$\sigma_{residual} = a \; X_m{}^b \tag{3}$$

where $a = 0.83$ and $b = 0.52$. The correlation coefficient is found to be $R = 0.93$. Since the exponent 0.52 is close to 0.5, the residual standard deviation is also quasi-Poisson in nature.

The above analysis shows that for the Tencor Surfscan 4000 used in the present study, the standard deviation in count can be represented by the equation,

$$\sigma^2 = X_m - \sigma_{residual}^2 = X_m - (a \; X_m{}^b)^2 \approx (1 - a^2) X_m \tag{4}$$

The coefficient, a, thus provides a quantitative measure of the degree of repeatability in count for the specific scanner--the larger the value of a, the smaller the actual standard deviation, σ , and hence, the more repeatable is the count. A comparison of the value of the coefficient a for different scanners can provide quantitative data on the relative repeability of

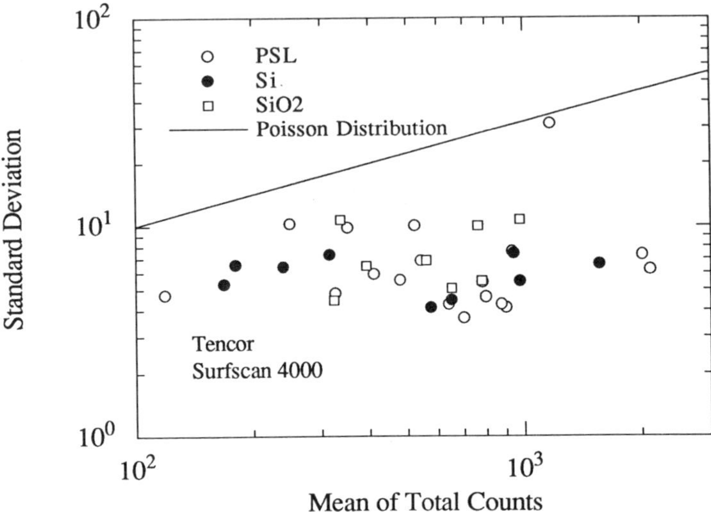

Fig. 9. Standard deviation of count as a function of total particle count on wafer showing count repeatability of the Tencor Surfscan-4000 for PSL spheres and Si and SiO_2 particles

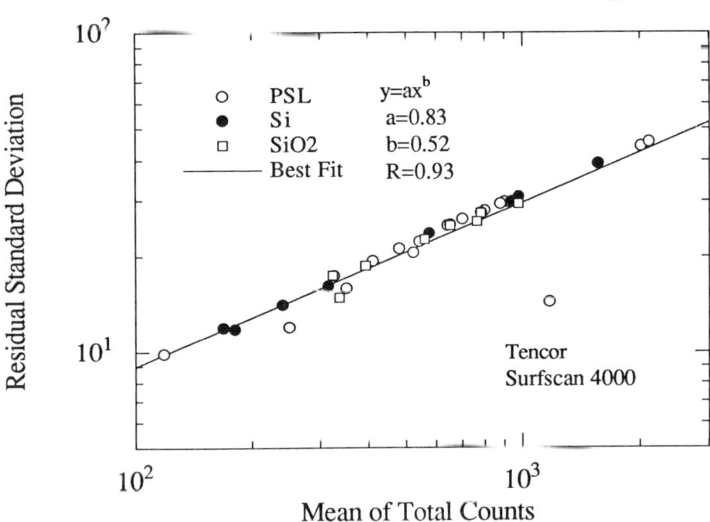

Fig. 10. Plot of residual standard deviation as a function of mean total particle count showing correlation between these two parameters for Tencor Surfscan-4000 with PSL spheres and Si and SiO_2 particles

different counters. This is an important measure of counter repeatability that needs to be studied for more commercial counters.

CONCLUSIONS

The important performance parameters for a Tencor Surfscan 4000 wafer surface scanner have been experimentally determined. They include the sizing accuracy, counting efficiency, lower detection limit, and count repeatability. Studies have been made using uniform sized PSL spheres and non-spherical particles of Si and SiO_2 deposited on bare silicon wafers. Particles in the 0.1 µm to 1.0 µm diameter range were used.

For PSL spheres (m=1.59 - i 0.0), an increase in particle size does not always lead to an increase in the scattering cross-section and the median scattering cross-section vs. particle size curve shows an oscillating behavior with two peaks and valleys in the size range between 0.261 µm to 2.0 µm. This oscillatory behavior has been predicted on the basis of numerical solution of the full Maxwell's electromagnetic equations and confirmed by approximate models based on Mie scattering and Fresnel surface reflection boundary conditions. The same oscillatory behavior in the scattering cross-section vs. particle size curve also applies to the theoretical results for Si (m=3.88 - i 0.02) and SiO_2 (m=1.46 - i 0.0) particles, with the former showing greater oscillation and larger number of peaks and valleys than PSL. However, the experimental result shows that this theoretically predicted oscillation based on a spherical particle shape has been largely damped out for real Si and SiO_2 particles of an irregular particle shape. This is believed to be due to the random orientation of the particles on the wafer surface and the influence of orientation on light scattering.

The counting efficiencies of the Tencor Surfscan 4000 for PSL spheres, and irregularly shaped particles of Si and SiO_2 particles have been determined by a statistical curve fit procedure. If one defines the lower detection limit of a scanner as the size at which 50% of the particles are counted, then the measured lower detection limit is 0.25 µm for PSL sphere, and 0.3 µm for SiO_2 particle as compared with manufacture stated lower detection limit of 0.3 µm. Further, the data for 0.1 µm Si particles shows that 90% of the particles are counted by the scanner. This is believed to be due to the high refractive index of the Si particles and its strong light scattering ability.

For the count repeatability study it has been found that the standard deviation of total count does not show any dependence on particle refractive index, being a function only of the total count indicated by the scanner. A residual standard deviation has been defined and found to be strongly correlated to the mean total count. The residual standard deviation shows a

quasi-Poisson distribution and mathematical corrections has been obtained
to provide a quantitative measure of the count repeatability for the scanner.

ACKNOWLEDGMENT

This research was supported by a grant to the University of
Minnesota Particle Technology Laboratory by the Particulate
Contamination Control Research Consortium. Current members of
consortium include Applied Materials, FSI International, GM/Delco
Electronics, IBM Corporation, Millipore Corporation, Nupro Company and
TSI Inc. Seung-Ki Chae's post doctoral appointment is supported by an
IBM Post-doctoral Research Fellowship from the Almaden Research
Center of IBM. A grant from the Minnesota Supercomputer Institute in
support of the theoretical calculations is also acknowledged.

REFERENCE

[1]   S. K. Chae, H. S. Lee and B. Y. H. Liu, *Proc. Institute of Environmental Science* 38th Annual Technical Meeting, Nashville, TN, May 1992, 1:271-279

[2]   D. Y. H. Pui, Y. Ye and B. Y. H. Liu, *Aerosol Sci. Technol.* **12**, 795-804 (1990)

[3]   S. K. Chae, "Light Scattering Applications for Microcontamination Control and Radiative Transport Properties Determination," Ph. D. thesis, Mechanical Engineering Department, University of Minnesota, Minneapolis, MN (1991)

[4]   G. L. Wojcik, D. K. Vaughan and L. K. Galbraith, *SPIE* **774**, *Lasers in Microlithography.* 21-31 (1987)

[5]   H. S. Lee, S. K. Chae, Y. Ye, D. Y. H. Pui and G. L. Wojcik, *Aerosol Sci. Technol.* **14**, 177-192 (1991)

[6]   D. W. Cooper and A. Neukermans, *J. Colloid Interface Sci.*, **147**(1), 98-102 (1991)

[7]   J. Pecen, A. Neukermans, G. Kren and L. Galbraith, *Solid State Technol.*, **30**, 149-154 (1987)

[8]   B. J. Tullis, "Handbook of Contamination Control in Microelectronics," (D. L. Tolliver, Ed.) Noyes, Park Ridge, NJ (1988)

[9]   D. Grant and B. Carlson, *Proc. Institute of Environmental Science* 38th Annual Technical Meeting, Nashville, TN, May 1992, **1**, 196-203

SENSORS FOR PARTICLE MEASUREMENT IN CORROSIVE
LIQUIDS AND GASES

Dr. Robert G. Knollenberg
Particle Measuring Systems, Inc.
Boulder, Colorado 80301

The development of a variety of new sensors to quantify
the particulate microcontaminants in corrosive liquids and
gases are described. Sensor designs are constrained by
choices in wetted material compatibility with aggressive
fluids to largely fluorinated polymers and sapphire. In spite
of the wide range of pressure, temperature, reactivity and
application, a diverse selection of **volumetric** and **in-situ**
sensor designs have emerged with sensitivities below 0.1μm in
most media.

1. INTRODUCTION

Semiconductor yield improvements can only be made by identifying and
eliminating small bits of foreign matter from all critical processing
steps. VLSI semiconductor firms have developed world class clean rooms
capable of removing particulates by factors of a million, and thus today's
manufacturing airspace is not the primary source of microcontamination.
The emphasis has switched to the process fluids (especially gases and wet
chemicals), and the various manufacturing processes as being the more
likely yield limiting contamination sources.

While large quantities of DI water and inert bulk gases are required
by semiconductor processes, chemically aggressive liquids and gases are
also used, oftentimes at high pressure and temperature. The establishment
of the cleanliness of these hazardous materials is of enormous importance.
Tables I and II list the important process liquids and gases. Most
instruments utilized to measure microcontaminants in these media rely upon
laser light scattering methods. Laser diodes are particularly important
because they offer higher power than conveniently sized gas lasers, and
laser cavities are not yet viable in liquid media. The general
application of such measurements is in support of semiconductor
manufacturing to improve material quality, reduce defects and increase
manufacturing yields.

The measurement of small particles in aggressive materials requires
numerous precautions unnecessary for inert material handling. For liquids
such as strong acids, material choices for fluid handling components
(largely fluorinated polymers) and sensor optics (typically sapphire) are
highly constrained. In addition, sampling procedures must not allow non-
compatible process materials to come in contact [1]. Thus, purging or
flushing and cleaning between sequential material runs must not be
compromised. The required sampling apparatus for gases is invariably
simplified; however, sensor design is often complicated by high pressures

80

as when assessing cylinder gas contaminants, or extremely low pressures
and flows as is found in addressing gas feeds and exhausts on process
plants. The wide dynamic range in gas volumes required to be sampled
often generates statistical sampling problems. Material permeability is
of obvious concern where toxic gases are used. The small gas volumes
often required to be sampled can generate statistical sampling problems
since gases can be filtered to quite clean levels. By contrast, most
process liquids are sufficiently contaminated that statistically
meaningful particle counts can be obtained in a reasonable length of time
utilizing relatively small volumes.

TABLE I
COMMONLY USED SEMICONDUCTOR PROCESS LIQUIDS

Liquids		Refractive Index
Hydrofluoric Acid (50%)	HF	1.29
Ammonium Fluoride (40%)	NH_4F	1.33
Ammonium Hydroxide (30%)	NH_4OH	1.33
Methanol	CH_3OH	1.33
Water	H_2O	1.33
Acetone	CH_3COCH_3	1.36
Hydrogen Peroxide (30%)	H_2O_2	1.36
Trichlorotrifluoroethane (Freon TF)	$CCl_2F-CClF_2$	1.36
Acetic Acid (Concentrated)	CH_3COOH	1.37
Methyl Ethyl Ketone	$CH_3COCH_2CH_3$	1.38
2-Propanol	$CH_3CHOHCH_3$	1.38
Butyl Acetate	$CH_3COOCH_2CH_2CH_2CH_3$	1.39
Nitric Acid (50%)	HNO_3	1.40
Hydrochloric Acid (38%)	HCl	1.41
Dichloromethane (Methylene Chloride)	CH_2Cl_2	1.42
Phosphoric Acid (85%)	H_3PO_4	1.42
Potassium Hydroxide (50%)	KOH	1.42
Sulfuric Acid (80%)	H_2SO_4	1.43
Sodium Hydroxide	NaOH	1.44
Trichloroethane	CH_3CCl_3	1.48
Trichloroethylene	C_2HCl_3	1.48
Tetrachloroethylene	$Cl_2CHCHCl_2$	1.50
Toluene	$C_6H_5CH_3$	1.50
Xylene	$C_6H_4(CH_3)_2$	1.50

TABLE II
<u>SEMICONDUCTOR PROCESS GASES</u>

<u>Silicon Gases</u>

Silane	SiH_4
Dichlorosilane	SiH_2Cl_2
Trichlorosilane	$SiHCl_3$
Silicon Tetrachloride	$SiCl_4$
Disilane	Si_2H_6

<u>Dopants</u>

Arsine	AsH_3
Phosphine	PH_3
Diborane	B_2H_6
Boron Trifluoride	BF_3
Silicon Tetrafluoride	SiF_4
Boron-11-Trifluoride	$B^{11}F_3$

<u>Etchants</u>

Methyl Fluoride	CH_3F
Halocarbon 116	C_2F_6
Halocarbon 23	CHF_3
Halocarbon 14	CF_4
Nitrogen Trifluoride	NF_3
Sulfur Hexafluoride	SF_6
Hydrogen Bromide	HBr
Boron Trichloride	BCl_3
Chlorine	Cl_2
Perfluoropropane	C_3F_8

<u>Reactants</u>

Hydrogen Chloride	HCl
Tungsten Hexafluoride	WF_6
Nitrous Oxide	N_2O
Anhydrous Ammonia	NH_3
Carbon Dioxide	CO_2
Hydrogen Fluoride – Anhydrous	HF

<u>ATMOSPHERIC/CARRIERS</u>

Argon	AR —— 2		Nitrogen	N_2
Hydrogen	H_2 —— 2		Oxygen	O_2
Helium	He			

Manufacturers of sensors to measure particulate microcontaminants in corrosive liquids and gases generally meet the needs of users by providing instruments that are capable of measuring sufficient numbers of small particles to characterize a clean media in a reasonable time (i.e., a few minutes). The ability to detect real short-term, out-of-spec fluctuations (episodic events) is often of nearly equal interest. Optical particle

82

counters (OPC) are the overwhelming choice while, for certain specialized applications, higher resolution particle size spectrometers are preferred. In recent years a new type of instrument has emerged which differs significantly from OPC's or spectrometers. It is a particle "monitor" [2]. The device has high sensitivity, high sample rate, but <u>quite poor</u> resolution. It represents a trade—off in design criteria to enhance the special needs of ultra—low levels of microcontamination (rare events). <u>It is the preferred instrument choice for measurements in many high—purity process fluids.</u>

Until most recently, all optical particle counting devices were of a **volumetric** type. A **volumetric** instrument is one in which all of the sample flow passes through the illuminated viewing volume. Such devices sample 100% of flow, and all particles at sizes slightly above threshold should be counted. The **volumetric** terminology is necessary to distinguish them from a second class of instruments —— **in-situ** instruments. "In—situ" instruments are a class of devices in which only a fraction (oftentimes less than a percent) of flow is illuminated.

2. VOLUMETRIC SENSORS FOR CORROSIVE LIQUIDS AND GASES

Volumetric sampling cells for particle size measurements in liquids and gases can be constructed using a transparent capillary (as shown in Figure 2.1). Since the flow path is always much smaller than the laser beam diameter illuminating the sample flow, all particles are illuminated and viewed, and the ratio of laser beam diameter to sample cell width largely determines the potential intrinsic resolution of such a sensor. Optical particle counters or spectrometers can be devised by meeting the appropriate requirements of size resolution and number of size channels. Liquid sensors with a $0.2\mu m$ lower size threshold and 20 ml/minute flow rates are possible. At $0.3\mu m$, 100 ml/minute is typical. Similar sensor designs can be used for cylinder gases at high pressure. Because of the small capillary inside diameter, pressure integrity is extremely good. Cylinder gas sensors (CGS) with $0.2\mu m$ sensitivity with 3 ℓ/minute sample flow at 3,000 psi can be configured. The liquid sensors differ only slightly from the gas sensors. The primary differences are in the capillary itself and the liquid sensor's need for several size ranges and flow rates. Capillaries of borosilicate glass and crystalline sapphire are used. (Sapphire is required for HF compatibility.) The reason that both types are used is that sapphire cannot be polished to the same level as borosilicate, and it has a higher refractive index. This generates a higher level of interfacial scatter (increasing background noise) which lowers the performance of the sapphire capillary as compared to borosilicate.

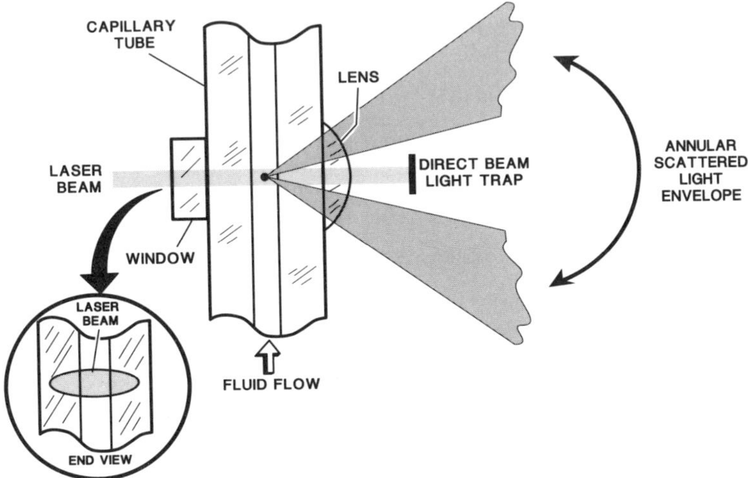

Figure 2.1   **Volumetric** Capillary Type Flow Cell. When used for liquids, this flow cell has close refractive index matching to reduce surface scattering at fluid/interface boundaries.

2.1   Corrosive Liquid Volumetric Sensors

An example of a liquid **volumetric** sensor is the PMS Integrated Micro-Optical Liquid **Volumetric** Sensor (IMOLV family) which can be combined with one of several Data Acquisition Systems (DAS) to provide spectrometer capabilities. Because the sensor is a **volumetric** device, the instrument is particularly useful for batch sampling; however, in-line monitoring is also possible when low flow (typically 100 ml/minute maximum) is required. The IMOLV sensors size particles suspended in liquid media using a laser light scattering technique as depicted in Figures 2.1 and 2.2.

These sensors have good size resolution as shown in Figure 2.3 for a 0.2μm IMOLV challenged with nominal 0.5μm polystyrene latex (PSL) microspheres. The particle produced pulses of radiant energy are sensed by a photodiode detector and sized with a sixteen channel pulse height analyzer to determine particle size. A scattering photodetector module in the sensor houses the photodiode detector and high gain (ultimate sensitivity of small particles) and low gain (for large particles) preamplifiers.

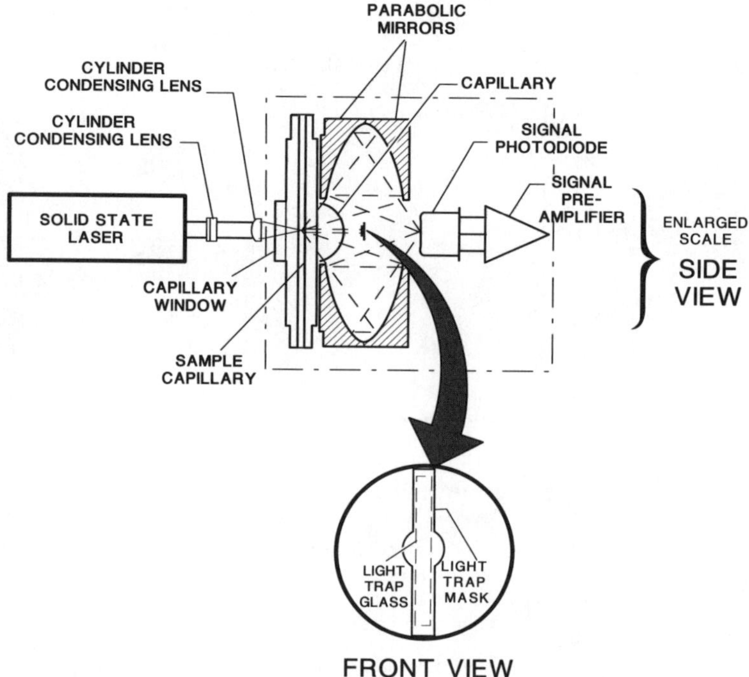

Figure 2.2   Optical Diagram for IMOLV Sensor Family.

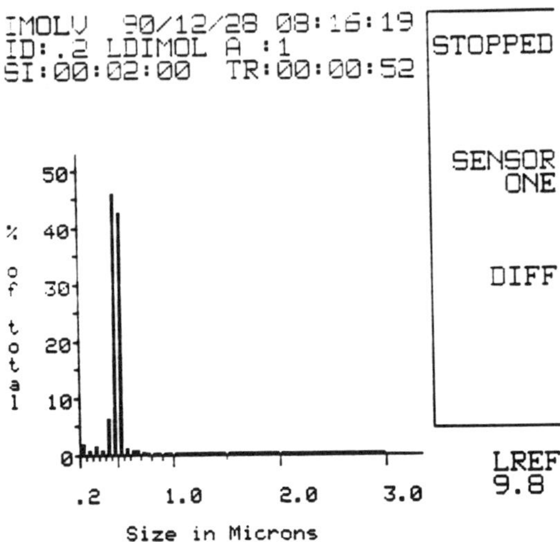

Figure 2.3   Histogram of 0.506μm PSL as Measured by a PMS 0.2μm **Volumetric** Sensor.   The resolution of this instrument is highest at smaller sizes where the response curve is steepest.

The laser used in the IMOLV sensors is a 40 mW semiconductor diode laser (SDL) derated to 30 mW.   The SDL is electrically driven by a constant power source to ensure constant laser output power, thus, eliminating any particle sizing errors due to laser power variation. Before illuminating the sample flow at the center of the sample capillary, the laser beam passes through two cylinder lenses (refer to Figure 2.2). The first of these lenses has a focal length equal to twice the distance to the capillary and acts as a 1:1 relay lens to allow adjustment of the laser beam in a direction perpendicular to the capillary tube axis.   The beam width is $\approx$ 3X the capillary diameter.   The second lens focuses the laser beam to a minimum dimension orthogonal to the capillary tube.   This focused dimension is typically 10–20μm.   The other dimensions of the illuminated "view" volume are determined by the cross–sectional area of the capillary tube.   The "view volume" dimension determines the probability of optical coincidence error.

Particles passing through the capillary tube and laser beam scatter light energy in all directions.   A portion of the light energy scattered from each particle is collected by the collecting optics (refer to Figure 2.2).   The first element of the collecting optics consists of a capillary lens which is cemented to the capillary tube.   This provides a glass–to–air optical interface corrected to ease collecting at a high numerical aperture. The second and third parabolic elements are the reflecting collecting optics. All the collecting optics are optimally coated for laser light at 780 nm. The amount of light collected is largely determined by the numerical aperture of the collecting optics, which is approximately 0.7 for the IMOLV. This large numerical aperture affords π steradians solid angle and is maintained throughout the entire depth of the sample view volume. A light trap mask located between the two parabolic mirrors is used to dump the primary laser beam.

The most recent addition to the existing **volumetric** family are the Liquilaz models shown in Figure 2.4. Optically, these instruments are essentially identical to the pre-existing IMOLV sensors, however, the pulse height analysis and particle size binning is accomplished using electronically variable thresholds controlled via a software package. This, in effect, provides an operator with threshold and size bin limit controls to accomplish calibration or compensation for sensitivity changes due to variations in liquid refractive index.

Figure 2.4 Photograph of a Liquilaz **Volumetric** Sensor.

The prime calibration is performed in two phases: first, the sensor must be optically aligned and the electronic functionality verified. The next step is to connect the sensor to a computer and the calibration software operated while supplying the sensor an appropriate concentration and size of monodispersed PSL particles in water. The actual calibration is then accomplished electronically. First, particles are supplied to the sensor and the optics are adjusted to get the specified particle sizing resolution. Next, various size particles are supplied to the sensor so that the response curve can be electronically determined. The proper curve fit parameters are calculated by the computer and then transferred to the Liquilaz to allow it to convert the digital signals to accurate particle sizes.

Compensation for different liquid refractive indices is accomplished by adjusting the prime calibration of PSL particles in water for a specific liquid refractive index other than water. With the high resolution threshold setting capabilities, any index contrast (ratio of particle to liquid refractive index) can, in reality, be accommodated. Because in most cases the liquid refractive index of chemicals is, with the exception of hydrofluoric acid, greater than water, the overall instrument sensitivity is reduced. This can have the effect of reducing the first channel threshold sensitivity. All other size bin limits are retained.

Fundamentally, for <u>homogeneous media</u> the relevant factor determining the amount of light scattered by a small particle is the ratio of the refractive index of the particle to that of the suspending medium. Historically, this ratio is referred to as the relative refractive index, but it is more descriptive to adopt the terminology <u>"index contrast"</u> (n_c) used in optical thin film technology. Since air and all gases have refractive indices close to unity, the index contrast for aerosol is essentially equal to the particle's refractive index. However, the liquids in Table I generally have refractive indices between 1.3 and 1.5 which are much closer to that of most solids (potential liquid contaminants) resulting in much lower index contrasts — even less than unity for some liquid suspended solid contaminants. The interplay of index contrast and particle size analysis in liquids has been explored rigorously in a previous work with the aid of Mie theory [3].

Figure 2.5 illustrates how latex particles of various size would scatter in liquids of various refractive index over a 10–60° collecting solid angle. Thus, a one micron latex PSL sphere in tricholorethane (n=1.48) would scatter the same amount of light as a 0.63μm PSL sphere in water and a 1.58μm PSL sphere in sulfuric acid (n=1.45) is equivalent to a 1.0μm PSL sphere in hydrofluoric acid (n=1.29). Clearly, the higher the liquid refractive index, the smaller the apparent particle size. But what about particles that have refractive indices lower than or equal to that of the liquid? In Figure 2.5, it is observed that a particle may scatter almost the same for an amount higher as for an amount lower in refractive index. Thus, a bubble in water (n_c=0.8) scatters the same as a particle of n=1.6; i.e., more than a latex sphere. The difficulty in estimating the true particle size is now apparent.

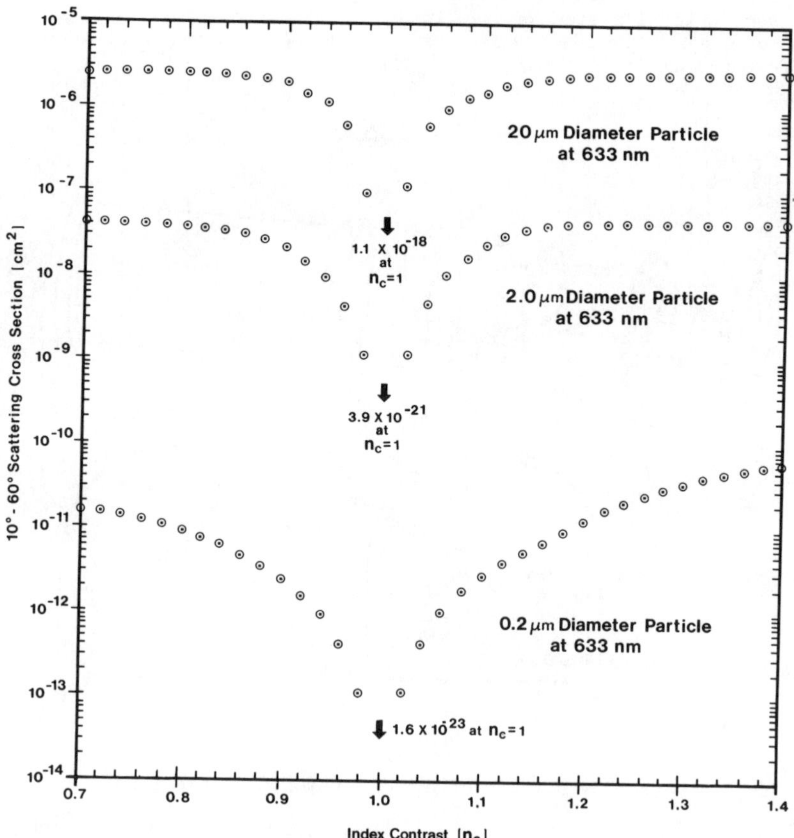

Figure 2.5   Scattering Cross Section as Function of Refractive Index
             Contrast.

2.2   Corrosive Gas Volumetric Sensors

An example of a **volumetric** sensor to measure microcontaminants in cylinder gases is the PMS Cylinder Gas Sensor, Model CGS–200. The CGS–200 sizes particles from $0.2 - 3.0\mu$m. This sensor is packaged in a cylindrical purge vessel with gas and purge connections in front and electrical connection on the rear as shown in Figure 2.6. Again, the entire sample flow passes through a uniformly illuminated portion of the

capillary providing good particle size resolution and all of the particles
in the sampled gas are measured and counted. This instrument also
operates as a particle size spectrometer. The CGS-200 has essentially
the same optical system as used in the IMOLV and as shown in Figure 2.2.
The SDL used in the CGS-200 is derated to 20 mW. Otherwise, no noteworthy
differences exist.

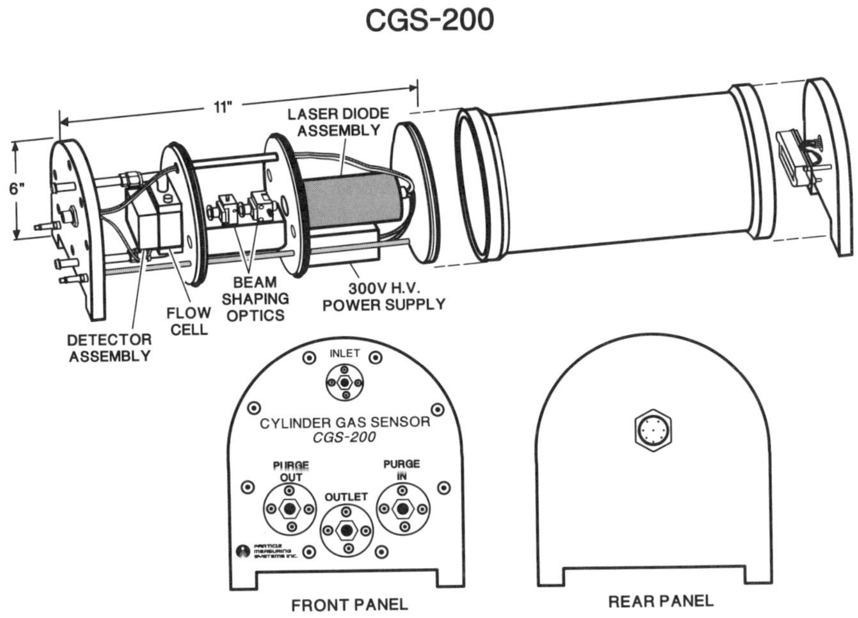

CGS-200

Figure 2.6   Assembly Diagram of a Model CGS-200 Cylinder Gas Sensor.

3. IN-SITU SENSORS FOR CORROSIVE LIQUIDS AND GASES

In-situ instruments are a class of devices in which only a fraction
of flow is illuminated and sampled. The need for in-situ devices arises
because of a fundamental limitation in the use of volumetric flow cells.
Since the entire volumetric sample cell cross section must be illuminated,
the interface between the cell wall and liquid/gas is also illuminated.
The liquid/cell-wall or gas/cell-wall interfaces are a source of stray
light due to surface defects or adhering contaminants. The noise
generated from this stray light limits ultimate sensitivity.

90

In properly designed **in-situ** instruments, the laser beam passes through entrance and exit windows as shown in Figure 3.1. The collecting optical system is typically oriented at 90° to avoid directly viewing the illuminating interfaces at the entrance and exit windows. This greatly reduces the collection of light scattered by surface defects, and the sensitivity approaches the fundamental limits imposed by noise from the media scatter alone. Since the particles are not constrained mechanically to intersect the laser beam in any special manner, the laser beam dimension becomes a "free" design parameter, and the beam diameter is ordinarily made extremely small to maximize sensitivity. Generating the smallest beams requires the use of Gaussian intensity distributions.

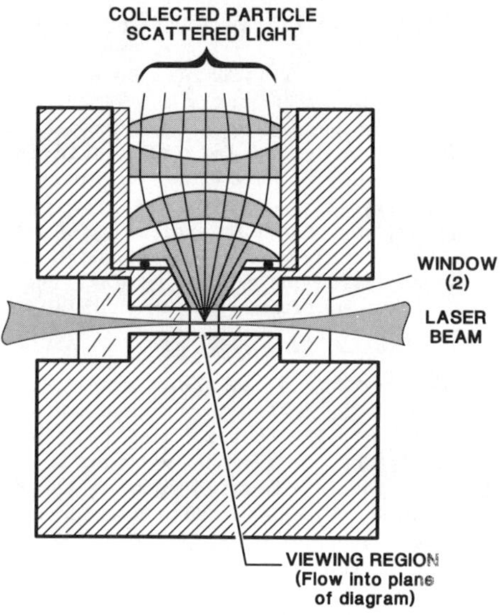

Figure 3.1 Liquid **In-Situ** Instrument Flow Cell Diagram. The wetted surfaces in a properly designed **in-situ** instrument are out of the field-of-view of the collecting optics.

In-situ instruments are generally designed as spectrometers or monitors. A spectrometer must be able to limit acceptable measurements to regions near the laser beam center where the intensity is high and fairly uniform. Liquid spectrometers with sensitivity to 0.05μm are available. A monitor instrument, unlike the spectrometer, provides no means of restricting, or otherwise sorting out, where particles pass through the laser beam. The light source may be an SDL or gas laser. The length of beam viewed is a function of detector size and imaging system magnification while the full width of the beam is used. This response depends on the particle's transect through the laser beam. Since no attempt is made to determine where a particle passes through the laser

91

beam, the same size particle has the opportunity to generate a range of scattering amplitude ranging from a maximum when traversing the beam center to minimum values at the beam extremities. Clearly, monitors are not high resolution devices; however, they fulfill a special niche when monitoring ultra-clean fluids in spite of their inherently poor resolution. They can provide relatively accurate cumulative size distribution information, greater sensitivity, and higher flow rates, at lower cost and complexity. In short, they provide the greatest statistical data base for the least cost. They are useful in both liquid and gas measurements. Monitors are the preferred instrument of choice for ultra-clean liquids and low pressure gases.

3.1 Corrosive Liquid In-Situ Sensors

To obtain a uniform region of illumination using a TEM_{oo} mode laser, one is restricted to working with regions near beam center. Such a requirement exists if one is to devise a spectrometer around an **in-situ** technique just as has been described for **volumetric** instruments. Since particles are not restricted in where they may pass through the laser beam, they are just as likely to pass through regions of weak intensity as regions of high intensity. At PMS our approach to solving this problem has been to map particle trajectories using high resolution imaging systems. The instruments are called High Sensitivity Liquid **In-Situ** (HSLIS) spectrometers. Two instruments with 0.05 and $0.1\mu m$ sensitivity are manufactured.

The optical system for the $0.1\mu m$ HSLIS spectrometer is a compound 10X system shown in Figure 3.2. (When configured for $0.05\mu m$ sensitivity, the magnification is 20X.) The laser used here is the same 40 mW polarized 780 nm TEM_{oo} mode SDL used in the **volumetric** sensors. The laser output is collimated and again re-focused forming the desired beam cross section at the sample volume. The beam cross section at the sample volume is elliptical with a $20\mu m$ minimum waist forming the ellipse minor axis positioned at the center of the sample volume.

Wide angle collecting optics are required to produce high sensitivities and obtain a monotonic calibration curve as shown in Figure 3.3. At small solid angles the response becomes ambiguous—particularly at larger sizes. The objectives of the HSLIS instrument collect light scattered from 50–130°. Each objective senses the same particle and has identical solid angles of collection and thus receives identical amounts of signal. Each objective relays its energy to independent photodetectors and pulse processing electronics at a 10:1 magnification.

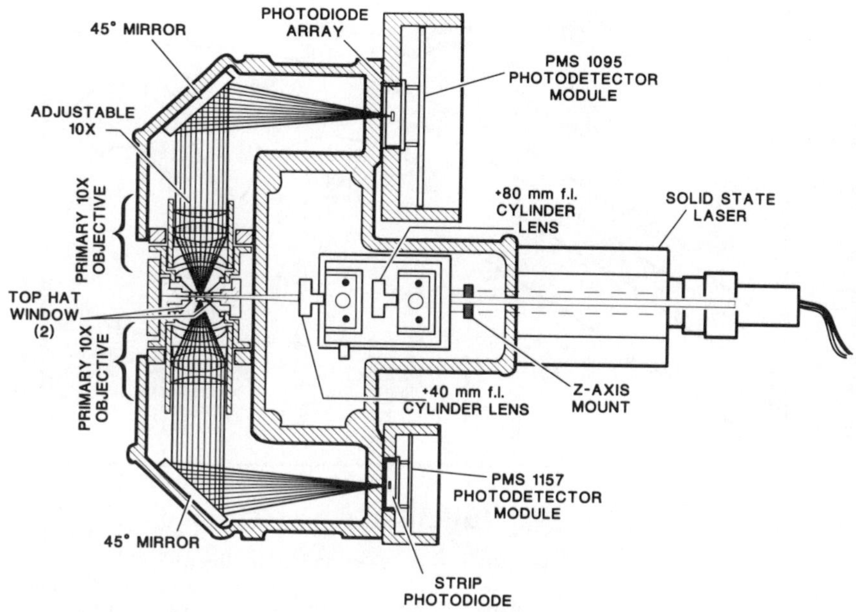

Figure 3.2   Optical System for PMS 0.1μm HSLIS Spectrometer. The laser shown is a He-Ne unit, however, current instruments use a solid state laser and a cast instrument housing.

The "sizing" objective shown in the lower optical train of Figure 3.2 focuses the beam onto a strip photodiode detector. The photodiode detector used with this optical train has a 10 mm X 1 mm active area. On the opposing side of the sample cell the "acceptance" objective produces the same magnification at its image plane. The photodiode used here is a linear photodiode array. The photodiode array has 20 active elements spaced on 0.5 mm centers. (Less than 20 are actually used in order to make the effective array length slightly less than that of the sizing strip detector.) Since the sizing optical train using the 10:1 magnification is viewed over a full 10 mm detector length, all events viewed by the array are simultaneously seen by the 10 mm strip photodiode. In addition to the photodiode array <u>width</u> being determined by the number of active elements and their spacing, a slit aperture is placed in front of the photodiode array to reduce its effective viewing <u>height</u> to approximately 0.25 mm. This, in combination with 10X magnification, results in an effective sample volume field height of only 0.025 mm.

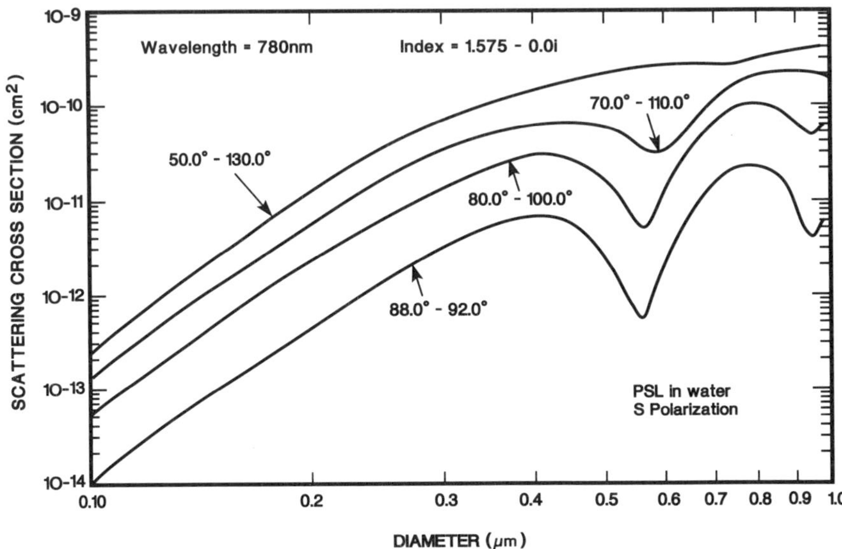

Figure 3.3   Scattering   Cross   Sections   for   PSL   in   Water   for   Various
Collecting Angles, at 780 nm.   These data indicate the need
for large collecting solid angles to produce a monotonic
calibration and maximize sensitivity.

The above optical systems, when combined with an elliptical laser
beam profile rotated 45°, allows one to define a region of suitably uniform
intensity to make reproducible measurements independent of the Gaussian
beam profile.   The overall situation is depicted in Figure 3.4.   When a
particle is in the preferred region, as defined in Figure 3.4, it is
simultaneously viewed with full amplitude by each detector; however, when
the particle is outside of this region, the energy received by the
aperture array detector is substantially reduced because of the vertical
shift in image position and the increase in image size with depth-of-field
position.   Pulse comparison circuitry is used to simultaneously compare
the amplitudes generated by the two detectors.   When the total signal
amplitudes (all summed elements) from the array is equal to or larger than
that of the strip photodetector, the particle is adjudged to be in the
preferred region of the uniform intensity and the pulse height measurement
is accepted.   When the amplitude produced by the array detector is less
than that observed using the strip photodetector, the measurement is
rejected.   In addition, signals from the individual elements on the array
photodetector are individually compared via parallel processing to a
common threshold.   The simultaneous presence of a signal generated by the
parallel processing circuitry in conjunction with the acceptance signal
determines   particle   validity.   In   this   manner,   high   resolution

94

measurements can be made and a region of uniform illumination is achieved **in-situ**. Because signals must be simultaneously observed by two independent detectors, random noise can also be more easily rejected.

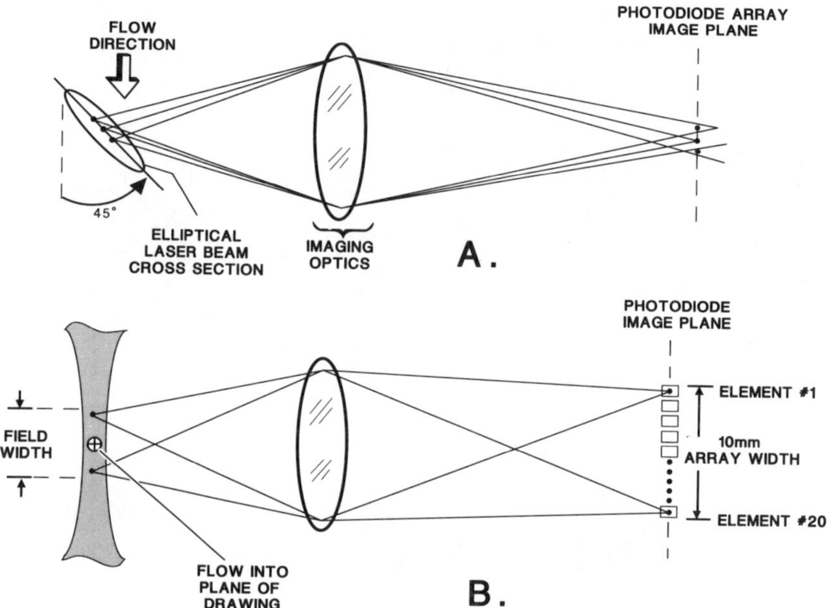

Figure 3.4   Ray Tracings of Particle Images in Viewing Volumes of HSLIS Spectrometers.   In (A), particle positions in the depth-of-field shift image position and image size.   In (B), particle position along beam axes generates image localized to individual array elements.

In the HSLIS spectrometers, the overall flow cross section in the sample region is approximately 10 mm^2. The sample volume definition scheme in combination with the flow profile generates a sample volume rate that is approximately 1 ml/minute for the 0.1μm sensor and 1 ml/10 minutes for the 0.05μm sensor with total liquid flow rates of 300 ml/minute and 100 ml/minute, respectively.

With the above sample volume definition scheme, one can utilize as small a portion of the Gaussian beam as desired. The trade-off is a reduction in sample volume in favor of improved size resolution. The HSLIS spectrometers provide the greatest size resolution at the smallest sizes because this is where the calibration response curve is steepest affording the greatest resolution opportunities (see Figure 3.3). The amplitude spread for a monodispersed particle can be quite high and still

provide good performance. For example, a ±50% uncertainty in intensity
generates less than 0.01μm error at 0.1μm. A good fraction of the laser
beam's power can thus be utilized. Figure 3.5 illustrates the resolution
of a 0.1μm HSLIS spectrometer for 0.157μm and 0.203μm PSL.

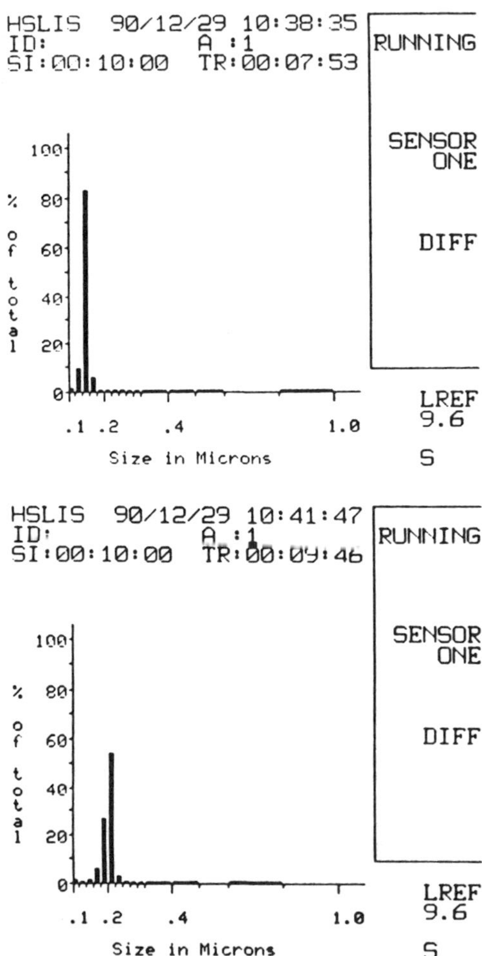

Figure 3.5   Histograms of 0.157 and 0.203μm PSL as Measured by a 0.100μm
HSLIS Spectrometer. The HSLIS can provide higher resolution
by further reducing the width of the laser beam for sizing
acceptance.

An example of a corrosive liquid monitor is the Model HSLIS—M65. It uses similar laser illumination and a high resolution optical system as the 0.1μm HSLIS but requires only one objective and viewing path and a single array detector. Sizing is accomplished in-situ via pulse height analysis (PHA) of the scattered light but the entire beam width is used. The HSLIS—M65 detects 0.065μm and larger particles using a 30 mW TEM_{oo} SDL. The flow rate is typically 100 ml/minute with 0.6% of flow or 0.6 ml/minute sampled. Typically four size classes are provided with thresholds at 0.065, 0.1, 0.15, and 0.2μm.

3.2 Corrosive Gas In-Situ Sensors

An example of an in-situ corrosive gas sensor is the Vaculaz model products designed to monitor microcontamination in the exhaust pump lines of semiconductor processes such as CVD reactors, oxide etchers, and furnace operations [4]. These sensors can also be used to monitor gas feeds to process equipment. The Vaculaz sensors use a viewing module that is an integral part of the plumbing as shown in Figure 3.6. The viewing module is a stainless steel cube with (3) windows. It can be supplied with heaters to maintain it at 150° C to help prevent condensation of certain volatiles. This sensor utilizes a simple refracting optical system, collecting scattered light at 90°.

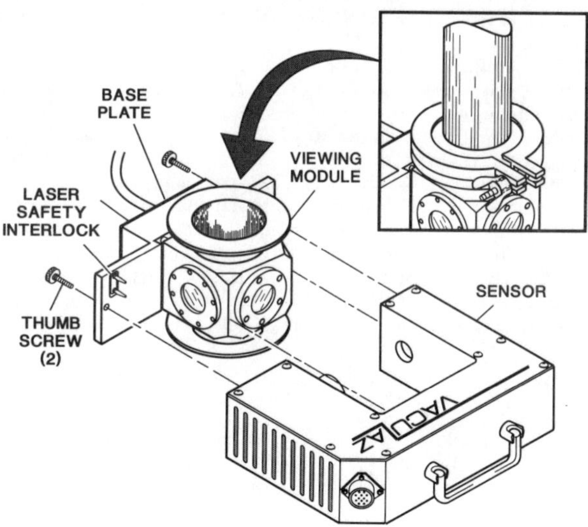

Figure 3.6 A Drawing Depicting a Vaculaz Sensor as Installed on a Viewing Module.

The primary installation required is that of the viewing module. The viewing module is a 5-way cube constructed of 304 type stainless steel. The sixth face is tapped for mounting to a plate. All three window faces are machined for o-ring seals. The three windows provide access for the laser light source, and transmit the collected light to the detector. The viewing module is intended to be installed as an integral part of the in-house plumbing. Connections are made at the inlet and outlet by use of standard vacuum flanges.

Two variations of Vaculaz sensors exist to accommodate line sizes 2" and smaller and 3-4" (see Figure 3.7). Like the liquid sensors, the Vaculaz sensors also use a 40 mW SDL that is derated to 30 mW. The SDL illuminates a sample area of 40 mm^2 by forming a 2 mm X 50μm beam cross section at the sample area which is viewed over a 20 mm beam length. This is accomplished by using a collimator, a 1/2-wave plate, and a cylinder lens combined into an astigmatic collimator assembly. The 1/2-wave plate is used to rotate the plane of polarization 90° so that the particles to be viewed scatter "S" polarization, maximizing signal at 0.2μm. After collimation, the cylinder lens is placed so that the large axis of the SDL is focused down to 50μm at the center of the sample area. The small axis of the beam is nearly collimated and has a width of 2 mm at the sample area. Alignment is accomplished through the adjustment of the laser diode assembly. Finally, an absorbing beam stop is used to capture the beam of light after passing through the viewing module.

The high levels of contamination produced in reactive processes combined with the highly reflective stainless steel sample cell make background light control an important issue. Photodiode arrays are employed to manage the background light. The refractive collecting optics design used for these sensors provide sufficient imaging performance for utilizing arrays with elements as small as 0.5 mm. The design generates an image field at a magnification of 1/2, thereby reducing the object field to an image field 1/4 the size of the sample area. This optical system is designed to localize as much as 90% of the particle's light scattering energy to a single array pixel, but does not have the spatial resolution of the HSLIS sensor design. However, this design does not require any of the imaging optics to be exposed to the vacuum environment, and the construction of the lenses is much less costly than with catadioptic designs which can achieve higher resolution.

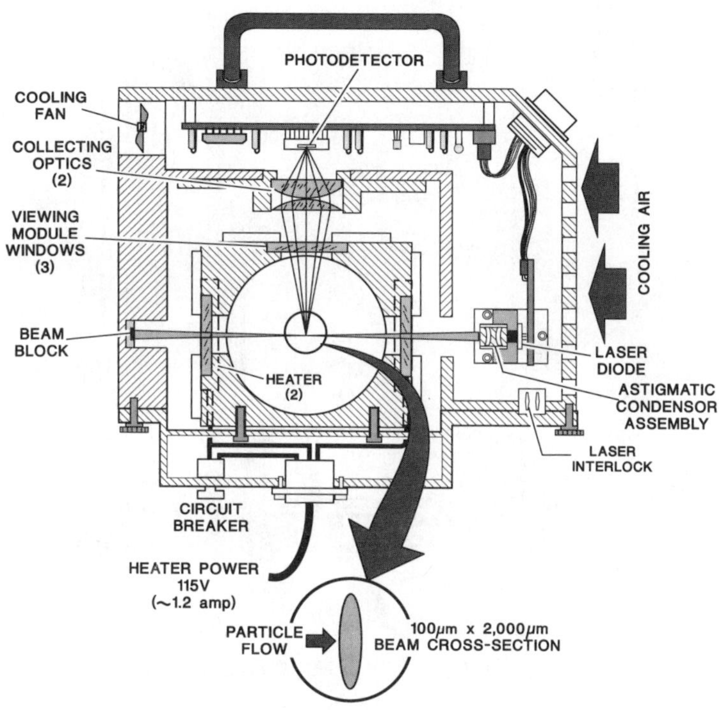

Figure 3.7a   Diagram of Vaculaz Sensor Utilized for Line Sizes up to 2".

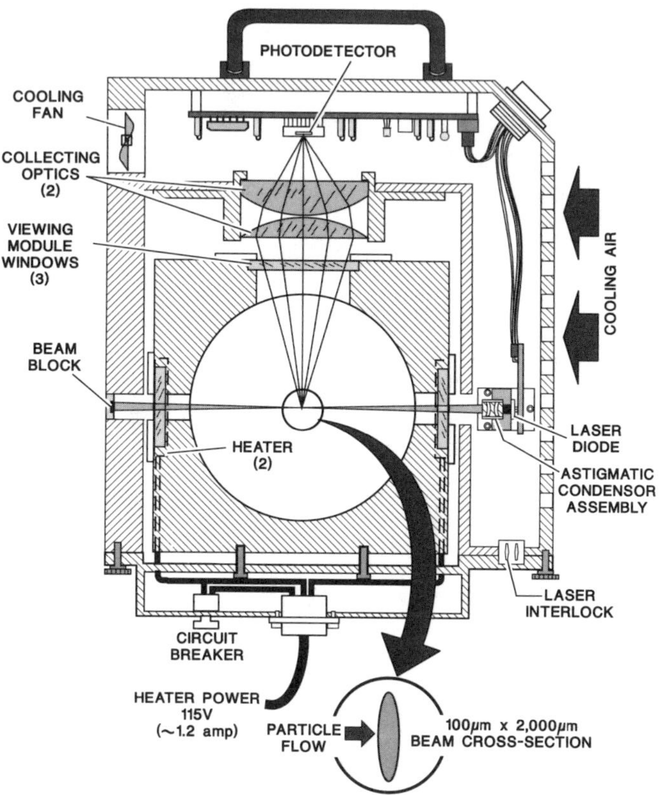

Figure 3.7b   Diagram of Vaculaz Sensor Utilized for Line Sizes up to 3"
and 4".

4.   SAMPLING CORROSIVE LIQUIDS AND GASES

Sensors are only part of the apparatus required to measure
particulate microcontaminants in corrosive liquids and gases. Clearly,
for liquid microcontamination a fluid delivery system is equally important
and invariably has considerable mechanical complexity, size and
engineering cost. Gas sampling, as mentioned in the introduction, may be
as simple as valves or flow controllers to regulate the flow discharged
from a pressure vessel or, in the case of process equipment, may be
attached in-line to a vacuum pumping or gas feed system. There are, of
course, certain liquid installations which are on-line but it is far less
common than with gas handling.

100

4.1 Corrosive Liquid Samplers

Corrosive liquid samplers are designed to sample liquids from non-pressurized containers such as baths, drums, or various sized bottles and deliver the liquid to a particle sensor. The CLS–600 shown in Figure 4.1 draws a sample from the container, compresses it to remove any bubbles — either pre-existing in the liquid or caused by the extraction process — then delivers the liquid to a sensor at a constant pressure and thus constant flow rate. Liquid is drawn into the sampler by a slight vacuum generated by an air/air aspirator. This process takes approximately ten seconds. The aspirator uses the same air supply that is used to operate the pneumatic valves in the system and to compress the liquid and purge the enclosure. This gas can be supplied from a house gas supply, compressed gas bottle, or a pump. The only requirements are that it be clean, compatible with the liquid being sampled, and deliver a pressure of from 60 to 150 psi.

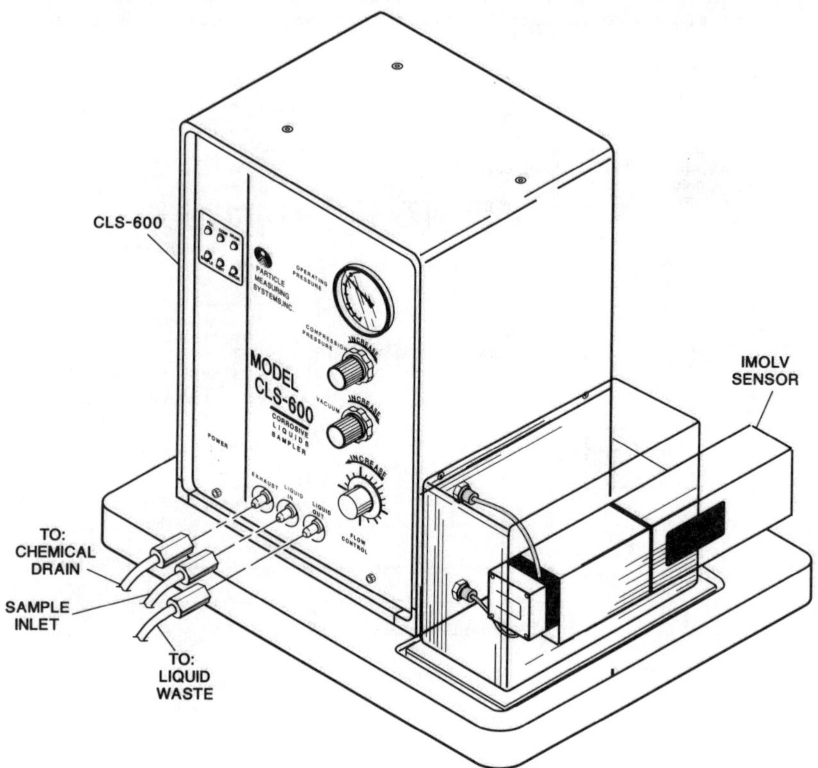

Figure 4.1   CLS–600 with IMOLV Sensor.

The CLS-600 is designed to minimize dead volume areas where
particles could become entrapped (refer to Figure 4.2). To minimize the
possibility of particle generation, all wetted surfaces of the CLS-600 are
PFA Teflon[1] with the exception of two Kal-Rez[2] o-rings and a Kal-Rez
diaphragm in the gauge protector. The sample buret is molded from PFA
Teflon. PFA Teflon's ability to clean up between samples makes it the
best material available. Two halves are accurately measured and welded
together, providing a strong and clean buret with a precise 45-milliliter
volume. The single pneumatic valve in the sample inlet line has little
dead volume and good demonstrated clean up qualities. The most difficult
particle trap to eliminate is the meniscus between the liquid and the gas
in the sampler. This meniscus is deliberately disposed of by overflowing
the sample buret. The size of the plumbing has been selected to keep flow
velocities high enough to facilitate clean up. The liquid does not come
in contact with threaded connections before it is sampled. All sample
handling valves are pneumatic valves which have shown a decided lack of
particle shedding, unlike electronically controlled valves. These
characteristics make the non-contaminating sample process possible.

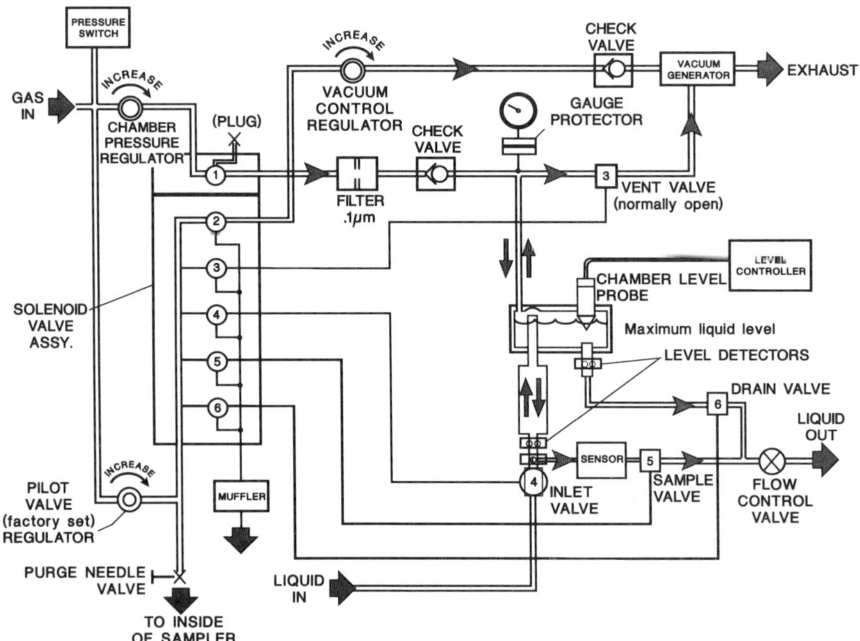

Figure 4.2   CLS-600 Fluidic Diagram

[1]Teflon is a registered trademark of DuPont Corporation.

[2]Kal-Rez is a registered trademark of DuPont Corporation.

The Corrosive Liquid Sampler, Model CLS–500, is a liquid feeding system which provides samples from open containers to any of the PMS IMOLV and HSLIS sensors. Figure 4.3 shows the CLS–500 with an IMOLV sensor attached. The CLS–500 accepts bottles of up to four liters in volume. It handles most liquids with viscosities less than 50 centipose, including corrosive and aggressive solvents and acids. The CLS–500 communicates with a Micro LPS data acquisition system for sample counting and control.

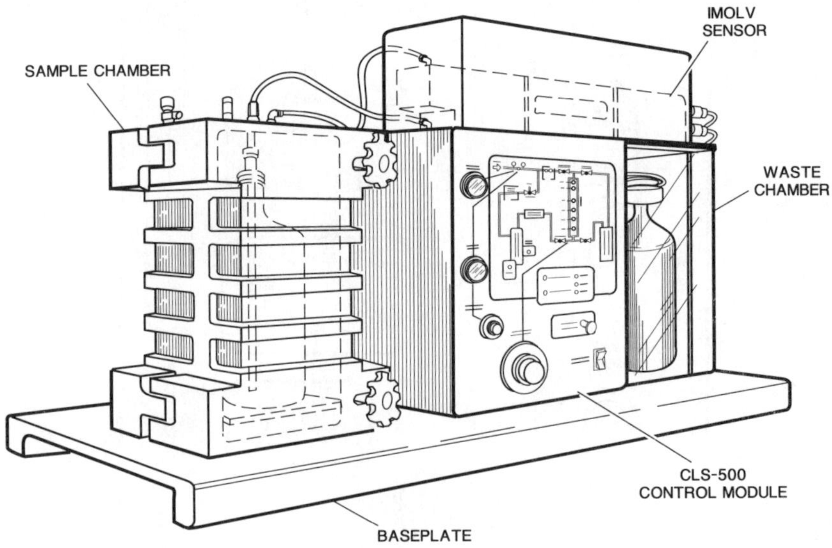

Figure 4.3   CLS–500 and IMOLV Sensor.

The CLS–500 uses a pressure vessel to force fluid directly from the sample bottle, through the sensor, and into a waste bottle. The system obtains the desired sample volume using either a direct buret measurement or a timed flow through a flow controller. The buret–measurement process provides precise sample volumes in standard amounts of 10, 25, 50, and 100 ml. The flow–controlled process allows selection of any amount of sample volume and minimizes contamination of the liquid. The data system determines the flow rate by timing buret–controlled samples at the current flow–controller setting.

All wetted surfaces of the CLS–500 are Teflon, Kal–Rez, Chemraz, or PVDF, and are resistant to essentially all liquids at ambient temperature. Therefore, acids (including hydrofluoric), and almost all solvents (including ketones) can be handled. The CLS–500 and sensor include all fluid–flow components necessary for operation; the user provides only the sample and waste bottles. All pressure–regulator knobs, switches and indicators, and the flow controller knob are located on the CLS–500 front panel (see Figure 4.4). A front panel flow diagram and LEDs provide continuous status for the valves, switches, and sensors in the CLS–500.

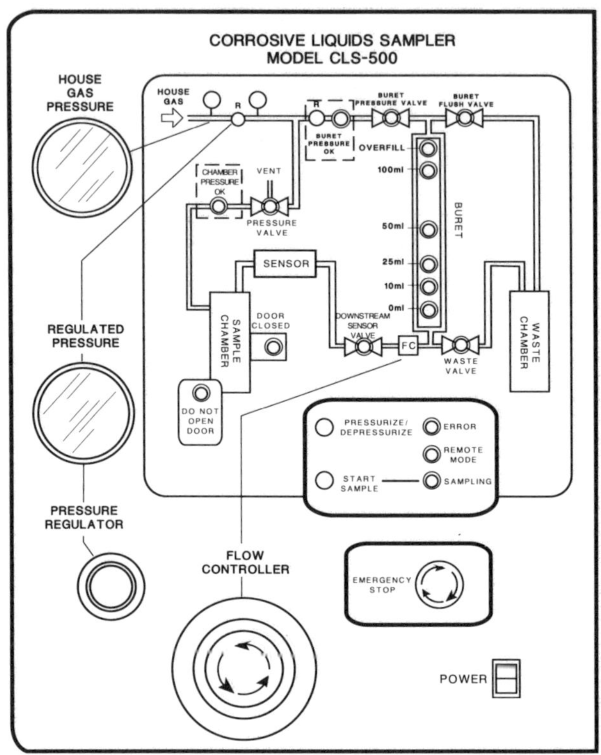

Figure 4.4   CLS–500 Front Panel.

Measurements in strong acids illustrate the use of the CLS–500.   In Figure 4.5, we have provided representative measurements performed at PMS for four <u>reagent</u> <u>grade</u> chemicals supplied by local vendors.   The measurements were performed using an IMOLV with a 0.3 to 15μm size range and a 0.1μm HSLIS.   Only the first ten channels of the HSLIS were used. To avoid bottle–to–bottle and batch–to–batch variations, samples from different lots were mixed together to represent an average.   The observed contamination levels are relatively high as might be expected.   Notice that sulfuric acid is the most contaminated of the group.   This is in spite of the fact that sulfuric acid would reveal the lowest n_c providing speculation that the high viscosity makes the material much more difficult to filter.

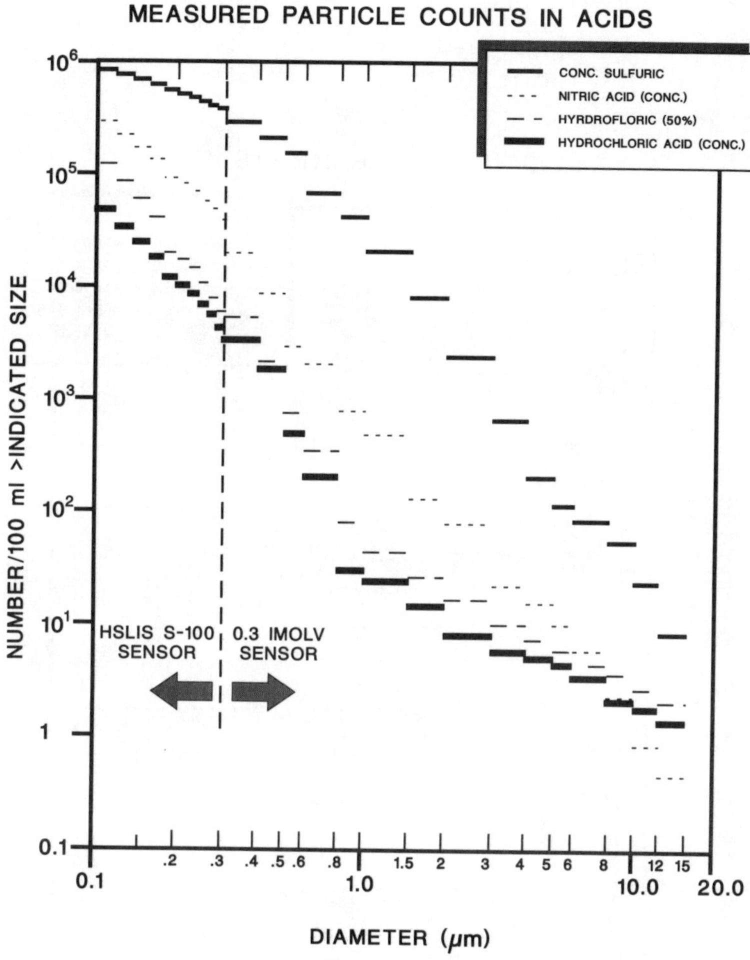

Figure 4.5   Measured Size Distributions in Semiconductor Process Strong Acids.   The   above   measurements   were   performed   without prefiltration.   The instruments used were an HSLIS S-100 (up to 0.3μm) and an IMOLV with a 0.3 to 15μm size range.

The above observations are to be contrasted with measurements made on the same acids in Figure 4.6 made with an M65 after undergoing reprocessing to remove particulates.   Such an acid reprocessor generally recirculates the acid through multi-pass filtration -- the end product

being reduced microcontaminants by orders-of-magnitude. The observed size
distribution is also much steeper suggesting that the filter approaches an
absolute cutoff at larger sizes. This latter data set illustrates the
need for high sensitivity when sampling the cleanest materials.

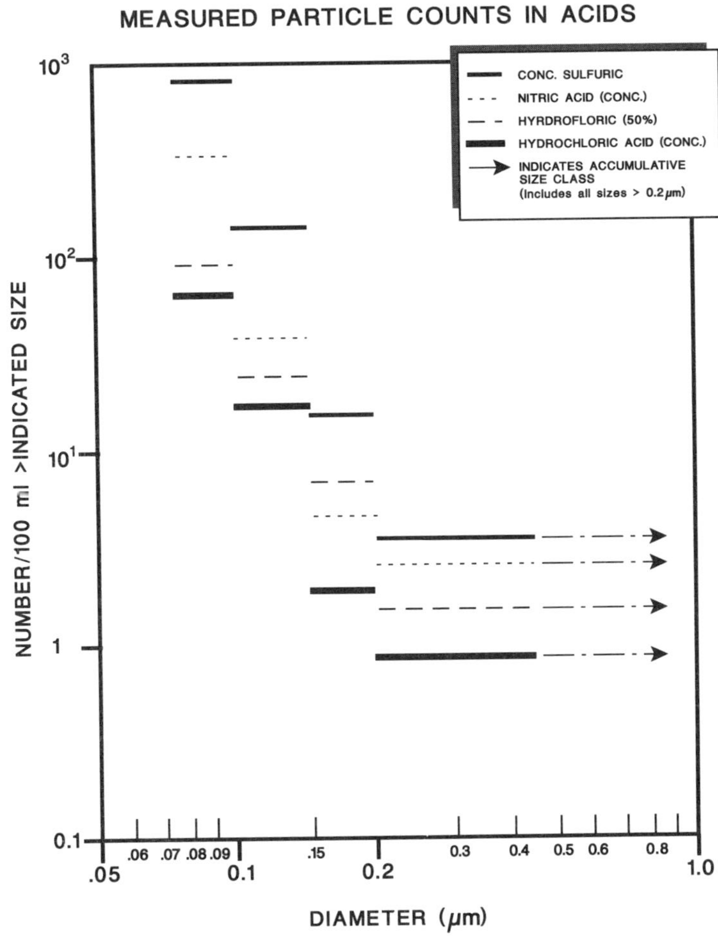

Figure 4.6   Measured Size Distributions in Recirculating Filtration Bath.

4.2 Sampling Corrosive Gases

When sampling corrosive gases from cylinders with the CGS, a suitable rotameter, mass flowmeter or mass flow controller can be used to measure total sample flow. Two basic sample flow systems that we have found reliable and accurate are presented in Figures 4.7 and 4.8. Figure 4.7 illustrates a method to be used for sampling at constant mass flow from a cylinder. The downstream valve (needle) adjusts the flow rate which is monitored by a flowmeter. Using this setup you are assured of minimum sample contamination and the sample flow is constantly metered. Replacing the needle valve and flowmeter in Figure 4.7 with a mass flow controller can provide automatic flow control while pressure varies. Figure 4.8 represents the constant **volumetric** flow method for sampling. This configuration lends itself to longer sampling periods where significant pressure drops occur. In this set-up, a mass flow controller is required to regulate the gas flow as a function of pressure. A pressure transducer with a 0 – 5 volt output calibrated to 0 – 3000 psi is also required. The DAS then computes the required control voltage for the mass flow controller as a function of pressure to deliver constant **volumetric** flow $\approx$ 20 cc/minute.

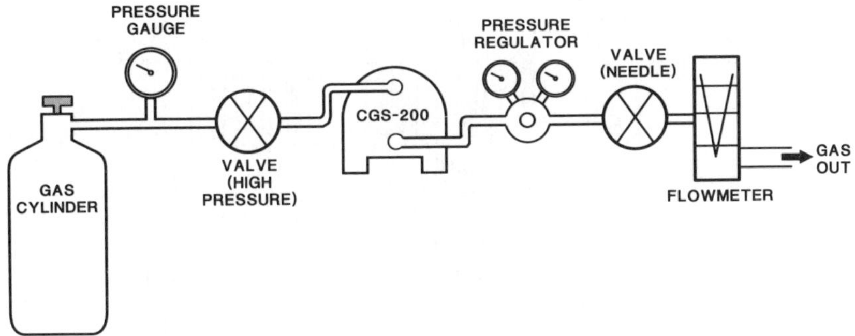

Figure 4.7   CGS–200 Constant Mass Flow.

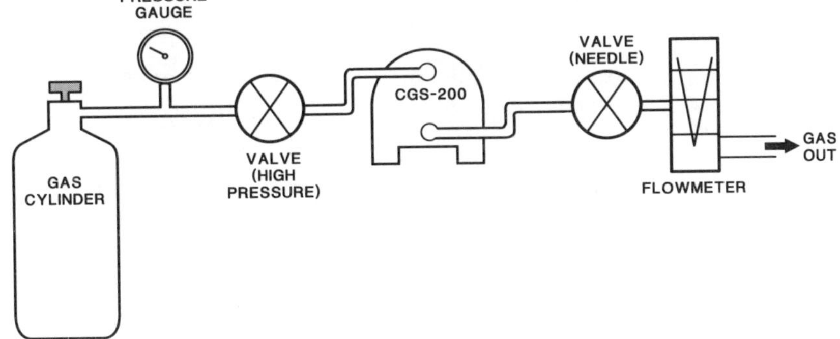

Figure 4.8   CGS–200 Constant Volume Flow (No Pressure Regulator).

Particle counts on process vacuum lines can be quite high. Data
samples using Vaculaz sensors are shown in Figures 4.9 and 4.10 for an
oxide etcher. Figure 4.9 shows particle counts exceeding 50,000 ℓ^{-1} at the
0.2μm level but only 100 ℓ^{-1} at 0.5μm. Each peak represents a single wafer
being processed. Every sixth peak is a "dummy" wafer. A much different
picture is revealed in Figure 4.10 where not only is the particle count
two orders-of-magnitude lower, but the ratio of the 0.2 to the 0.5 counts
is only about a factor of 10 different. Probably more important than the
absolute magnitude of such particle counts (oxide etchers are obviously
not clean processes) is the repetitive signature associated with each
process.

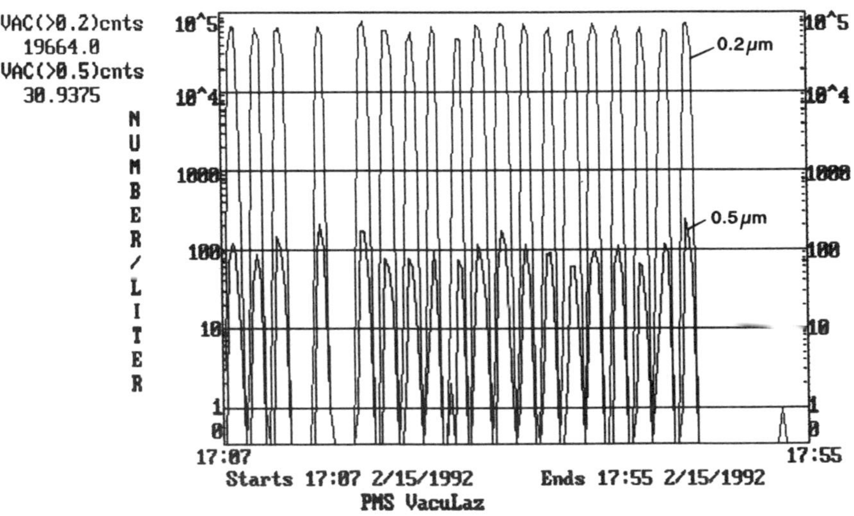

Figure 4.9   Particle Concentration Profiles Observed for an Oxide Etcher.
Each peak represents a separate wafer.

108

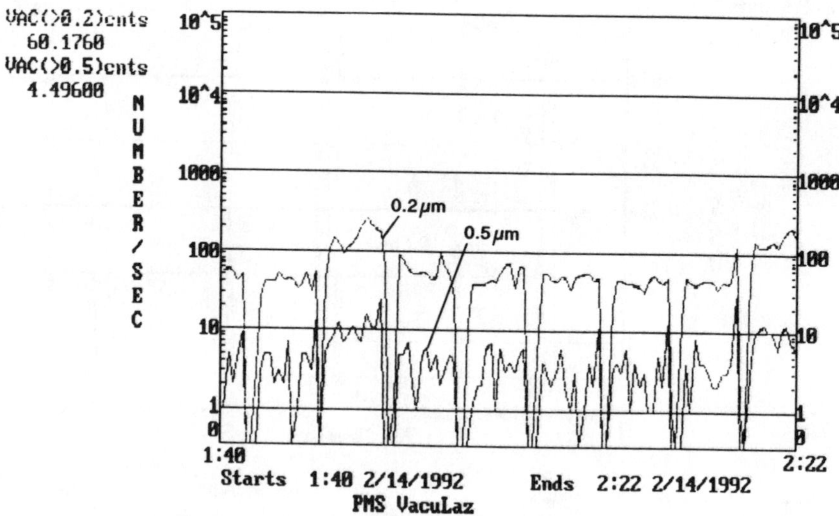

Figure 4.10   Particle Counts Observed for an Oxide Etcher. The process
is different from that shown in Figure 4.9. Every sixth
wafer is a "dummy" wafer.

Data for a CVD process are shown in Figure 4.11. The particle
activity at the 0.2μm level revealed here is nucleation mode particle
generation and should not be construed as microcontamination. There is
very little activity at the 0.5μm level and, in fact, it is only observed
at the end of each process cycle where we believe trapped particles are
released during the plasma field collapse.

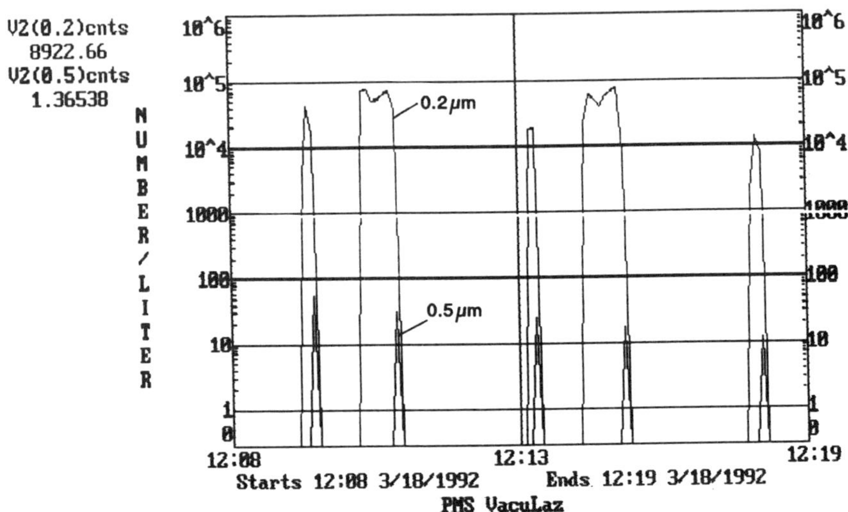

Figure 4.11   Particle Concentrations Observed on a CVD System.   The particle concentrations observed at 0.2μm represent nucleation mode activity.

5.   CONCLUSIONS

During the last several years, a variety of new sensors have been developed to quantify the microcontaminants in corrosive liquids and gases.   Sensor design requires considerable care utilizing a limited selection of materials and safety must be of ultimate concern.   In spite of the wide range of reactivity, pressure, temperature and application, a host of sensor designs have emerged with sensitivities well below 0.1μm in most media.   Both **volumetric** and **in-situ** sensors have found application with a compliment of simple-to-complex supporting sample delivery systems.

REFERENCES

[1]   Tolliver, D., <u>Handbook of Contamination Control in Microelectronics</u>, Noyes Publications, pp. 257–300, 1988.

[2]   Knollenberg, R. G., and Veal, D. L., "Optical Particle Monitors, Counters and Spectrometers: Performance Characterization, Comparison and Use", presented at the Institute of Environmental Sciences 37th Annual Technical Meeting, San Diego, CA, May 1991.

[3]   Knollenberg, R. G., "The Importance of Media Refractive Index in Evaluating Liquid and Surface Microcontamination Measurements", presented at the Institute of Environmental Sciences 32nd Annual Technical Meeting, Dallas, TX, May, 1986.

[4]   Knollenberg, R. G., and Knollenberg, S. C., "Vacuum Process Monitoring at the Sub-Tenth Micron Level", to be presented at the International Confederation of Contamination Control Societies, London, England, September 1992.

THE CURRENT STATE OF THE ART OF DEFECT DETECTION USING LASER SCANNING SYSTEMS

Armand Neukermans

Tencor Instruments
2400 Charleston Road
Mt. View, CA 94043

The fundamental principles of defect detection by laser scanning are briefly reviewed. For very small particles (0.1-0.2µm), photon noise usually becomes dominant in their detection. Methods are described that deal with the statistical variations observed in detecting particles near their detection limit. For particles on bare wafers 0.1µm particles can now be routinely detected with 0.05µm particles expected to be detectable in the future. For patterned wafers dark field spatial filtering is a very powerful technique, capable of resolving 0.2µm defects and below in periodic array patterns.

A. DETECTION OF PARTICLES ON BARE WAFERS

In laser scanning systems for particle detection, we try to detect the light scattered by the particle. The scattered electric field is caused by a perturbation of the incident electric field. The stronger the local field, and the perturbation, that stronger the scattered field will be. In general, any discontinuity will cause scatter: this may include particles, voids, subsurface dislocation, grain boundaries, or any thin contaminating layers. Looking at Figure 1, we see why it is appropriate to discuss scattering in terms of perturbation theory. There exists an incoming wave and a reflected wave, and the sum of these two give rise to a standing wave. Note that the field intensity above the surface of the wafer is highly non-uniform and repeats itself with a period of $\lambda/2$. The particle is subjected to the standing wave field, and this is what causes the scattering by the particle. In general, for normal incidence, we notice that the field never goes to zero, there is always a finite value, and it is also non-zero at the surface of the wafer because the reflection coefficient of silicon (-.63) is substantially less than -1. The particle experiences a highly non-uniform electric field giving rise to a very complex radiation pattern.

If a spherical particle is put into a uniform field, it becomes a dipole, and it radiates like a dipole. The characteristics of dipole radiation are illustrated in Figure 2. The dipole radiates mostly perpendicular to its own axis, with an intensity proportional to its dipole moment squared, and the fourth power of the frequency(1). A particle subjected to a uniform electric field acquires a dipole moment proportional to the applied electric field, the difference in indices of the media, and the volume of the particle. Therefore, the power radiated by a particle is proportional to its radius to the sixth power, or in other words, to its volume squared, and to the square of the difference in dielectric

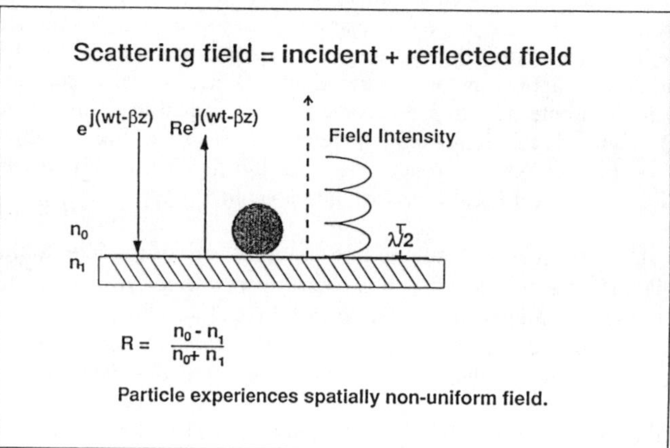

Figure 1

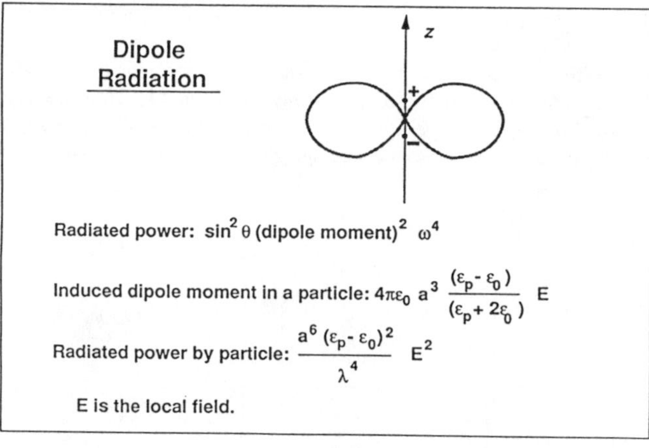

Figure 2

constants between the particle and the surrounding medium. It is also inversely proportional to the fourth power of the wavelength. The above formula is only correct for a single dipole, and in fact, it only applies when the particle is smaller than the wavelength, which we will call the Rayleigh regime.

For larger particles, we have to talk about collective dipole radiation, which is the sum of the radiation of various dipoles added together. As we can see in Figure 3, when we have an assembly of dipoles illuminated by an incident electric field, the forward waves are always in synchronism and therefore tend to reinforce themselves. However, for the back scattered radiation, we find that the phases of the various radiators are not in step. Hence, the back scattered radiation is a very strong function of the orientation and the position of the dipoles in the electric field. When we do a calculation for four dipoles, (2) for example, we do see that the forward radiation is highly intensified but that the back scattered radiation is only moderately intensified. For two dipoles the same phenomena occur, but not as pronounced. As the number of dipoles increases, the general tendency of forward scattering becomes amplified.

Figure 4 shows the Mie scattering in free space for various particles at the wavelength of 6328A (3). The particle size which is present here varies from 2μ to $.1\mu$. Over that size-range, the scattered intensity is seen to vary by a factor close to a million. It is also apparent that the larger particle, 2μ, is very much forward scattering, while for the smaller particles, the angular distribution becomes much broader and more uniform.

If we were to know the radiation patterns of various particles, the task at hand of detecting them would be substantially simplified. Unfortunately, there are no good analytical models to calculate the scattering from particles which are not suspended in free space. Most analytical models which have tried to describe quantitatively the scattering of a particle on a surface tend to give somewhat disappointing results.

We have, therefore, used a finite element analysis in modeling the scattering from a particle on a substrate. This work was performed by John Mould and Gregg Wojick of Weidlinger Associates (4) in cooperation with Tencor Instruments. The sphere is divided into finite elements, and the substrate can be given periodic or random roughness, if necessary.

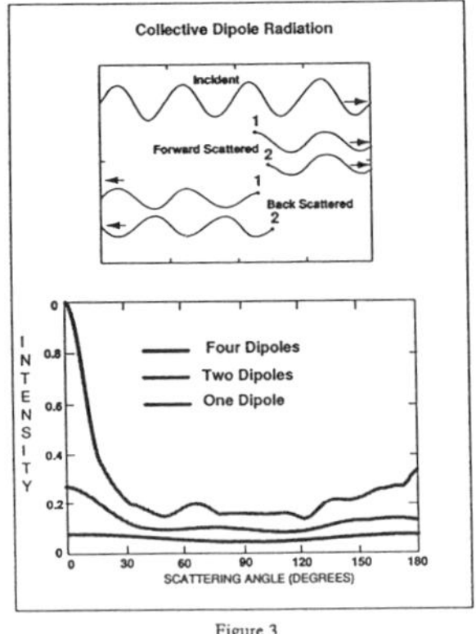

Figure 3

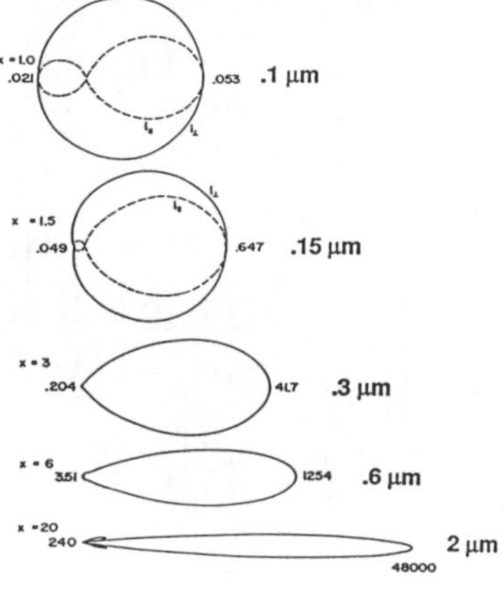

Figure 4

Figure 5 illustrates the calculated radiation pattern of a .5μm particle on silicon, under normal incidence illumination, illuminated with .48 μm radiation. (Ar laser)

Figure 6 illustrates the measured relative response of various polystyrene spheres on substrates under normal incidence, as measured on a Surfscan® 6200.

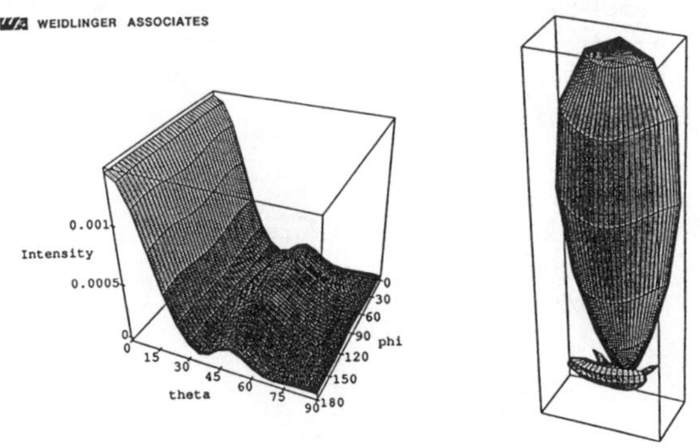

Scattering from 0.5 micron polystyrene sphere on smooth silicon
Plane Wave: 0° incidence angle, s-polarization, 0.48 micron wavelength

Figure 5

115

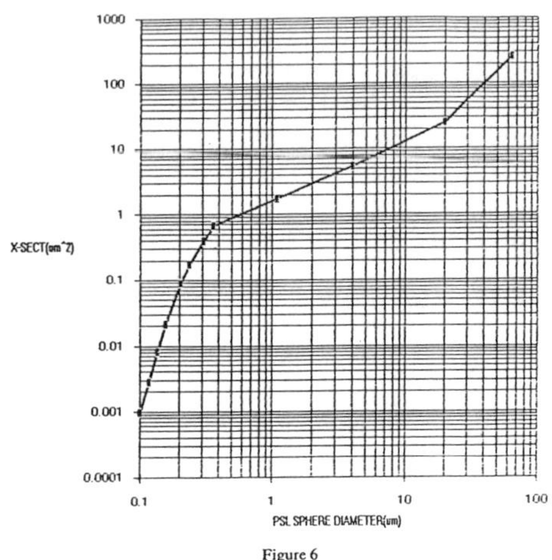

Figure 6

From the graph of the particle scattering cross-section versus particle diameter, we can see that below .3µ diameter, the particle appears to follow the Mie scattering regime. The power law which is followed here, is closer to the fifth than to the sixth power, but in general, there is good agreement.

Most of the above results were derived using latex spheres on silicon. We can see on the next graph (Figure 7), which is from the work from Fujino, et al. (5), that even using non-spherical particles, we get relatively good agreement with the laws for latex spheres.

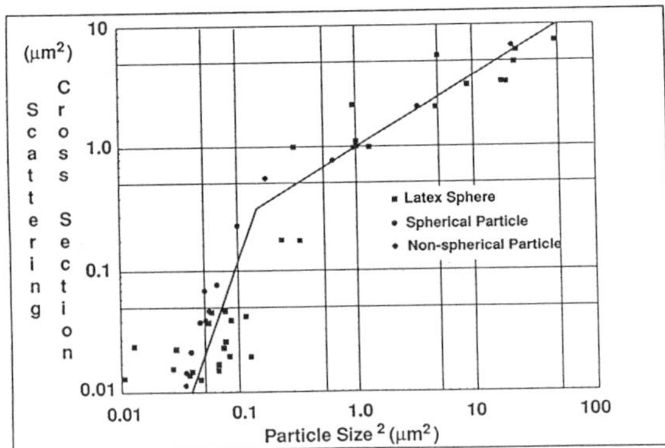

Figure 7

116

We have, so far, only talked about the scattering of the particle. We must also consider the scattering from the surface on which the particle is situated. Even a very well polished silicon wafer (with 1A RMS surface roughness) will scatter a fair amount of light and eventually it is this random scattering background which to a large degree will determine the size of the smallest particle which we can find or detect on this wafer.

In scattering from a surface, we usually define a quantity which is called TIS, which stands for Total Integrated Scatter. It is defined as the ratio of the scattered light to the reflected light from the surface. Early studies (6) have determined that the TIS is related to the surface roughness and the laser wavelength by the following formula: $TIS = (4\pi\sigma/\lambda)^2$. In this particular expression, σ is the RMS surface roughness and λ is the laser wavelength of the light we are using. Since we are interested in the scattered light, we can say that the scattered light is equal to the reflected light times $(4\pi\sigma/\lambda)^2$. As we will find in many cases, it is of interest to be able to express this formula as a function of the electric field at the surface,and we find, indeed, that this formula can be rewritten as follows: Scattered light $= (4\pi\sigma/\lambda)^2 * E^2 * ((n_0 - n_1)/2n_0)^2$. In this particular formula, E is the electric field which exists at the surface and which we will find to be a very sensitive function of the film thickness if a film is present. The quantity $(n_0 - n_1)$ is the difference in refractive index between the various media. We will find this formula most useful when applying it to the scattering from dielectric films.

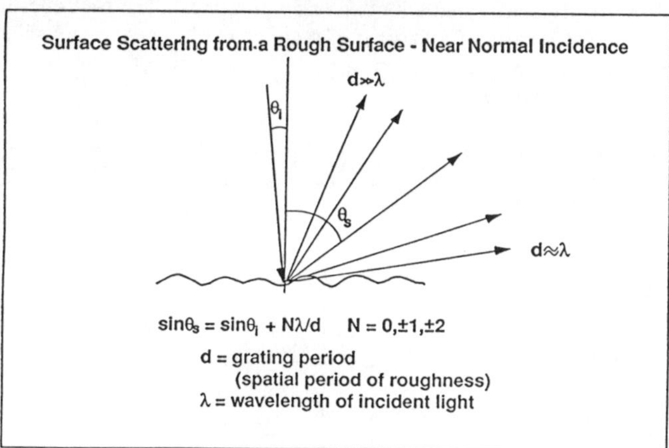

Figure 8

The distribution of the scattered light depends on the wavelength, the polarization, and the surface spatial frequencies. The exact distribution of the scattered light has been the object of very many studies and tends to be rather complicated. However, we can take a relatively simplified look at it, as is illustrated in Figure 8, where we treat the surface scattering from a rough surface under near normal incidence as a grating. We find $\sin\theta_s = \sin\theta_i + N\lambda/d$, where d is the grating period. In this particular case, it happens to be the spatial period of the roughness that is present on the wafer. Upon inspection of this formula, we see that grating periods much longer than the wavelength, tend to scatter very close to the specularly reflected beam. Grating periods which are equal to the wavelength tend to scatter at 90° from the incoming light and, therefore, tend to graze the surface. Hence, the angular distribution of the scattered light tells us what the various spatial components are of the roughness, present on the wafer, at least within a given spatial frequency band. The same is equally true under oblique incidence (Figure 9). Again, spatial gratings which have periods which are much longer than the wavelength scatter very close to the reflected light. Surface roughness, which has a period that is equal to $\lambda/2$, tends to scatter in the backward direction. So, in principle, oblique incidence is capable of resolving spatial frequencies which are twice as high as those under normal incidence.

This simplified discussion gives us a summary understanding of the scattering that is present on the surface on which the particle itself is situated. The analysis is the same for all wafers, bare, rough or patterned, but the magnitude of the surface scattering varies by five to six orders of magnitude among them.

In returning now to bare wafers, which is the simplest of the scattering problems, we find that we are usually confronted with two basic problems. The first question is: what is the probability of detection of a particle of a given size? The second: given a certain number of counts or particles detected on a wafer, how many of them are real and how many of them are false alarms?

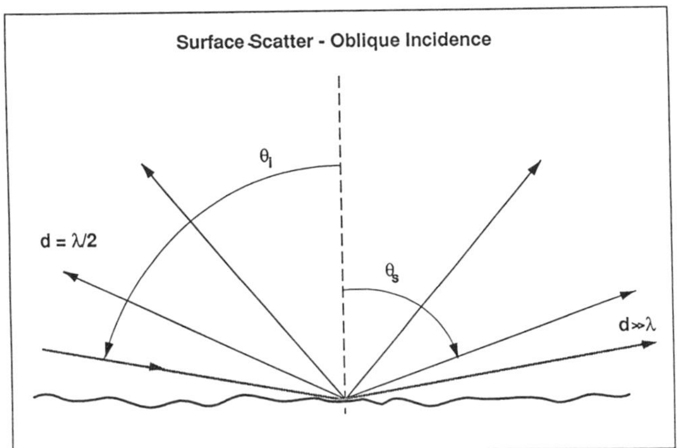

Figure 9

The detection of particles, in general, can only be expressed by statistical laws. The detection of a particle must therefore by expressed as a probability of detection, rather than a certainty. Trying to analyze this particular problem, we usually make use of monodisperse PSL spheres. When PSL spheres are put down on a wafer, nominally all of the same size, we find that the response which we get in terms of scattering cross section is a Gaussian distribution (Figure 10). The smaller we make the particles, the wider we will find the scattering distribution to be. There are various factors which affect the width of the histogram which we obtain. Instrument errors contribute to this finite width, and among them is the finite sampling grid, which is established by the signal processing. By and large, for small particles (<.2μm) the width of the histogram is defined by photon noise characteristics. Because of the very high speed of the scanning beam that is used in these laser scanners (>100 m/sec), we find that we have relatively few scattered photons and therefore, we have to start dealing with quantitization effects of these photons.

The light scattering is governed by Poisson statistics. We find that if we repeat a measurement on a particle, that the standard deviation of the distribution is equal to the square root of the number of quanta which the particle scatters. In this case, the dominant source of photon noise is at the input of the photo detector, where the number of quanta is smallest. Similarly, the background noise from the haze varies statistically, governed by the same process. In order to guarantee no more than a few background counts in the detection of particles on the wafer, the threshold has to be set judiciously. Both of these criteria have been well described in the paper of Leslie, et al. (7). Once both of these effects are taken into account, we can move on to the description of the count accuracy based on these statistical effects.

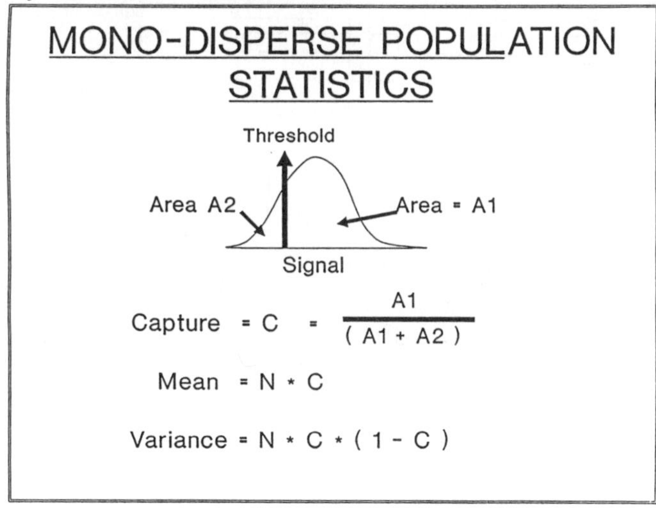

Figure 10

When we take a monodisperse population, as illustrated in Figure 10, we find that the capture rate C (or the probability of detection) is given by the ratio of the areas: $A_1/(A_1+A_2)$. If there happen to be N particles in the distribution, then the mean number that we will catch is actually N*C, the number times the capture rate. And the variance which we find, when repeating the measurement is N*C* (1 - C). This was first postulated by Cooper (8), recognizing that binomial statistics are involved. So, for monodisperse distributions by measuring the variance, in principle, we are able to tell what the capture rate is of a particular particle. Experiments were performed with monodisperse distributions, and the agreement between the theory and the measurement is quite accurate, as can be seen in Figure 11. In the analysis of real defect distributions, we need to look at the particle distribution that is present and describe them as a set of monodisperse particles, to each one of which the above theory is applied, as illustrated in Figure 12. The mean count, then, is the sum of the means, and the variance is the sum of the variances, because these are independent processes. This theory, then, was compared with data from natural detects found on a prime wafer, and one can see in Figure 13 that agreement between theory and actual data is quite respectable. The theory of Leslie et al. gives a good description of the count respectability which may be expected.

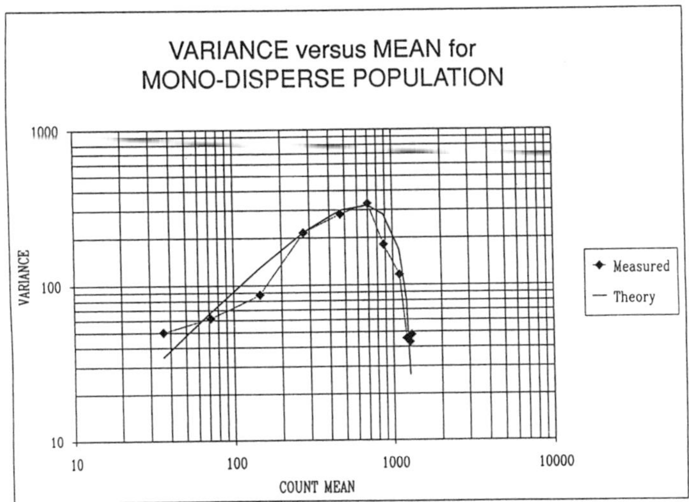

Figure 11

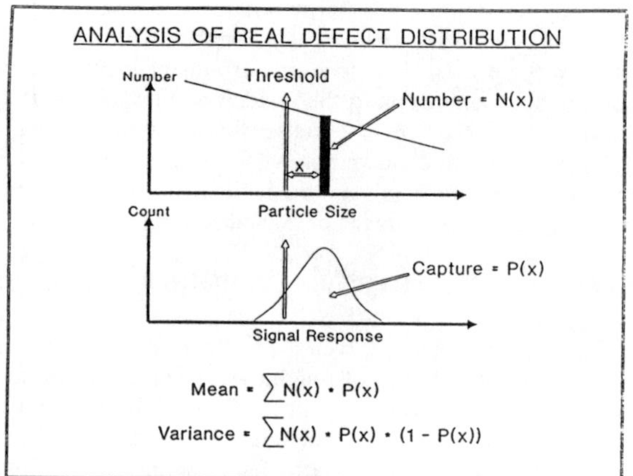

$$\text{Mean} = \sum N(x) \cdot P(x)$$

$$\text{Variance} = \sum N(x) \cdot P(x) \cdot (1 - P(x))$$

Figure 12

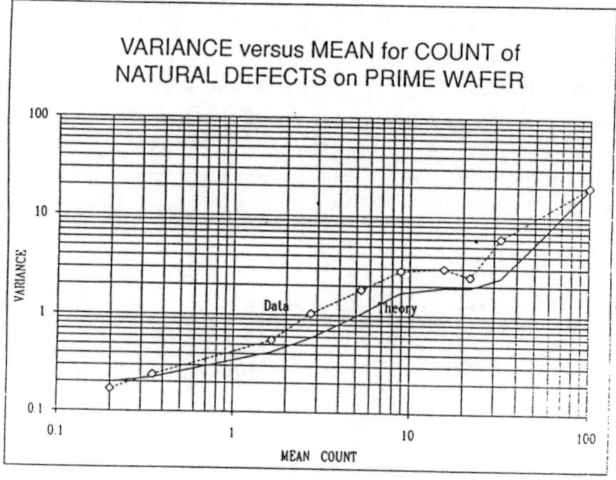

Figure 13

The detection limits which we experience on bare polished wafers are actually set by practical considerations. The more laser power we can apply, in general, the better; particularly, if we can apply it in a small beam size. However, if we make the size of the scanning beam very small, then it takes us a relatively long time to scan the whole wafer. And so we face a compromise between wafer throughput and sensitivity.

Clearly photon noise is the dominant concern. People can relatively easily detect .1 micron particles now on wafers with on the order of 1ppm haze. Detection of particles half this size, .05 micron, is a definite possibility with an increase in laser power, some improvement in the signal processing such as fully matched filtering, and various other effects. On bare wafers there exist the capability of enhancing the detection through various condensation methods analogous to the condensation nucleation counter. With this particular effect, one can probably extend the sensitivity an other order of magnitude. However, in practice there is some concern about the contamination that this detection method would imply and secondly it is very difficult to do it in such a way that the nucleated distribution truthfully represents the original distribution.

B. DETECTION OF PARTICLES ON ROUGH SURFACES

The second class of wafers which we would like to address is that of unpatterned wafers, which have an appreciable degree of roughness. In principle we face exactly the same problems as in the bare wafer case, except that the scattering from the surface is substantially higher, often by several orders of magnitude. However, in general, we find that we can do better than we would anticipate by using oblique angle illumination. As illustrated in Figure 14, under oblique illumination, at angle ϕ_0 where ϕ_0 is typically on the order of 75-80 degrees, the expression for the Fresnel reflection coefficient tells us that we almost have total reflection ($r = -1$). As a result, the field at the surface becomes zero and we then have the intensity distribution as illustrated. Notice now, that the apparent period of the intensity distribution is much longer, at least in the vertical direction above the water. In fact, it is $\lambda/2\cos\phi_0$ and this is about five to six times larger than $\lambda/2$ for the usual angles of incidence. Just above the surface the intensity increases as a quadratic function in z.

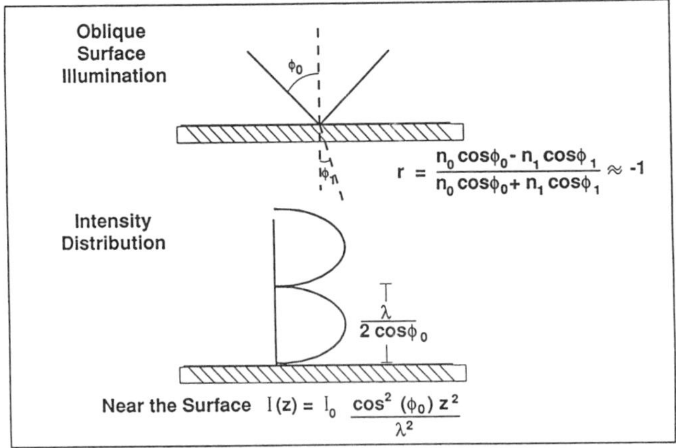

Figure 14

By putting a null of intensity at the surface, the surface roughness becomes substantially less important. If there is no field, the surface roughness cannot give rise to scatter. This artifact removes the scatter from the surface very dramatically. A particle, sticking out above the surface roughness experiences a finite field, and does scatter.

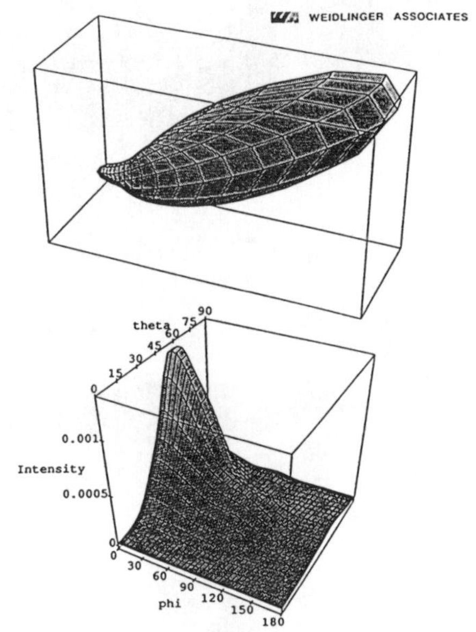

Scattering from 0.3 micron polystyrene sphere on smooth silicon
Plane Wave: 75° incidence angle, s-polarization, 0.48 micron wavelength

Figure 15

The scattering from particles with oblique illumination also gets dramatically modified and although still mostly in the forward direction, there are some changes that take place. The next figures (15, 16, 17) show the radiation patterns under 75 degree illumination for .3, .5 and 1 micron particles again calculated by finite element analysis, with s polarization.

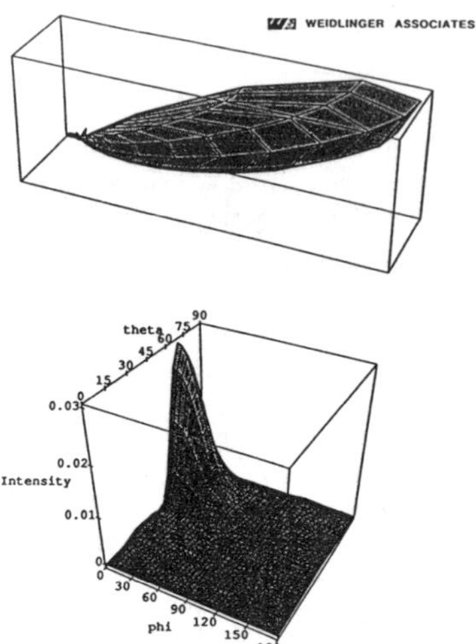

Scattering from 0.5 micron polystyrene sphere on smooth silicon
Plane Wave: 75° incidence angle, s-polarization, 0.48 micron wavelength

Figure 16

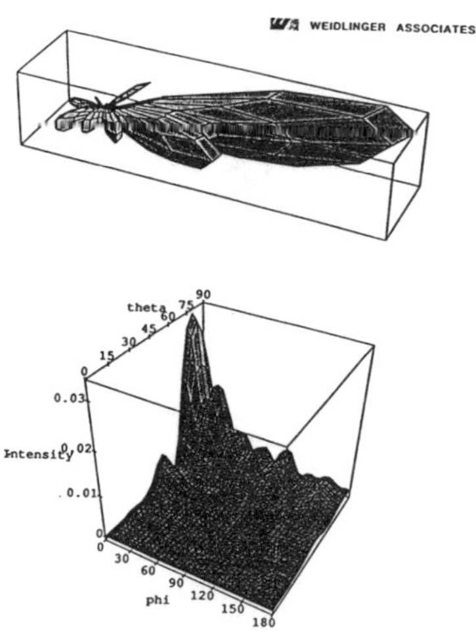

Scattering from 1.0 micron polystyrene sphere on smooth silicon
Plane Wave: 75° incidence angle, s-polarization, 0.48 micron wavelength

Figure 17

124

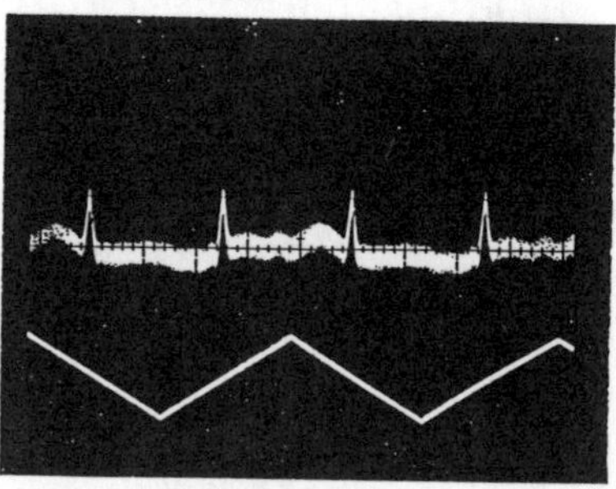

Figure 18

For rough surfaces an increase in laser power does not necessarily result in improved detectivity of the particles. Decreasing the spot size usually helps, but eventually speckle effects on the surface become more pronounced, as the number of scattering centers in the beam decreases. The real difficulty is to find optimum orientations which provide good particle scattering but have drastically reduced surface scatter. Figure 18 for example, shows the detection in an experimental Surfscan 7000 type configuration of a .26μ particle on polysilicon. Under normal incident illumination, particles ten times this size can barely be detected on this substrate.

There are some simple ground rules which can be defined, which are very helpful for optimizing the detectivity of particles on dielectric films. The surface roughness (haze) of the film affects the detectivity of particles very substantially. For a given surface roughness of the film, we also find that the minimum particle which can be detected is a strong function of the film thickness.

By looking at the electric field distributions in the film, we find that films, $N\lambda/2$ thick (maximum reflectivity) present minimum haze for a given surface roughness, particles above the film are optimally detected while defects which are in the film, will be detected at minimum sensitivity. On the other side, films which are an uneven multiple of $\lambda/4$ thick (minimum reflectivity), will present maximum haze, the detectivity of the particle above the surface will be minimized, but there will be maximum sensitivity to defects in the film itself. For film thicknesses in between, we need to use the formula derived above. All of these effects are really caused by the distribution of the electric field inside, at the boundary, and above the film.

FOUR QUARTER WAVES OF OXIDE

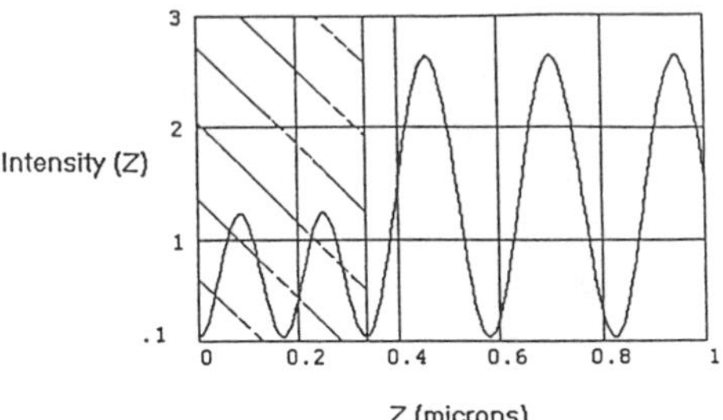

Figure 19

Let us look at an oxide layer on silicon, which is two half waves thick (Figure 19). Above the film, the field is just the same as when no film is present. On the inside, the fields are relatively small; and right at the interface between the oxides and the air we see that the electric field intensity is very low. It is this local electric filed at the surface that determines the amount of scattering caused by surface roughness, and as one can see, multiple half wave thick films will cause the scattering to be minimal because the field is extremely low. Because the intensity inside the field is low, the detectivity for defects in the film is also quite low. Because the electric field distribution outside the film is high, we will have optimum detectivity for particles above the surface. At this particular film thickness, we note that the reflectivity of the film is equal to that of the bare film In general, the particle scattering cross-section also happens to be equal to those for the bare wafer, or very close to it (save for image charges in the dielectric film).

FIVE QUARTER WAVES OF OXIDE

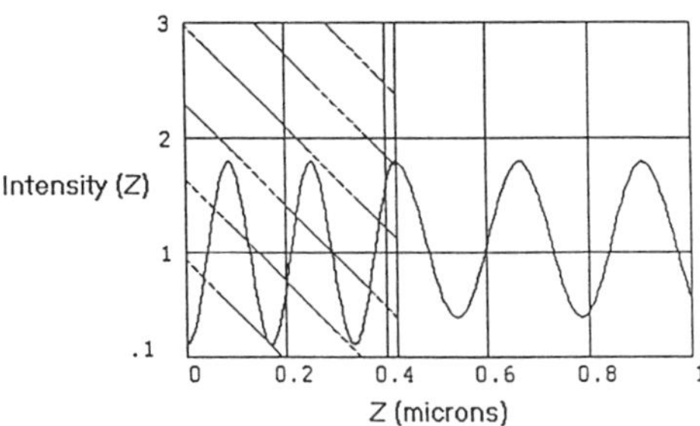

Figure 20

126

When we make the film 5λ/4 thick, (Figure 20), we see now that the inside field and the outside field are not vastly different and therefore, we might expect equal sensitivity for defects in or outside the film. The scattering cross-section for particles above the film has dropped by a factor of 2 and so has the intensity. We also notice that the field at the interface between the oxide and the air is now very substantial. We expect that any surface roughness at this location will be highly visible as haze, and this is indeed what we find experimentally.

When we use nitride layers instead of oxides, these same phenomena take place, but they are even enhanced. For an uneven quarter wave thick film (Figure 21), we see that the field on the outside becomes quite low; experimentally the scattering cross-section for particles drops dramatically (by a factor of 10), much more so that would be expected from the field. The almost constant amplitude of the field above the surface indicates that very little energy is reflected, hence this particular film acts as a very good anti-reflection layer. The forward scattered radiation of the particle is hence mostly absorbed in the substrate. Inside fields are very high. One can see that putting a λ/4 layer of nitride over silicon enhances the scattering on the inside of the film and of the surface very dramatically.

Figure 22 shows how this enhancement of surface roughness varies as a function of the film thickness. We see that we almost have a 9 to 1 increase of surface scatter, due to these interferometric effects. even though the actual surface roughness stays the same. Experimental results, illustrating the above theory for resist and oxide, are illustrated in Figures 23 and 24. In order to detect particles on film, it is clear that one would prefer to use films that are a certain number of half wavelengths thick.

SEVEN QUARTER WAVES OF NITRIDE

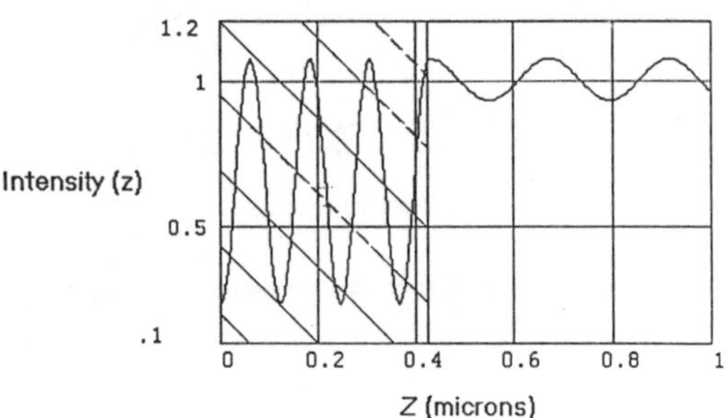

Figure 21

127

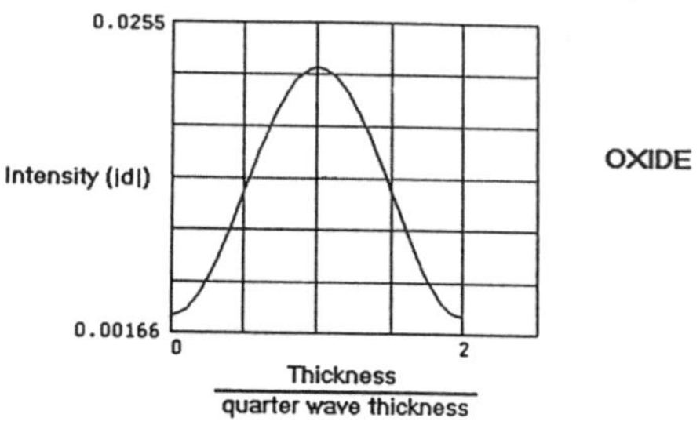

Figure 22

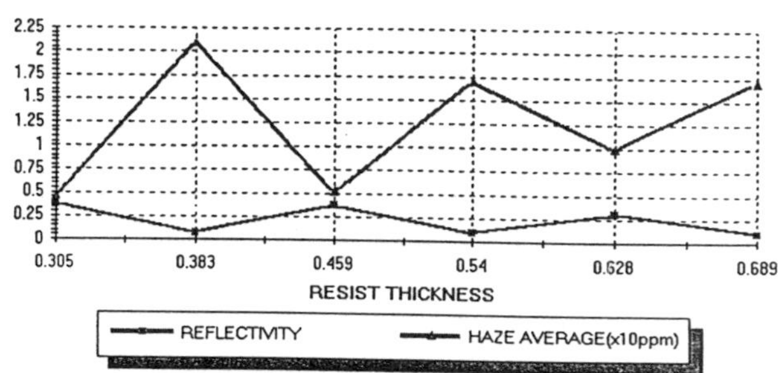

Figure 23

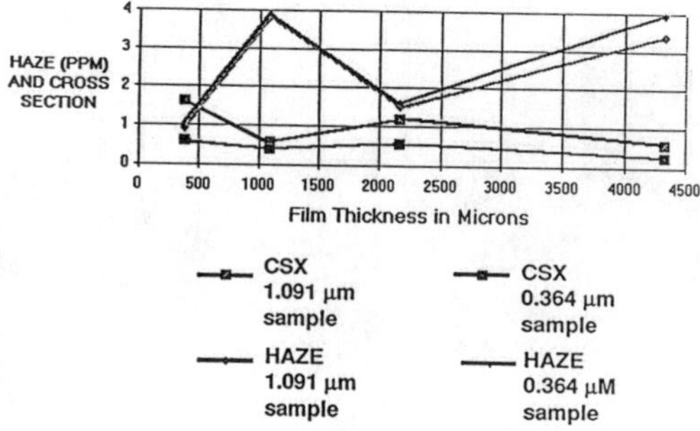

Figure 24

C. PATTERNED WAFER INSPECTION

Many of the discussions above, carry over into patterned wafer inspection. For the same reason as we have discussed under rough surfaces, patterned wafers are most effectively illuminated under oblique angle, in order to get the scattering from the surface roughness down. However, we not only have random rough surfaces, but we also have the presence of periodic structures such as memory cells. Under oblique illumination, these periodic structures give rise to diffraction patterns as illustrated in Figure 25. The folding in one direction comes because of the oblique illumination. However, for purely periodic structures, these diffraction spots can be readily removed with spatial filtering. Spatial filtering, in the Fourier plane of the image, can remove these diffraction spots, as illustrated in Figure 26.(9) Only the non-periodic defects and the scattering due to surface roughness are left over in the filtered image.

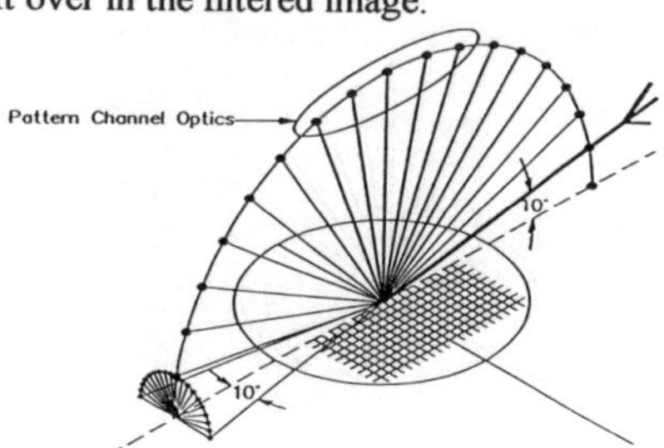

DIFFRACTION PATTERN OFF WAFER
("MANHATTAN GEOMETRY")

Figure 25

129

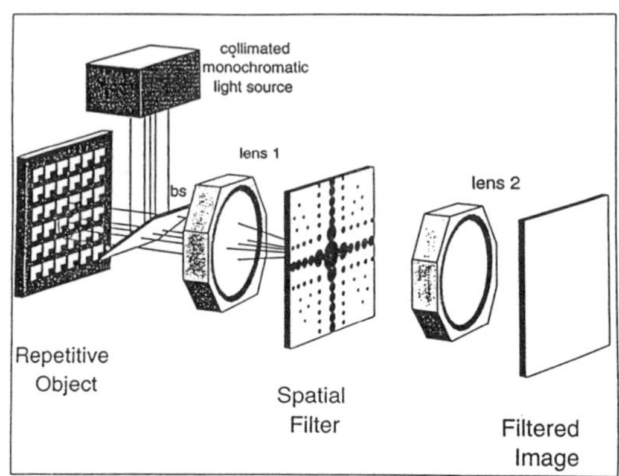

Figure 26

3-D Raw Data

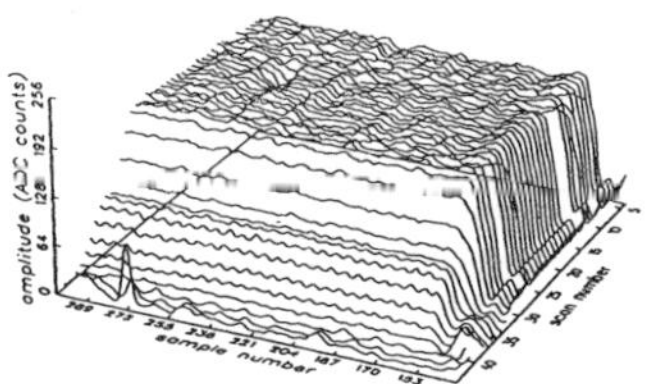

3-D Raw Data

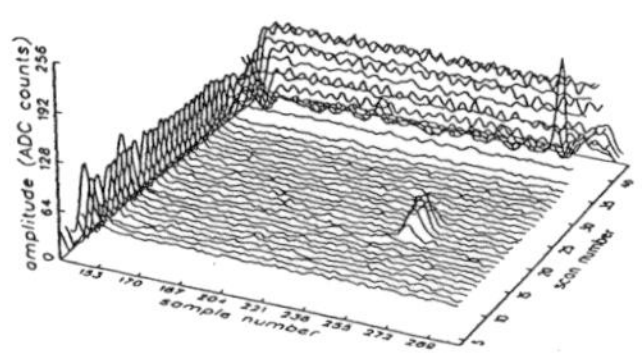

Figure 27

Hence, at least in wafers with periodic structures, many of the difficulties arise not so much from the pattern, as from the random surface roughness. At the later stages of processing, the pattern itself can add substantial roughness, giving rise to additional scattering. We feel that in the future, combinations of dark field illumination providing the advantages listed above, together with spatial filtering, will provide systems which are both sensitive and fast. The effects of spatial filtering are illustrated in Figure 27, which gives the raw data with and without spatial filter in a dark field illumination system. Spatial filtering is just one example of background scatter reduction. This background is very regular with memory arrays, but more random in other product wafers.

The darkfield detection scheme can be further enhanced to give better background reduction through improvements in its collection optics. This is illustrated in Figures 28 to 30, which show substantial improvement in particle detectivity and background reduction, due to judicious changes in the collection angle.

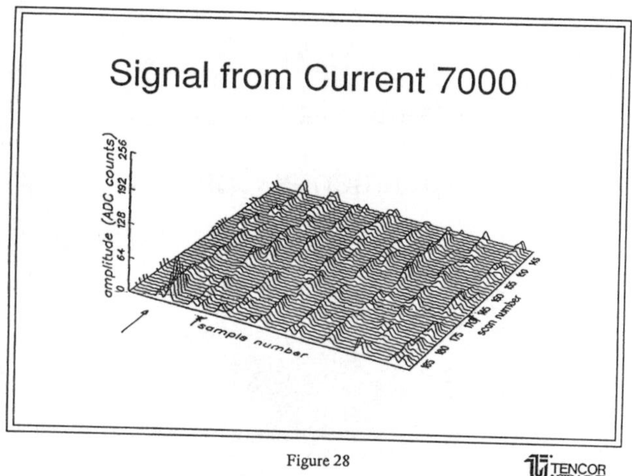

Figure 28

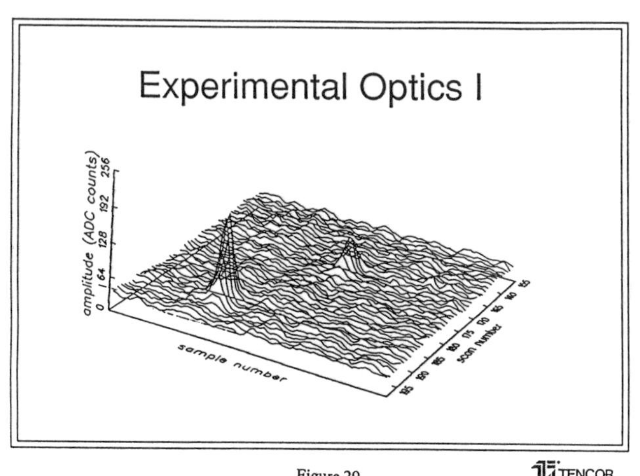

Figure 29

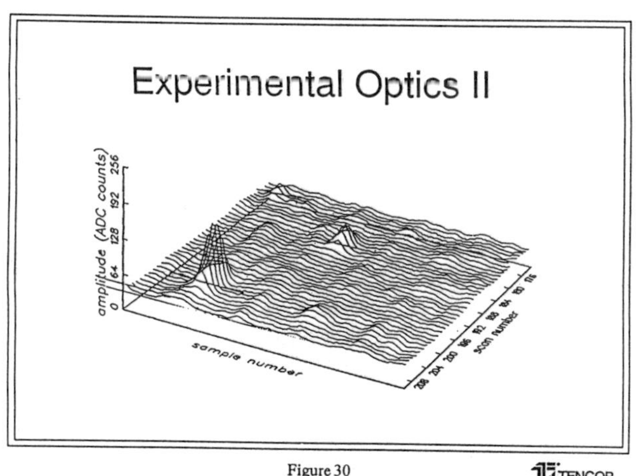

Figure 30

There is an inherent advantage in the use of these darkfield spatial filter systems because such systems do not really need to resolve the defects in order to detect them. With the spatial filtering, for example, a .1μ particle can be detected with 1μ optics resolution, although it is not resolved, while microscopy systems often need to resolve the defect in order to detect it. In addition, spatial filtering gets simpler, as the cell size keep shrinking, because there are fewer diffraction spots. We think that in the not too distant future, systems will become available having .1 - .2μ detectivity for front end processes, .5μ detectivity on metal, with throughput on the order of 12-15 wafers per hour. This is for wafers which have periodic features associated with them such as DRAMS. For wafers that do not have as many periodic features, the problem is not as simple, and inspection speed will be somewhat slower.

In conclusion, just as optical lithography has often been able to push its limitations further out, we believe that much life is left in optical inspection systems, in particular, those discussed here using laser-scanning systems.

REFERENCES

1. R. Plonsey, R. Collin "Principles and Applications of Electromagnetic Fields", McGraw Hill, 1961.

2. C. Bohren, "Clouds in a glass of beer", J. Wiley, 1987.

3. Bohren and Huffman, "Absorption and scattering of light by small particles," J. Wiley, 1983.

4. Weidlinger Associates, El Camino, Los Altos, CA 94022.

5. N. Fujino, S. Miyazaki, H. Horie, M. Takeshita, S. Sumita, T. Shiraiwa, 19th Annual Meeting of the Fine Particle Society, Santa Clara, 1980.

6. H. Bennett, J. Porteus, Journ. Opt. Soc., 51, 123, (1961).

7. B. Leslie, J. Pecen, K. Gross, A. Neukermans, Proc. 1990 Microcontamination Conf., 158, Santa Clara, 1991.

8. D. Cooper and A. Neukermans, J. Colloid and Interface Science, Vol. 147, #1, 1991.

9. M. Taubenblatt, J. Batchelder, "Patterned Wafer Inspection Using Spatial Filtering", IBM Res Report, RC 17150, submitted to Applied Optics.

IN SITU PARTICLE MONITORING IN AN
IN-LINE METAL DEPOSITION SYSTEM

Hoang K. Nguyen
High Yield Technology, Sunnyvale, California 94086

James Hunter
Cypress Semiconductor, Round Rock, Texas 78664

A simple and effective method of in situ particle monitoring in an in-line metal deposition system was developed and implemented in manufacturing. The in situ particle sensors were placed in the deposition chamber and the load lock pump line to monitor the particle level continuously. The in situ particle monitors were found effective at detecting the particle trends and periodic particle excursions such as those caused by peeling or flaking of the deposited film. The in situ particle monitors have since replaced the traditional method of particle monitoring with silicon test wafers. The use of in situ, real-time monitors has lead defect reductions up to 97% in the deposition chamber, an increase in tool utilization, and as a result, a major decrease in cost-of-ownership.

INTRODUCTION

Process and equipment generated defects have been recognized as a major source of contamination and can severely affect wafer yield [1]. This class of defects is even more critical at the metal deposition process because of the high value of product wafers that are near the end of their process cycle. Therefore, there is a need to monitor the particle level at metalization accurately and, if possible, continuously. The most common method of particle monitoring (PM) in an in-line metal deposition process is cycling of bare silicon wafers through the process chamber without deposition and then measuring the number of particles added. The inability to run process leads to inaccurate results. Furthermore, this process displaces production, and can only be performed periodically. Any reasonable PM sampling technique could not monitor all periodic and short particle excursions. Clearly, it is necessary to develop a simple and effective method of particle monitoring to prevent further processing of valuable product wafers when the particle level is high.

A simple and effective method of in situ particle monitoring for an in-line deposition system was developed and implemented in manufacturing. Since replacing the traditional method of particle monitoring using wafers, the use of in situ, real-time particle sensors placed inside the deposition system has lead to a significant reduction in defect density, an increase in tool utilization and a reduction in monitor wafer usage, and a significantly lower cost-of-ownership.

EQUIPMENT SETUP

The in situ particle sensors were evaluated in a pair of in-line single wafer metal deposition systems (Varian 3290). This deposition system includes a process chamber and a load lock. The process chamber has four processing stations: plasma etch conditioning, aluminum deposition, titanium deposition and titanium tungsten (TiW) deposition all separated by isolation shields. The process chamber is kept at low pressure to minimize gaseous contamination of the deposited film [2]. A load lock is used to transfer wafers in and out of the process chamber. A maximum of five wafers can be held in the chamber on a rotating transfer plate at any one time. The transfer plate holds the wafers vertically using metal clips and ferries them from the load lock past the four stations, and finally back to the load lock for removal.

It was expected that significant particle generation would occur both in the load lock and the process chamber. Since mechanical contacts such as metal to metal and wafer to metal (wafer holding clips) have been known to generate particles in the load lock and these particles are stirred up during pumping, good in situ particle sensor location is in the load lock pump line. Because the process chamber operates at low pressure (low militorr range), it is assumed that particles fall freely in this low pressure environment [3], another good sensor location is at the bottom of the chamber directly below the TiW sputtering station. This station was known to contribute to wafer defects because the deposited TiW film is under high stress and prone to flaking and peeling. The configuration of the in situ monitoring system for the in-line metal deposition system is shown in figure 1.

The in situ particle sensors use laser light scattering to detect particles as they pass through the controlled laser beam. These sensors are compact so that they can be placed inside the deposition chamber or in the pump line. Because the sizes of film flakes inside the deposition chamber are typically large, an in situ particle sensor with 0.5 micron minimum detectable size was used there (HYT model 15A). For the load lock pump line, an in situ particle sensor with 0.38 micron minimum detectable size (HYT model 20) was used.

Fig 1: HYT sensor configuration for Varian 3290

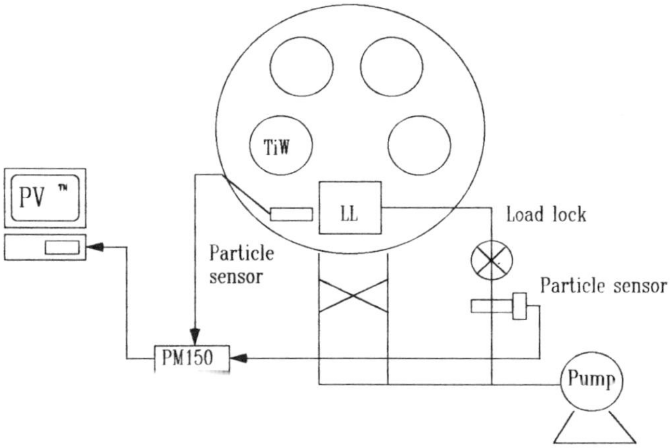

Initially the particle data was collected continuously on one minute intervals for the deposition chamber. After the deposition chamber's defect level was reduced by the defect reduction efforts, the time interval was increased to two minutes to enhance the particle count rate. This enhanced count rate made the particle baseline and any excursion more visible. In the load lock, the particle data was collected per pump down cycle. A trigger box received input from the load lock ram forward switch. When the ram was forward (load lock sealed from the deposition chamber), data acquisition began. After the load lock was pumped and the ram was pulled back to allow wafer transfer, the particle data acquisition stopped and posted the value. A personal computer stored and displayed the particle data for the deposition chamber and load lock in real-time. The proprietary software package provided with the in situ particle monitoring system provided the capability for closed loop statistical process control.

The conventional method of particle monitoring was to cycle a 6-inch diameter test wafer through the deposition system without deposition and measure the particles $\geq$ 0.3 micron added to the wafer with a surface defect scanner. Each test took approximately 30 minutes to complete and was performed four times a day.

DATA AND CORRELATION

The in situ particle counts for both sensors correlated well with system events that were known to cause high particle levels such as wafer clip changes, broken ceramic ring (the rings that support the wafer clips), and particle build-up in the load lock isolation valve. Figures 2 and 3 show the time synchronized data that portray these events along with data from the wafer PMs. Although the correlation between the in situ particle counts and wafer PM was poor, the in situ particle counts were considered more accurate because the in situ counts agreed with each other and correlated to known system events such as those shown in figures 4 and 5. The wafer monitors were not effective in detecting increased particle levels during these system events. Furthermore, the in situ monitors were able to detect the flaking trend inside the deposition chamber as shown in figure 6.

The conclusion that the wafer monitor counts were poor indicators of system particle conditions is based on two factors. First, many particle generating events are short in duration and very sporadic. Periodic particle monitoring can easily miss events that could have a significant impact on the defect level of product. Second, the particle monitors are run without deposition on non-product wafers. This eliminates most process related particle effects. The combination of the insensitivity of the particle wafer monitors and infrequent test intervals resulted in failure to detect out-of-control particle events as those shown in figure 2. During this period shown in figure 2, there were seven particle wafer monitor runs with five out of control events where the in situ particle level exceeded the control limit of 20. Because the frequency of out-of-control conditions roughly equals the frequency of wafer particle tests, the periodic monitors cannot detect most, or even all out-of-control events.

Experiments were conducted to test the sensitivity of the in situ monitors in detecting flaking and mechanically induced defect events. It was noticed that significantly higher particle counts were seen during post TiW processing when the station was shut down and the wafer heater was turned off. The high particle counts seen even during system idling (no mechanical activity) indicated that the particle generation was a thermal effect from the cooling of

the highly stress film. This effect was easily reproduced
in an experiment where the TiW station was heated up and ran
for some specified time and then abruptly shut down to cool.
The results of this experiment are shown in figure 7.   The
in situ monitors were also effective at showing mechanically
induced defects such as scraping shutters.

Similarly, the in situ monitor for the load lock was
equally effective at detecting particles generated by the
deposited film flaking off the wafer holding clips.   A
portion of each clip became coated with the sputtered metals
and the clips themselves were bent back and forth
continuously during the loading and unloading of wafers.

Fig 2: Load Lock Particle Data

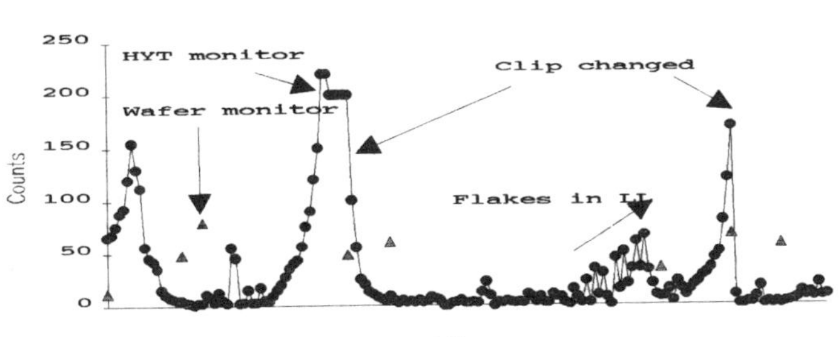

Fig 3: Sputter Chamber Particle Data

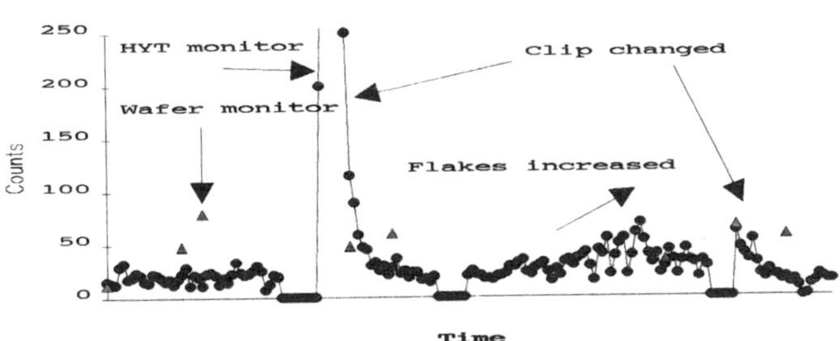

This combination required the clips to be changed at regular
intervals (the clips are disposable) since the metal build-
up eventually began to flake. One distinct advantage with
in situ monitoring is the ability to predict the need for
clip changes. Clip related particle problems are shown in

138

figures 2,3,4 and 5. These figures also show that the
particle levels declined more rapidly in the load lock than
in the deposition chamber after the clip changes. This is
because the clips actuated or flexed in the load lock, and
held their positions throughout the deposition chamber. The
in situ monitor was also effective at showing that the clips
at a single station on the transfer plate were flaking.
This is confirmed by every fifth count being high as shown
in figure 8. Standard wafer monitor techniques could have
easily missed this isolated problem. All sensitivity tests
and correlations were repeated to confirm the results before
implementing the in situ particle monitoring method into
manufacturing.

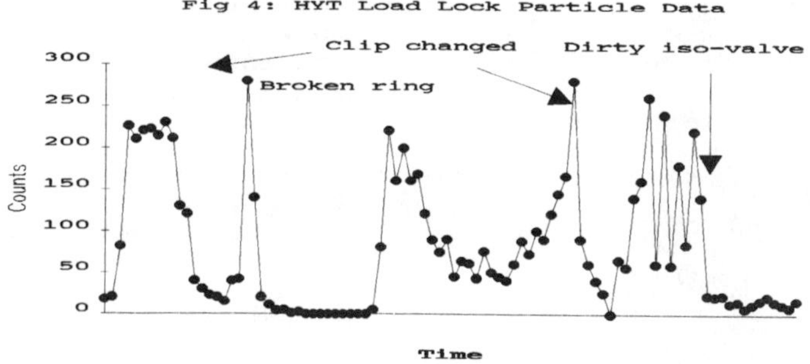

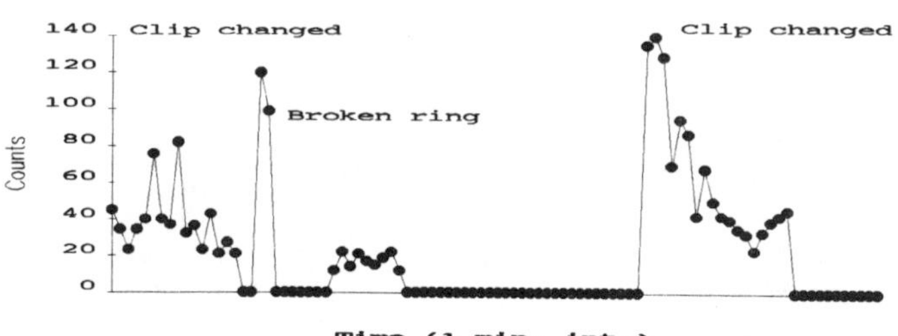

STATISTICAL PROCESS CONTROL AND IMPLEMENTATION

Once the particle baseline was established for normal system operation and confidence was established that excursions in the particle counts would be detected, the particle control limits were implemented into production. Obviously, the aim was to prevent further processing of product if a high level of particles was detected or an upward trend in the particle level was seen. The software was configured to trigger an alarm and stop the machine during processing (a non-fatal interruption) when a single count exceeded a high limit defined by 1 point $\geq$ X standard deviations and/or a trend defined by 2 of 3 points $\geq$ X standard deviations was detected, as shown in figure 6. This technique was found to be reliable and effective at detecting brief and periodic particle excursions as well as increasing particle trends. Because the in situ particle monitoring is continuous, it has increased the awareness of defect levels and has been an effective tool for driving down defect baselines.

Fig 6: Sputter Chamber HYT Trend

Fig 7: Cooling Induced Flaking

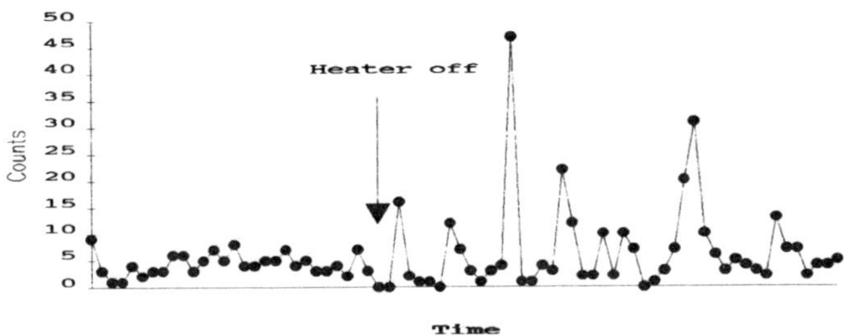

140

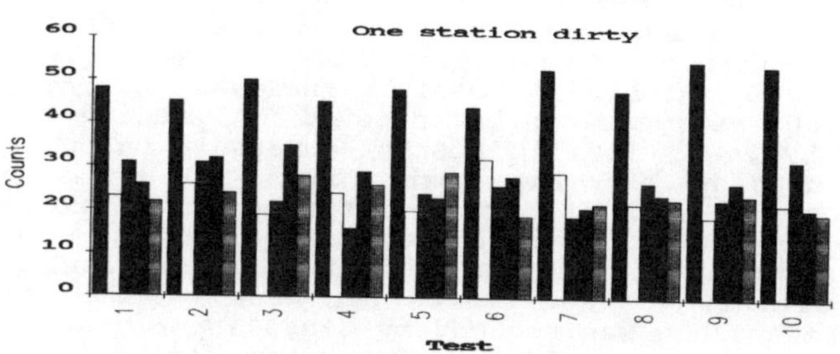

COST OF OWNERSHIP REDUCTION

With the increased awareness of defect levels, the defect reduction followed soon after production implementation. Using a trial and error method while continuously monitoring the in situ defect level, the levels for the chamber and the load lock were reduced by 93 and 27 percent, respectively, as shown in figure 9. The biggest gains in reducing the chamber defect levels were made by modifying the chamber cleaning and conditioning procedures as well as maintaining the temperature of the TiW deposition station as much as possible. In chamber cleans, it was confirmed that there were no substitutes for clean and properly handled parts. Also, attaining a set base pressure prior to initiating TiW deposition was found to be critical. Proper conditioning could extend the number of wafer processed per clean cycle from 1000 to 1500. By maintaining TiW station temperatures, flaking due to cooling stresses was minimized.

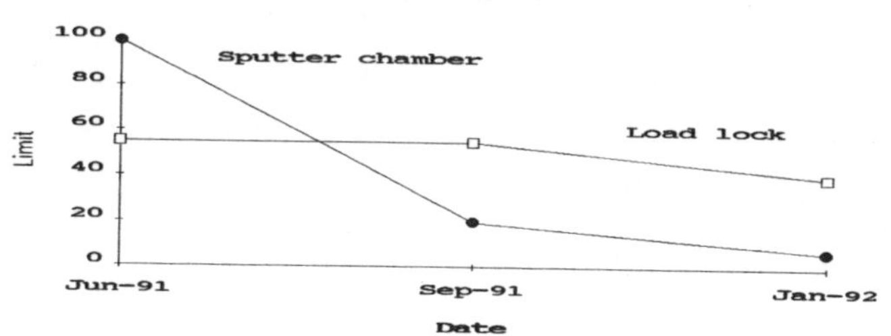

141

    With the reduction of defects and elimination of
silicon particle monitor wafers, the potential cost savings
with the in situ particle monitors were calculated with the
cost-of-ownership model developed at SEMATECH [4], as shown
in figure 10 and 11. A possible increase in profitability
of $500,000 per year can be achieved by reducing the defect
density by 20%. In this cost-of-ownership calculation the
defect density has the most impact in terms of cost
reduction even with a conservative assumption that only 17%
of total defects cause a failed die, a die size of 1 cm^2,
the wafer value at this step is $1,000, and final value is
$3,000 with 2000 wafer starts per week. The calculated
cost-of-ownership was reduced by $70,000 per year by virtue
of eliminating particle monitor wafers (assuming $15 per
wafer and 28 test runs per week). In addition to cost
savings from eliminating wafer particle monitors, there are
also cost savings due to a 2% increase in utilization, an
improvement attributed to reduced downtime from eliminating
wafer particle monitor tests. The in situ particle monitor
could also be used to pinpoint the particle sources, thus
resulting in less system down time.

Fig 10: Cost of Ownership Reduction

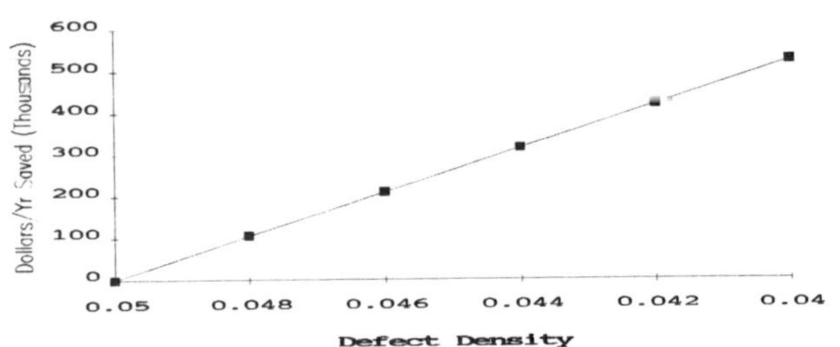

Fig 11: Cost of Ownership Reduction

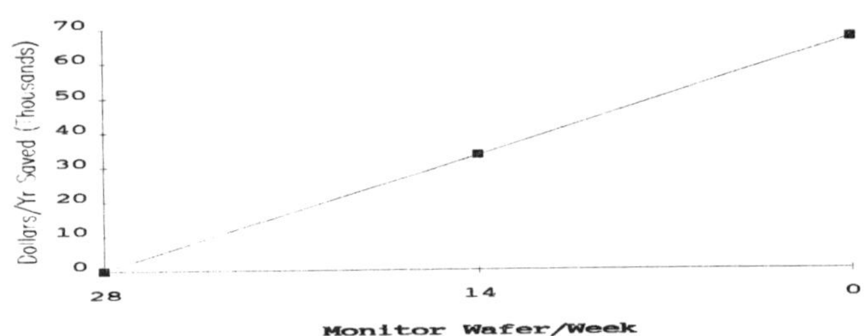

CONCLUSION

The in situ, real-time particle monitor has shown to be a reliable and effective method of particle monitoring in a single wafer metal deposition system. The in situ monitors installed in the deposition chamber and the load lock pump line were effective at detecting particle excursions and particle trends generated by film flaking, metal build-up and mechanical effects, whether normal or abnormal. A particle baseline for this metal deposition system was established, making it possible to set threshold limits with statistical guidelines. In situ monitoring has replaced the traditional wafer particle monitoring method. The continuous and sensitive nature of the in situ particle monitors has raised awareness for defect levels and has driven the defect levels down by 97% and 27% for the deposition chamber and load lock, respectively. With the full implementation of the in situ monitoring in manufacturing, the calculated annual cost-of-ownership was reduced by $70,000 by virtue of eliminating the wafer particle tests. Although the particle reductions of 97% in the deposition chamber and 27% in the load lock chamber were achieved, it is difficult to translate these gains directly into cost-of-ownership benefits. However, with a calculated cost-of-ownership reduction of $500,000 for 20% of defect density reduction, it may be assumed that large gains were made by defect density reduction.

ACKNOWLEDGMENTS

The authors would like to thank Chuck Stone, Greg Hamilton, Robert Ortiz and Marc Thompson of Cypress Semiconductor, and Dr. Peter Borden, Jim Stoltz of High Yield Technology for their support in the project.

REFERENCES

[1]   B.Y.H. Lin, presented at the 37th Annual Meeting of the IES
[2]   S.M. Sze, VLSI Technology, 2nd edition, 1988, pg. 390
[3]   Peter Borden, Microcontamination, May 1991
[4]   JoAnn Trego and Daren Dance, SEMATECH Cost-of-Ownership documentation, December 1991

CORRELATION BETWEEN *IN SITU* GAS ANALYSIS AND THE REACTION CHEMISTRY FOR LPCVD SILICON NITRIDE PROCESS IN VERTICAL THERMAL REACTORS

Mansour Moinpour, and Mohsen Shenasa*
Fairchild Research Center
National Semiconductor Corp., Santa Clara, CA 95052-8090

Azita Sharif, Dennis F. Brestovansky and George C. Guzzo
LINDE Division of PRAXAIR Inc., Tarrytown, N.Y., 10591

and

Richard J. Markle and Richard Wilkes
SEMATECH, Austin, TX 78741

Recently, silicon nitride Low Pressure Chemical Vapor Deposition (LPCVD) process which is widely used in VLSI technology has been studied in terms of in situ gas purity analysis and contamination control. The emphasis of previous studies has been mainly on the detection, measurement and reduction of moisture and oxygen levels within the process reactor. In this paper, we report on the correlation between the silicon nitride reaction chemistry and the in situ analysis of vapor phase species during deposition and post deposition steps. Several Si-Cl-H and Si-Cl-O compounds were detected which could be intermediate species formed as a result of the overall reaction between NH_3 and $SiCl_2H_2$ to deposit silicon nitride. In addition, several Si-O-H species were detected which may suggest a vapor phase interaction between moisture and oxygen and dichlorosilane species.

INTRODUCTION

A current focus of contamination control efforts in the semiconductor industry is on the identification and elimination of undesirable contaminants which may exist within manufacturing process equipment during wafer processing. Low Pressure Chemical Vapor Deposition (LPCVD) of silicon nitride film is widely used in VLSI processing. With increase in process complexity and decrease in circuit geometry, both uniformity and defect density of the LPCVD nitride film play an important role in high wafer yield and die reliability. The focus of previous studies has been mainly on improving the process capability and film uniformity through investigation of the effects of process parameters such as deposition temperature, total pressure and reactive gas flow and reactive gas ratio[1,2]. More recently, silicon nitride (LPCVD) process was studied in terms of in situ gas purity analysis and contamination control[3-5].

For a vertical (LPCVD) silicon nitride system, an axial profile of moisture and oxygen levels along the wafer load was obtained using Linde's Low Pressure Reactor Analysis (LPRAS) methodology[5]. The emphasis of these studies has been mainly on the detection, measurement and reduction of moisture and oxygen levels within the process reactor. The level of H_2O and O_2 within a LPCVD silicon nitride reactor is believed to play a direct role in the formation of defects. Moisture and O_2 could react with the process gas(dichlorosilane, $SiCl_2H_2$) to form particles or other contaminants that could be incorporated into the film. Oxygen and moisture could also compete with the intended reactions between the process gases($SiCl_2H_2$, and ammonia, NH_3) and the wafer surfaces. Additionally, the process chemistries were believed to contribute to potential wafer defects. Residual reactive gases could be present after the intended processing was completed and cause subsequent undesirable chemical reactions with the wafer surfaces, or lead to the formation of particles which could be incorporated into the film[6]. Ammonium chloride(NH_4Cl) by-products could be formed as a result of the silicon nitride deposition process. Ammonium chloride has been suspected as a possible cause of silicon nitride film "starry night" defects[6,7]. The purpose of this study is to complement the previous in situ contamination studies of silicon nitride process by elucidating on the correlation between the silicon nitride reaction chemistry and the in situ analysis of vapor phase species during deposition and post deposition steps.

EXPERIMENTS

The Silicon Valley Group (SVG)/Thermco Systems' Vertical Thermal Reactor (VTR) 6000 which is a vertical, multi-wafer LPCVD reactor was used for silicon nitride deposition. A schematic of the vertical reactor is shown in Figure 1. Process gases were introduced via quartz injection ports located at the top of the quartz process tube. The furnace gases were exhausted from the bottom/door end of the process tube through the vacuum lines. Heating coils maintained the desired temperature setpoints. The gas flows and temperature were controlled through the use of mass flow controllers and thermocouples, respectively. The furnace had three temperature zones. The LPCVD silicon nitride process employed a temperature-gradient with the highest temperature setpoint at the bottom zone to compensate for any reactant gas depletion effects down the process tube. In a typical nitride deposition experiment, 150 wafers are positioned in slots 1 through 150 of the 160-slot quartz boat which is subsequently loaded into the LPCVD furnace. The Si_3N_4 film deposition was achieved by reacting dichlorosilane($SiCl_2H_2$, DCS) and ammonia(NH_3) gases at pressure of 200 mTorr. The $NH_3/SiCl_2H_2$ ratio was kept at 3:1. Typical temperature setpoints were 765°C on the top, 780°C in the center and 790°C at the bottom of the reactor. Base pressure on this reactor was typically 3-4 mTorr.

The Linde LPRAS gas analysis system consisted of a quadruple mass spectrometer(residual gas analyzer) with the ability to detect trace vapor phase species with fragment sizes up to 300 atomic mass units(amu). It employed a rapid-response sampling system in order to monitor gas purity inside the LPCVD reactor in a real time manner. This way, any changes in gas composition or purity could be correlated to process events. A schematic of the gas analysis system and the sampling configuration are also presented in Figure 1. Gas was withdrawn from the vicinity of the wafers using the sampling probes as shown in Figure 1. Gas sampling probes were used along the length of the process tube to obtain gas composition profiles. The LPRAS also provided the calibration of H_2O and O_2 concentrations[4].

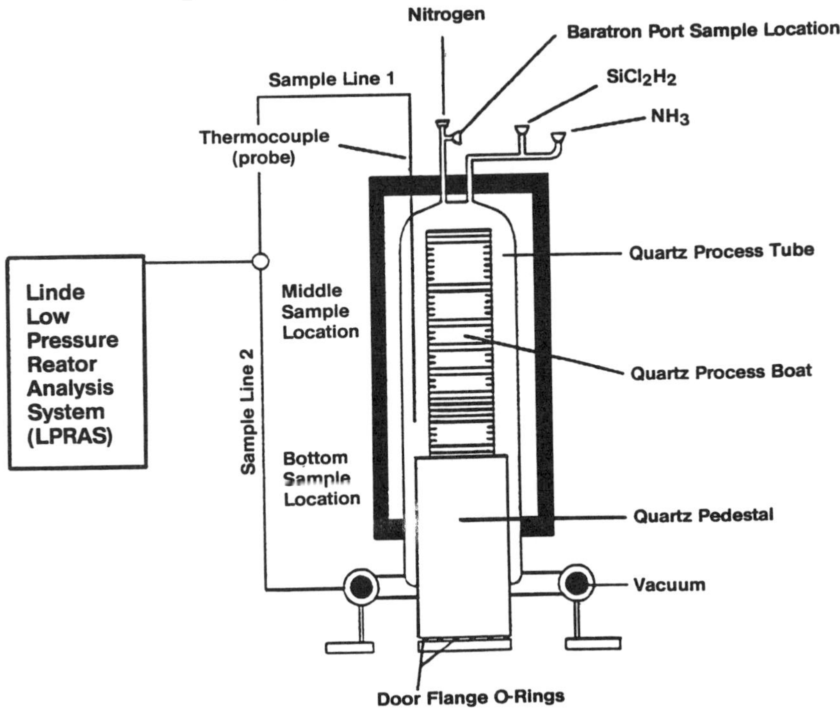

Figure 1. Schematic of VTR 6000 (LPCVD silicon nitride) and Linde LPRAS Systems

RESULTS AND DISCUSSIONS

The sequence of silicon nitride deposition process steps is outlined in Table 1. Gas phase species were sampled during the deposition and post-deposition purge steps(Steps 10 and 13 in Table 1, respectively). Spectrum scans at the beginning, middle and the end of the actual deposition interval were taken as samples were measured from both the bottom of the reactor zone and the exhaust line. Spectral data was also collected from the beginning, middle and end of the post-deposition purge(Step 13) during several experiments. In all experiments, ion peaks were detected and recorded up to the limits of the mass analyzer, i.e., 300 amu. Figure 2 shows all the species detected during temperature stabilization(Step 7). This was used as a reference point for detection of gaseous species in the reactor so that ions resulting from impurities present in the system could be distinguished from the those resulted from process gas interactions. There only existed a one-minute NH_3 flow and a 10-Sec DCS flow between this step and the onset of silicon nitride deposition. Species detected 3 minutes into deposition and 33 minutes into deposition are shown in Figures 3 and 4 respectively. These species(ions) were presumably derived from molecular intermediates existing in successive stepwise reactions leading to the formation of Si_3N_4 and/or formed as a result of interaction with contaminants(mainly oxygen and moisture) present in deposition reactor .

Table I. Standard 1100 A silicon nitride process

Temperature	=	782°C at the center
Process Pressure	=	200 mTorr
Base Pressure	=	< 5 mTorr

Interval	Description
0	STANDBY (N_2, 700°C, 200 mTorr)
1 (N/A)	Boat loaded with N_2 purge
2 (8 min)	Evacuate to base pressure (< 5 mTorr)
3 (10 min)	Vacuum pump check (N_2, 700°C)
4 (02 min)	Evacuate to base pressure
5 (01 min)	Leak check (5-10 mTorr/min)
6 (10 min)	Temperature cycle (N_2, 200 mTorr, from 700°C to 782°C)
7 (25 min)	Temperature stabilization (N_2, 200 mTorr)
8 (01 min)	NH_3 on (200 mTorr)
9 (10 sec)	$SiCl_2H_2$ on (200 mTorr)
10 (57.5 min)	Deposition ($SiCl_2H_2$, NH_3), 200 mTorr, and 782°C
11 (01 min)	$SiCl_2H_2$ off, maintain NH_3 off
12 (03 min)	Evacuate to base pressure, NH_3 off
13 (01 min)	N_2 purge (700°C, 200 mTorr)
14 (05 min)	Slow backfill (to 5 Torr, N_2, 700°C)
15 (05 sec)	Full backfill (to 760 Torr, N_2, 700°C)
16 (02 min)	Pre-lower boat hold (N_2, 650°C)

Comparing spectra scans during the temperature stabilization step(Figure 2) and the deposition step(Figures 3 and 4), it is seen that several Si-Cl-H and Si-Cl-O compounds were detected only after the deposition started which could be intermediate species formed as a result of the overall reaction between NH_3 and $SiCl_2H_2$(DCS) to deposit silicon nitride. These are SiOH, SiHOH, SiO, SiCl, $SiCl_2$ and SiH_2ClNH_2. The chemical reactions for the deposition of Si_3N_4 and the formation of ammonium chloride can be written as follows:

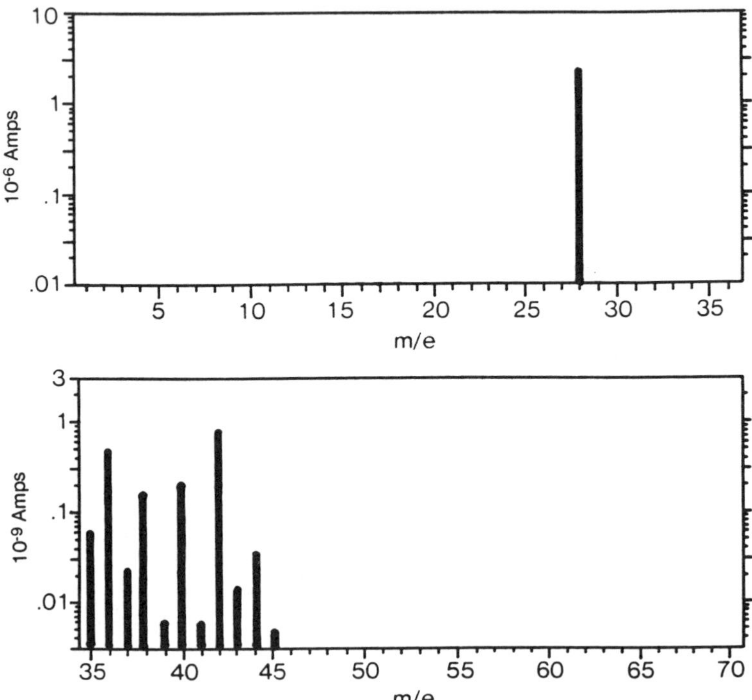

Figure 2. Si_3N_4 process, scan during temperature stabilization

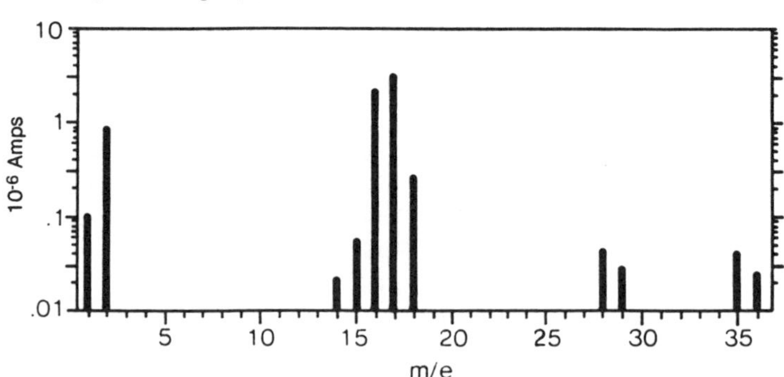

Figure 3. Si_3N_4 process, spectrum scan of deposition, T= 3 min. into deposition

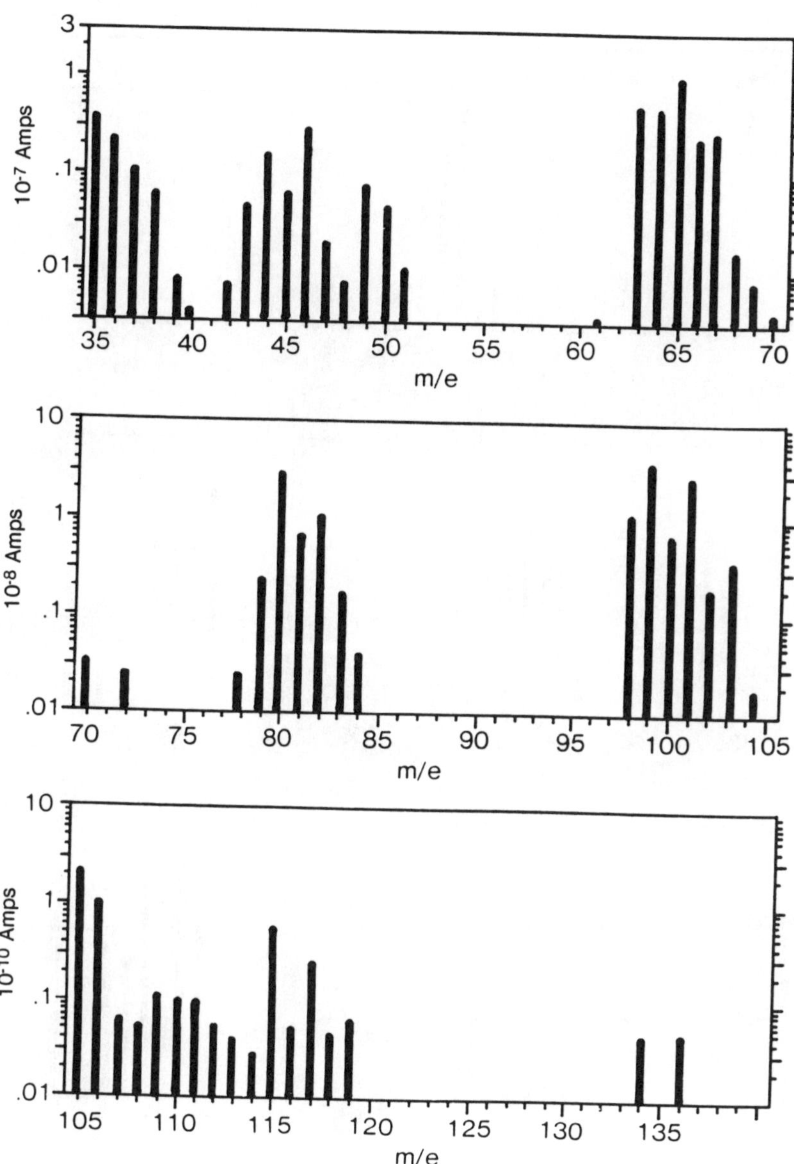

Figure 3 (Cont.). Si_3N_4 process, T= 3 min. into deposition

149

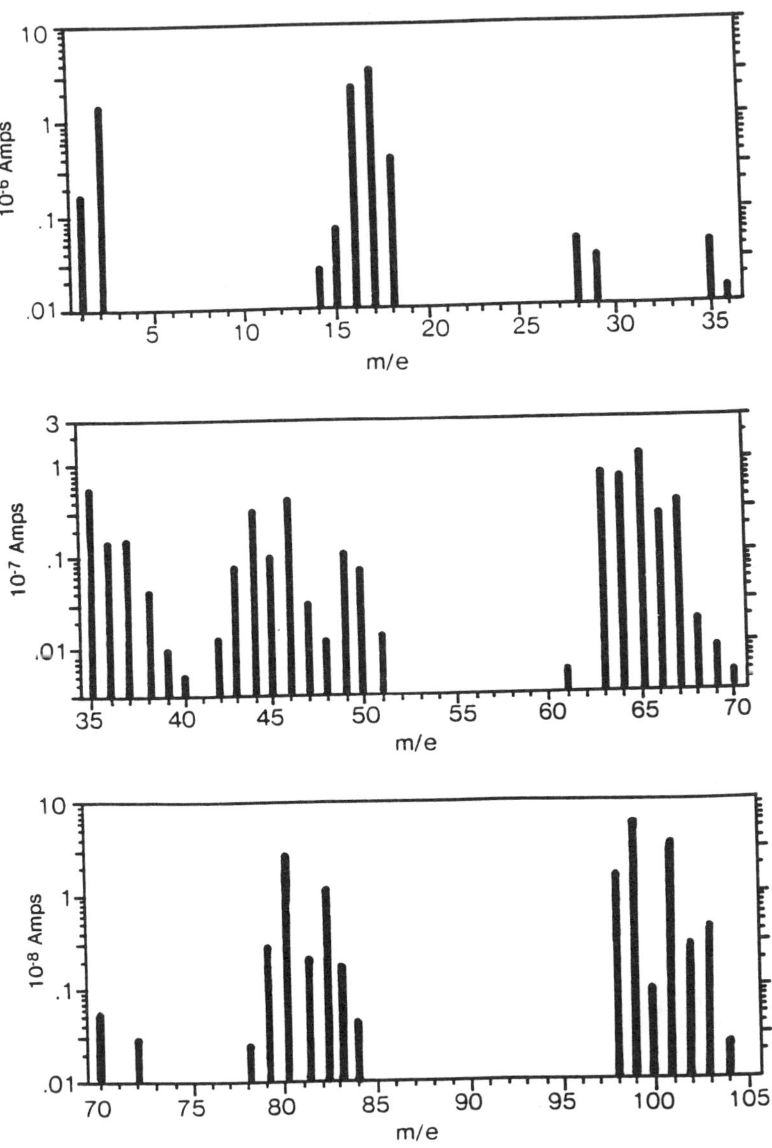

Figure 4. Si$_3$N$_4$ process, spectrum scan of deposition, T= 33 min. into deposition

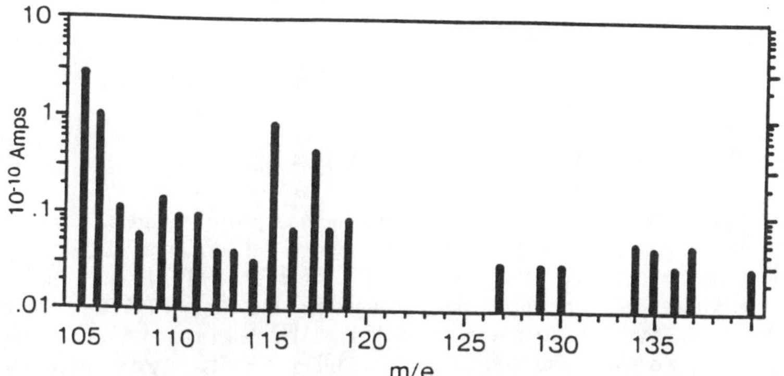

Figure 4 (Cont.). Si₃N₄ process, T= 33 min. into deposition

$$SiCl_2H_2 + 4/3NH_3 = 1/3Si_3N_4 + 2H_2 + 2HCl \tag{1}$$

$$HCl + NH_3 = NH_4Cl \tag{2}$$

The net reaction is:

$$SiCl_2H_2 + 10/3NH_3 = 1/3Si_3N_4 + 2NH_4Cl + 2H_2 \tag{3}$$

One reaction mechanism which involves gas-phase formation of $SiCl_2$ and NH_3 decomposition is as follows[8]:

$$SiH_2Cl_2 = SiCl_2 + H_2 \tag{4}$$

$$SiCl_2 + 4/3NH_3 = 1/3Si_3N_4 + 2HCl + H_2 \tag{5}$$

$$NH_3 = 1/2N_2 + 3/2H_2 \tag{6}$$

For the sake of clarity, we will call this the Reaction Mechanism I. This model suggests that the excess ammonia decomposes to nitrogen and hydrogen according to reaction (6). However, Lin[9] has detected by mass spectrometry very small amounts of N_2 in a flow of ammonia through a silicon nitride deposition reactor at 800°C and a pressure of 50-200 Torr. On the other hand, thermodynamic studies of Si-H-Cl system reveal that the $SiCl_2$ intermediate could be a prominent equilibrium species at LPCVD conditions[10]. The other possible reaction mechanism(Reaction Mechanism II) involves direct gas-phase DCS-NH₃ reaction, formation of the silylamine and its surface reaction to form nitride[9]:

$$SiH_2Cl_2 + NH_3 = SiH_2ClNH_2 + HCl \tag{7}$$

151

$$SiH_2ClNH_2 + 1/3NH_3 = 1/3Si_3N_4 + HCl + H_2 \tag{8}$$

Reaction (7) can be written as:

$$SiH_2Cl_2 + NH_3 = ClSiH_2Cl \dots HNH_2 = SiH_2ClNH_2 + HCl \tag{9}$$

As it can be seen from reactions (7) to (9), these stepwise reactions leading to the formation of the nitride are suggested[9,11] to involve the formation of one or more intermediate species and the detachment of HCl from a highly unstable molecular complex. Roenigk and Jensen[8] have suggested that for LPCVD conditions $NH_3/SiH_2Cl_2<10$, the deposition can be described by reactions 4-6 involving the formation of the gas-phase intermediate species, $SiCl_2$. Results of mass spectrometric analysis of the gas phase species during nitride deposition in the vertical reactor, shown in Figures 3 and 4, which indicates relatively higher abundance of $SiCl_2$(amu=99) compare to SiH_2ClNH_2(amu=82) supports their hypothesis. However, the existence of SiH_2ClNH_2 intermediate in the mass spectra also suggests that the Reaction Mechanism II can not be completely overruled.

Figures 5 and 6 show the peak intensity as a function of time during deposition step. The gas sampling was performed at 3 minutes into deposition(t=3), 22 minutes into deposition(t=22) and 33 minutes into deposition(t=33).

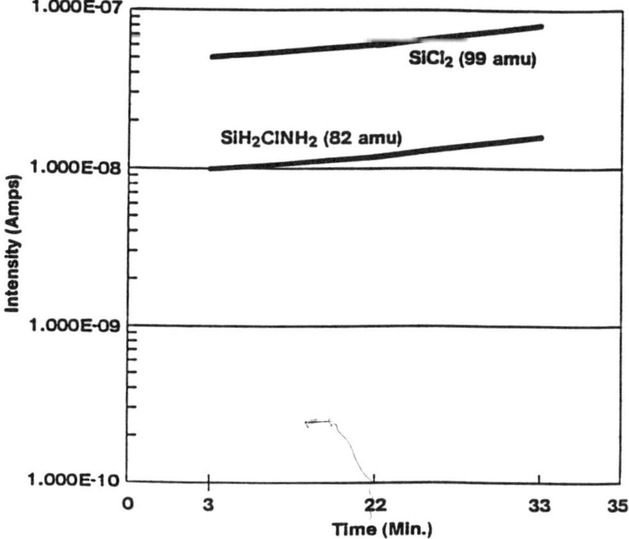

Figure 5. Concentration of species vs. time, deposition step

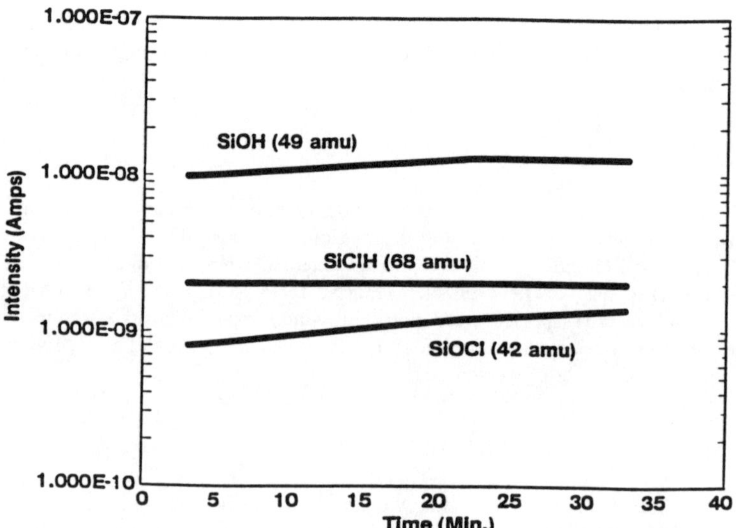

Figure 6. Concentration of species vs. time, deposition step

It can be seen from Figure 5 that both peaks for SiH_2ClNH_2 and $SiCl_2$ show steady increase over time. This may suggest that both reactions (4) and (7) take place during the progress of nitride deposition. Hence, there could be more than one set of independent chemical reactions leading to the formation of various intermediate species which eventually results in the nitride formation. The change in concentration(peak intensity) of several Si-O-H and Si-O-Cl is depicted in Figure 6. It is seen that the peak intensity for these species remain relatively constant. The presence of SiCl/SiHCl species(63-69 amu grouping) suggests that in addition to intermediate species formed as a result of chemical reactions (4)-(9), other reaction paths may be plausible.

Species SiO, SiOH and SiHOH(21-23 amu and 44-46 amu groupings) which were detected after the onset of deposition might have formed as a result of direct interaction of H_2O and/or O_2 with DCS or with intermediate species depicted in reactions (4), (7) and (9). In addition, species at 47 and 48 mass/charge(m/e) were observed in the spectra scan that could be identified as $SiOH_3^+$ and $SiOH_4^+$. This may also suggest a vapor phase interaction between moisture and oxygen and dichlorosilane species. Water from atmosphere could also react with ammonium chloride formed per reaction (2) as follows:

$$NH_4Cl + H_2O = (NH_3).(H_2O)(ad) + HCl \tag{10}$$

The basic ammonia and water complex can attack the surface and provide nitrogenous Si nucleation sites. Hence, the presence of oxygen and moisture within the reactor environment during deposition further complicates the overall reaction path leading the Si$_3$N$_4$ formation and contributes to formation of "starry night" defects per reactions with the ammonium chloride.

CONCLUSIONS

The correlation between the silicon nitride reaction chemistry and the in situ analysis of vapor phase species during deposition and post deposition steps were described. The Si-Cl-H and Si-Cl-O compounds detected where believed to form as a result of the overall reaction between NH$_3$ and SiCl$_2$H$_2$ to deposit silicon nitride. In addition, several Si-O-H species were detected which may suggest a vapor phase interaction between moisture and oxygen and dichlorosilane species. The formation of "starry night" defects was briefly discussed.

ACKNOWLEDGMENTS

The authors would like to acknowledge the contributions of Mike Bizak of National Semiconductor Corp. to the setup of the sampling probe.

REFERENCES

[1] R. S. Rosler, Solid State Technol., **20(4)**, 63, (1977).

[2] G. Depinto and J. Wilson, IEEE/SEMI Advanced Semicon. Manufact. Conf., 47 (1990).

[3] B. A. Sacha, R. L. Rodgers, D. F. Brestovansky and G. C. Guzzo, Proc. of IES Conf., 373 (1990).

[4] D. F. Brestovansky, R. M. Vadjikar, G. C. Guzzo, D. Denning and O. K. Reynolds, Proc. of Microcontamination Conf., 74 (1988).

[5] M. Moinpour, M. Shenasa, A. Sharif, G. C. Guzzo, D. F. Brestovansky, R. J. Markle, and R. Wilkes, Symposium B, MRS Conf. Proc., April/May 1992.

[6] M. Simard Normandin and B. A. Oliver, Proc. of Microcontamination Conf., 94 (1989).

[7] D. F. Brestovansky, G. C. Guzzo, R. J. Markle and M. T. Schroth, Proc. of IES Conf., Nashville, TN, May 1992.

[8] K. F. Roenigk and K. F. Jensen, J. Electrochem. Soc., **134**, 1777 (1987).

[9] S. S. Lin, J. Electrochem. Soc., **124**, 1945 (1977).

[10] F. Langlais, F. Hottier, and R. Cadoret, J. Cryst. Growth, **56**, 659 (1982).

[11] U. Wannagat, *Advances in Inorganic Chemistry and Radiochemistry, Vol. 6, 225,* Academic Press, New York (1964).

** Mansour Moinpour is currently at Intel Corp., California Technology Development Center, Santa Clara, CA 95052.*

ANALYSIS OF DEFECTS ON THE SURFACE OF BARE
UNPATTERNED SILICON WAFERS BY SEM-EDS

Erik J. Mori and Larry W. Shive

MEMC Electronic Materials, Inc.
501 Pearl Dr., P.O. Box 8
St. Peters, MO 63376

ABSTRACT

Contamination control and defect reduction are essential to obtaining higher yields and reliability in semiconductor manufacturing. This work demonstrates the detection and morphological characterization of sub-0.2 micron particles and metallic residues deposited onto bare, unpatterned silicon wafers using Scanning Electron Microscopy (SEM) and Energy Dispersive Spectrometry (EDS). Results have shown that polystyrene latex spheres, polishing and lapping materials of 0.1 micron dimensions are readily observed on polished wafers. However, the components have been mostly found to agglomerate. The location of the particles are dependent on sample deposition conditions. Metallic residue collected by Vapor Phase Decomposition (VPD) have been examined for morphology, size, and composition. Intentionally contaminated wafers with iron, copper, nickel, and zinc were used for this study. The VPD residue characteristics have been shown to differ between metallic contamination.

KEYWORDS: silicon, particle, defect, VPD, TXRF, SEM, EDS

INTRODUCTION

The drive towards smaller device geometries and higher yields in ULSI processing requires strict process control and reduction of particle deposition and metallic contamination since these defects severely affect electrical device characteristics. Particles are known to cause a host of processing defects ranging from thin film nonuniformity to circuit patterning imperfections. Metallic impurities aggregates or residues can also lead to device failure either through mechanical or chemical interference. In order to understand the nature, origin, and source of these defects SEM-EDS was employed to detect and characterize the contamination source at a greater than 0.2 micron level [1,2]. For quantitative analysis of metallic impurites, Vapor Phase Decomposition-Total X-ray Reflection Fluorescence (VPD-TXRF) has been used to increase sensitivity of the instrument [3]. However, the morphology of the VPD residue, which is key to understanding the residue and plating characteristics with respect to x-ray emission profiles, has not been studied. This work will also describe the analysis of iron, nickel, copper, and zinc contaminated metallic residues by SEM-EDS on Si wafers.

EXPERIMENTAL

Both 100 and 150-mm n- and p-type Czochralski grown Si wafers were used for this study. The wafers were subsequently cleaned using a modified-RCA process. Polystyrene latex spheres (Particle Measuring Systems), alumina powder and colloidal silica polishing compounds (Fujimi Corp.) were obtained and deposited on wafer from an aqueous solution at a spin station. Table I lists the particle nominal size and applications:

TABLE I

Particle Type	Particle Size (micron)	Applications
Alumina	0.08	lapping
Silica	.074	polishing
PSL	.102	particle counter
	.203	calibration

The aluminum solution consisted of adding 5 grams of alumina powder dispersed in 100 ml of DI water. After 10 minutes of ultrasonication, a 25 microliter aliquot was taken and diluted into another 250 ml of DI water. Finally, a 10 microliter solution was dispensed during the rinsing cycle at the spin station. For the silica slurry, a 25 microliter aliquot was added to 250 ml of DI water of which 10 microliters was also dispensed during the rinsing cycle. Three drops each (ca. 100 microliters) of 0.1 and 0.2 micron sphere concentrated solution were diluted in 40 ml of DI water. A 100 microliter solution was dispensed during spinning.

Metallic contamination consisted of dispensing a 10 ml solution of 1 ppm wt. Fe, Ni, Cu, and Zn standard onto a wafer on a spin coater. The solution was allowed to stand on the wafer for 1 minute prior to spinning of the liquid at 3500 rpm for 30 seconds. The residue and oxide underlayer was subsequently stripped using VPD [3]. A 100 microliter DI water droplet was utilized to collect the condensates from the VPD process. Wafers were subsequently dried using an IR lamp in a clean air glove box. Particle counting and mapping before contamination was performed using an ESTEK WIS 150 automated wafer inspection system. An Amray 186OFE, JEOL 6400F, Zeiss DSM 960, and Hitachi S4000, S4100 SEM; and Kevex Quantum, Superdry, APD; Noran Voyager and Z-MAX 30 detectors were evaluated for applicability in this study.

RESULTS AND DISCUSSION

SEM-EDS was utilized to determine whether individual particulates of ≤ 0.2 micron dimensions could be observed on a polished Si surface. The latex spheres, alumina, silica, and metallic VPD residues were detected and characterized for morphology, size, and composition. The characteristic physical shapes and x-ray emission were used to distinguish against the Si background and used to verify the presence of contaminants.

Particles:

Particles of nominal diameters ≤ 0.2 microns were used to demonstrate defect detection on bare unpatterned and polished Si wafer surfaces using SEM-EDS. Although individual particles of silica, alumina, and PSL were found, most particles were aggregates. Polystyrene latex sphere nominal diameters size agreed with manufacturer specifications. The PSL solution was injected near the center of a wafer during the spin cycle. Due to the centrifugal force from the spin station, these spheres were located near the edge of the wafers.

From Figure 1 below, colloidal silica particles were also readily observed.

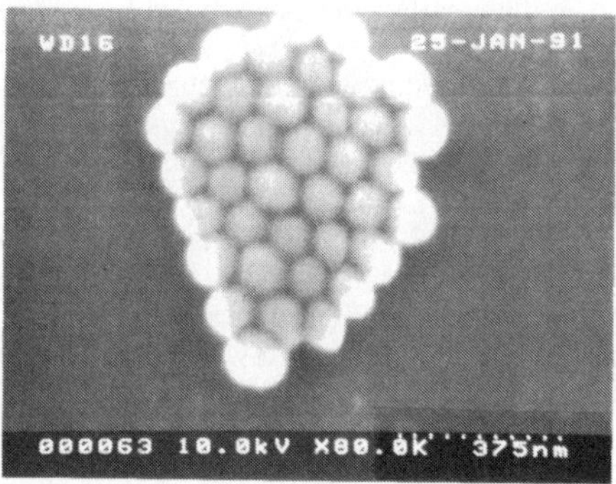

Figure 1. Colloidal silica particles on a bare polished Si wafer surface.

Clusters of colonies consisting of at least three spheroid particles were found scattered throughout the wafer surface. The particulates were highly charged against the featureless silicon background, rendering it amenable to physical and chemical characterization.

The alumina solution which was ultrasonicated to produce a suspension in DI water was not completely dispersed as shown in Figure 2 below. However, few individual particles were scattered and the composition of a 0.1 micron particle was verified by EDS as shown in Figure 2 below:

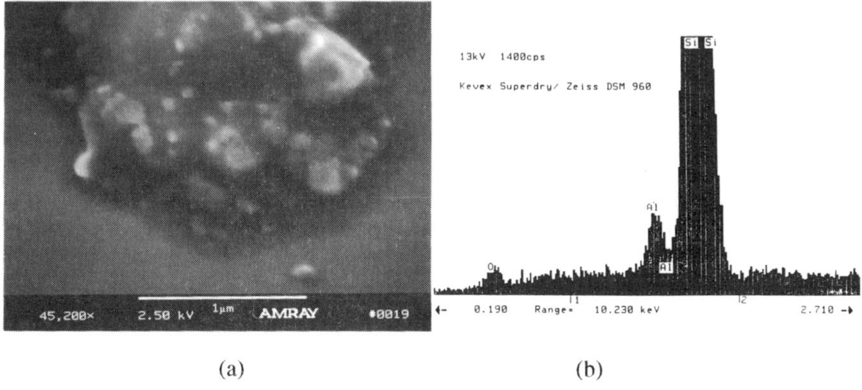

(a)   (b)

Figure 2. a) Scanning electron micrograph of alumina aggregates and b) an x-ray spectrum of a 0.1 micron alumina particle on a bare polished Si wafer surface.

The primary difficulty with this work was the accurate assessment of particle location for SEM work with particle counter coordinates. In addition, a permanent feature on the Si surface was required to initially contrast and focus the beam to the particle size of interest in order to maintain an appropriate field depth for detection throughout the entire wafer surface. Furthermore, long integration times and low accelerating voltages were used to obtain detectable x-ray emissions and minimize the deformation, decomposition and volatization of materials. Future work will include a completely automated integration unit between a wafer inspection system with an SEM-EDS to facilitate detection efficiency of <0.1 micron particles on large bare unpatterned Si wafers.

VPD Metallic Residue:

The morphology, size, and composition of a VPD residue varied with the type of metallic contamination. These parameters are essential in establishing optimal conditions in terms of the usage of glancing angles in TXRF for rough surfaces. An Fe-, Zn-, Ni-, Cu-contaminated and uncontaminated VPD residue was used for this study. A representative control wafer with a dried droplet with no contamination under low magnification are presented in Figure 3a and b below:

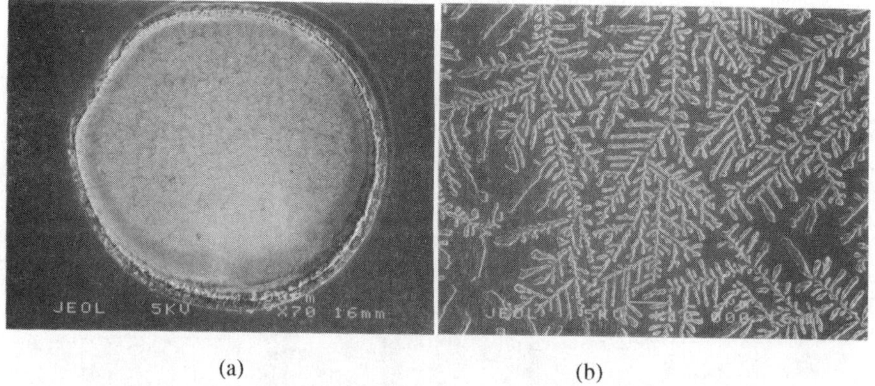

(a)   (b)

Figure 3. a) Low and b) high magnification of a typical uncontaminated VPD residue on a bare polished Si wafer surface.

The size and shape of the ring pattern observed depend on the initial condensate volume, drying time, air flow conditions, surface roughness and tilt, and heating intensity. From secondary and backscattered images, a typical oxide ridge (height >1 um) at/near the perimeter, and a dendritic structure approaching the center of the residue are observed. However, the morphology and composition changes with contamination. A closer examination of the Fe residue lying at the inner portion of the concentric ring reveals a plate-like morphology as shown in Figure 4a and b below:

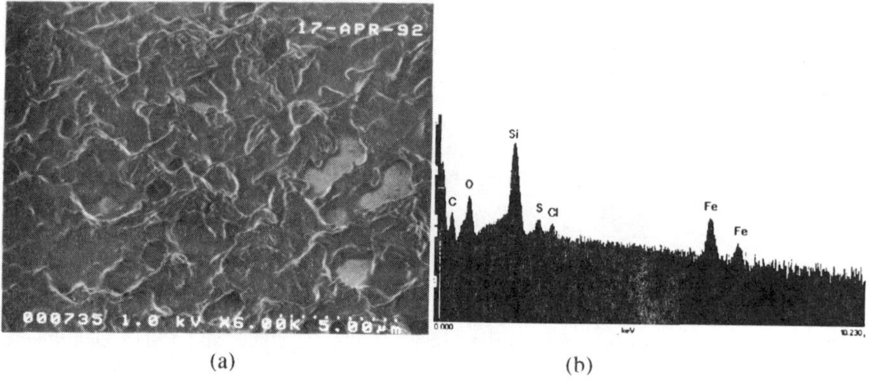

(a)   (b)

Figure 4. a) Scanning electron micrograph and b) x-ray spectrum of an Fe-contaminated VPD residue on a bare polished Si wafer surface.

159

As shown in Fig. 4b above, iron contamination was confirmed using EDS. The origin of residual chlorine and sulfur observed in iron may be from the modified RCA cleaning process.

Zinc- and nickel contaminated residue revealed a similar dendritic structure to the uncontaminated wafer. Nickel displayed a discontinued dendritic pattern. However, characteristic x-ray emission for Zn and Ni were either weak or not observed. Low accelerating voltages (<5 keV) were required to minimize dendrite volatization and thus enabling imaging of Zn and Ni residues.

Unlike the previous contamination, Cu residue was not directly observable to the naked eye. However, several small circular features of approximately 15 microns in diameter were observed scattered throughout the wafer and are shown under low and high magnification in Figure 5a and b below:

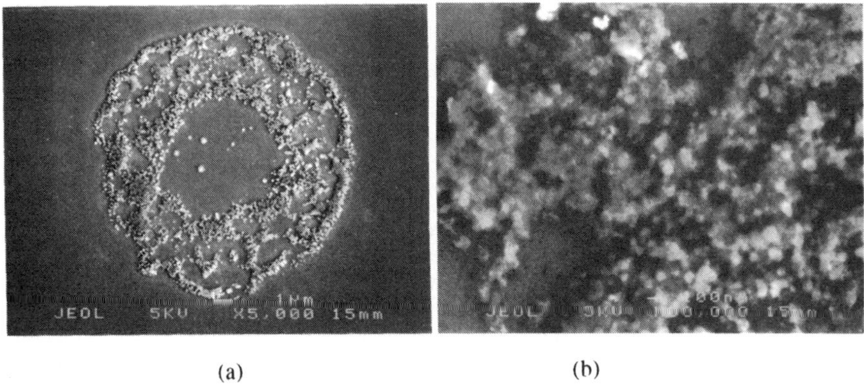

(a)   (b)

Figure 5. a) Low and b) high magnification of a Cu-contaminated VPD residue on a bare polished Si wafer surface.

Copper exhibits a fine powdery structure with particulate diameter of approximately 50-100 nm. Scattering across the wafer indicates Cu plates and precipitates out during the drying stage of the VPD droplet due to the electronegativity difference between Cu and the Si surface [5].

CONCLUSION

SEM-EDS has been demonstrated to be capable of detecting sub-0.2 micron size particles on a bare polished Si wafer surface. Alumina, silica, and PSL spheres were examined for this study. In addition, the morphology of the VPD residue changes with metallic contamination. Typical dendritic structures formed in the case of an uncontaminated wafer is essentially unchanged with Zn and slightly altered with Ni. No characteristic x-ray emission peaks were observed. However, VPD residue morphology is drastically altered with Fe and Cu. Fe reveals a more platelette-like structure while copper exhibits a fine powder-like morphology. Scattering of Cu residues across the wafer agree with plating mechanism invoking electronegativity differences between Si and the metal.

ACKNOWLEDGMENTS

The authors wish to thank Amray, JEOL, Hitachi, Zeiss, Kevex, and Noran for utilization of the instruments. In addition, assistance from Kenneth Ruth, Todd Hedges, Phil Schmidt, and Sharon Brumer are greatly appreciated.

REFERENCES

[1] S.J. Miller, E.J. Baker, K.-H. Ahn, *Microcontamination*, 10, 1989.

[2] T. Hattori, S. Koyata, *Solid State Technology*, 9, S1, 1991.

[3] P. Eichinger, H.J. Rath, H. Schwenke, *Semiconductor Fabrication: Technology and Metrology*, ASTM STP 990, ed. D.C. Gupta, 305 (American Society for Testing and Materials), 1989.

[4] H. Kondo, J. Ryuta, E. Morita, T. Yoshiniki, Y. Shimanuki, *Jpn. J. Appl. Phys.*, 31, L11, 1992.

[5] T. Imaoka, T. Kezuka, J. Takano, I. Sugiyama, T. Ohmi, *Proc. IES*, 1, 466, (IES, Mt. Prospect, Ill.), 1992.

PHYSICAL CHARACTERIZATION LIMITS
IMPOSED BY RANDOM SAMPLING

T. J. Shaffner
Materials Science Laboratory
Texas Instruments, Incorporated
Dallas, Texas 75265

ABSTRACT

Characterization technology has managed to keep a step ahead of ULSI requirements for smaller spot size analysis and increased sensitivity to impurities and defects. This pace cannot continue as we approach atomic dimensions. In sampling a random distribution of trace impurity atoms for example, we must include at least one in the analytical volume before detection is possible. Smaller and more surface sensitive probes emphasize this issue. We provide a quantitative framework for understanding and dealing with this fundamental problem.

INTRODUCTION

The shrinking of ultra-large scale integration (ULSI) linewidths and stringent requirements for higher purity materials continue to challenge our ability to analyze submicron point defects and sense parts-per-trillion levels of impurity atoms in silicon. Until recently, modern characterization technology has kept pace with these requirements, but this becomes increasingly difficult as analytical probes reach atomic dimensions and impurity levels relevant to manufacturing become vanishingly small.

The problem is illustrated in Figure 1. To sample a distribution of impurity atoms, we must include at least one in the probe volume before detection is possible. The probability that this will happen depends on the concentration, probe size, and depth of the escaping electron, photon, or particle. As these diminish, so do our chances for a successful analysis. In effect, the phenomenon is an extension into the atomic realm of the more familiar sampling problem we encounter when selecting a small number of representative specimens from many [1].

The fact that we are encroaching on this limit is documented in the literature [2,3] for modern tools like the first three listed in Table I. In particular, the scanning tunneling microscope, which is capable of imaging single atoms, poses the tricky issue of deciding which atoms are relevant to the problem at hand. We also discover that the simple act of finding a sparsely populated species in a sea of matrix atoms is often an unrealistic endeavor. Transmission electron microscopists have struggled with the problem for years.

While these points are self evident, a quantitative definition of the problem for tiny analytical probes and the boundaries it imposes remain absent in the literature. The following will provide a framework that incorporates the sampling issue with existing and more mature treatments of signal detection. We begin by adopting a working definition for the *relative concentration* of an impurity atom in silicon. However, it is sufficiently general to include other micro-anomalies, such as vacancies, interstitials, atom clusters, and lattice defects.

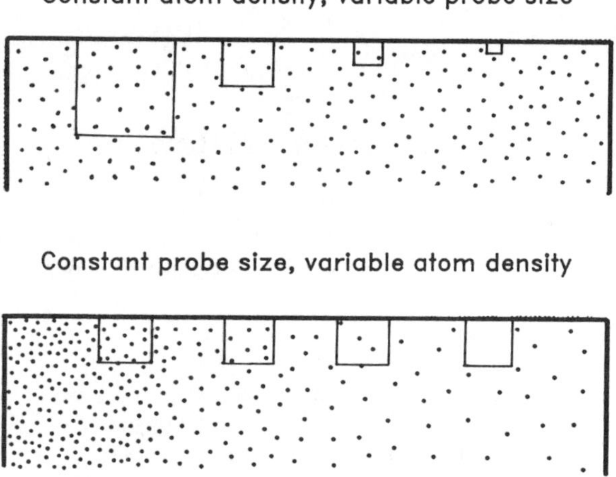

Figure 1. An analytical probe can be so small or the impurity concentration so low, that none of the atoms or defects of interest are contained within V.

Table I. Estimates of Total Atoms in the Probe Volume, V

Analytical Technique	Diameter	Depth	$V\ (cm^3)$	Atoms
Scanning Tunneling Microscopy	10 $\mathring{A}$	1.6 $\mathring{A}$	$2\mathrm{x}10^{-21}$	$1\mathrm{x}10^2$
Scanning Auger Microscopy	200 $\mathring{A}$	10 $\mathring{A}$	$7\mathrm{x}10^{-19}$	$4\mathrm{x}10^4$
Transmission Electron Microscopy	200 $\mathring{A}$	200 $\mathring{A}$	$6\mathrm{x}10^{-18}$	$3\mathrm{x}10^5$
Auger Electron Spectroscopy	25 μm	10 $\mathring{A}$	$5\mathrm{x}10^{-13}$	$3\mathrm{x}10^{10}$
Photoluminescence Spectroscopy	1 μm *Sphere*		$5\mathrm{x}10^{-13}$	$3\mathrm{x}10^{10}$
Electron Microprobe	1 μm *Sphere*		$5\mathrm{x}10^{-13}$	$3\mathrm{x}10^{10}$
Secondary Ion Mass Spectrometry	10 μm	100 $\mathring{A}$	$8\mathrm{x}10^{-10}$	$4\mathrm{x}10^{10}$
X-ray Photoelectron Spectroscopy	150 μm	10 $\mathring{A}$	$2\mathrm{x}10^{-11}$	$1\mathrm{x}10^{12}$
Rutherford Backscattering	1 mm	100 $\mathring{A}$	$8\mathrm{x}10^{-9}$	$4\mathrm{x}10^{14}$
Total Reflection X-ray Fluorescence	14 mm	200 $\mathring{A}$	$3\mathrm{x}10^{-6}$	$2\mathrm{x}10^{17}$
X-ray Fluorescence	1 cm	20 μm	$2\mathrm{x}10^{-3}$	$9\mathrm{x}10^{19}$
Instrumental Neutron Activation	5 cm *Cube*		$1\mathrm{x}10^{+2}$	$7\mathrm{x}10^{24}$

QUANTITATIVE FRAMEWORK

Definitions

We assume impurities that are randomly distributed throughout a silicon volume of dimensions X, Y, and Z, and write the *relative concentration* as

$$C = N_i/N_s \tag{1}$$

where N_i is the total number of impurity atoms in the sample, and N_s is the number of matrix silicon atoms. Implicit in the definition is the assumption that each impurity replaces a silicon atom, which is not always the case. A more general treatment would include interstitial impurities as well ($C = N_i/(N_i + N_s)$), but the matter becomes inconsequential at the trace levels ($N_i << N_s$) treated here.

We consider a probe volume V, expressed by the total number of atoms amenable to detection. The z boundary is defined by either penetration depth of the exciting radiation, or escape depth of the signal electron, photon, or particle. For simplicity, we approximate V with a rectangular box of dimension d in x and y, and z in depth. Details of the probe shape become insignificant later when we allow sample dimensions (X and Y) to be large by comparison.

We lose no generality in thinking only of the surface slab defined by X, Y, and z, since the remaining bulk is not susceptible to analysis. In effect,

$$z = Z \tag{2}$$

Sampling Distribution

We wish to express the probability α, that a single impurity atom distributed at random will be included in the probe volume, V. For one atom fixed within XYz, the number of available choices for probe positions, only one of which includes the atom, is given by

$$Number\ of\ Probe\ Volumes\ =\ \frac{XYz}{d^2 z} = \frac{XY}{d^2} \tag{3}$$

α is the inverse of (3):

$$\alpha = d^2 / XY \tag{4}$$

The probability that N impurity atoms are included in V is given by the Binomial distribution

$$D_1(N) = \frac{N_i!}{N!\,(N_i - N)!}\ \alpha^N\,(1 - \alpha)^{N_i - N} \tag{5}$$

We impose a useful approximation to D_1 for large samples (large X and Y) and small probe area (d^2). In these limits as $N_i \to \infty$ and $\alpha \to 0$, (5) reduces to the Poisson distribution

$$D_1(N)\ =\ \frac{(\alpha\,N_i)^N}{e^{\alpha N_i}\,N!} \tag{6}$$

The roles of N_i and α are no longer independent, and their product reduces to

$$\alpha N_i = \left(\frac{d^2 z}{XYz}\right)\left(\frac{N_i}{N_s}\right) N_s = CV \tag{7}$$

using (4) and (1). Substituting into (6), we obtain the simplification

$$D_1(N) = \frac{(CV)^N}{e^{CV} N!} \tag{8}$$

which confirms it is unnecessary to know the exact number of impurity atoms, or probability that a given one falls within the probe volume. We need know only the total matrix atoms encompassed by V and the relative impurity concentration, C.

Examples of $D_1(N)$ are shown in Figure 2 for the scanning tunneling microscope and scanning Auger microprobe. The average value of each distribution is CV, and these broaden with large N in proportion to the standard deviation, $\sqrt{CV}$.

The curves express the obvious fact that impurities must be included in the probe volume before detection can take place. Given their presence, the act of producing counts on a video display, detector, or recording device evokes yet another distribution that depends on details of the physics underlying the measurement. Although a full description of the techniques in Table I cannot be dealt with here, it is possible to incorporate the concept of detectability into the present framework.

Signal Detection

Considering a single impurity atom in V, we express the probability p of recording one count during course of the measurement as

$$p = \beta I t \tag{9}$$

where t is the counting time and I the intensity of the incident exciter. β is a proportionality constant that expresses the cumulative effect of excitation and emission processes from the sample, collection and detection efficiency of the electronics, and a host of details peculiar to the technique at hand. Frequently it is small in value, and must be compensated by high intensity, long duration measurements for successful detection to take place.

It is beyond the scope of this work to attempt first principles estimates of p for each technique in Table I. However, it is evident that it is near unity for a

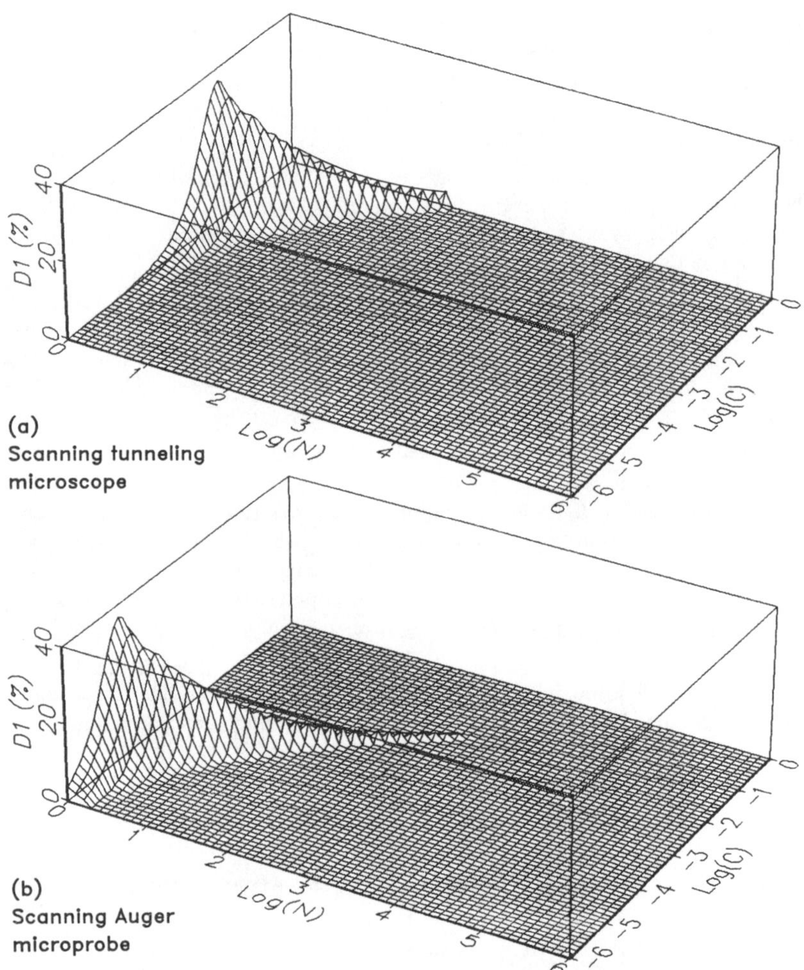

Figure 2. D_1 is the probability distribution that N impurity atoms are included in the probe volume as a function of relative concentration C. Values in Table I are assumed for the scanning tunneling microscope (a) and the Auger microprobe (b).

scanning tunneling microscope operated in atom imaging mode. Estimates for the remaining methods will be close to familiar experimental sensitivities, provided the distribution D_1 at a given concentration is not sharply peaked near $N = 1$. This condition guarantees the probability is small that only a single atom (also, zero atoms) resides in the probe volume.

It is assumed in the following for example, that p for the Auger microprobe is close to its measured detection limit (about 0.01 *atomic %*). Peak of the D_1 distribution in Figure 2(b) at $log(C) = -2$ is sufficiently shifted ($log(N) = 2.58$) that its tail at $log(N) = 0$ is vanishingly small.

An important point of definition is that β does not include counts from instrument fluctuations or background originating from competing physical processes within the sample. In effect, these diminish our ability to label which counts are correct, and require an independent approach for extracting signal from noise in the measurement.

Literature dealing with this issue is abundant, but the generalization provided by L. A. Currie [4] is the author's preference. Currie proposes three limits that describe 1) the critical level for anticipated detection (L_C), 2) the condition for qualitative determination (L_D), and 3) the level for which quantification is possible (L_Q). Either definition requires the availability of more impurity atoms than the ideal noise-free and zero-background scheme developed so far. We will impose Currie limits later when specifying the number of atoms n required in a real experimental situation.

Given that N impurity atoms fall within the probe volume, the probability distribution for producing n counts at the console during the experiment is given by the Binomial distribution

$$D_2(n) \; = \; \frac{N!}{n!(N-n)!} \, p^n \, (1-p)^{N-n} \tag{10}$$

Examples of $D_2(n)$ for scanning tunneling and Auger microscopes are shown in Figure 3.

Until the last decade, it was characteristic of analytical techniques that N is large, and p small. Notable examples are instrumental neutron activation, x-ray fluorescence, and the others listed at the bottom of Table I. In these cases, (10) reduces to a Poisson form that that underpins the familiar expression for the standard deviation

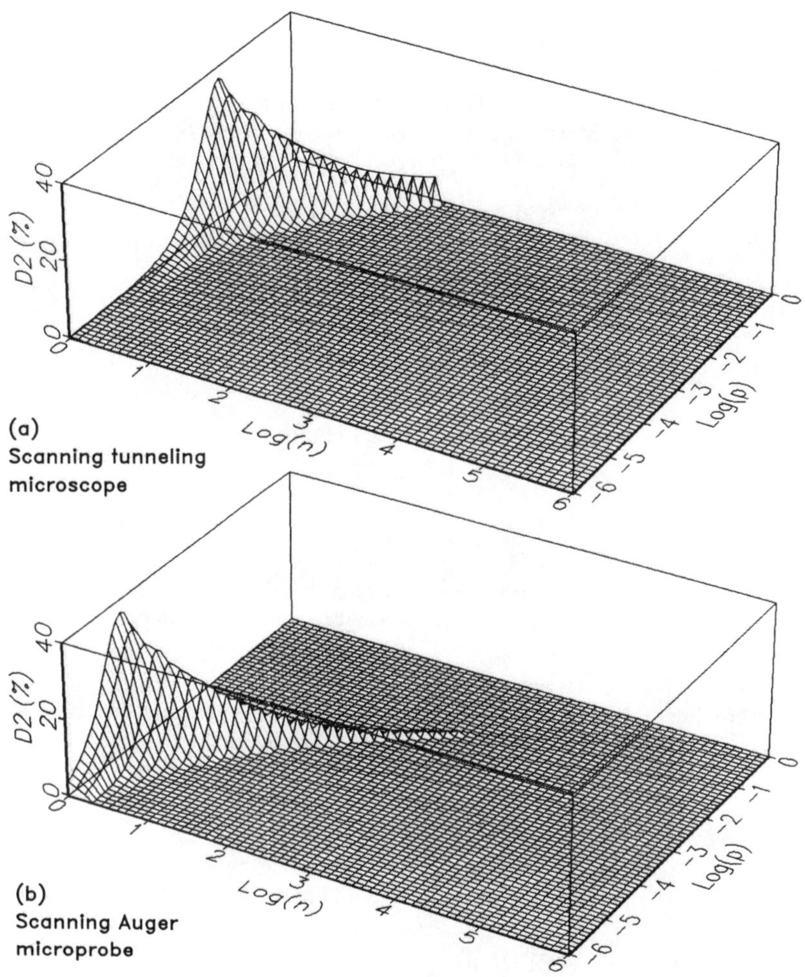

Figure 3. D_2 is the probability distribution that n atoms produce counts during course of the measurement as a function of p, the probability that one atom produces a count.

$$\sigma \propto \sqrt{N} \tag{11}$$

It appears that (11) can no longer be valid for the non-Poisson case when N is small, but we demonstrate in the following that the proportionality is preserved when D_1 and D_2 are combined into a composite distribution.

Composite Distribution

To derive the composite distribution $D_C(n)$, we sum the product of the sampling and detection components over all values of N from n to ∞:

$$D_C(n) = \sum_{n}^{\infty} D_1(N)\, D_2(N) \tag{12}$$

From (8) and (10), it can be shown that this reduces to

$$D_C(n) = \frac{(CVp)^n}{n!\, e^{CVp}} \tag{13}$$

which is itself a Poisson distribution of mean value CVp and standard deviation $\sqrt{CVp}$. A plot of $D_C(n)$ for the scanning tunneling and Auger microscopes is shown in Figure 4.

While (13) gives the probability of detecting precisely n atoms during course of the experiment, it is the sum of those above one of the Currie detection limits that is relevant to the measurement. For example, if we are able to detect 100 atoms given by the probability segment at $n = 100$ ($log(n) = 2$) in Figure 4, we should also include the increments representing more than 100 atoms (101, 102, *etc.*) when determining how many counts are recorded during the experiment.

The composite probability P_C that we detect n or more atoms (assuming one count per atom) is determined from the sum of (13) from n to ∞

$$P_C = \sum_{n}^{\infty} \frac{(CVp)^n}{n!\, e^{CVp}} = 1 - \sum_{0}^{n-1} \frac{(CVp)^n}{n!\, e^{CVp}} \tag{14}$$

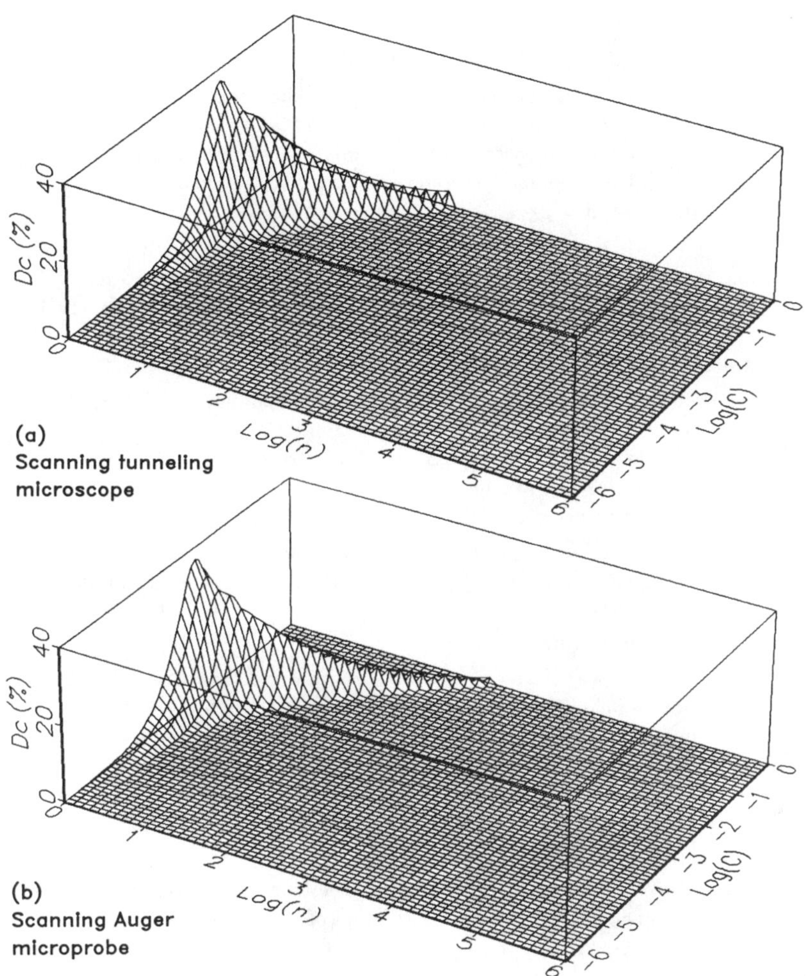

Figure 4. The composite distribution $D_C(n)$ is derived from D_1 and D_2. It gives the probability that that n atoms in the probe volume produce counts, as a function of relative concentration C.

SUMMARY

The salient feature of the composite P_C plots shown in Figure 5 is the plateau of probability values near unity. Such *detection plateaus* give a visual and intuitive feeling about the overall sensitivity of a technique, when constraints of sampling with tiny analytical probes and inherent sensitivity to low concentration levels are both taken into account.

The plateaus are bounded on the n axis by the product Vp, because there can be no more impurity substitutions than the total number of silicon atoms within the probe. Hence, the largest n possible for the scanning tunneling microscope ($p = 1$) in our example is 100 atoms ($log(n) = 2$). The remaining boundaries are defined by equation 14.

Detection plateaus portray an ideal measurement regime which is free of instrument noise and background counts that originate from competing physical processes in the sample. A strict interpretation of 5(b) for example, leads one to believe that a single Auger electron can be reliably identified provided the concentration is 0.1% ($log(C) = -1$) or greater.

It is here that we must impose the Currie detection limit of choice (L_C, L_D, or L_Q). Either defines a reference point n_o on the n axis, below which detection in real-life situations cannot occur. In effect, we zero the plateau between the origin and a parallel plane passing through n_o.

The plateau in 5(a) is governed by the sampling distribution D_1, and that in 5(b) more by an inefficient detection process (D_2). Although p is near unity for the scanning tunneling microscope, the tool is inefficient at quantifying a one-part-per-billion impurity level, for example. Even though we may stumble on the *one part* by chance early in an analysis, we must ultimately establish a baseline by measuring at least a billion silicon atoms before the quantification is complete.

Although atomic scale sampling and detection efficiency can be decoupled in special cases like this, they are indistinguishable in most experimental situations. We can expect to encounter the ambiguity more often as the purity of silicon material increases, and our ability to apply analytical probes within the atomic regime matures.

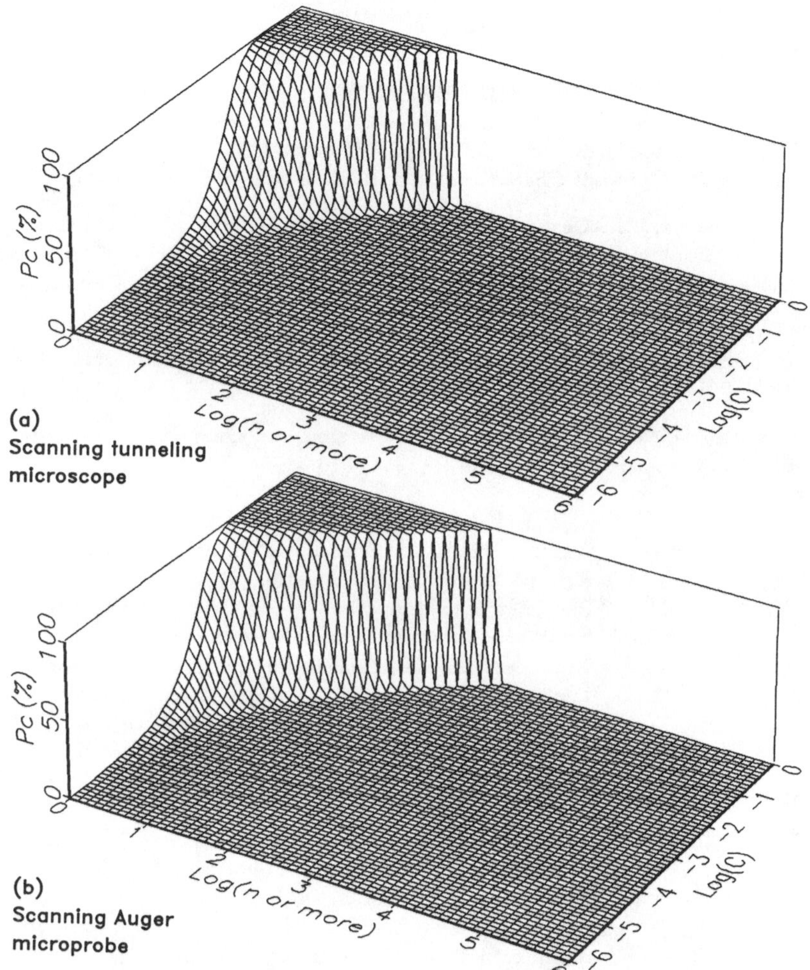

Figure 5. Total probability plots $P_C(n)$ for detection of n or more atoms in a sample of impurity concentration C reveal *detection plateaus* characteristic of the technique. These are similar for the scanning tunneling microscope (a) and the Auger microprobe (b), but for different reasons.

REFERENCES

1. J. K. Taylor and B. Kratochvil, "Sampling," *Metals Handbook Ninth Edition, Volume 10, Materials Characterization*, R. E. Whan, ed., ASM (1986) 12.

2. D.E. Passoja, L.A. Casper, and A.J. Scharman, "Analytical Approaches and Expert Systems in the Characterization of Microelectronic Devices," *Microelectronics Processing: Inorganic Materials Characterization*, L.A. Casper, ACS Symposium Series **295** (1986) 1.

3. T.J. Shaffner, "Trace and Surface Analysis for ULSI," *Semiconductor Silicon 1990*, **90-7**, Electrochem. Soc., Pennington, NJ (1990) 891.

4. L.A. Currie, "Limits for Qualitative Detection and Quantitative Determination," Analytical Chemistry, **40**(3) (1968) 586.

LIST OF SYMBOLS

C	Relative Impurity Concentration
X, Y, Z	Specimen Dimensions
N_i	Number of Impurity Atoms in Specimen
N_s	Number of Silicon Atoms in Specimen
d	Probe Width
z	Smaller of Probe Penetration or Escape Depth
V	Probe Volume ($d^2 z$)
α	Probability One Impurity Atom is in V
N	Number of Impurity Atoms in V
$D_1(N)$	Distribution of Impurity Atoms in V
p	Probability of One Impurity Count from One Atom
β	Proportionality Constant (p/It)
$L_C,\ L_D,\ L_Q$	Currie Detection Limits
n	Number of Counts from Impurity Atoms
n_o	Atom Number Equivalent of Currie Limit
$D_2(n)$	Distribution of Impurity Atom Counts
$D_C(n)$	Distribution Composite of D_1 and D_2
P_C	Probability of Detecting n or More Impurity Atoms

MEETING FUTURE DEFECT DENSITY REQUIREMENTS FOR MICROELECTRONICS

Daren L. Dance
SEMATECH
2706 Montopolis
Austin, Texas 78741

Defect density reduction is required to maintain competitive yield as the complexity of semiconductor design and technology increases. Historical trends indicate that for every three generations of technology, a defect density reduction of an order of magnitude is required. Meeting these requirements will require teams of manufacturers, researchers, and suppliers to address defect reduction, removal and tolerance in all areas of manufacturing technology. This report investigates the factors driving defect density requirements, summarized published defect density forecasts, and applies those trends to forecast defect density requirements for future technologies.

INTRODUCTION

As microelectronic technologies advance from generation to generation, the manufacture of advanced semiconductor devices requires stringent defect control. Defect density reduction is required to maintain competitive yield as the complexity of semiconductor design and technology increases. Many defects are due to particles on the wafer. Processes and equipment must be designed for minimum particles, assembled in a contamination free environment, installed correctly, and operated to minimize defect density. Manufacturers, researchers and suppliers must form effective teams for continued success.

More than 75% of the wafer probe yield loss in an advanced semiconductor fab may be attributed to random structural defects. Most of these defects are due to particles on the wafer [1]. These particles come from the semiconductor process and equipment, materials, people, and the cleanroom environment. As manufacturers address cleanroom and people issues, the focus on particles is shifting to the process and equipment [2]. The semiconductor industry is mastering cleanroom design and construction, reducing contamination through better garments and practices, and improving the cleanliness of gases, chemicals, and materials. But defect reduction efforts must continue to focus on processes and equipment.

Efficient defect reduction efforts require a road map of defect density requirements for future technologies. This paper investigates the factors driving semiconductor defect density requirements and summarizes published defect density forecasts. Those trends are used to forecast defect density requirements for future technologies and outline areas where additional contamination and defect density control efforts are needed.

DEFECT DENSITY DRIVERS

Competitive pressures are driving the requirement for ever lower defect densities in semiconductor manufacturing. Manufacturing equipment cost, die sizes, integration levels, and process complexity are increasing while design rules and minimum critical defect sizes are shrinking. Smaller geometries, larger devices, and larger wafer sizes require equipment with better process control and uniformity. These factors increase the cost of manufacturing equipment. Zieber of Dataquest estimates that the cost of a new wafer fab will increase by 60% for each new product generation. Product development costs are also increasing. For example, development costs for the Intel 8086 microprocessor were about \$25 million, the 386 was developed for about \$100 million and the 486 cost about \$250 million [3]. Increased cost must be offset by increased productivity and yield to remain competitive. Larger wafers increase productivity but yield must be improved by reducing process variance and defect density. Thus increasing development and equipment cost drives the requirement for reduced defect density.

Similarly, die sizes, integration levels, and process complexities are increasing as customer applications require greater levels of functionality on each integrated circuit. Figure 1 shows several die size forecasts [4, 5] and published reports of Japanese DRAM die sizes [6]. Increasing integration level also increases process complexity, as illustrated in Figure 2 which compares the number of masking levels to feature size [5, 7]. Increased die size, integration level, and process complexity increases the probability of a particle causing device failure and drives the requirement for reduced defect density. The same competitive forces increasing size and complexity lead to smaller feature sizes. Minimum critical defect size, the size of a defect that may cause failure, is a function of feature size. The relationship between feature size and minimum critical defect size is often estimated by rules of thumb. Fisher [8] refers to a tenth rule (minimum critical defect size is one-tenth

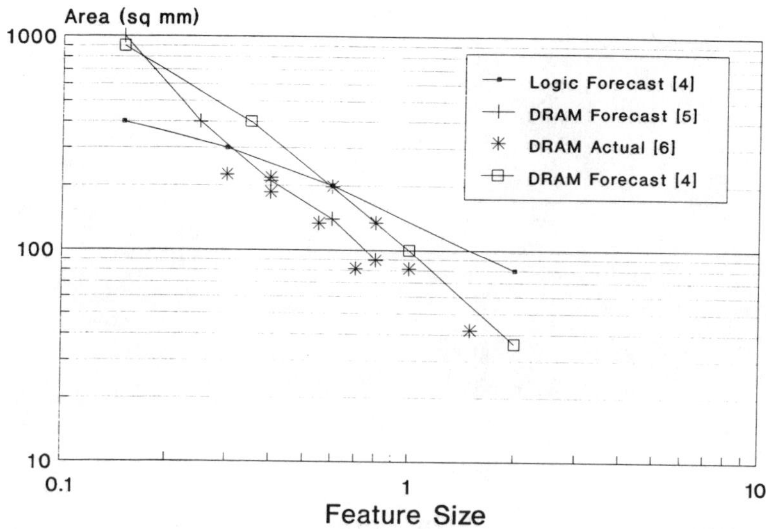

Figure 1:   Die Size Trends

circuit feature size) while others use a one-fifth rule for similar geometries [5]. Even the tenth rule may be too conservative for vertical dimensions, such as gate oxide thickness. Thus a defect significantly smaller than device feature size can cause failure.

Defect size distribution follows an inverse exponential relationship ($1/x^k$) with defect size, which indicates that small defects are much more likely to occur than large defects [11]. For the same overall level of cleanliness, defect density appears to increase as minimum defect size is reduced. Thus, not only are particles more likely to cause failure at smaller feature sizes, the number of particles causing failure increases to drive the requirement for reduced defect density. Assuming a $1/X^3$ distribution of particle sizes indicates that there are more than 16 times as many potentially damaging particles threatening a 64 Mb DRAM as affected a 256K DRAM. Defect size distributions, smaller feature sizes and increasing equipment cost, die size, integration level, and process complexity drive requirements for reduced defect density as semiconductor manufacturers compete in advanced technologies.

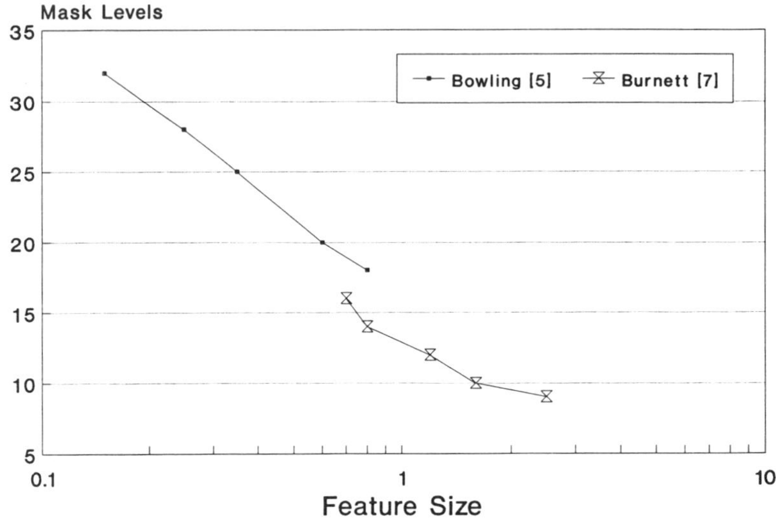

Figure 2:   Process Complexity Trends

DEFECT DENSITY TRENDS

Wafer yields and defect densities are among the most guarded information in semiconductor manufacturing, however, historical defect density information has been published for several unidentified companies.   While actual defect density information is sparse, authors are more forthcoming with forecasts of future defect density requirements. Figure 3 shows historical defect density information [12, 13] and several defect density forecasts in relation to the actual data. Bowling [5] provided estimates of defect density requirements for future technologies to the National Science Foundation (NSF). The Shioya data [14] are from published defect density versus yield curves assuming 80% yield. Die size reported in Japanese press reports [6] was used to estimate a defect density using a Poisson model ($Y = e^{-AD}$) also assuming 80% yield without redundancy. These data show the learning curve effect as experience is gained through volume production. Fitting a trend line to

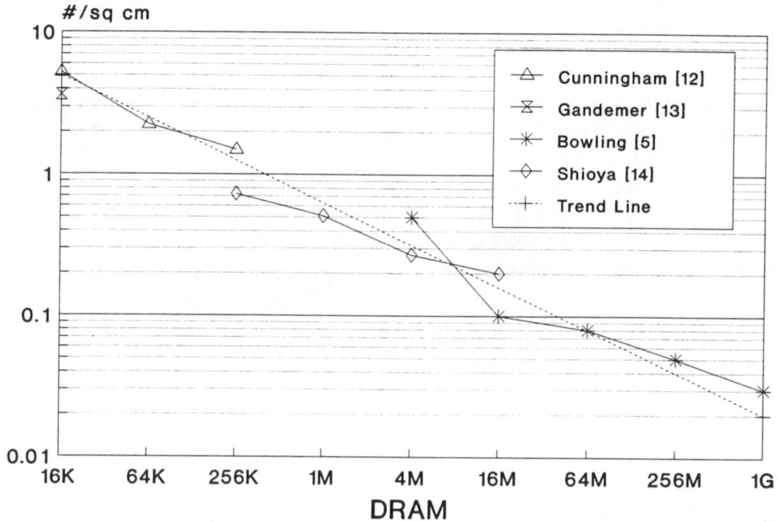

Figure 3:   Defect Density Trends

these forecasts indicate that increasing complexity by three technology generations requires defect density reduction of at least one order of magnitude. For example, the 256M DRAM will require a defect density of about 0.05 defects per cm^2 which implies a 3 fold improvement over today's 4M and 16M DRAM requirement of 0.15 to 0.3 defects per cm^2.

FUTURE DEFECT DENSITY REQUIREMENTS

Defect density reduction trends and marketing requirements for advanced technologies indicate that stable manufacturing processes must be established on a generally well defined schedule. These trends drive the need for continuous defect density improvement along a clearly defined path. Knowing the required defect density reduction allows estimating equipment particle requirements for future technologies. Particle densities may be scaled from existing tool contamination levels to future technologies using expected yield and estimating the required number of processing steps. This scaling assumes existing tool contamination levels correlate to actual defect-limited yield and requires estimating a relative degree of difficulty for future technologies.

Electrical yield is the product of several components: defect-limited yield, parametric-limited yield, and other yield limiters. Defect-limited yield loss is due to random defects introduced during manufacturing. While not all defects will cause failure, all defects have a probability of causing failure at some point in manufacturing. Several formulae exist for modeling defect-limited integrated circuit yield as a function of process defect density and device area. The negative binomial model is one of the most common [15, 16]:

(1)

$$Y = \frac{1}{\left(1 + \dfrac{A_c D_0}{C}\right)^C}$$

where:

     Y = percentage yield
     A_c = critical area
     D_0 = defect density
     C = degree of clustering

Critical area is the area in which a defect must occur in order to cause device failure. If $C=1$ then the above formula becomes the Seeds formula:

(2)

$$Y = \frac{1}{1 + A_c D_0}$$

If $C = \infty$ then we have the Poisson formula:

(3)

$$Y = e^{-A_c D_0}$$

While the literature refers to C as a clustering parameter, it may be more properly referred to as the shape parameter for a gamma distribution, since defect-limited yields tend to follow a gamma distribution. Different semiconductor companies and manufacturing areas have different yield characteristics and use different yield models. Thus selection of the proper yield model formula is best left to the user.

Yield formulae for parametric limited yields have not been as clearly defined. However, many device parameters have an upper and lower bound. Outside of this process window, yields may be negatively impacted. Thus if all of the processing parameters are centered, then parametric yield is maximized. Other components of yield loss include design limited yield and gross yield loss. Design limited yield is a function of the suitability of the product design to the design rules and device parameters specified. Gross yield is the loss of product due to processing errors such as missed process steps or wafer breakage. Since the large majority of yield loss in an advanced semiconductor fab is due to defect-limited yield, other sources of yield loss will be neglected in considering defect density requirements for future technologies. The correlation between tool contamination and yield requires separating defect-limited yield from other yield limiters. This may be done through failure analysis of completed devices and by using electrical defect monitor test devices such as interdigitated comb and meander structures.

Knowing the required defect density reduction allows estimating future equipment defect density requirements. Using current information and estimating future feature sizes and process complexities allows forecasting equipment defect density requirements shown in Table I. This table compares defect density per masking level requirements for several memory technologies. Defect density is from the trends shown in Figure 3 and the number of masking levels is estimated from Figure 2. Defect density is shown for both the minimum feature sized of the mature technology and for a standard minimum defect size of 0.2 μm.

DRAM Tech	Feature Size	Defect Density #/cm²	Mask Levels	Defects /Level #/cm²	Defects /Level @ 0.2μ
64K	2.50	2.50	9	0.2778	43.4028
256K	1.60	1.00	10	0.1000	2.5000
1M	1.20	0.50	12	0.0417	0.2604
4M	0.80	0.30	16	0.0188	0.0422
16M	0.60	0.15	20	0.0075	0.0042
64M	0.35	0.08	25	0.0032	0.0005
256M	0.25	0.05	25	0.0018	0.0001
1G	0.15	0.02	32	0.0006	0.0000

Table I: Defect Density Forecasts

The smaller feature sizes of future technologies leads to smaller minimum defect sizes. Estimates of future defect density must be normalized to the present minimum defect size. By assuming defect size distributions follow a $1/X^3$ relationship, the following equation can be used to normalize the defect density estimates:

(4)

$$D_2 = D_1 \left(\frac{X_{C1}}{X_{C2}} \right)^2$$

where:

D_1 = Defect density 1
D_2 = Defect density 2
X_{C1} = Minimum Defect Size 1
X_{C2} = Minimum Defect Size 2

The relative difficulty of the future technology compared to the present technology is estimated by the number of masking levels required.

Forecasting equipment defect density requirements has several limitations:
- The forecast is only as good as the existing correlation.
- The forecast uses only horizontal scaling.
- The forecast scales uniformly for all equipment.

If significant non-defect related problems are limiting the actual yield, then the correlation between equipment particle measurements and yield is weakened. Any forecast based on a weakened correlation will also be less accurate. The horizontal scaling limitation does not comprehend the greater sensitivity of thin oxides to vertical defects. Thus the thinner dielectrics of future technologies may have a greater effect on defect-limited yield than horizontal scaling implies. Finally the method assumes uniform equipment improvement and does not consider shifting paradigms. A significant process or equipment improvement may make other defect reduction efforts less important. While these forecasts have limitations, they are useful in pointing the direction of defect density reduction efforts.

STRATEGIES FOR MEETING FUTURE DEFECT DENSITY REQUIREMENTS

Meeting the future defect density requirements outlined in Table I is a difficult challenge for the microelectronics industry. Strategies for meeting future defect density requirements include defect prevention, defect removal, and defect tolerance. Microelectronics manufacturers, researchers and suppliers must form effective teams to apply these strategies for continued success in meeting defect density reduction requirements.

Defect prevention is the first line of defense, including all aspects of contamination reduction in the fab environment, processing equipment, chemicals and materials. Current fab environment activities include investigation of better-than-class 1 cleanrooms, application of minienvironments around processing tools, and the use of automated material handling systems to reduce particle sources in the cleanroom environment. Much work must be done to reduce contamination from processing equipment, chemicals and materials. Process modeling to understand particle dynamics in processing chambers is contributing to cleaner equipment designs. Research is also needed in the materials and components used to make the manufacturing equipment. The benefits of ultra-high purity chemicals and materials also need to be clearly identified. Addressing the aspects of defect prevention will contribute to meeting future defect density requirements.

If defects on the wafer cannot be prevented, then contamination removal is the next line of defense. Many different methods of cleaning wafers are being used including wet chemical bathes, mechanical brush cleaners, and low pressure plasmas. Most cleaning processes do not actually remove particles but are used to modify wafer surface characteristics. Particle removal requires adding sufficient energy to overcome bonds between the particle and the surface. Thermal, mechanical, sonic, and phase change kinetics are sources for energetic particle removal. While wafer cleaning is generally effective with larger particles, additional work remains to develop particle removal methods for sub-micron sized particles.

In addition to defect prevention and contamination removal, the process and circuit must be designed to tolerate remaining defects for acceptable circuit yields. DRAM manufacturers provide redundancy and error correcting circuitry to provide defect tolerance in their products. Defect tolerance can also be accomplished by using design rules which minimize critical area or the area in which defects must fall to cause failure. Processes may also be designed for defect tolerance. For example, most organic defects are transparent to X-ray lithography, cold traps in vacuum systems may trap defect inducing moisture away from wafer surfaces, and adding chelating agents to wet bath processes may cause small particles to agglomerate, facilitating removal.

SUMMARY

Historical information and forecasts of future defect density trends can be used to estimate defect density requirements for advanced microelectronics. Meeting these requirements will require teams of manufacturers, researchers, and suppliers to address defect reduction, removal and tolerance in all areas of manufacturing technology. Additional research is needed in equipment, processes, gases, and chemicals for contamination free manufacturing. Research is also needed in the materials and components used to make the manufacturing equipment. Cost effective development of new tools and processes needs to be addressed. Usage of equipment and process modeling needs to be increased to reduce trial and error methods of development. Finally capability of the tools for detection, measurement, and analysis of defects need to be extended to smaller defect sizes. With this knowledge, equipment and processes can be designed, assembled, installed, and operated to meet future defect density requirements.

ACKNOWLEDGEMENTS

This report would not be possible without the direction and assistance of Paul Calhoun, Richard Jarvis, and Venu Menon. Management support from Mohammad Ibrahim and Ed Hall is also greatly appreciated.

REFERENCES

1.	T. Hattori, "Contamination-Control Engineering in Wafer Processing: Problems and Prospects," *Technical Proceedings SEMICON/Japan 1989*, pp. 244-255.

2.	V. Menon, quoted in "Industry News: Front-end tools are the new front line in contamination war," *Microcontamination*, Jan. 1991, p. 10-14.

3.	F. Zieber, quoted in "Industry News: Fab Costs Escalating 60% for Each Device Generation," *Semiconductor International*, Jan. 1991, p. 13.

4.	A. S. Oberai, "Lithography -- Challenges of the Future," *Solid State Technology*, September 1987, pp. 123-128.

5.   A. Bowling, "DRAM Device Manufacturing Trends," from a talk presented to the NSF Panel on determining research needs in sub-micron particles, November 19-20, 1990.

6.   Japanese press reports abstracted and translated in *Today's Headlines from Japan*, June, 1990 to May 1992.

7.   J. Burnett, "World Class Contamination Control Practices," *Semiconductor International*, April 1988, pp. 160-170.

8.   W. G. Fisher, November 1988, Private Communication.

9.   R. J. Kopp, "Forecast 1991: Timing is the Key," *Semiconductor International*, Jan. 1991, pp. 40-45.

10.   V. Ramakrishna and J. Harrigan, "Defect Learning Requirements," *Solid State Technology*, January 1989, pp. 103-105.

11.   D. M. H. Walker, *Yield Simulation for Integrated Circuits*, Kluwer Academic Publishers, Hingham MA, 1987, pp. 37-41.

12.   J. A. Cunningham, "The Use and Evaluation of Yield Models in Integrated Circuit Manufacturing," *IEEE Trans. Semiconductor Manufacturing*, May 1990, pp. 60-71.

13.   S. Gandemer, B. C. Tremintin, and J. Charlot, "Critical Area and Critical Levels Calculation in I.C. Yield Modeling," *IEEE Trans. Electron Devices*, Vol. 35, No. 2, February 1988, pp. 158-166.

14.   Y. Shioya, A. Abiru, and K. Yanagida, "Possibility and Expected Date of Shift to 8-inch Wafer," *Semiconductor World*, Vol 10, 1989, pp. 98-102.

15.   C. H. Stapper, "Fact and Fiction in Yield Modeling," *Microelectronics Journal*, 1989, Vol. 20, No. 1-2, pp. 129-151.

16.   C. Kooperberg, "Circuit Layout and Yield," *IEEE Journal of Solid-State Circuits*, Vol. 23, No. 4, August 1988, p 887.

HOLOGRAPHIC WAFER INSPECTION AS AN EFFECTIVE DEFECT REDUCTION METHODOLOGY FOR ULSI WAFER FABRICATION

Michael J. Satterfield
Advanced Products Research and Development Laboratory
Motorola, Inc., Austin, TX 78721

ABSTRACT

The application of a holographic defect detection tool has been used as an effective weapon to locate and identify process induced defects (PIDS) in a sub-micron wafer fabrication area. Whole wafer inspection has provided knowledge of process defect signatures that is critical for solving defect problems. Stripping and partitioning software capabilities have shown themselves to be essential in finding the sources of PIDS. Finally, the concept of process induced defects per wafer pass (PIDPWP) will be discussed.

INTRODUCTION

Reducing process induced defects is very important in a ULSI wafer fabrication area in order to achieve high yields so that die cost is minimized. In a ULSI research and development laboratory it is important to reduce process induced defects so that a maximum number of learning cycles on new processes and devices can be accomplished in the shortest time possible, and that these new processes and products can be transferred to a high volume manufacturing fabrication area in a short period of time. The use of automatic patterned wafer inspection tools is essential to find, characterize, and monitor process induced defects in an SRAM, DRAM, or microprocessor process flow. A holographic, whole wafer, automated patterned wafer inspection system was purchased for this purpose. The holographic tool works on the principle of two dimensional Fourier transform filtering. This enables the highly repetitive patterns from DRAM and SRAM arrays to be filtered out leaving only the defect information in the re-constructed Fourier Transform filtered image[1]. Holographic technology is well suited for detecting low contrast defects and defects that reside in vias, contacts, and trenches as compared to other systems that utilize optical imaging and laser scattering technology.

PROCEDURE

Once the holographic defect tool was installed and functionality established, several SRAM lots were selected from the pilot line to serve as "test vehicles" for the new defect tool. A subset of wafers from each lot was selected as the monitor wafers. These wafers were randomly chosen at the first step, but were used for subsequent wafer scans throughout the process flow. The subset of wafers from the lots were scanned at certain process levels, the defects were classified, and the lots were then sent on to either the next process step, or they were allowed to go several more steps in the process before they were re-scanned. The partitioning software was used between two steps or several steps in order to detect new defects which the intervening step or steps had added to the wafers. In addition, short flow electrical defect monitoring test lots were also examined using the holographic defect tool.

RESULTS

Several killer process induced defect types and sources were detected in a short period of time. After corrective measures were taken to eliminate these defects, the holographic defect tool was utilized to see if a reduction in defects at the step in question had been accomplished. If no reduction was apparent, other corrective measures were taken until the defect was either eliminated or reduced by at least one order of magnitude.

During this time, several defects were detected that proved to be rather subtle using normal bright field methods. Figure 1 shows one of these defects after a certain etch step is performed. This defect was typically easy to spot after performing the etch as is evident in the photograph. However, prior to the etch it was rather difficult if not impossible to see. When the holographic tool first found this defect prior to the etch, there was some skepticism as to the validity of the data. However, according to the holographic defect tool, the intensity readings for this defect were extremely high (their levels were at 255 which is maximum). Readings like these were commonplace for large particles, but no large particles existed at these particular coordinate sites. However, when the defect was reviewed from the two dimensional Fourier transform filtered image, commonly called the holographic reconstructed image, the defect was immediately visible and its classification was well known from previous work. Figure 2 shows a photo of this low contrast defect prior to etch. The defect was finally detected by placing the microscope at a low magnification, adjusting the focus through several iterations, and finally literally walking across the room to review the CRT display on the off-line review station. After all of this, it was still difficult to see the defect! Figure 3 shows the holographic image of the defect, and there was no mistaking that this defect was a water spot caused by a poor dry. The "water spot" identification was further supported by the wafer map shown in Figure 4. This shows the defects to cluster at the center of the wafer. The drying process under scrutiny was performed in an on-axis spin rinse dryer. A physical assessment of the on-axis dryer shows that no force is present toward the center of the wafer since the linear velocity is directly proportional to the radius. During the work with the holographic tool, it has excelled in finding low contrast defects like the one described above in Figures 1-4.

DISCUSSION

From the work with the SRAM and snake wafer lots, several interesting and useful features on the holographic defect tool were used to make well founded decisions about the data obtained from the tool. Also, these features and combinations of these features and attributes of the tool were used to maximize its potential at finding the sources of process induced defects. These items are discussed in the following paragraphs.

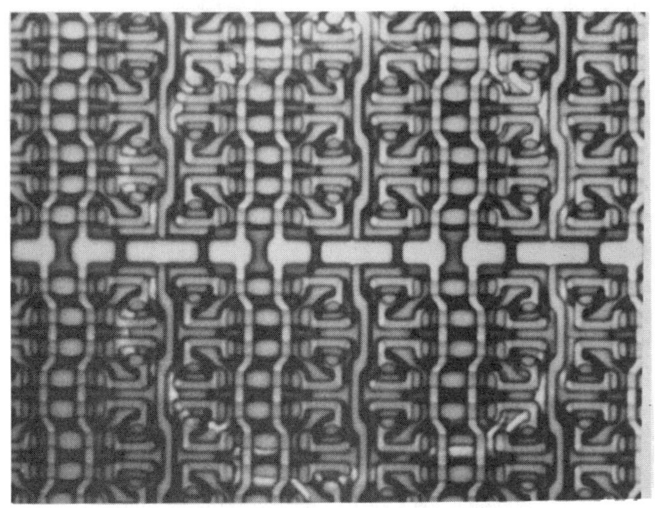

Figure 1: Water Spot Defect After Etch

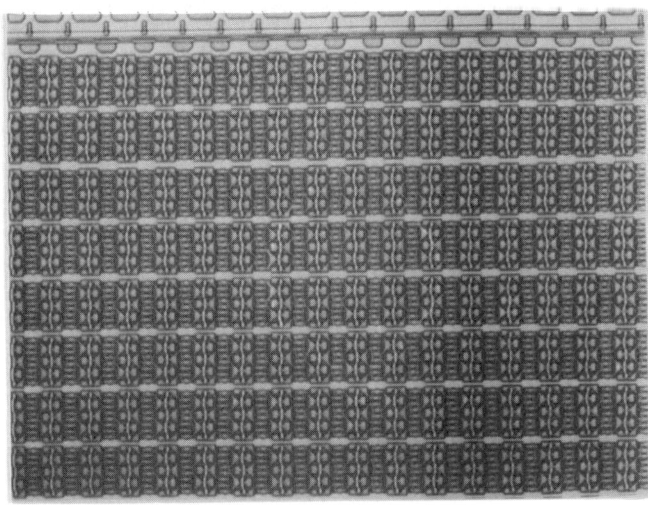

Figure 2: Water Spot Defect Before Etch

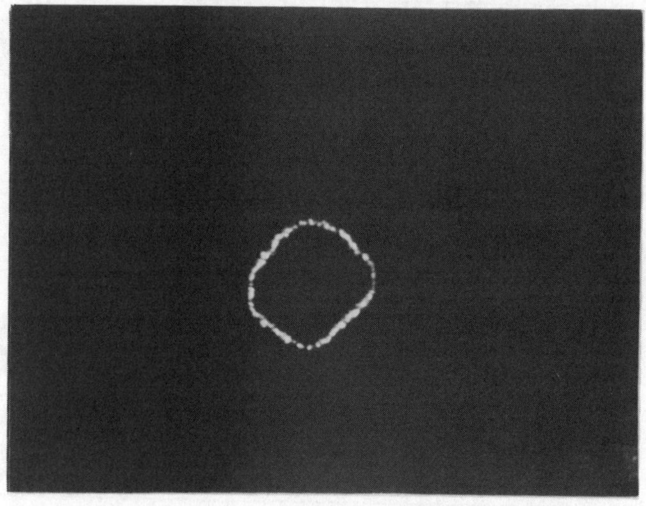

Figure 3: Holographic Image of Water Spot Defect Before Etch

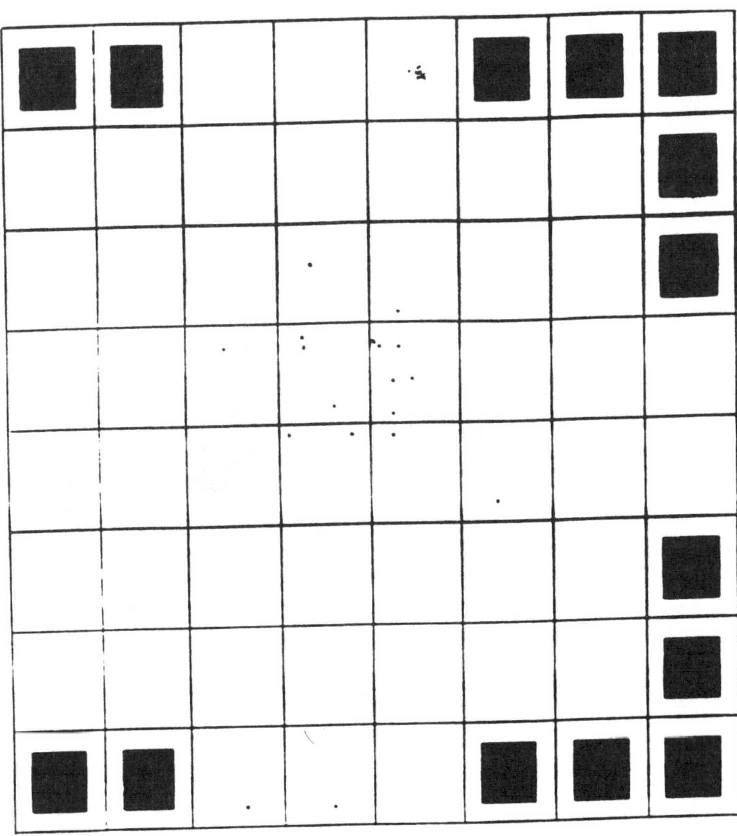

Figure 4: Full Wafer Map Showing the "Process Signature" of the Water Spot Defect.

190

Partitioning

The concept of partitioning is relatively simple, yet it is a very powerful tool available to be utilized in finding the source of process induced defects (PIDS)[2]. Partitioning enables one to find the source of a PID relatively easily, and more important, the data obtained from partitioning is usually more convincing than other defect finding methodologies. A simple partition study is performed by doing a pre-scan on a set of wafers at a certain step in the process flow, inspecting the wafers, and classifying the defects found. This creates a set of unique data files for the set of wafers at this certain process step. For simplification, assume this set of files, or a single file, is called "A." The wafers are then allowed to continue to the next process step and then the same set of wafers are scanned again after the next process step is completed. This again creates a unique set of data files, or a data file, that can be referred to as "B." Once the final scan is completed, the partitioning software on the holographic defect tool's review sequence or its accompanying off-line review station, can be employed to create new files, or a single file, that can be labeled as "C." The partitioning software simply takes the data points, i.e. the defect coordinates, in file(s) "B" and subtracts the defect coordinates in the file(s) "A" from file(s) "B" and automatically creates a new file, or set of files, which were labeled earlier as "C." Mathematically, the partition function can be expressed as:

$$C = B - A$$

The file "C" is a new file created from files "A" and "B", with "C" containing only the defects added by the last process step. Figure 5 shows a developer wetting defect that was found during the inspection of an active area developed photoresist pattern on a one micron SRAM process; this is the first highly repetitive pattern applied in the SRAM process. This defect is caused by a bubble forming during the photoresist developing process. The bubbles can originate from the manner in which the developer is applied or from the chemical reaction of the developer with the exposed photoresist or both. The positive resist process normally produces nitrogen gas from the chemical reaction of the developer with the exposed resist. If the developer solution is sprayed incorrectly on the wafer, or if it is applied too forcefully to the wafer, foam can result as a major source of bubbles. Having the coordinates of this particular defect was very important in proving it was indeed caused by a developer bubble. With the assistance of Photo Engineering, a defect like the one shown in Figure 5, was examined using a microscope that contained a yellow filter. The microscope was on another review station that would accept generalized defect coordinates. Upon examination with this microscope, a latent image could be seen in the bridged resist pattern indicating the resist had been exposed but not developed! When the wafer was re-developed, the latent image was cleared away and the "bubble defect" disappeared! In addition, the wafer was loaded into a scanning electron microscope (SEM) and examined further. The SEM examination showed a circular ridge in the area where the resist had been bridged. It is assumed that the edge of the bubble created this ridge due to the re-deposition of developed photoresist around the edge of the bubble. A careful examination of the developer tracks by Photo Engineering found several problems with the developer dispense method. After altering the develop dispense, further inspections of the active photoresist pattern at this step showed a dramatic decrease in these developer wetting defects.

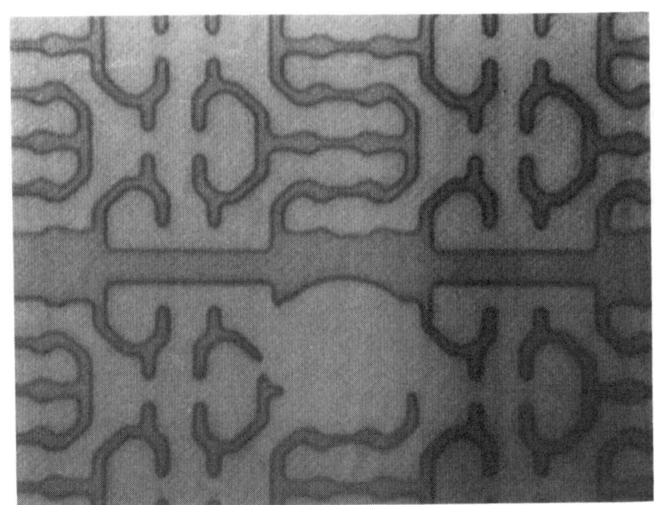

Figure 5: Developer Bubble Defect

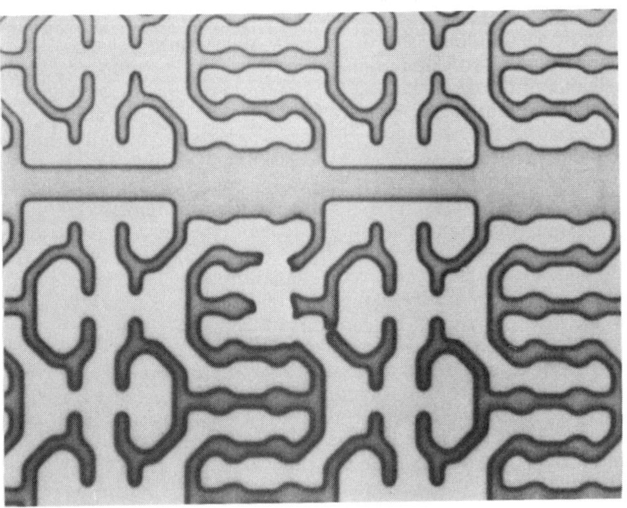

Figure 6: Etch Particle Type Defect

The next step in the process involved anisotropically etching the substrate beneath the developed active photoresist pattern and then removing the photoresist. After this step, the wafers were re-scanned on the holographic defect analysis tool and the data files were partitioned using the partitioning software. The wafers were then loaded onto the review station and examined using the newly created partition files. The new defects found in the partitioned files resembled the defect pictured in Figure 6. This defect was recognized as an "etch particle defect" and had been observed at other steps in the process using other defect detection methodologies[3 & 4]. An etch particle defect is defined as a particle of unknown origin, usually originating in but not limited to the etch process, that falls onto the patterned resist and blocks the highly selective and anisotropic reactive ion etch (RIE). These defects are characterized by their planar nature, irregular shape, and anisotropic profile. The shape of etch particle defects is determined by the shape of the particle that blocks the etch; the perimeter of the particle defines the shape of the etch particle defect. The etch particle defects were randomly distributed across the wafer and were not found prior to the etching of the wafer. Therefore, the partitioning study showed that the etch had indeed added these defects to the wafers. Some of the new defects exhibited a "ghost" active pattern as shown in Figure 7. This further re-inforced the theory of a fall-on particle during etch and was easily explained by the particle falling onto the resist pattern after the etch had begun but before the etch had completed.

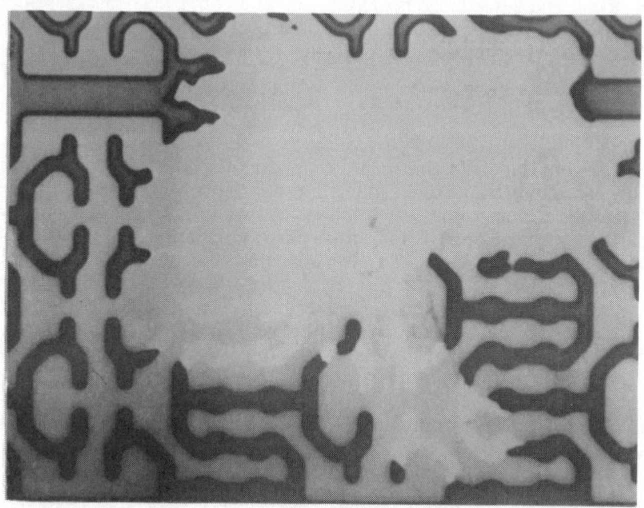

Figure 7: Etch Particle Defect with Ghost Pattern

Etch Engineering solved this defect problem by altering the cleaning process on the etcher where this process was performed. Finally, these wafers were examined using the original file ("A"); the etch particle defects were not present and their coordinates were very different from defect coordinates shown in file "A." Also, during the inspection of the wafers with file "A," the circular regions of undeveloped resist where found to be permanently transferred to the active pattern. Having the coordinates of previously detected defects is very powerful tool to see how a defect changes as the process moves from step to step. In this case, the defects were shown to be permanent after the etch which greatly increased the likelihood of causing yield loss and this in turn was very helpful in getting this serious problem fixed. When the wafers were probed, failing bits were traceable, via failed bit map analysis (FBMA), to both the developer bubble defects and the active pattern etch particle defects. Therefore, the information from the partitioning study was very useful in eliminating these two killer defects.

Another method of partitioning is available using the multi-level overlay software available on the holographic defect detection tool and its accompanying off-line review station. This software allows maps of any source to be stacked onto one another in order to make intelligent decisions concerning wafer process signatures[5]. However, by overlaying the data from file "A" and file "B", a "quick" partition study may be done, and it can be used to verify the normal partitioned file. The multi-level overlay software on the review station maps the first file in the color red. The second file is mapped in the color green. If two defects align, i.e. they have the same coordinate values in both files, then these coordinate sites will be colored yellow on both the review station high resolution CRT display and the colored map hard copy printouts. Therefore, the defect sites that are colored green are the defects that were added by the last process step. By reviewing just these sites, the added defects can be scrutinized and classified. The classified defects can then be stripped out using the stripping software and new maps containing just the new defects added by the last process can be made.

Manual Partitioning

Partitioning can take one last form with the holographic tool. This form of partitioning has been designated "manual partitioning." Use of the two dimensional Fourier transform filtered images is important when using manual partitioning. With manual partitioning, it is very important to save all filters and holograms of each scan performed during the partition study. The concept of manual partitioning involves taking the most recent wafer scan file and using this file to examine the previous level's holographic plates. This method will determine if a "seed" defect was present at the previous process step. Seed defects tend to be very small and may not be detected due to the low signal to noise ratio. Seed defects tend to grow larger as the wafers move through the rest of the process flow. This is especially true in the last parts of the process where chemical vapor deposition (CVD) dielectrics are used between conductors. Such dielectrics are widely used on devices that contain more than one level of metal. It is important to find a previous defect on the initial scan of a manual partition study, that will be easy to spot and identify at the next wafer scan and subsequent inspections. Photo and etch defects, as well as a large particle, make good "reference" defects. This defect should be classified at the initial scan and its die coordinates written down for future use. This defect will be used to correct stage error (offset) on both the wafer and the holographic plates during the manual partition. A manual partition study is performed by loading the previous level's holographic plates, and reviewing the defects on the holographic tool, not the review station. The present defect scan file is selected and reviewed using the two dimensional Fourier transformed filtered images of the defects. This review takes place after going to the known defect and offsetting the defect coordinates on both the wafer and the holographic image. After the offsets are accomplished, each defect from the new scan file can be examined and easily recognized as a previous defect by simply looking for a small dim point of light. Previous defects will show a dim point of light. New defects will have no visible point of light; i.e. the holographic image will be completely black.   However, since the review process must be used to perform a manual partition, the defects in the latest scan file can be classified as "P" for previous and "N" for a new defect added (no visible light detected). After each defect is classified either "P" or "N," then the latest defect file can be stripped using the stripping software, into two new defect files that contain just the defects classified as "P" and the defects classified as "N"; i.e. the previous defects, and the defects classified as "N" which are the new defects added by the last process step.

The above methodology is quite tedious, but it can yield exciting and revealing information about the process. One way to employ the concept of manual partitioning is to attempt to "create" known defects. By creating a known defect, a good deal of understanding on how the defect occurred is learned[6]. For example, a double deposition of two different dielectrics is sometimes used to separate conductive polysilicon from the first layer of metal on a double level metal device.[7] The two dielectrics are usually low temperature oxide (LTO) and/or plasma enhanced tetraethylothrosilicate glass (PTEOS) and heavily doped borophosphosilicate glass (BPSG). BPSG is commonly used dielectric material, and is often used as a "flow glass."[8] In Figure 8 these defects were found after a PTEOS deposition onto a conductive polysilicon film. This step was followed by a deposition of BPSG on top of the PTEOS. The initial defects grew quite large after the deposition of BPSG as shown in Figure 9.

After doing a partition study, many of the large defects as shown in Figure 9, were displayed as "new" defects that had been added by the last process step; i.e. the BPSG deposition. However, routine particle monitoring wafers from this system, which are formed by depositing BPSG onto a clean silicon wafer, did not show the large particles like the ones found during the partition study. Manual partitioning was employed to see if the defects originated in the PTEOS deposition. The old plates used for the PTEOS scan were examined using the data file obtained from the BPSG level scan and the majority of the new large defects showed evidence that a very small "seed" defect was present after the PTEOS deposition as verified by a very dim but visible point of light on the holographic image from the old plates! The dim point of light is analogous to observing a very faint star in the clear night sky. Therefore, by using manual partitioning, this particle problem was shown not to originate at the BPSG level, but at the preceding PTEOS deposition process level!

Again, this method was used to pinpoint large defects that occurred in the dielectric separating metal 1 and metal 2 on a double level metal device. This interlevel dielectric (ILD) process involves planarization of a CVD PTEOS deposition which is followed by another CVD PTEOS deposition.[9] By attempting to create a well known "killer" defect as shown in Figure 10, that was known from previous work to originate in the complete ILD process, two different and separate defect origination points were found to be the origin of the killer defect. The first was due to small spherical glass beads forming during the first PTEOS deposition; these were due to gas phase nucleation in the PTEOS process. The second source of the killer defects was found to be from the planarization process which left particles behind. These defects were greatly reduced by cleaning up the process equipment where these two processes were performed, and by introducing an effective and economical clean after the planarization process.

Figure 8: LTO Defects Prior to BPSG Deposition

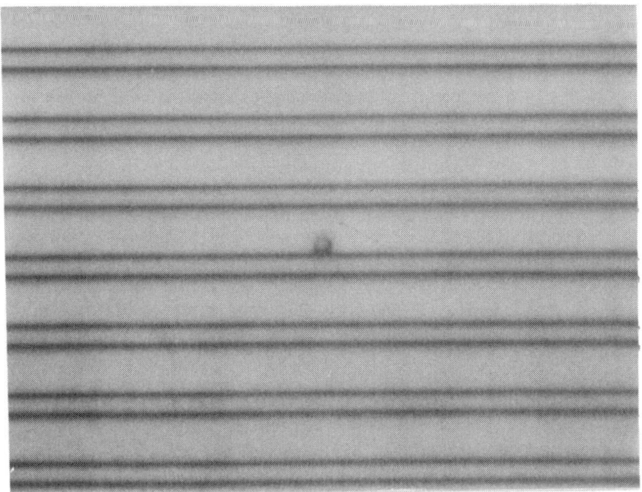

Figure 9: LTO Defects After BPSG Deposition

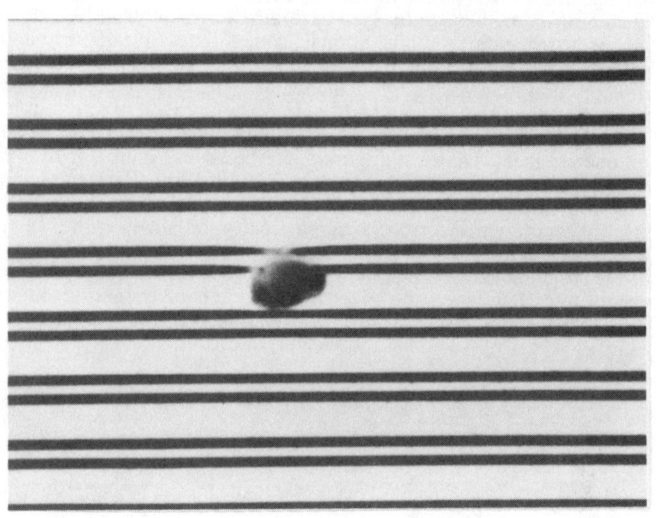

Figure 10: Example of an ILD Defect

Defect Stripping and Multi-Level Overlay

The holographic defect tool and its accompanying off-line review station have two very useful functions in their software that aid greatly in determining process defect signatures on patterned wafers. These two functions are known as "defect stripping" and "multi-level overlay". Defect stripping allows the use of a fully classified defect file to make new wafer maps from that file that contain sub-sets of the classified defects in them. In particular, one routine allows for the stripping of individual defects via their classification scheme. Therefore, it is possible to create new maps containing certain defects of interest. The other feature, multi-level overlay, simply allows one to stack several different wafer maps on top of one another and view the outcome. When these two features are combined, the results can be dramatic. For example, during a routine examination of a line monitor snake vehicle, a new defect was detected as depicted in Figure 11. During the engineering inspection and classification, it appeared the new defect manifested itself along the periphery of the wafer. The new defect was classified with a two letter code and then the stripping software was employed to strip out just this defect code and create new wafer maps from it as shown in Figure 12. All of the new maps were overlaid on one another. The resulting map, as shown in Figure 13, showed the defects did indeed cluster around the edges of the wafer and in particular were located where the flat, crown, and minor flats were located. These defects were found to be from electrostatic discharge caused from a magnetically enhanced etch. Interestingly, the magnets are located at the periphery of the wafers where the defects clustered. A physical analysis of the situation showed that the Lorentz force [10] was pushing electrons toward these magnets and that when the electric field density increased beyond dielectric breakdown, an electrostatic discharge resulted which literally blew a hole in the dielectric.

Figure 11: Electrostatic Discharge Defect

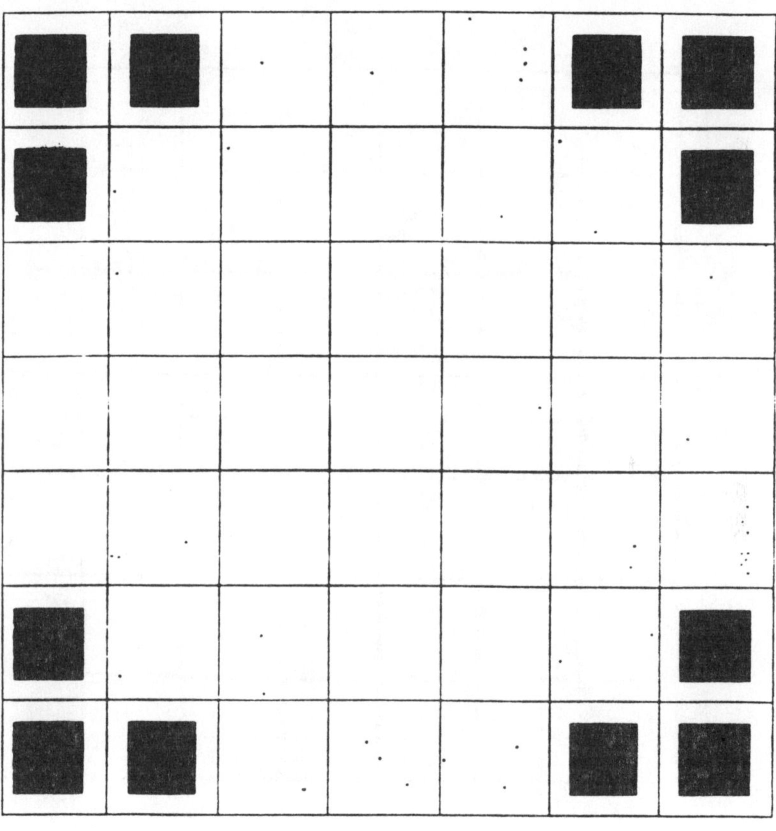

Figure 12: Single Wafer Map Containing Only ESD Defects

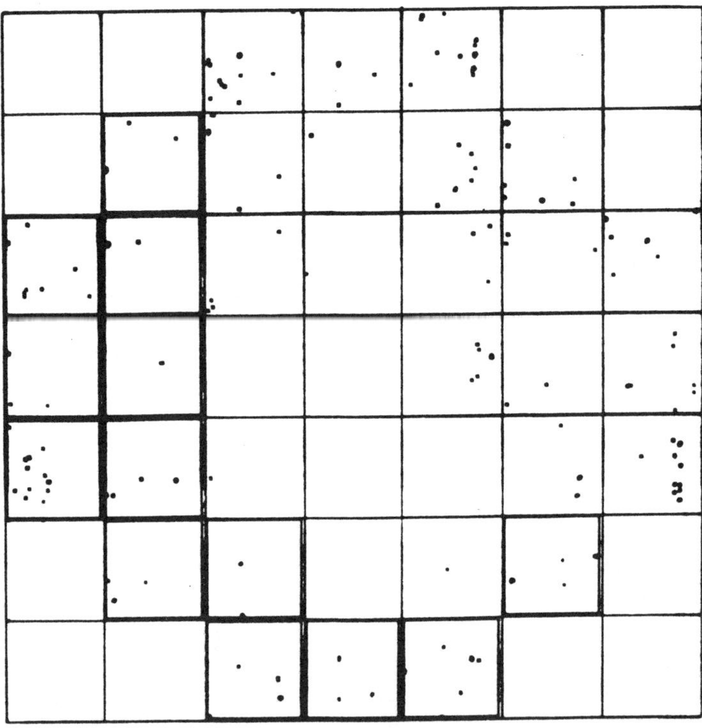

Figure 13: Multi-Level Overlay Map of Stripped Maps Containing Only ESD Defects.

The Concept of Process Induced Defects per Wafer Pass

For many years wafer processing equipment has been qualified using bare silicon and the concept of particles added per wafer pass (PWP). This methodology is well documented in the technical literature[11]. However, this method usually involves compromise and usually does not represent the true process. For example, to monitor an etcher, a bare wafer is pre-counted, inserted into the etcher, the etch chamber is pumped down, process gases are allowed to flow, the chamber is then vented and the wafer is removed and re-counted. The number of particles detected before the check is subtracted from the number of particles obtained after the test and a delta is obtained for particles added per wafer pass and is plotted on a one-sided control chart. Unfortunately, this does not represent the true process. The RF energy and resultant plasma is not turned on during the test because it usually roughens the silicon surface and makes a bare wafer surface scanner count many false particles due to surface scatter. However, by using a simple highly repetitive pattern and the holographic defect detection tool, it is possible to set up intelligent defect monitors that examine the exact process in question. For example, to monitor the aforementioned etcher, create a bare substrate, such as a blanket deposited film, of the material that is normally etched, process the wafer through the normal photolithography step, substitute the highly repetitive pattern mask instead the current level mask, and expose and develop the resist with this pattern. Once the highly repetitive resist pattern is developed, this wafer is now the initial scan wafer (or the "before" wafer). This wafer is scanned on the holographic tool and all defects found are classified. Small defects that are not important can be eliminated by sieving the data via the threshold sieve routine on the holographic defect tool and its accompanying review station. Etch the wafer with the exact recipe and process that product wafers would see, and re-scan the wafer on the holographic defect tool. Perform a simple partition between the last scan and the first scan to find the defects added by the etch process. These defects are the process induced defects that were added to the wafer. This methodology and concept can be referred to as process induced defects per wafer pass (PIDPWP), of which particles per wafer pass (PWP) are a subset. An example of a PIDPWP is the anisotropic etch defect that was discussed above (see Figure 6). A bare surface scanner cannot detect such a defect unless it detects the particle that causes the defect. Without striking a plasma during a simulated etch, the particle that blocks the etch may not be dislodged or created.

Process induced defects added by the process can be charted on a one-sided control chart and statistical process control (SPC) methods can be performed to keep the piece of process equipment in control from a PIDPWP standpoint. The system could be checked for PIDPWP each shift (8 hours) or each half shift (4 hours) to ensure it is performing correctly. Thinking from a process standpoint, the concept of PIDPWP utilizing a highly repetitive pattern and the holographic defect tool, can be accomplished for a large number of the process steps in a SRAM or DRAM flow. For example, to monitor a photo process, simply print a highly repetitive resist pattern on a clean prime silicon wafer and examine it for PIDPWP with the holographic defect tool. Deposition or sputtering tools can be monitored by using a wafer with a "hard" pattern (such as an oxide pattern) and scanning the oxide pattern before and after the deposition. Figure 14 shows the wafer map for an oxide pattern wafer scan prior to a CVD PTEOS deposition. Figure 15 shows the same wafer after the PTEOS deposition. Figure 16 shows a representative photo of the defects detected after the deposition. These defects are small spherical glass beads that are due to gas phase nucleation during the PTEOS deposition. These beads were being detected on

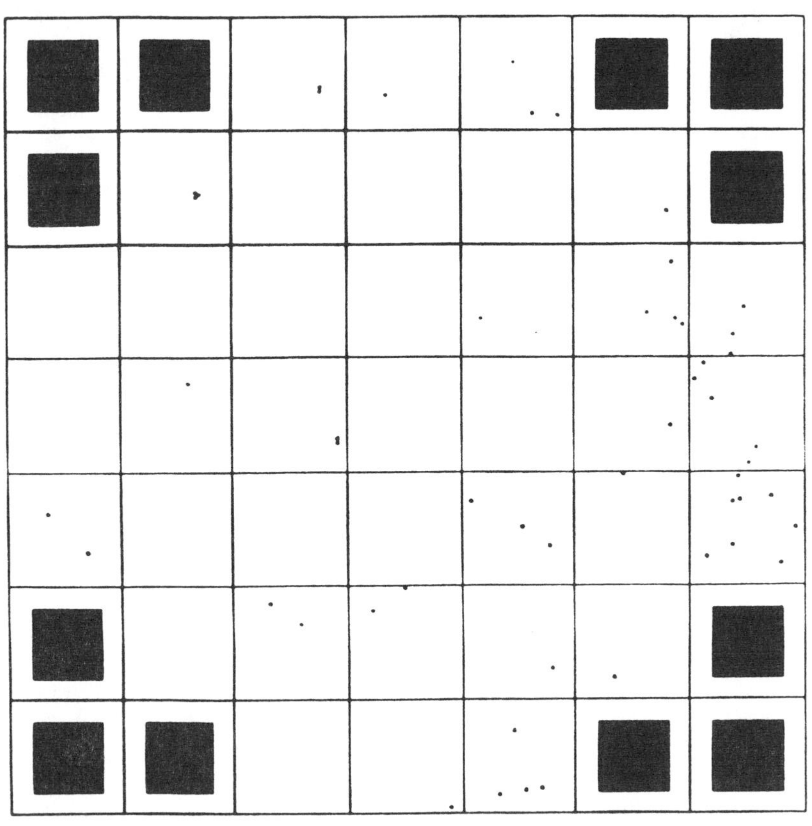

Figure 14: Wafer Map Showing Initial Scan of Oxide Patterned Wafer

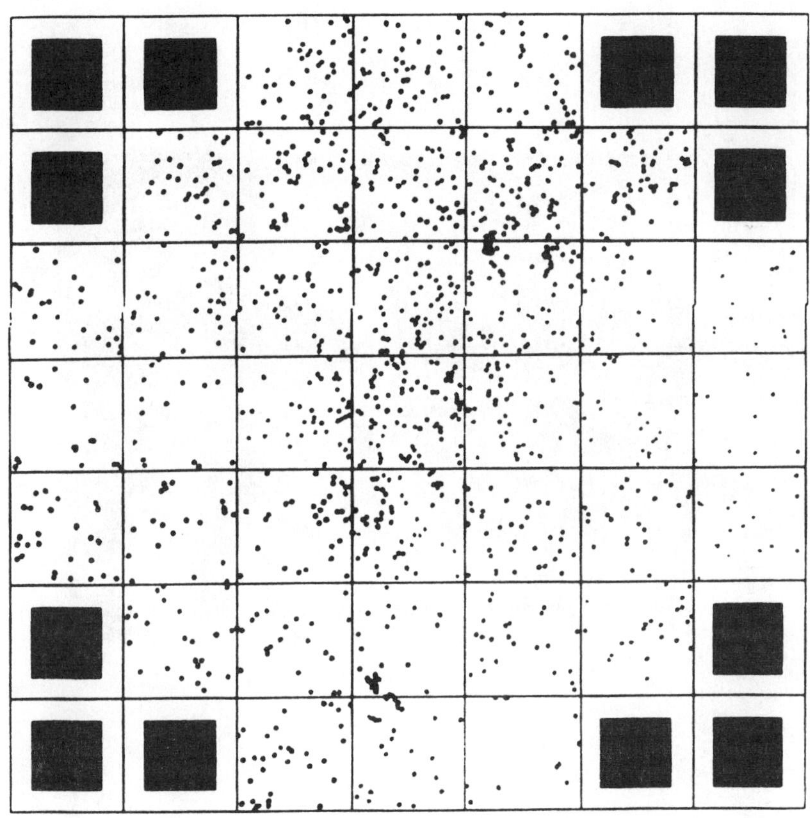

Figure 15: Wafer Map of Oxide Patterned Wafer After PTEOS Deposition.

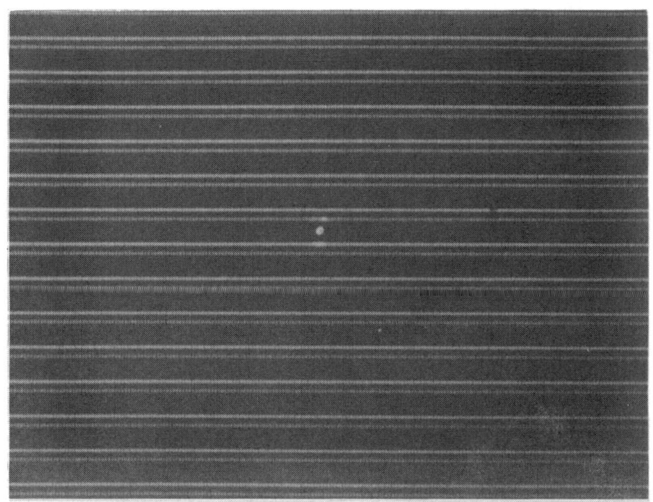

Figure 16: Example of Spherical Defect Added By PTEOS Deposition.

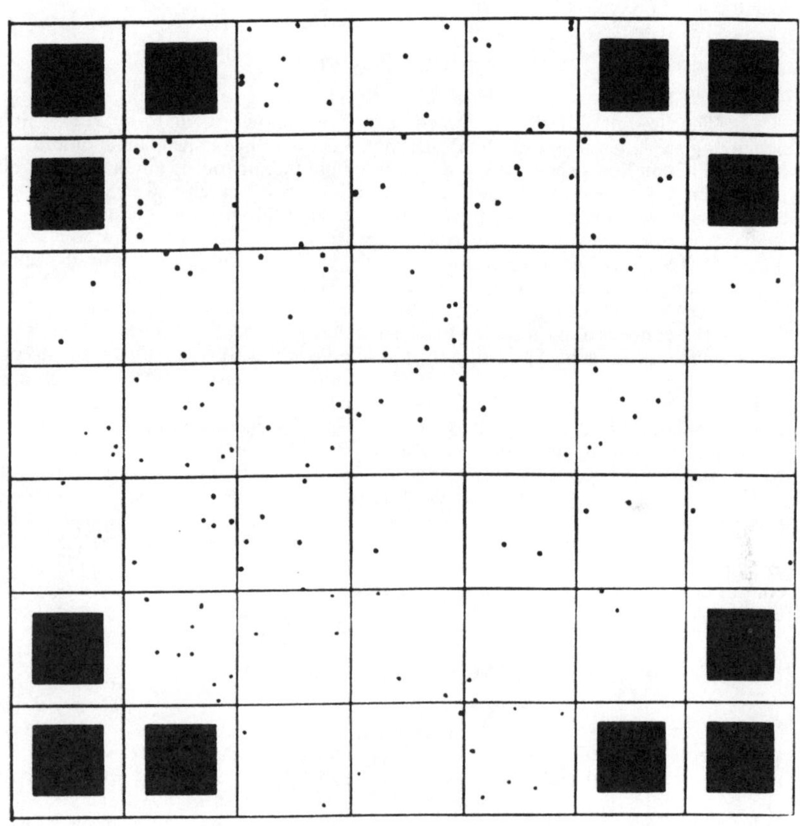

Figure 17: Wafter Map of Oxide Wafer After Alternate PTEOS Process Deposition.

unpatterned substrates, but it was difficult to find them under a microscope. This tends to make the data obtained from a bare surface scanner less convincing. The data obtained from this experiment was much more convincing because the defects could be easily reviewed with the review station and their shape and size could be directly obtained. This method can be used to compare different pieces of equipment that perform the same process. Considering the same PTEOS deposition, historic snake yields had shown another PTEOS process to yield better than the earlier PTEOS process shown in Figures 14-16. Figure 17 shows the oxide pattern wafer map after the PTEOS deposition. Please note that the number of defects from this process are at least one order of magnitude lower than the other process (Figure 15)!

CONCLUSION

The holographic defect detection tool has shown itself to be very valuable in debugging sub-micron and sub-half micron processes from a defect standpoint in the ULSI pilot production line. The speed of the current holographic tool requires that the system be used as an engineering tool that is used primarily for "search and destroy" type defect elimination. However, the next generation of this tool will give greatly increased scan speeds and higher resolution on patterned wafers[12]. This will allow for "search and destroy" as well as line monitoring via product wafers, and it will allow the concept of process induced defects per wafer pass (PIDPWP) to be fully realized.

The concept of partitioning has been extremely beneficial in finding the sources of process induced defects. The stripping and overlay software available on the holographic defect tool and its accompanying review station have been instrumental in detecting process signatures for certain types of defects. Manual partitioning has shown itself to be beneficial in finding defects in the later processes of an SRAM process, and has been very helpful in "creating" defects in order to increase defect learning and understanding. Finally, combining the concept of partitioning and by using a simple highly repetitive pattern, intelligent monitors that represent the true process can be developed to understand the process induced defects added per wafer pass. The concept of process induced defects per wafer pass (PIDPWP) can be utilized to monitor process equipment by performing the exact process rather than mimicking it. The data obtained from PIDPWP can be plotted on a control chart and used to control the process equipment via well known statistical process control (SPC) methods.

ACKNOWLEDGMENTS

I would like to thank the management, especially my supervisor John Alvis, of the Advanced Products Research and Development Laboratory for providing me with the necessary equipment and resources that made this work possible. I would like to thank Henry Gonzales for the countless number of wafer scans he has done for me on the holographic tool. I would also like to thank all of the process engineers in APRDL who helped me with this project. I would also like to thank Jeff Wetzel, and Ted White for providing me with line monitor SRAM lots for partitioning studies, and allowing me ample time to thoroughly study the results obtained from these lots before sending them to the next process steps.

REFERENCES

[1]   D.L. Caven, L. H. Lin, R. B. Lowe, R.E. Graves, and R. L. Fusek, "Patterned Wafer Inspection Using Laser Holography and Spatial Frequency Filtering", Journal of Vacuum Science Technology, B6 (6), Nov/Dec 1988.

[2]   Susan P. Billat and Prasanna Chitturi, "Defect Partitioning: A New Methodology," Proceedings of the SPIE, Volume 1087, Integrated Circuit Metrology, Inspection and Process Control III, p 207, February 1989.

[3]   Charles F. King, G. Peter Gill, and Michael J. Satterfield, "Electrical Defect Monitoring for Process Control," Proceedings of the SPIE, Volume 1087, Integrated Circuit Metrology, Inspection and Process Control III, p. 76 (see Figure 7), February 1989.

[4]   Keith Dillenbeck and G. Peter Gill, "Using Snake Patterns to Monitor Defects and Enhance VLSI Device Yields", Microcontamination, February 1989, p. 23, March 1989, p 33.

[5]   K. Radigan, B. Sheumaker, and Neil Heller, "Using Full Wafer Defect Maps as Process Signatures to Monitor and Control Yield", Proceedings of the Third Annual International Semiconductor Manufacturing Science Symposium, May 1991.

[6]   Skip Mencio, Motorola Internal Presentation, July 1991.

[7]   L.C. Parrillo, et. al., "A Versatile, High-Performance, Double-Level-Poly, Double-Level-Metal, 1.2-Micron CMOS Technology, 1986 IEDM, p 244.

[8]   G. L. Schnable, et. al., "Devitrification in Borophosphosilicate Glass Films Used in VLSI", Journal of the Electrochemical Society, Vol. 127, No. 12, December 1990.

[9]   J. Hayden, et. al., "A High-Performance Sub-Half Micron CMOS Technology For Fast SRAMS", 1989 IEDM, p. 417.

[10]   D. Halliday and R. Resnick, <u>Fundamentals of Physics</u>, 2nd Extended Edition, John Wiley and Sons, Inc., 1981, p. 539.

[11]   B.J. Tullis, "A Method of Measuring and Specifying Particle Contamination by Process Equipment", Microcontamination, p. 67, November 1985.

[12]   M. Slama, M. Bennett-Lilley, and P.W. Fletcher, "Advanced In-Line Process Control of Defects", Proceedings of SPIE Conference on Microlithography, March 1992.

A METHODOLOGY FOR CONTINUOUS DEFECT REDUCTION IN A HIGH VOLUME SUB-MICRON CMOS FACTORY

Mark Berry, Gary DePinto and Joseph Steinberg
Motorola Semiconductor Products Sector,
Microprocessor & Memory Technologies Group,
3501 Ed Bluestein Blvd., P.O. Box 6000, Austin, TX., 78762

Continual improvements in yield, reliability & manufacturability measure a fab and ultimately result in Total Customer Satisfaction. A new organizational and technical methodology for continuous defect reduction has been established in a formal feedback loop, which relies on yield and reliability, failed bit map analysis, analytical tools, inline monitoring, cross functional teams and a defect engineering group. Payoff was a 9.4 X reduction in defectivity and a 5 X improvement in reliability of 256K fast SRAMs over 20 months.

INTRODUCTION

Keys to the success of a wafer fabrication area are organizational structure and dynamics. Continual improvements in yield, reliability and manufacturability are the measures of success in the fab which ultimately result in Total Customer Satisfaction. Within Motorola an organizational and technical methodology has been integrated into a sub-micron manufacturing facility which relies on fast feedback, team work and trust among all functional groups for continual defect reduction. In this paper, the typical organizational structures in wafer fabs are reviewed and contrasted with a new structure. The concept of *nemawashi* is introduced as a means of enhancing trust and buy in.

Previously, in our facility, each functional area developed goal sets with little overlap in terms of defect reduction. A new approach was developed within each functional group that linked defect reduction goals. This systematic approach to continuous improvement and feedback has been institutionalized. Continuous defect reduction relies on a blend of yield and reliability data, failed bit map analysis, in line monitoring and subsequent process module partitioning to identify the sources of yield loss.

Yields are tracked and significant failure modes are studied in Pareto form. External support groups also supply feedback on any failures from reliability stressing and/or customer returns. Together with our customers we have clearly demonstrated a link between defect reduction and reductions of burn-in failure rates and field returns. Failed bit map analysis (FBMA) is performed on a regular basis. FBMA consists of meticulous, layer by layer deprocessing and inspection of predominant defects from predominant failure modes. Mechanisms identified by FBMA are put in Pareto form and shared with all groups in the factory. This focuses resources on "killer defects". Others have reported similar methodologies [1,2,3]. At process modules where killer defects are suspected, inline monitoring can be increased or the module can be systematically analyzed with progressive partitioning. Once the source of the defect is identified and the problem is corrected, inline monitors are then refined into statistical process control (SPC) tools.

To aid in the transition from FBMA to process/equipment isolation, cross functional teams are formed. These teams are more effective than traditional, management edicted "task forces". They formulate their own strategy and goals to eliminate defects and improve process robustness. Such teams are emerging as required resources in today's work force. This approach has been highly successful and has resulted in reductions of 9.4 X in defect density and a 3.4 X for production burn-in fallout over a 20 month period.

Continuous defect reduction is invaluable to both the manufacturer and the customer. Internally, lower defectivity can help reduce cycle time, increase capacity, reduce cost and testing requirements and ultimately improve profitability. To the customer it means improved on time delivery and higher quality ICs going into their end products resulting in fulfillment of our overall objective: Total Customer Satisfaction. A methodology for continuous defect reduction has been institutionalized. Details of the approach are founded in a new organizational structure and a technical feedback system relying on a blend of failure analysis, defect detection equipment and empowered cross functional teams.

ORGANIZATIONAL STRUCTURES

All business groups whether it be a service, manufacturing or support function have structures that theoretically allow them to operate with efficiency, speed and effectiveness. Sometimes the structure comes from a creative brainstorm, but frequently is based on history and function of each group. Integrated circuit manufacturing has specific tasks that are divided into different disciplines, the main function being manufacturing with several support groups. The support groups include Process Engineering, Equipment Services, and Device Engineering. These elements are necessary for a state of the art high volume integrated circuit factory to compete effectively.

Within manufacturing organizations, different functions can be divided to most efficiently serve the factory. Efficient service includes high volume and yield, all delivered at the lowest possible cost with the best possible quality. In any situation where an organization is separated into departments, the communication between these departments is critical to the success of the entire operation. A typical integrated circuit manufacturing organization (fig.1) can be designed with three main operational groups: 1) Manufacturing, 2) Process Engineering and Equipment Services, and 3) Device Engineering. Each group has their own responsibilities but must service the entire manufacturing organization. Manufacturing is responsible for productivity improvements, cycle time reduction, scrap reduction, procedural documentation, training and personnel issues. Process and Equipment Engineering maintain responsibility for equipment availability, process development, particle and defect reduction, SPC and tightening of the process distributions. Device Engineering has the task of identifying the sources of yield loss and feeding back the data to the process and equipment groups to find the root cause source of the defect. They are also responsible for the process integration, and are the communication focal point for working with internal and external customers and fab support groups. These job descriptions are not exclusive, on the contrary, the success of each group is contingent upon cooperation from the others.

Each section is responsible for their own performance, as well as the entire organization's goals. Each functional group has its own internal goals that address the areas over which they are responsible, have the most influence, or where the manager has the most personal interest. Only with coordinated effort among all sections can the entire organization meet its goals.

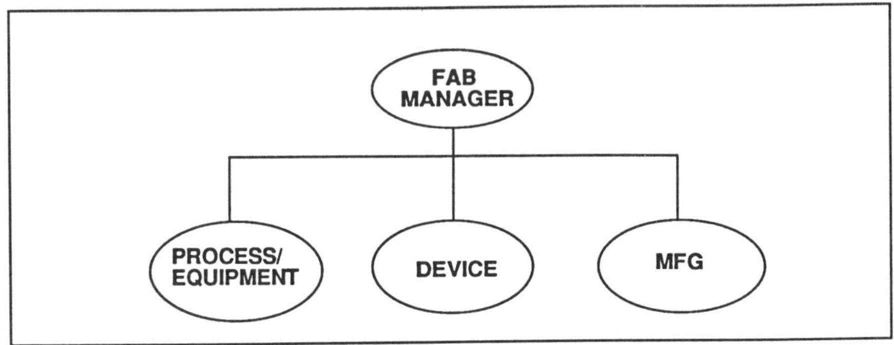

Fig. 1. Traditional fab organization chart

An organization can be viewed as a processing system to accomplish tasks or goals. This process system can be very complex and with human factors as one of the inputs, the system can fail due to poor communication and incomplete understanding. It is possible for departments within an organization to concentrate so intensely upon their own functions that they become defensive and lose sight of the operation's overall goals.

White space is a term which defines the area between groups within an organization. Typically a paper organizational chart has boxes containing functions. The area on the chart between the functions is literally white space. The organization must be set up in a fashion that recognizes that these gaps exist, and has the ability to bridge the white space. A typical organization chart that looks rather innocuous can actually be hiding severe problems. Each group may actually be building a silo of self defense (fig.2).

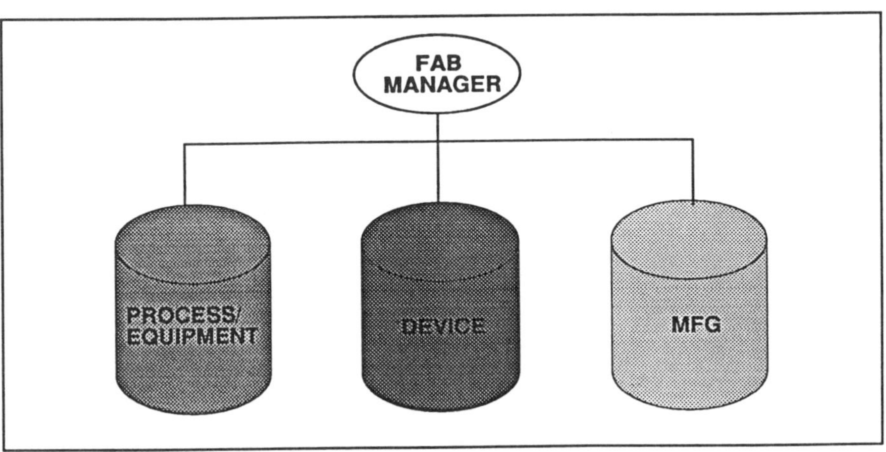

Fig. 2. Defensive structure

These silos could also be used as rocket launching pads for the more aggressive managers. To cover weaknesses or take attention from their problems, bombs can be tossed into other groups territory. This causes them to be defensive and collect data proving that the problem is elsewhere. This lack of relationship and trust between groups can be very destructive, caused by complete breakdown in communication, lack or respect for other individuals and having one's own interests larger than the interests of the entire organization. It is the responsibility of the operation's managers to recognize the problem and build a organization that can bridge this white space for total communication and teamwork.

An effective organization must see the entire operation as a series of handoffs between functions. The different groups must be linked with common goals and objectives, and must be continuously conscious of the responsibilities of each group in essence making them mutual responsibilities. If one group fails, the entire organization fails.

Effectively managing continuous defect reduction requires support from all parts of the organization. Manufacturing must accept monitoring techniques and give Process Engineering time on equipment for characterization. The Process Engineering group must take ownership of the defect levels for specific equipment, and Device Engineering must supply good feedback identifying the areas of the process to work on. Despite good intentions, it is quite natural for there to be separations between these groups. This Ven diagram (fig.3) shows how these groups may operate with no complete overlap.

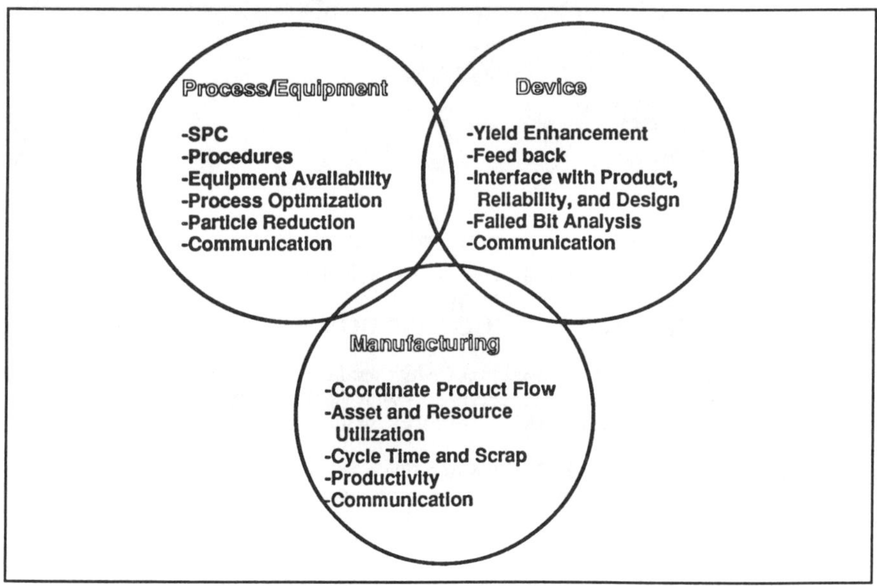

Fig. 3. Three section Ven diagram

The creation of a Defect Engineering group provides the needed overlap of all three groups to enhance communication, tie together goals and be a focal point for team-oriented problem solving. To maintain focus on defect reduction, the Defect & Device groups report to the same manager. It is very important to staff the defect group with people from each of the groups that they support, to have a complete knowledge base, and to help enlist the buy-in of the entire organization. The following Ven diagram (fig.4) shows how the defect group links the organization. The key objective of the defect group is to be a focal point for communication to bridge the white space that hinders organizational effectiveness.

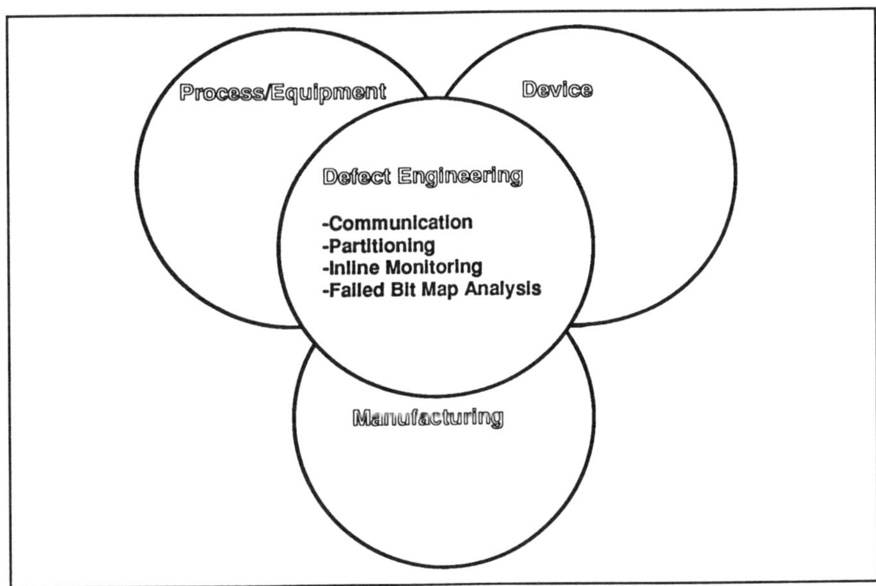

Fig. 4. Four section Ven diagram

METHODOLOGY FOR CONTINUOUS IMPROVEMENT

To reach the goals of a sub-micron CMOS facility an organizational structure has been set up with a common defect reduction goal with the necessary people and equipment resources allocated to achieve this goal. Proper blending of people and equipment is required to make these goals a reality. Defect reduction work is a sustained, systematic effort and cannot be accomplished with a step function effort over a short period of time. Common goals between all groups in an organization provides the foundation needed for success. Goals must be set aggressively enough to enhance the performance of individuals in the organization. Finally, top management must visibly support the goal by providing the necessary organizational resources.

In order for the defect group to help the organization reach its defect reduction goals total buy-in is necessary. Horizontal buy-in across functional areas including manufacturing, process, equipment, device, plant services and vendors is needed. Vertical buy-in down the management chain from operations manager, staff, section heads, supervisors, engineers, technicians and operators is also necessary. Total buy-in and integration of the defect reduction goal makes success imminent.

The key ingredient for defect reduction is pre-work and early communication of the data collected. Presenting new data in a public forum that is controversial without prior knowledge is a direct route to failure. *Nemawashi* is defined as pre-discussion of information leading to agreement before a formal or public forum. This allows disagreements to be worked out prior to entire group meetings. Problems or grievances with opposing positions can be resolved in a private forum. Practicing nemawashi leads to trust and team work. The time spent going over data with involved groups avoids public confrontation and long non-productive meetings.

An integrated circuit manufacturing facility can be thought of as a feedback loop, as shown in fig. 5. Basic inputs are processing equipment, process recipes and procedures to run and monitor the equipment. Results of fabrication are judged by process and probe yield, as well as, reliability. Other outputs include cycle time and cost. Yield and reliability failures are broken up into failure modes and failure analysis identifies the mechanisms. These results are used to establish where to place and what type of inline monitors to use. Both the inline monitor results and failure mechanisms lead to changes and optimizations of the basic three fabrication inputs.

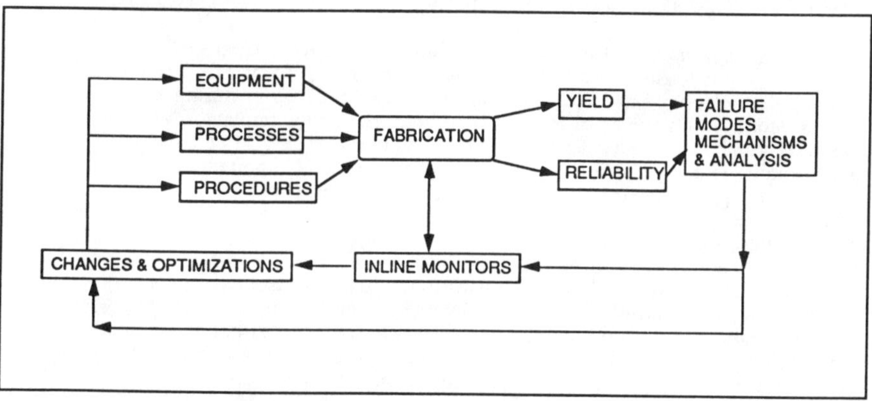

Fig. 5. Model for continuous improvement

Methodology

There are five key components of our continuous improvement methodology. From the model there are 1) yield and reliability feedback supported by 2) failed bitmap analysis and sometimes leading to 3) inline monitors. These results lead to the formation of 4) cross functional teams and 5) weekly defect meetings. In the following paragraphs each component is detailed.

Yield and reliability feedback - Yield and reliability feedback are the first step in identifying "killer defects" to eliminate for continuous improvement. On a daily basis all wafer probe yields (circuit test) are collected and summarized. Abnormal lots are investigated within 24 hours of their probe test. Yield variation is categorized as to whether its within wafer (spatial distribution), wafer to wafer or lot to lot. Yields are accumulated into trend charts and converted to defect densities through a modified Murphy's model. Failure modes are put in Pareto form and compared over time to separate random from systematic yield loss . Reliability failures from internal qualifications and customer returns are also collected and put into pie chart format to give a perspective and provide direction.

Failed bitmap analysis - Failed bitmap analysis is performed on a weekly basis on 0.8um products including a 32 bit microprocessor (MPU) with cache memory and 256K density fast SRAMs [4]. The MPU employs twin well, single poly, double level metal while the fast SRAM adds a second poly layer. This architectural difference results in better visibility into interconnect defects on the the MPU and double poly / dielectric defects on the fast SRAM. The selected wafers are probed and their failure modes put in Pareto form. From the predominant failure modes, the bitmaps are inspected to give another failure mode Pareto chart more related to the yet unknown failure mechanism. More precisely, the bitmap Pareto's are made up of categories of failures (i.e., double bit, row, column, etc.). From there, layer by layer deprocessing and inspections (optical and scanning electron microscopy) take place until all potential layers have been examined. The failure mechanisms found are then put into a final Pareto format which identifies the hierarchy of killer defects. This information is then shared among all levels and areas of the fab.

In-line monitors - Occasionally, the failure mechanisms from FBMA are so clear that their source is readily identifiable. This is the exception, not the rule. Take for instance an interpoly dielectric particle. It could have landed on the wafer during any number of process steps spanning much of the process (photo, etch, poly, implant, CVD oxide, etc.), yet all the bitmap failure mechanism can say is that its under poly 2 but atop poly 1. This is an opportunity to apply inline monitors and partitioning. It is important to use all the tools available and appropriate for partitioning and analysis. These may include light scattering defect detection systems, comparitors or holographic inspection. These tools can be applied to partition the process flow with multiple inspections at different levels to isolate the source of a defect to a specific process or equipment set. Again this is another idealistic case. Often the common reality is that partitioning does not isolate a defect source that precisely. Frequently, problems are caused by interactions between multiple processes. It is this combination of failure analysis, partitioning, and Pareto chart perspective that determines whether a cross functional team will be formed. Once a failure mechanism's source is identified procedures or changes must be put in place to prevent its reoccurrence. To make continuous improvement, SPC must be put in place to insure that yield loss does not occur due to out of control situations. In parallel with this inline product inspection effort, inline monitoring with nonpatterned defect detection equipment, utilizing SPC control, is continually performed on all equipment. Additionally, patterned wafer inspections are used to monitor the line at key levels in the process.

Cross functional teams - To attack elusive failure mechanisms, cross functional teams can be formed after failure analysis, Pareto diagrams and partitioning clearly illustrate modules in need of attention. These same tools are employed by the team. The team's members typically consist of people from the defect group, device, process, manufacturing and equipment. Wherever possible they are the experts on the process module(s) in question. Resources and support provide group cohesion and increases the chances of success in achieving the group's goals. A strategy to solve the problem is developed by the individuals on the team. This type of environment along with group problem solving builds trust among team members so that the source of the yield loss can be identified and solved. Goals and strategies are defined by the team, not by management (i.e., not an appointed "task force"). These teams are provided with the necessary support and resources to eliminate sources of yield loss. Teams disband when a combination of inline monitoring and FBMA illustrates the sought after improvement.

Weekly meetings - A large manufacturing facility can devote massive amounts of time to meetings regarding cycle time, scrap and productivity. We added a weekly meeting solely to focus on continual improvements through defect reduction. It serves as a forum for fast feedback and immediate problem response. The meetings are attended by all levels and disciplines. FBMA, inline monitor data and team updates are presented. Device Engineering also has weekly yield enhancement meetings to address product/process specific issues. Meetings are used to provide the organization with knowledge so that coordinated effort and decision making occurs to make continuous yield improvements. This meeting is a good forum for teams to present their progress.

CONTINUOUS YIELD IMPROVEMENT FLOW CHART

When introducing new products or processes into a fabrication area, the majority of the yield loss and reliability problems can be attributed to a few obvious problems. Once the obvious problems are resolved, the key to continual yield improvement is to have a systematic pre-determined approach of defect reduction. A continuous yield improvement flow chart has been developed that incorporates the five key components of our continuous improvement methodology. This flow chart is shown in fig. 6 on the following page.

There are strong arguments that in line inspection should determine the appropriate starting place to work on defect reduction. This approach of reacting to in line defect data, whether obtained visually or through defect inspection equipment, is widely used, but it has one flaw; the defects found in line may not be your highest source of yield loss. A great deal of effort may go into eliminating a defect problem without any major increase in yield. For this reason, the flow chart as illustrated in fig. 6 uses wafer probe as a starting point of focus.

A Case Study

Using this flow chart of continuous improvement many defect problems have been identified and corrected. Using the steps outlined in this flow chart, an example of how a major defect was identified and eliminated will now be reviewed. Fig. 7 is an illustration of composite (normalized) yield on a 256K fast SRAM for 625 wafers. All die that have a normalized yield in excess of 90% are highlighted. It is clear that two areas of the wafer have suppressed yield, the edge and a strip down the center of the wafer. The lower yield on the outer edge of the wafer was expected, but the low yield down the center of the wafer was not.

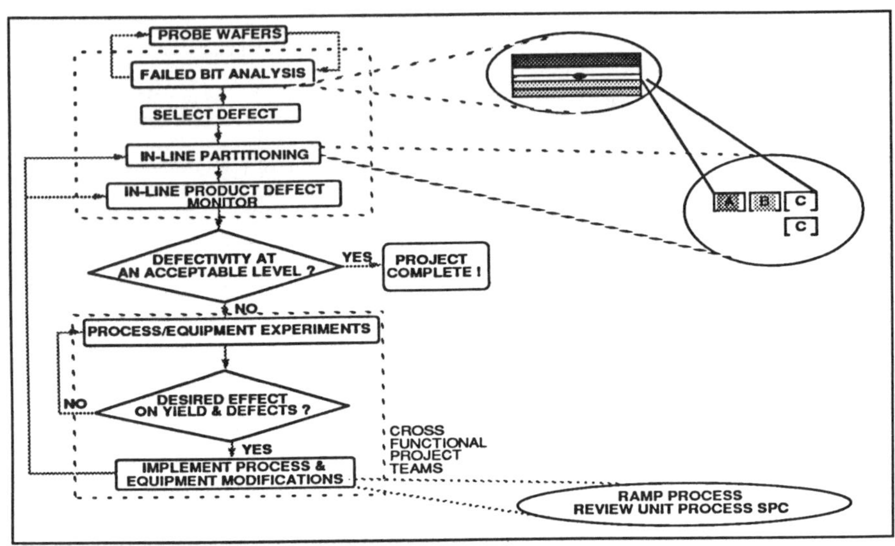

Fig. 6 Continuous improvement flow chart.

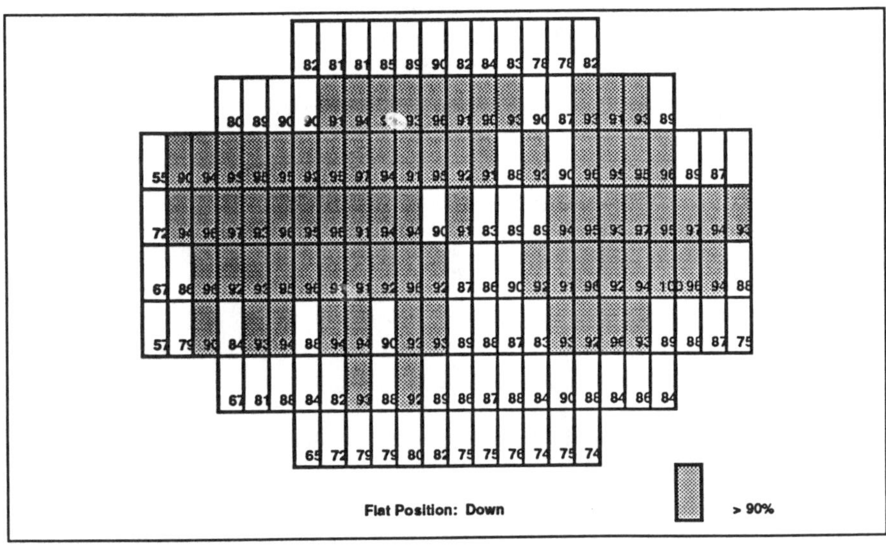

Fig. 7 Normalized cumulative yield map of 625 wafers.

As discussed previously, wafers are bit mapped and deprocessed, and all of the identifiable failures are collected and added to a composite Pareto chart. Pareto charts are continually updated for both fast SRAMS and the MPU. A Pareto chart for one of the products indicated that a specific defect type was a major source of yield loss. This defect is illustrated in fig. 8.

Fig. 8 SEM photomicrograph of defect.

Once the defect above was selected to be worked on, in line inspections to partition the line to identify the source of the problem were set up. The holographic inspection tool was identified as the tool of choice for identifying this defect. Although this system is slow, it can scan any portion of a wafer that has a repetitive area. For a fast SRAM, this is effectively the entire area of a wafer. Scanning the entire wafer provides insight into whether or not a problem is spatially dependent. The narrow width of the defect (.2 to .5 microns) made light scattering tools ineffective in identifying the source of the problem. As we would later find out, the localized nature of this problem made die to die comparison also ineffective as an initial screening tool.

Several iterations of partitioning using the holographic inspection tool identified the source of this defect to be between gate oxide and the second poly deposition of the gate poly (I) layer. The holographic inspections showed that the distribution of the defect was not random, but had a spatial dependance. All of the defects identified on ten wafers from ten different lots using holographic inspection were compiled into a composite map. The results of this composite map comprising 76 defects are shown in fig. 9. Also

217

superimposed in fig. 9 is the composite yield map of all die that yield greater than 90%
from fig. 8. It is clear that the identified defects are the source of the suppressed yield
down the center of the wafer.

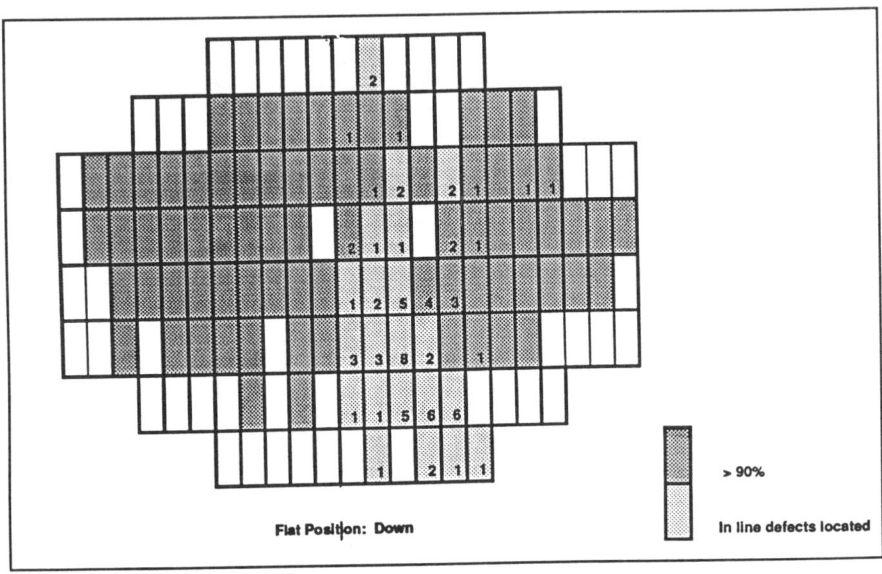

Fig. 9 Normalized yield/ holographic inspection defect map overlay.

Once the spatial dependance across the wafer was determined, a die to die
comparison tool was set up to monitor the problem and establish a baseline, and a cross
functional team was set up to solve the problem. It has been found that establishing a
baseline is necessary so that progress in solving a problem can be measured and a goal can
be set. It is also important to establish a baseline because certain types of defects can not be
totally eliminated. The baseline forms the quantitative basis for a team to track defect
reduction and continue work until the point where the goal is met, or a point of diminishing
returns is encountered.

After several carefully planned experiments, the cross functional team isolated the
problem to one particular process. Identification of the problem process was difficult due to
the interaction of poly deposition and the true defect source. The cross functional team
identified that the defect was not a poly defect at all, but was enhanced by depositing poly
over it. This enhanced defect without going through poly deposition is so small that it can
not be found by using die to die comparison or holographic inspection. Yet once poly is
deposited over the defect it grows to a size such that it is able to short contacts to poly lines.

Once the specific problem process was identified, the holographic inspection tool was used to resolve the problem. Typical results of the standard process screened using the holographic inspection tool resulted in a spatial pattern like that in fig. 10. Typical defect counts from this process were in the range of +60 defects with 70% of them being the defect in fig. 8. A very simple, yet subtle, process change lowered the defect count to less than 6 at this process with all targeted defects being eliminated.

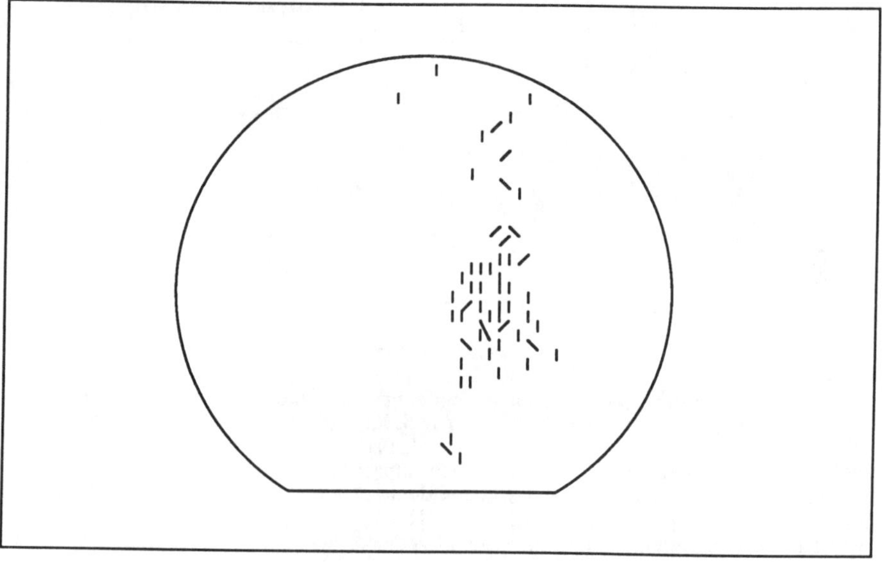

Fig. 10 Spatial defect pattern using holographic inspection tool

QUALITY VERSUS EFFECTIVENESS

Use of this problem solving approach has shown that cross functional teams are more effective in an organizational setting and provide a higher amount of quality as measured by Total Customer Satisfaction than do individual efforts and task forces. It is envisioned that the next continuum beyond cross functional teams is a higher performance organization (fig.11).

Individuals can have a big impact upon solving problems, but when groups of people are incorporated many more ideas can be generated. A task force focusing upon a problem is a very effective method of fixing a given major problem, but goals and methodology defined by management limits the number of problems that can be worked on at a given time. Because a team is self guided, many difficult problems can be addressed simultaneously which decreases the cycle time of learning. This creates a much stronger team leading to a high performance organization.

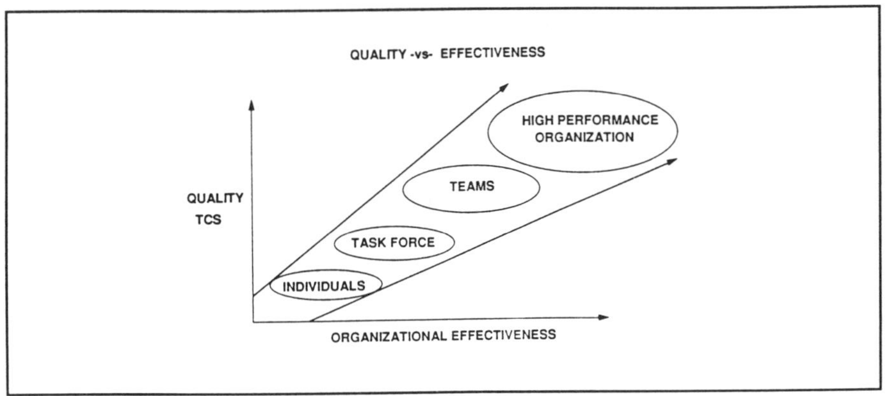

Fig. 11 Quality versus effectiveness

RESULTS

The continuous improvement methodologies and the efforts of cross functional teams described in this paper have paid off with impressive results on 1.0um 256K fast SRAMs over a 20 month period. Many non value added steps have been consolidated or removed. Many process modules have been improved to known better equipment and processes. Defect reduction work has forced the implementation of SPC methods so that the major problems do not continually recur. These changes took place while the factory was running at capacity. Defect density improvements were not traded for increased cycle time or inline scrap.

Three quantitative measures testify the effectiveness of our methodology. In fig. 12, the 20 month trend of defect density on 1.0um 256K fast SRAMs is shown. This defect density is a normalized version of actual defect density derived by Murphy's model from wafer probe yields. Overall defect density improved by 9.4 X. There is a strong correlation between low defectivity at wafer level probe and low fallout at production burn-in which showed a 3.4 X reduction in fallout. These improvements propagate even further into product lifetime. All failures from extended reliability tests and field returns in the first 12 months have been incorporated into an "incoming + 12" calculation to measure field reliability which improved 7.2 X. In fig. 13, production burn-in and field reliability are plotted along with defect density. Note all three are closely correlated.

Continuous improvement and defect reduction positively impacts many aspects of a product which benefit the customer, the manufacturer and society. For the wafer fab it means reduced cycle time, increased capacity and reduced costs. Higher yield and quality require less testing. Overall financial performance and ability to compete in the market are improved. Ultimately, continuous improvement and defect reduction is one of routes to Total Customer Satisfaction in terms of on time delivery, performance and quality.

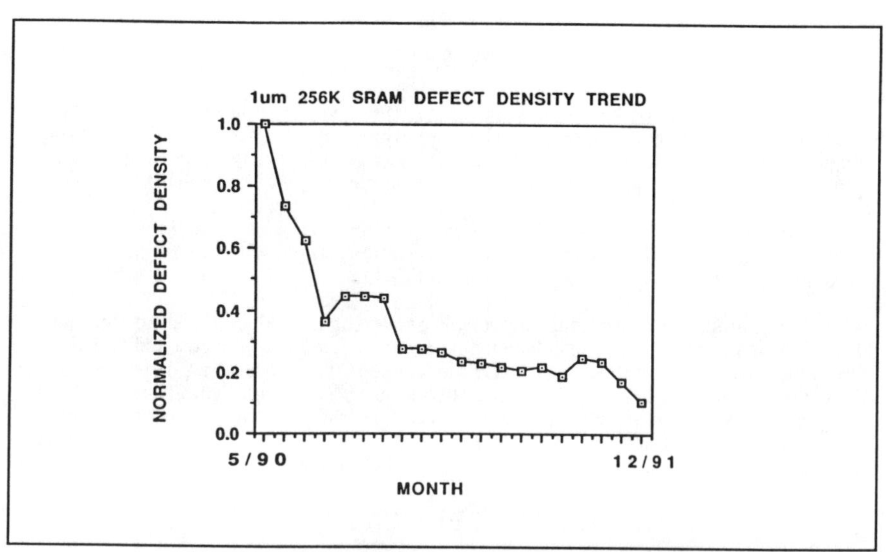

Fig. 12 1.0um 256K fast SRAM defect density trend

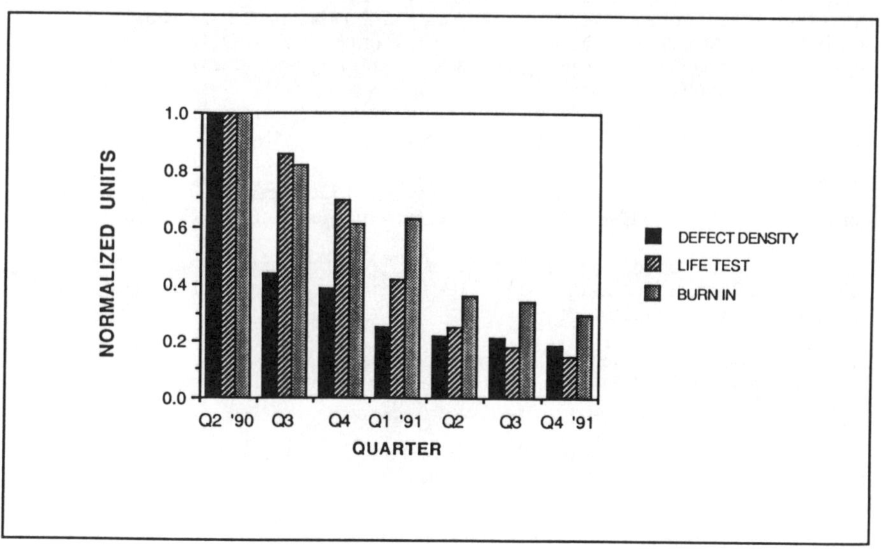

Fig. 13 Correlation of defect reduction to reliability improvement

SUMMARY

Every organization has white space, and it is the responsibility of the organization to determine how to facilitate communication between groups. Cross functional teams enhance communication and create buy-in between groups through representation and input. This helps bolster teamwork and increases the speed of problem solving.

Nemawashi is the key ingredient in building trust and removing sources of confrontation. This methodology will help organizations reaching for six sigma results in improving yield, reliability and total customer satisfaction.

Continuous improvements impact all phases of a product. Within the wafer fabrication area improvements include reducing cycle time, reducing die cost and effectively increasing capacity by increasing the good die shipped. There is also a true trickle down effect, the test area will see a reduction in test time and burn-in can be decreased with higher reliability. The company will see improved financial performance. Most important of all, the customer will receive HIGH QUALITY PRODUCT with ON TIME DELIVERY. This produces the direct result of Total Customer Satisfaction.

ACKNOWLEDGEMENTS

Mark Hall, George Kong, Felix Leonardi, Mike Alperin, Harold Cartey and Pat Rister have supplied the failed bit map analysis necessary to identify defects. Tim Chesnut, Michael Phillips, Dean Dreier, Patrick Billings, Clark Shepard and Chris Allen have supplied key probe and reliability support, and to Don Horst, Scott Morgan, Robert Broyles, Errol Moore and the entire Process Engineering, Manufacturing, Device Engineering and Defect Engineering teams for supporting and believing in the results possible with Empowered Teams.

REFERENCES

[1] H.G. Parks; "Fast Turn Around Post Process Yield Enhancement for Custom VLSI Foundries" 1990 IEEE / SEMI Advanced Semiconductor Manufacturing Conference, p.82

[2] E. Hall, D. DiMarco, P. Berndt, A. Zandi ; "A Structured Approach for Failure Analysis of a 256K BiCMOS SRAM" ISTFA International Symposium for Testing and Failure Analysis, 1989, p.167

[3] F.T. Agricola, K. F. Anderten ; "Failure Mode Analysis Methodology on High Density CMOS SRAMs" ISTFA International Symposium for Testing and Failure Analysis, 1989, p. 389

[4] S. T. Flannagan, et. al., "8-ns CMOS 64K x 4 and 256K x 1 SRAM's" IEEE Journal of Solid-State Circuits, Vol. 25, No.5, October 1990, p.1049

COMPUTER DEPROCESSING OF PRODUCT WAFER DEFECTIVITY DATA

Steven W. Daniel

Semiconductor Products Sector
Motorola Inc.
2200 W. Broadway Road, Mesa, AZ. 85202

The purpose of this work was to define a method for identifying and prioritizing defect sources in an integrated circuit manufacturing line, based on data generated using an automated laser scanning patterned wafer inspection system. Defectivity levels at the final process block of the manufacturing flow are known to correlate to yield. The objective is to understand how previous process blocks contribute to the defectivity observed at the final process block.

INTRODUCTION

With the current level of wafer processing cleanroom technology and protocol, the most significant defect sources affecting the product come from the wafer processing equipment. Most manufacturing areas address this issue by developing a particle monitor wafer program where unpatterned test wafers are routinely run through specific process equipment and added particles are counted and sized using a scanning laser surface particle measurement system. While this is a useful and necessary technique for maintaining particle control in process equipment, it fails to replicate the dynamic interaction of the real product with the process.

Within the past few years, the introduction of automated patterned wafer inspection systems utilizing laser scanning defect detection [1] has made possible the collection of a statistically significant amount of data, in real time, directly from product wafers. The efficiency of these systems allows for the evaluation of 15 - 20 wafers per hour (complete wafer surface) and creates the challenge of how to effectively manage the large amount of data so as to extract defect information that can be used in the manufacturing line [2].

Early application of the patterned wafer inspection system involved establishing a routine monitoring program to partition the manufacturing line into manageable blocks of process operations. Sampling points were established with regard to both the number of process steps between sampling points and the critical nature of particular process steps to the functionality of the product. All starting material lots of a particular product flow are designated for inspection at one of five points across a flow containing 70 major process levels. At each inspection point five wafers are randomly selected for inspection. The average defects per wafer as well as the range are tracked on control charts and reaction mechanisms are established for material outside of the normal distribution. Although useful for monitoring defect trends at these designated inspection points, this application alone was only moderately effective at identifying prior sources which could be driving defect levels at these inspection points. At every process level inspections are performed, a new set of defects are revealed and seldom do clear points of focus emerge. Attempts to gain further information about defect sources by microscope inspection of detected defects for visual classification of types also proved to be only moderately effective as most defect types were transformed into generic lumps after processing through a few harsh processes.

The goal of defect inspection directly on product wafers is to characterize the link between successive processes and to understand which processes are responsible for contributing the defects which will ultimately remain on the product wafers. This enlightenment was not forthcoming prior to the development of relational database software for the comparison of defect sites through multiple, sequential inspections on wafers as they progress through the process flow.

EXPERIMENTAL APPROACH

A typical use of a laser scanning patterned wafer inspection system is to perform bracketing inspections on product wafers before and after a single process, or block of processes. The current level of relational database software allows for comparison of the data files from two inspection points by comparing the x,y coordinates (for a defined radius) of the defect sites and classify defects as "Added", "Common" or "Removed". This gives a very limited picture of what can be a very long process flow.

The next step involves defining a series of inspection points to bracket numerous processes, or blocks of processes, then applying the current level of relational database software to analyze each bracketed process for "Added" defects (Figure 1).

"Added" Defects

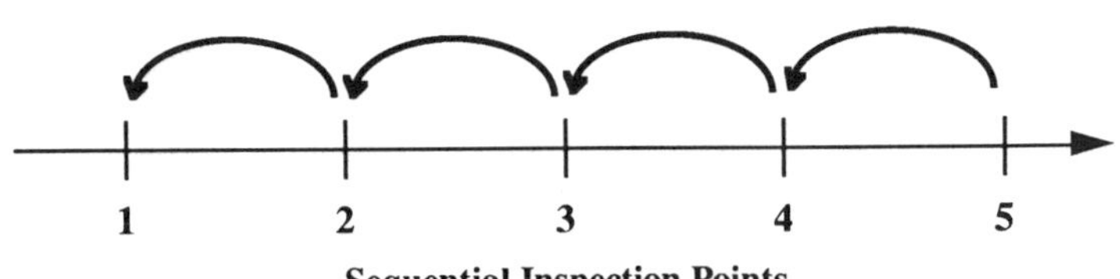

Sequential Inspection Points

Initial Determination of "Added" Defects

Figure 1

Analysis of data in this manner often fails to lead to any points of focus for identifying defect sources which are having a detrimental impact on the finished product wafers. Typically, analysis in this manner will show a significant number of "Added" defects for each inspection point and the sum of these will be much greater than the total defects detected at the final inspection point. This tends to indicate that every bracketed process, or block of processes, is a problem and fails to give any real direction as to where improvement efforts need to be focused.

This lack of direction led to analyzing the total defect population at the final inspection point from the standpoint of determining in which previous process block they had their origin. This was accomplished by comparing the final total defect population to the defect population at the initial inspection point and to the "Added" defect population for each intermediate inspection point (Figure 2).

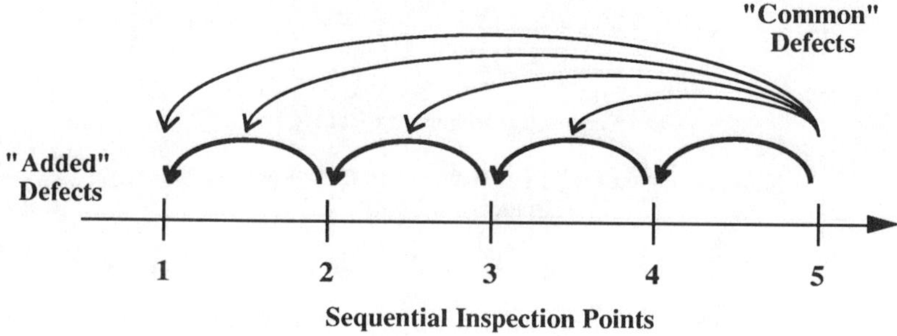

Sequential Inspection Points

Initial Determination of "Remaining" Defects

Figure 2

Data analyzed in this manner always showed that the sum of all the defects contributed from the various process blocks was always greater than the total defect population detected at the final inspection point. The greater the number of inspection points, the more exaggerated the difference. This discrepancy was subsequently determined to be caused by variations in attainable sensitivity at different process levels. This caused defect sites to fade in and out of detection during inspections at sequential process levels and resulted in individual defect sites being detected as "Added" at multiple inspection points.

The solution to this problem involved developing an algorithm to define a complex matrix of data overlay comparisons to track individual defect sites to the first inspection point at which they were detected. To accomplish this, the data from every successive inspection point must be resolved against the data from every preceding inspection point to identify the defect sites which are unique to every point in the series of inspections. The data files containing the defect sites resolved in this manner will now be referred to as "True Added" defects as opposed to the data files created in Figure 1, which will be refered to as "Simple Added" defects. The resolution of the "True Added" defects for the final inspection point for a series of 5 sequential inspection points is represented by the overlay sequence in Figure 3.

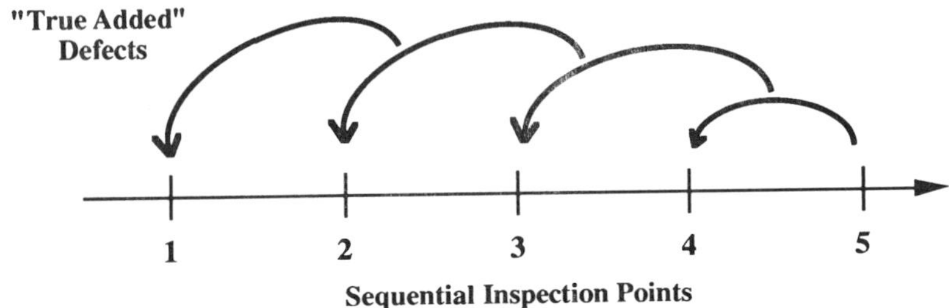

**Determination Of "True Added" Defects For The Final
Inspection Point**

Figure 3

In a similar fashion for the fourth inspection point in a series of five sequential inspections, the "True Added" defects must be resolved against every prior inspection point. Next the "True Added" defects file from the fourth inspection point must be resolved against the the total defect population at the final inspection point for "Common" defects to derive the contribution of the fourth process level to the total remaining defect population at the final inspection point (Figure 4).

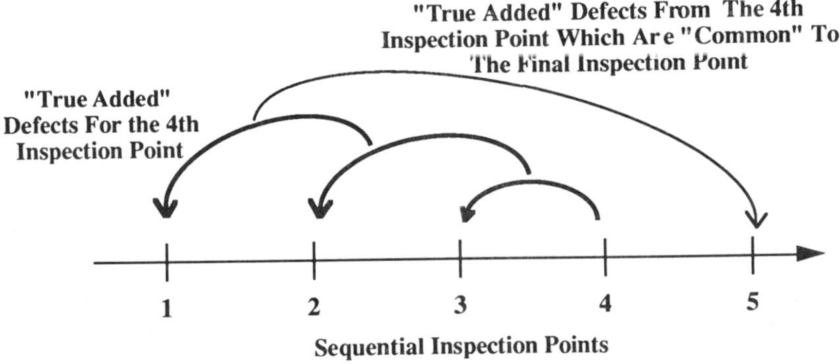

**Contribution Of Defects From The 4th Inspection Point
To The Total Defects At The Final Inspection Point**

Figure 4

This overlay sequence is repeated all the way back to the first inspection point, then the initial defect population is resolved against the final inspection point for "Common" defects to determine the contribution of defects incoming to the inspection sequence to the total remaining defect population at the final inspection point (Figure 5).

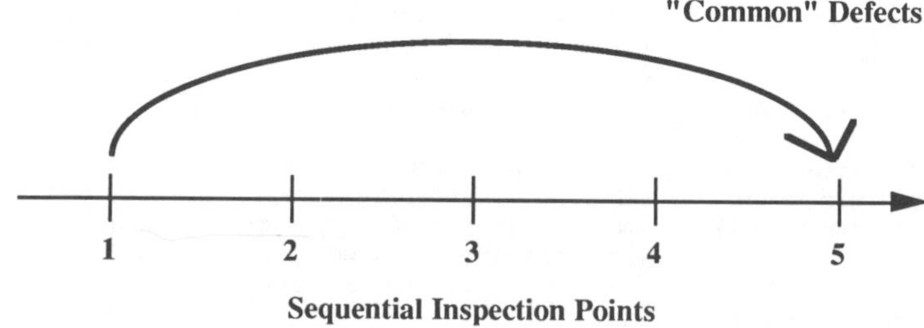

**Contribution Of Defects Incoming To The Inspection
Sequence To The Total Defect Population At The Final
Inspection Point**

Figure 5

The number of overlay comparisons required to accomplish this for a series of n inspection points is defined by:

$$\sum_{i=2}^{n} n_i$$

Applying this complex matrix of overlays to any give series of inspection points yielded data where the sum of the individual process defect contributions equalled the total defect population detected at the final inspection point.

RESULTS AND DISCUSSION

Throughout the semiconductor industry, a widely used technique for identifying defect sources involves deprocessing of finished product. To accomplish this, single process levels are chemically stripped back a layer at a time. Microscope inspections are performed initially and after each process level is removed to track defects back to their origin. The data analysis technique described here is the computer equivalent of deprocessing in the sense that defects at the final inspection point are stripped back to correctly identify their source. This is especially important in that the vast majority of defect sites which appear in any given sequence of inspections are transitory in nature and must be sifted out of the data to reveal the process levels responsible for contributing hard defects (those which ultimately remain).

Defect data which is analyzed from the standpoint of what defects get onto the product wafers tends to be very random in nature, indicating nearly all processes as significant contributors. Defect data analyzed from the standpoint of what defect sites ultimately remain on the product wafers tends to be very systematic and correctly identifies the processes which are significant defect contributors.

By performing this procedure on numerous product lots in a varying series of inspection points, major defect sources were successfully mapped out and identified. This led to very specific points of focus in the manufacturing line for concentration of yield improvement efforts. With the understanding of which processes are responsible for contributing hard defects, Defect Engineering can utilize this information to refine reactions for the routine monitoring program.

This data analysis technique has been successfully applied to resolve the data generated on up to 22 sequentially inspected process levels. Using the current level of relational database software, this required performing 256 overlay comparisons for each wafer in the inspection matrix. While the current level of software makes this type of data extraction possible, it is far from a straightforward application. The techniques and results discussed in this paper have been provided to the developers of the relational database software who are incorporating this multilevel data extraction technique into the second generation of the system.

SUMMARY

This paper has described a methodology for integrating automated patterned wafer inspection into a high volume manufacturing line. Equally as important as the capability of detecting defect sites on product wafers is a relational database for the comparison of defect sites from multiple inspection points as the product wafers progress through a complex process flow. The development of an algorithm to define a matrix of overlay comparisons for successive inspection points has resulted in a nearly complete characterization and understanding of defect sources affecting product wafers across a flow containing 70 major process levels.

ACKNOWLEDGEMENTS

For his continuous support and help as this work evolved and in the preparation of this paper, I would like to acknowledge Gavin Woods.

REFERENCES

1) R. Browning, I. Lincoln and P. Stonestrom, in "Integrated Circuit Metrology, Inspection and Process Control III," Vol. 1087 Society of Photo-Optical Instrumentation Engineers, Bellingham, WA (1989).
2) G. P. Woods in "Defects In Silicon II," W. M. Bullis, U. Gosele and F. Shimura, Editors, Vol. 91-9, The Electrochemical Society, Pennington, NJ (1991).

MEASURMENT OF PERIPHERAL DAMAGE CAUSED BY SCRATCHES ON SILICON WAFERS

Michael K. Kocsis

Intel Corp., M/S CH2-59

5000 W. Chandler Blvd.

Chandler, Arizona 85226

Scratches on Silicon wafers have long been known to degrade integrated circuit yield by disrupting the circuitry of the scratched die. However, this work shows that scratches also create particulate which migrate and kill surrounding die. A factorial experiment was run which put reproducible scratches at four different steps in the fabrication process on serpentine-comb electrical defect monitor wafers. The number and distribution of shorts and opens on the electrical defect monitors as well as in-line particle readings were used to quantify the clustering relationship between the primary (scratch) defect and the secondary particle defects. The susceptibility of each process layer to damage as well as the mechanics of particulate migration are discussed.

INTRODUCTION

Defect reduction is a major concern in the production of semiconductor devices. At Intel's Fab 6 in Chandler, AZ, there are several different processes and many different products currently in production. Each process/product combination has its own set of defects. Locating the source of these defects can be a very difficult and time consuming task. Consequently, only one type of defect is pursued at a time. To help choose which one to work on, a pareto chart is made of the six most common killer defects found by end-of-line failure analysis. The top categories are usually particle related with scratches lower in the list. However, it has been suggested that many of these particle related defects are in fact created by particles coming from scratches, and that scratches should receive a higher priority. An experiment was designed to determine if a scratch can create particles which then migrate across the wafer and kill surrounding die, causing peripheral damage. The experiment was also designed to determine the mechanism of any particle migration, whether scratches at different steps in the fabrication process have different effects, and if different types of scratches have different effects. The results from this experiment could also be used to further understand and model defect clustering.

EXPERIMENTAL METHODOLOGY

This experiment consisted of a 4x2x2 factorial using 3 repetitions of 16 wafers each, with 2 control wafers in each group. The three variables were:

1) Step in the fabrication process where the scratch was applied
2) Type of stylus used to make the scratch
3) Location on the wafer where the scratch was applied

Electrically testable defect monitor wafers were chosen as the device to apply the scratches to. The results from the end-of-line electrical tests as well as in-line Surfscan 7000 (SS7000) measurements were used as the response in the analyses.

Four places in the process were selected to receive scratches:

1) After Al/Si Deposition/Before Photoresist (PR) Coat
2) After PR Coat/Before Expose
3) After Expose/Before PR Develop
4) After PR Develop/Before Plasma Etch

The experiment focused its attention on the lithography steps, because of a history of problems with equipment causing scratches in that area.

There are two types of scratches predominantly seen in Fab 6, 1) scratches caused by vacuum wands, and 2) scratches caused by process equipment (i.e. oven doors, robotic handlers, etc.). To duplicate these two types of scratches, two different styluses were used. A vacuum wand paddle made of a hard thermal plastic was used to duplicate vacuum wand scratches, and a diamond tipped scribe was used to duplicate the worst case equipment scratch.

Scratches were placed at the center and edge of the wafer to investigate the positional dependency of any particles generated by the scratches (This could be important during resist spin for example).

Electrically testable defect monitor wafers were chosen as the vehicle for this experiment. These monitors detect defects that cause opens/shorts and give their distribution across the wafer [1][2]. The wafer consists of test die that contain Al/Si lines in a serpentine-comb pattern with a pitch of 3.5 microns (Fig. 1). Each wafer has 620 die with each die having 40 individually testable sections. Each die can detect up to 40 individual killer defects at the end-of-line electrical test. Because of their tight geometry and high percentage of wafer coverage, these monitors are much more sensitive than product wafers in detecting defects. Also, they have an advantage in that they only go through a small part of the processing that production wafers do, so that the noise coming from other defects is greatly reduced. This allows a clear signal to be seen from the variables in this experiment.

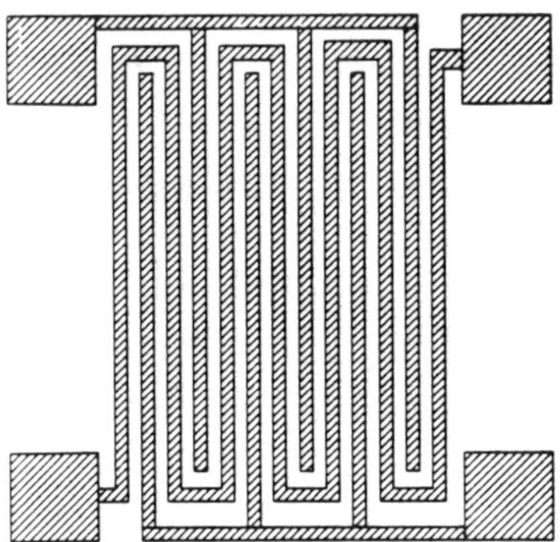

Fig 1. Electrically Testable Serpentine-Comb

Besides the end-of-line electrical tests, Surfscan 7000 measurements were taken in-line on all the wafers at each process step where a scratch was applied. The SS7000 uses a laser to scan and detect particles on patterned wafers. It then prints out a wafer map showing the location of all the particles detected. This allowed the particle distribution after each process step to be determined.

The experiment proceeded according to the flow in Fig 2. At each of the four chosen steps in the process the selected wafers were scratched using one of the styluses. Each wafer received only one scratch (edge or center) during the experiment. All the wafers were then measured on the SS7000. The SS7000 measurements along with the end-of-line electrical tests were used as the response in the analyses.

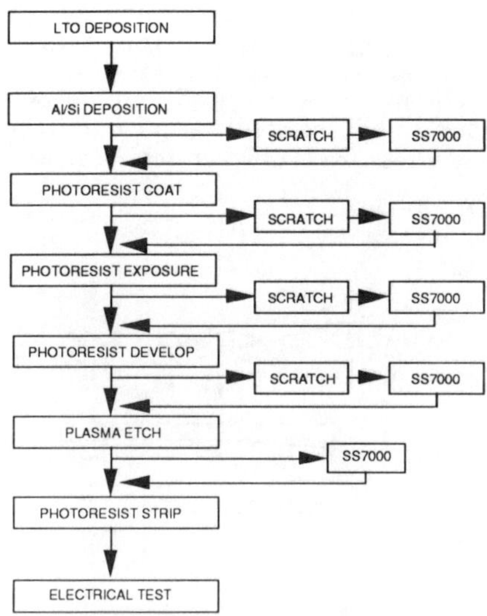

Fig 2. Experiment Flow

The most critical part in this experimental procedure was in the method of applying the scratches. Previous experiments done on scratches have had problems in the analysis of their results because all the applied scratches were different and unique and could not be duplicated. In order to achieve valid results in this experiment, a way to repeatable apply scratches to wafers at any step in the process had to be developed. This was done by using a Kinetek programmable motorized stage with an automatic flat finder, a robotic autoloader, an adjustable spring loaded tool holder, and a stylus to make the scratch. To make a repeatable scratch, the wafers were first placed in the Kinetek stage input cassette and flat found so all the wafers had the same orientation. This was done so the wafers would all make contact with the autoloader at the same wafer location, and therefore all be placed at the same location on the stage. The robotic autoloader then loaded a single wafer onto the stage. The Kinetek stage has an automatic flat finder which then oriented the wafer with the flat towards the operator. This first part of the procedure insured that all the wafers were repeatedly placed and oriented on the stage.

A magnetic base with a rigid arm, at the end of which was a spring mounted tool holder, was placed at the base of the stage (Fig 3). This tool holder is the same type of device that machinists use for dial indicators. The particular stylus to be used (vacuum wand paddle or diamond tipped scribe) was placed in the tool holder. The Kinetek stage, which can be programmed to move to very precise coordinates at very precise speeds, was then driven to its first programmed position. The stylus, attached to the spring mounted tool holder, was then lowered onto the wafer with a specific amount of pressure. The stage was then driven at a set speed, in a straight line, to the next set of programmed coordinates, allowing the stylus to scratch the wafer. The stylus was raised and the wafer unloaded. This setup allowed scratches to be repeatedly applied with respect to position on the wafer, length, and pressure of the scratching instrument.

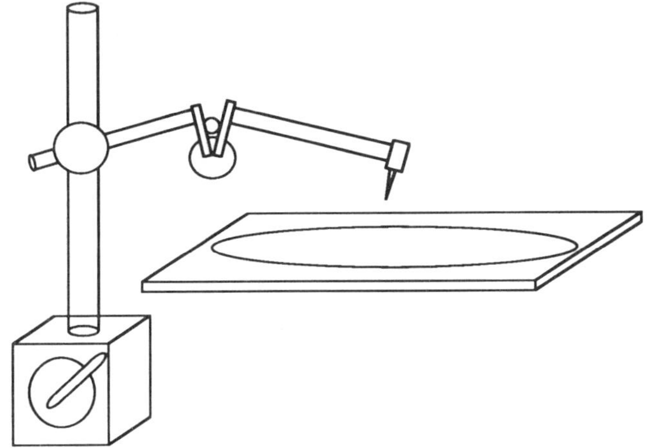

Fig 3. Magnetic Base and Tool Holder with Stylus

RESULTS/DISCUSSION

Scratch Applied After Al/Si Deposition/Before PR Coat

The vacuum wand paddle created no peripheral damage (Damage to the surrounding die from particles generated by the scratch) for either the center or edge scratches. The SS7000 measurements showed a very low level of particulate surrounding the scratch (Fig 4), and did not detect any of these particles migrating across the wafer at any point latter in the process (Fig 5). The end-of-line electrical tests showed that only the die which contained the actual scratch failed, while the surrounding die all came out good (Fig 6).

The diamond tipped scribe created peripheral damage for both the center and edge scratches. The SS7000 measurements taken directly after the scratches were applied showed a high level of particulate directly around the scratch, but none in the surrounding die (Fig 7). The wafers were then put through PR Coat and again measured on the SS7000. These measurements showed a stream of particles going directly from the scratches to the edge of the wafer for both the edge and center scratches (Fig 8). There were no particles going from the scratches towards the center of the wafer. Subsequent SS7000 measurements taken after PR Develop, showed an even higher level of defects going from the scratches to the edge of the wafer (Fig 9). The end-of-line electrical tests show a region of failed die following the particle distribution from the scratches to the edge of the wafer, with ~20 failed die due to peripheral damage and 2 failed die due to the actual scratch (Fig 10).

The vacuum wand paddle scratches gouged the metal surface of the wafer, but did not generate any particles that migrate. The diamond tipped scribe scratches generated a large number of particles which then migrated towards the edge of the wafer during the Photoresist Coat Spin. The particles that did not get pushed over the edge of the wafer were then covered with photoresist. Since the photoresist planarized the surface, many of these underlying particles were not seen by the SS7000. The particles locked in the photoresist then blocked the exposure, causing defects. These defects caused 10 times the damage of the actual scratch. After develop, the exposed photoresist was washed away leaving all the extra/missing patterns created by the particles, which were subsequently detected as an increased level of defects by the SS7000.

Scratch Applied After PR Coat/Before Exposure

Both the vacuum wand paddle and the diamond tipped scribe created peripheral damage for both the center and edge scratches. The SS7000 measurements taken directly after the scratches were applied showed a high level of particulate directly around the scratch, but none in the surrounding die (Fig 11). The SS7000 measurements taken directly after Exposure showed that the particles had migrated into the surrounding die (Fig 12). The particles had a circular distribution 4 to 6 die in diameter with its center located at the scratch. The particles were evenly distributed and showed no directional preference either toward or away from the center of the wafer. The end-of-line electrical tests showed a region of failed die following the pattern of the particle distribution, with ~75 failed die due to peripheral damage, and 2 failed die due to the actual scratch (Fig 13).

The vacuum wand paddle and the diamond tipped scribe both created a high number of photoresist particles due to the resist s soft nature. These particles, which were sitting on the surface, then migrated across the wafer during its passage through the exposure stepper. The scattered particles blocked the exposure and caused extra patterns. These defects caused 30-40 times the damage of the actual scratch. The particles migrated in all directions so the pre-alignment rotation does not appear to be the underlying mechanism. The most likely cause of the particle migration is the stepper focus air probe which uses jets of nitrogen to determine if the wafer is in focus. Experiments to verify this hypothesis are currently ongoing.

Scratch Applied After Expose/Before PR Develop

Neither the vacuum wand paddle nor the diamond tipped scribe created any peripheral damage for either center or edge scratches. The SS7000 measurements taken directly after the scratches were applied show a high level of particulate directly around the scratch, but none in the surrounding die (Fig 14). Subsequent SS7000 measurements did not detect any of the particles migrating across the wafer at any point in the process. However, measurements after develop showed that many of the particles were no longer there (Fig 15). The end-of-line electrical tests showed that only the die which contained the actual scratches failed, while the surrounding die all came out good (Fig 16).

The scratches left particles of exposed and unexposed resist on the surface of the wafer next to the scratch. The exposed resist particles dissolved in the develop process and most of the unexposed resist particles were washed away. There were then few particles left on the surface of the wafer which could cause defects at etch.

Scratch Applied After PR Develop/Before Plasma Etch

Neither the vacuum wand paddle nor the diamond tipped scribe created any peripheral damage for either center or edge scratches. The SS7000 measurements taken directly after the scratches were applied show a high level of particulate directly around the scratches, but none in the surrounding die (Fig 17). Subsequent SS7000 measurements did not detect any of the particles migrating across the wafer at any point in the process (Fig 18). The end-of-line electrical tests showed that only the die which contained the scratches failed, while the surrounding die all came out good (Fig 19).

The particles generated by the scratches never went through a piece of equipment that would cause them to migrate. The particles remained concentrated around the scratch and did not damage the surrounding die.

Findings For Each Of The Three Variables Used In The Experiment

Step where Scratch was Applied - Four steps in the process were chosen to receive scratches in order to determine if they had different susceptibilities to damage. Scratches applied before Exposure caused peripheral damage that was 10 to 40 times greater than the damage the actual scratch caused. None of the scratches applied after exposure caused peripheral damage. Scratches that were applied after the wafer had photoresist on it generated a higher number of particles due to the resist s softness. All of this indicates that scratches occurring before Exposure have a high yield leverage. Also, particular attention should be given to scratches occurring both in the Photoresist Coaters and the Steppers.

Type of Stylus Used - Two types of styluses (vacuum wand paddle, diamond tipped scribe) were used to duplicate scratches commonly seen in the fab. The vacuum wand paddle showed that it did not generate many particles when the wafer was not covered with photoresist. When there was photoresist on the wafer, the vacuum wand paddle scratch generated large amounts of particles due to the resist s softness. The diamond tipped scribe, which was used to duplicate equipment scratches, generated large amounts of particles at every step in the process. This suggests that vacuum wand scratches should receive a lower priority than equipment scratches on non photoresist layers. When there is photoresist on the wafer, any type of scratch can generate a large number of particles.

Location of Scratch on Wafer - Scratches were made on both the edge and center of the wafer, to determine if the differences in particle distribution would give an indication of the type of mechanisms which cause migration. The diamond tipped scribe scratch, applied After Al/Si Deposition/Before PR Coat, created particles which migrated from the scratches to the wafer edge. This indicated that the particles were moved during the photoresist coat spin. The scratches applied After Coat/Before Exposure created particles which migrated in all directions. This indicated that a wafer spin was not responsible for the particle migration and lead to the theory that the stepper focus air probes are responsible.

Table 1 gives a summary of all the results from this experiment.

Table 1

STYLUS	LOCATION ON WAFER	PERIPHERAL DAMAGE (YES/NO)	SS7000 # DIE WITH PARTICLES/ # DIE WITH SCRATCHES	ELECTRICAL TEST # FAILING DIE WITH PARTICLES/ # FAILING DIE WITH SCRATCHES
AFTER Al/Si DEPOSITION/BEFORE PHOTORESIST COAT				
WAND	EDGE	NO	0/2	0/2
	CENTER	NO	0/2	0/2
SCRIBE	EDGE	YES	15/2	15/2
	CENTER	YES	23/5	23/5
AFTER PHOTORESOST COAT/ BEFORE EXPOSURE				
WAND	EDGE	YES	83/2	39/2
	CENTER	YES	172/2	70/2
SCRIBE	EDGE	YES	142/2	56/2
	CENTER	YES	210/2	83/2
AFTER EXPOSURE/ BEFORE PHOTORESIST DEVELOP				
WAND	EDGE	NO	0/2	0/2
	CENTER	NO	0/2	0/2
SCRIBE	EDGE	NO	0/2	0/2
	CENTER	NO	0/2	0/2
AFTER PHOTORESIST DEVELOP/ BEFORE PLASMA ETCH				
WAND	EDGE	NO	0/2	0/2
	CENTER	NO	0/2	0/2
SCRIBE	EDGE	NO	0/2	0/2
	CENTER	NO	0/2	0/2

DEFECT CLUSTERING

Accurate yield models and control charts are dependent on knowledge of the defect distribution across the wafer. Many of the early yield models and most current control charts make the assumption that defects fall in a Poisson distribution. However, it has been widely proved that many defects fall in clusters and therefore violate one of the key assumptions of the Poisson distribution, namely that all individual defects are independent of each other. Yield models created since the 70's have tried to take this defect clustering into account. In developing these models, the authors often used die yield data on product wafers, and made the assumption that die that failed next to each other for the same test, failed for the same defect [4]. This is not always a good assumption, and it also ignores the fact that defects at opposite ends of the wafer may be related. As this experiment clearly shows, particles from a scratch can migrate across a wafer and kill die a large distance from the original defect. According to Maley et al, The accuracy of yield modeling depends on both the accuracy of the yield models and accuracy of defect characterization used by these models [5]. The method of experimentation presented in this paper can be used to give a much clearer picture of the mechanics and characterization of defect clustering for scratches. This , in turn, may help future yield models to better take into account effects due to defect clustering.

CONCLUSIONS

Scratches occurring between Deposition and Photoresist Exposure can create particles which then migrate across the wafer and kill surrounding die. The peripheral damage caused by these particles can be 10 to 40 times greater than the damage caused by the scratch alone. The two mechanisms found that cause particle migration are the Photoresist Coat Spin, and possibly the Focus Air Probes in the Exposure Stepper. Scratches occurring on photoresist covered wafers generate more particulate due to the resist s softness, and equipment scratches cause more particulate than vacuum wand scratches. Because of the greater amount of peripheral damage caused, scratches occurring between Deposition and Photoresist Exposure have a high yield leverage.

ACKNOWLEDGEMENTS

The author wishes to acknowledge the technical advice and support of Christopher W. Teutsch.

REFERENCES

[1]   Teutsch C.W., Drain D.C., Combining Electrical Defect Monitors with Automatic Visual Inspection Systems, in Integrated Circuit Metrology, Inspection, and Process Control , K.M. Monahan, ed., Proc. SPIE 1089, 189 (1989).

[2]   Teutsch C.W., Miller B., Fournier C., Isolation and Characterization of Particle Induced Defects from the Lithography Process Using an Electrical Defect Monitor, in Particles on Surfaces 3: Detection, Adhesion, and Removal , K.L. Mittal, ed., pp173-190, Plenum Press, New York (1191).

[3]   Dean R.L., Implications of particulet contamination in E-beam lithography, in Particles on Surfaces 2: Detection, Adhesion, and Removal , K.L. Mittal, ed., pp253-265, Plenum Press, New York, (1989).

[4]   Maly, Moore, Stojwas, Yield Modeling and Defect Tolerance in VLSI, in Inter. Workshop on Design for Yield , Oxford 1-3 July (1987).

[5]   Frieman D.J., Albin S.L., Clustered Defects in IC fabrication: Impact on Process Control Charts, in IEEE Transactions on Semiconductor Manufacturing , Vol 4. No. 1 Feb (1991).

VACUUM WAND SCRATCH APPLIED
AFTER Al/Si DEPOSITION/BEFORE PR COAT

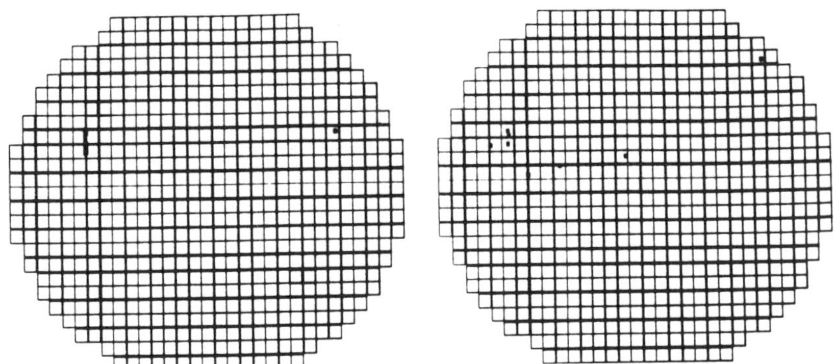

Fig 4. SS7000 Map of Scratch
After Al/Si Dep/Before PR Coat

Fig 5. SS7000 Map of Scratch
After PR Coat/Before Expose

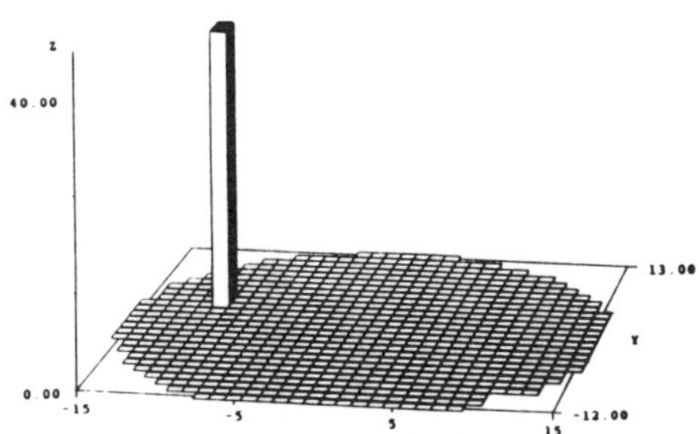

Fig 6. Number of Shorts/Opens Per Die.
Filled in Squares Represent
Location of Scratch.

DIAMOND TIPPED SCRIBE SCRATCH APPLIED AFTER Al/Si DEPOSITION/BEFORE PR COAT

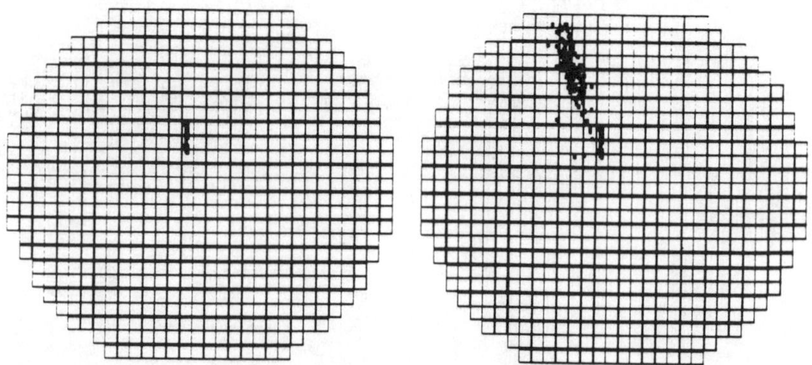

Fig 7. SS7000 Map of Scratch
After Al/Si Dep/Before PR Coat

Fig 8. SS7000 Map of Scratch
After PR Coat/Before Expose

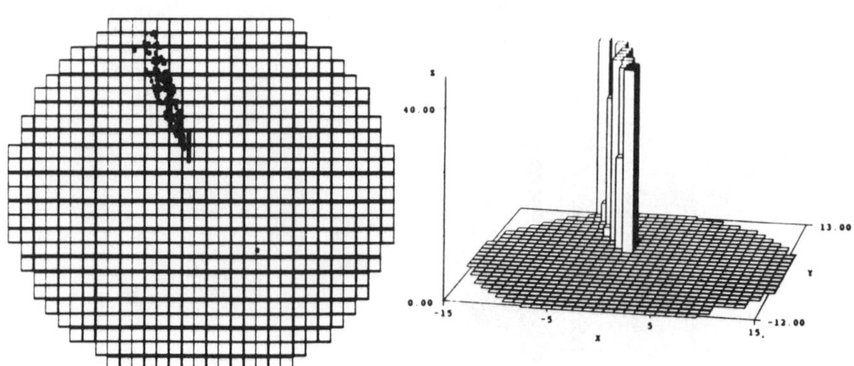

Fig 9. SS7000 Map of Scratch
After PR Develop/Before Etch

Fig 10. Number of Shorts/Opens Per Die. Filled in Squares Represent Location of Scratch

DIAMOND TIPPED SCRIBE SCRATCH APPLIED AFTER PR COAT/BEFORE EXPOSURE

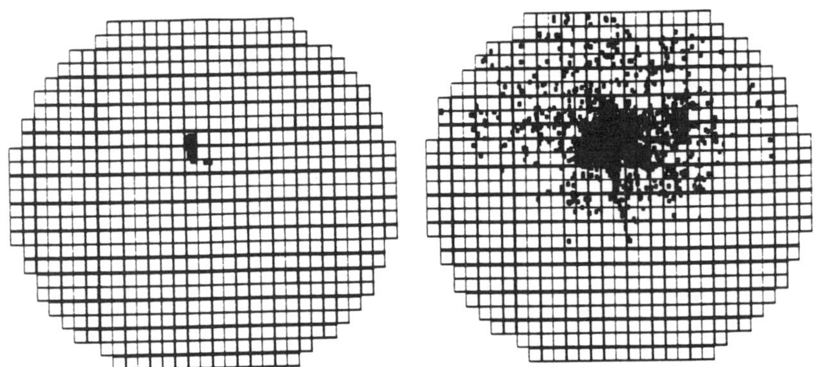

Fig 11. SS7000 Map of Scratch After PR Coat/Before Expose

Fig 12. SS7000 Map of Scratch After Expose/Before PR Develop

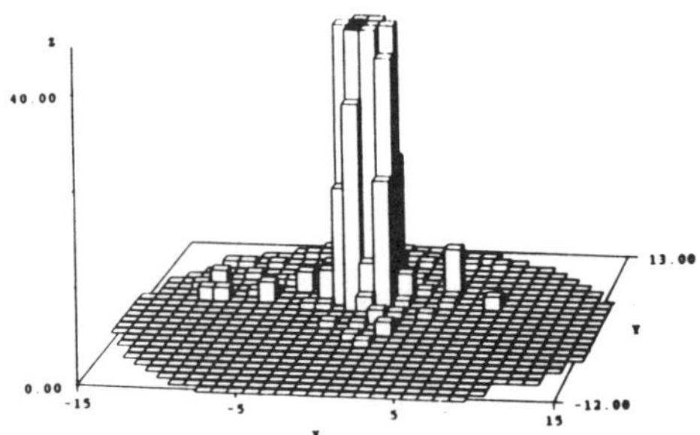

Fig 13. Number of Shorts/Opens Per Die. Filled in Squares Represent Location of Scratch

VACUUM WAND SCRATCH APPLIED AFTER EXPOSURE/BEFORE PR DEVELOP

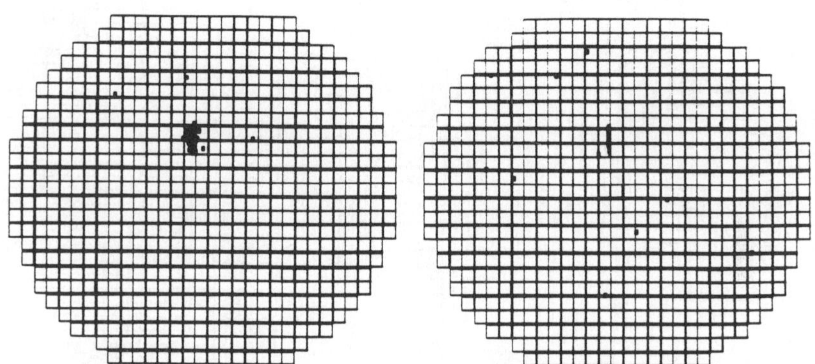

Fig 14. SS7000 Map of Scratch After Expose/Before Develop

Fig 15. SS7000 Map of Scratch After Develop/Before Etch

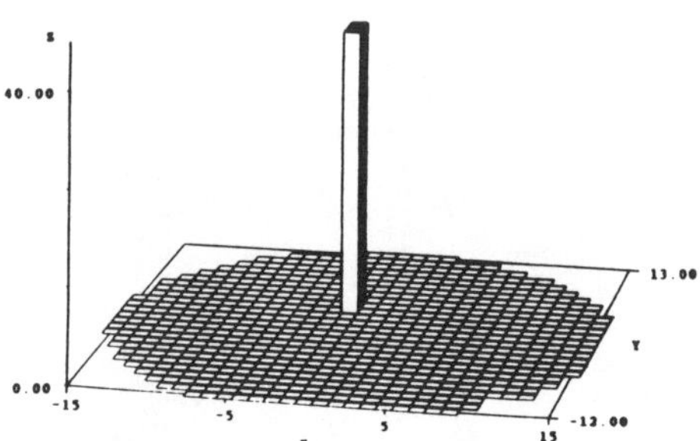

Fig 16. Number of Shorts/Opens Per Die. Filled in Squares Represent Location of Scratch

DIAMOND TIPPED SCRIBE SCRATCH APPLIED AFTER PR DEVELOP/BEFORE PLASMA ETCH

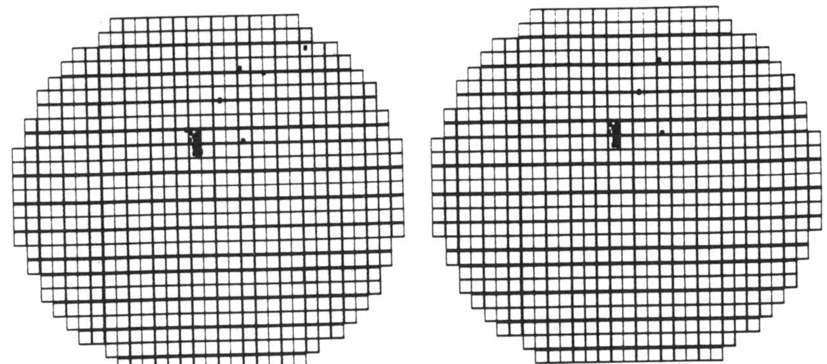

Fig 17. SS7000 Map of Scratch After Develop/Before Etch

Fig 18. SS7000 Map of Scratch After Etch/Before PR Strip

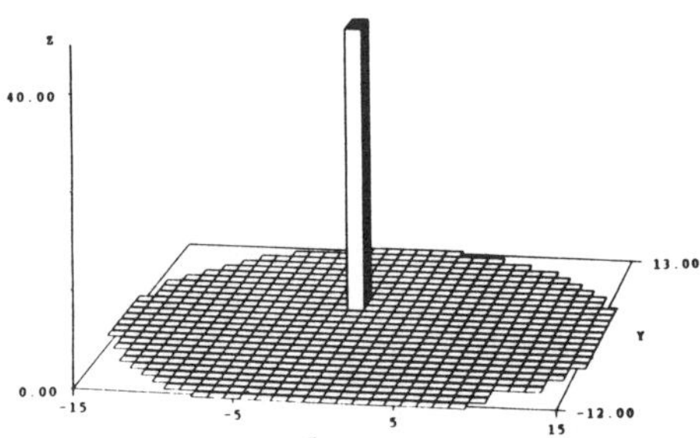

Fig 19. Number of Shorts/Opens Per Die. Filled in Squares Represent Location of Scratch

ADVANCED ULTRA-HIGH-PURITY WATER SYSTEMS FOR SEMICONDUCTOR MANUFACTURING

Glen T. Nickerson, P.E.
MOTOROLA Inc.
Plant Services Department
6501 William Cannon Drive West
Austin, Texas 78735-8598

ABSTRACT

Methods for producing ultra-high-purity deionized water for semiconductor manufacturing have evolved to a higher level of sophistication to stay ahead of increasing device complexity. Contaminants in water must be reduced from the parts per billion levels to parts per trillion levels. This paper describes the major issues and trends in ultrapure water production systems in order to support submicron product manufacturing. Seven major categories are discussed:

* Pretreatment considerations and RO membrane systems
* Degasification and water storage technology
* Ion exchange and regeneration systems
* Filtration and ultrafiltration methods
* Analytical methods and testing programs
* System sanitization considerations
* Component material testing

These methods become significant for the design of ultrapure water (UPW) systems to support the factories of the future. They also apply for upgrading and expanding existing UPW systems.

INTRODUCTION

The two most important factors in construction and operation of an ultrapure water system are the capacity and the quality required. Microprocessors and memory product manufacturing demands that certain water quality parameters must be met. Low particle counts, bacteria, organic and ionic levels, sodium, silica, and oxygen levels must be achieved continuously for successful yields. Each semiconductor device type has a set of critical dimensions (1.0 μ, 0.8μ, 0.5μ, etc) that cannot be violated by D.I. water contaminants. These parameter specification levels are usually available from the process engineers along with quantities of ultrapure water required for each process tool. A study must then be performed to determine the diversity of flow requirements to all tools, so that the water make-up supply system and the distribution loop supply and return network will not be grossly oversized (or undersized). Another factor in the distribution loop design is to be able to maintain sufficient and stable pressure levels, to avoid surges that can produce particle shedding when tools are cycled on/off. A controlled temperature is also desirable.

In the design and construction of a complete ultrapure water system, it is a tremendous challenge to overcome all of the potential areas for contamination. Many purification steps and many materials are utilized to manufacture the water: chemicals, filters, membranes, pumps, storage tanks, linings, piping, control valves, pressure vessels, ion exchange resins, ultraviolet sterilizer units, filter housings, probes, sample valves and tubing. All elements of the system must be scrutinized for their individual contamination potential. Materials of construction must be chosen carefully to reduce the possibility of organic or ionic leaching. An emphasis is made to remove contamination potential as early as possible in the system. Every detail counts when tight water quality parameters are required.

PRETREATMENT CONSIDERATIONS

Water source quality varies with location, some sites have well water, others use treated surface water supplied by a municipality. Most new systems today utilize reverse osmosis membranes in combination with ion exchange resin to supply water into a storage tank. Based on the composition of the source water, usually tested by an independent laboratory, a pretreatment system is designed. A chemical injection program is defined to precondition the feedwater: for example, reducing the pH by injecting sulfuric or hydrochloric acid, reducing the scaling potential by injecting a scale inhibitor, and reducing the chlorine level present.by injecting sodium bisulfite or sulfur dioxide gas. Recent systems have used two-stage reverse osmosis machines for the earliest possible removal of contaminants. At Motorola for example, the level of TOC is reduced by almost two orders of magnitude (from 2.3 ppm to 26 ppb) by the use of a two-stage RO unit (see Figure 1). Silica levels can also be reduced by two orders of magnitude (622ppb to 5.8 ppb) by the use of thin film membranes (refer to Figure 2).

DEGASIFICATION AND WATER STORAGE TECHNOLOGY

Degasification is utilized primarily to remove carbon dioxide and dissolved oxygen from the water, typically through a packed tower by means of a forced draft fan or by using a vacuum pump system. Forced draft degasifiers are used to reduce the elevated CO_2 level formed by pH adjustment previously described. Ambient air is pushed through a prefilter section into a fiberglass tower filled with polypropylene saddles. The saddle shapes act to break up the water as it falls through the tower into a clearwell basin, and the air flow reduces the CO_2 level out through the exhaust. The water is then re-pressurized by a transfer pump to the next treatment step.

Vacuum degasifiers are used similarly for the reduction of dissolved oxygen. For example, the oxygen level can be reduced from the saturation level of 8.38 ppm to less than 260 ppb (see Figure 3). Single stage or two-stage vacuum systems can be used, depending on the specified oxygen removal level desired. There are several newer methods of membrane separation degasification, and also a method to bubble nitrogen into a tower to effectively combine with oxygen for removal (1). These newer methods, however, have been limited to small scale laboratory systems.

Water storage tank materials have been improved as water quality specifications have become more stringent. Standard water storage tanks in a DI water system are fiberglass reinforced plastic material. High purity distribution loop water tanks have been upgraded to include fluoropolymer lining applied to the inside of a fiberglass structural shell. This is typically a 90 mil thickness of thermally welded PVDF sheet material. This inert material is also compatible with the use of ozone that is being injected into some of the newer plant systems for the purpose of bacteria control. In addition, a pressurized nitrogen gas blanket can be introduced into the top dome of the tank to prevent atmospheric contaminants from entering the high purity storage tank.

ION EXCHANGE SYSTEMS

Primary and polishing deionizers have been installed in series to reduce ionic contaminants to trace levels. For example, sodium can be reduced to less than 30 µg/liter (refer to Figure 4). A variety of ion exchange resins are available for use, depending on the conditions of the specific application. Where higher physical stability is desired, such as in an application where resins are transferred or where the potential for oxidative attack by ozonated or chlorinated compounds exist, macroreticular resins might be utilized. In applications where an increase in ion exchange capacity (longer run length between chemical regeneration) is desired, gel type resins might be employed.

Multiple ultraviolet sterilizer units have been utilized in the downstream piping sections of deionizer systems for bacteria reduction, since the resin vessels are known to have a high potential for microbial growth (cool, dark, low velocity, large surface areas. In addition, ultraviolet sterilizers are used upstream of DI tanks for the purpose of ozone destruction to protect the resins from oxidation. At Motorola, ultraviolet sterilizers with 185 nanometer wavelength bulbs are also used to reduce total organic carbon in the polishing loop from 10 mg/l to less than 2 mg/l.

FILTRATION METHODS

Series filtration steps have also been employed, including sub-micron filter cartridges and ultrafilter membranes to reduce particle contaminant levels. Many different filter materials, support materials, and membranes have been utilized to filter ultrapure water to meet the specification levels required (2). It is becoming more common to install 0.1μ or 0.04μ cartridge filters constructed of nylon or PVDF materials. A variety of ultrafilter modules are also available, at ratings as low as 10,000 molecular weight cutoff. Spiral wound and hollow fiber membrane modules constructed of polysulfone or polyamide are among the most popular types used. A careful selection of gasket and o-ring materials is critical to minimize contamination.

COMPONENT MATERIAL TESTING

The construction phase of building an ultrapure water system is subject to a variety of contamination sources, and this must be managed properly as well. Each component is manufactured to a set of specifications at the supplier's shop, but one aspect is often given inappropriate attention: clean manufacturing methods. Some components are not packaged or cleaned prior to delivery, particularly large vessels and pumps, where grease and dirt can become embedded. Machining, welding and lining techniques are critical, as well as final cleaning methods used on the components. It is useful to perform a source inspection of all major components to witness the performance testing and final packaging for shipment. This prevents having to flush and clean up equipment and components at the site, and also prevents the flushing of contaminants downstream into the distribution loop and point of use. Another key to a successful startup is to fabricate all high purity piping in a class 10 clean room with strict protocol measures concerning handling, cleaning, and bagging, before the pipe spool sections are taken into the factory for installation.

Test methods for UPW components are currently being generated by a SEMATECH task force. This endeavor focuses on developing methods to test UPW system components such as valves, piping, fittings, filter cartridges and o-rings that are manufactured for use in the polishing distribution loop. Four areas of test methods being developed include chemical, mechanical, surface analysis, and particles.. The intent is to provide both the manufacturers and the end users with standard methods to evaluate the performance of product A versus product B in order to make better purchasing decisions regarding cost versus quality.

ANALYTICAL METHODS AND TESTING PROGRAMS

Systems are being installed with more on-line instrumentation and better data acquisition tools. This helps in tracking the performance of unit operations and for troubleshooting problems (3). At Motorola, utilizing an on-site laboratory has proven to be beneficial in performing routine quality control checks, troubleshooting the performance of UPW components, and helping reduce response time. Methods are being developed to achieve lower limits of detection for ions, silica, and organics. The laboratory function also serves to provide data for evaluating system efficiency. Employing skilled operations personnel is critical in order to maintain efficiency and stability of the ultrapure water system.

SYSTEM SANITIZATION CONSIDERATIONS

There are a variety of system sterilization methods available (4). A shutdown period is required to sanitize ultrapure water systems in order to control the level of bacterial growth to a manageable level. The frequency of shutdown must be determined based on bacteria test samples taken from strategic areas in the system. Hot water, ozone, hydrogen peroxide, quaternary ammonium and peracetic acid chemicals have been used with varying degrees of success. Three important considerations are the impact of thermal expansion on the piping distribution system, the level of ionic contaminants in the sterilizing solution, and the ability to completely rinse out the residual sterilant chemical before putting the distribution system back on line. Hot water is an effective method, but there are strains of bacteria that can survive above 100 degrees C. Chemical sterilizations such as hydrogen peroxide are common, however there can be as much as 500 ppb of trace metallics present in the supplier's material. Depending on the layout of the distribution piping network, there may be low point traps that allow for chemicals to become trapped. This can result in a slug of chemical releasing when the hydraulics of the loop change after the loop is put back into service. Ozone is perhaps the most effective sterilant since it is quite powerful in low concentrations (25-50 ppb), and it returns to H2O without requiring a flush to drain before going back on line. At Motorola, a two hour ozonation contact period has shown to be successful in controlling bacteria.

SUMMARY

In summary, there are many facets of an ultrapure water system design that can impact the final semiconductor product quality. It is essential to take a global approach and consider methods for reducing each system component's potential for contamination. Yet a wide gap continues to exist between the application of purification steps available and understanding the actual levels of contaminants that cripple manufacturing. Further research is required by process engineering teams to study the impact of these trace level contaminants on submicron manufacturing yields.

ACKNOWLEDGMENTS

I would like to recognize and thank the ultrapure water operations team, the manufacturing team, and the Facilities support team for their inputs and hard work on this project. Special thanks to Bob Patton for preparing the charts.

REFERENCES

(1) H. Sato, N. Hashimoto, T. Shinoda, and K. Takino, "Dissolved Oxygen Removal in Ultrapure Water for Semiconductor Manufacturing", Proceedings of Tenth Annual Semiconductor Pure Water Conference, pp. 147-164, February (1991)`

(2) Kevin Pate, "DI Water Filters-Ozone Compatibility Study", Proceedings of ULTRAPURE WATER EXPO '90-EAST, pp. 137-153, April (1990)

(3) Glen Nickerson, "A Continuous High Purity Water Improvement Program for Submicron Device Manufacturing", Proceedings of ULTRAPURE WATER EXPO '89-WEST, pp. 35-38, November (1989)

(4) Marc Mittelman, "Biological Fouling of Purified Water Systems: Part 3, Treatment", Microcontamination, pp. 30-40, January (1986)

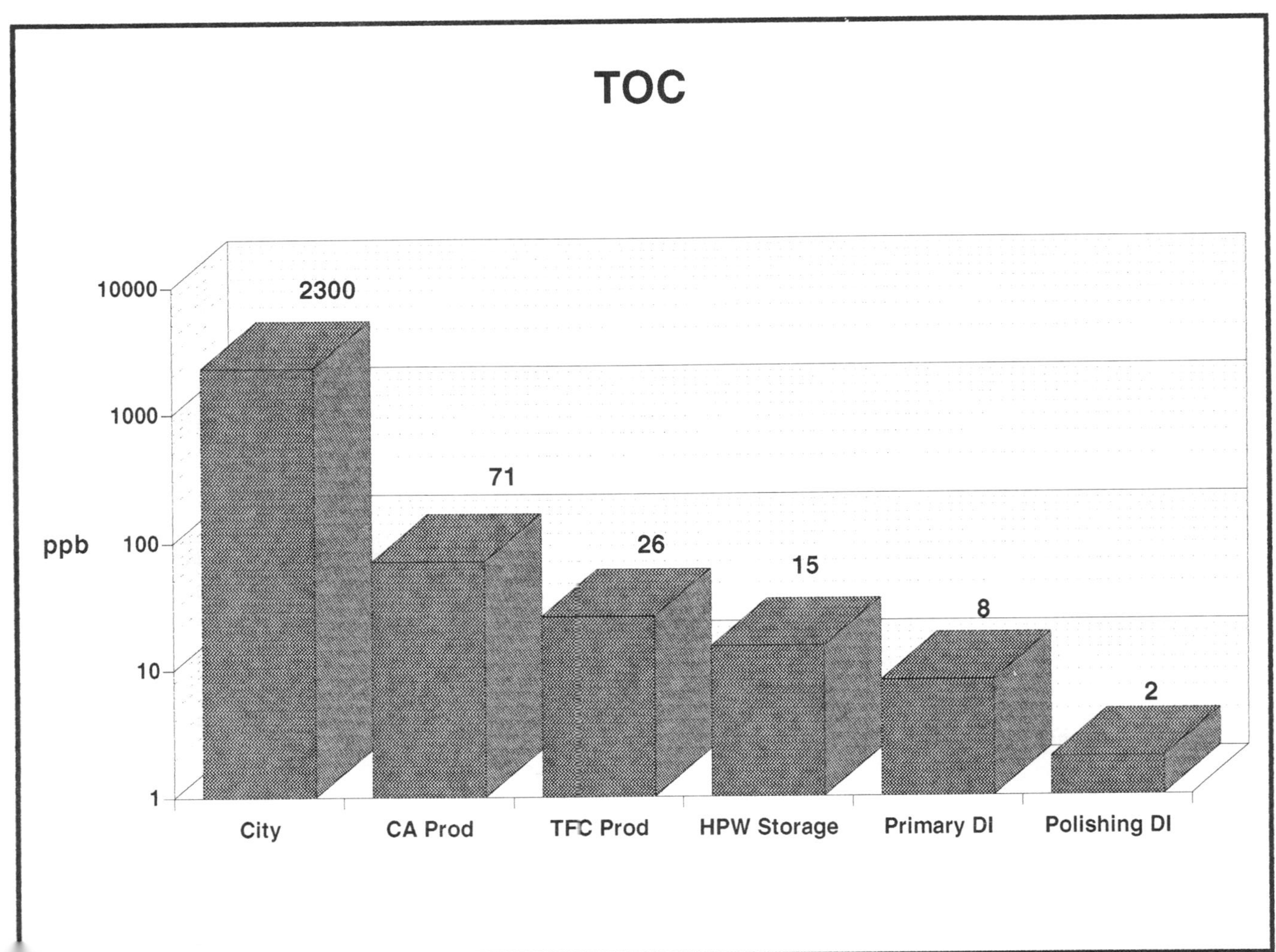

TOC
10000
1000
ppb
100
10
1
2300
71
26
15
8
2
City
CA Prod
TFC Prod
HPW Storage
Primary DI
Polishing DI

SILICA

SILICA
ppb
10000
1000
100
10
1
0.1
7500
622
5.8
2.9
1
0.8
City
CA Prod
TFC Prod
HPW Storage
Primary DI
Polishing DI

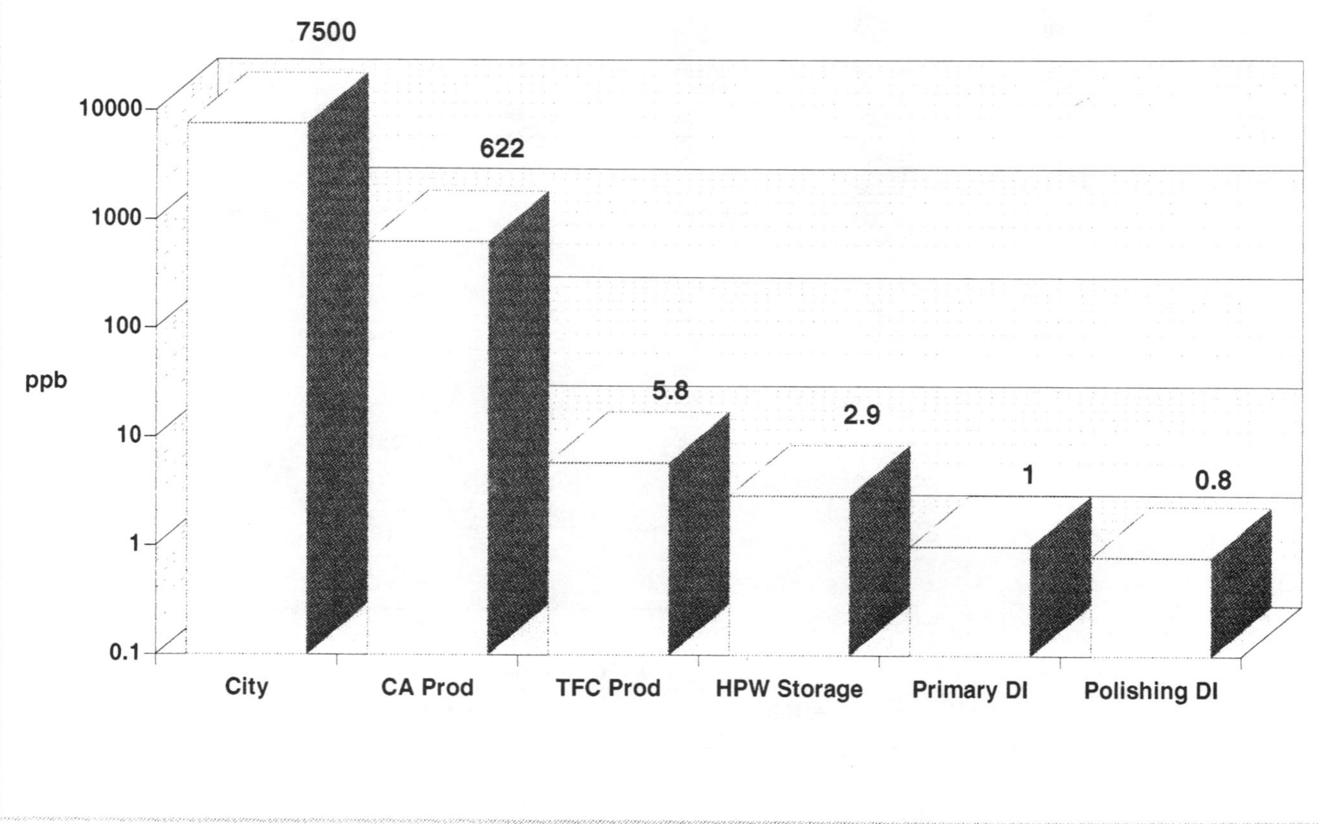

251

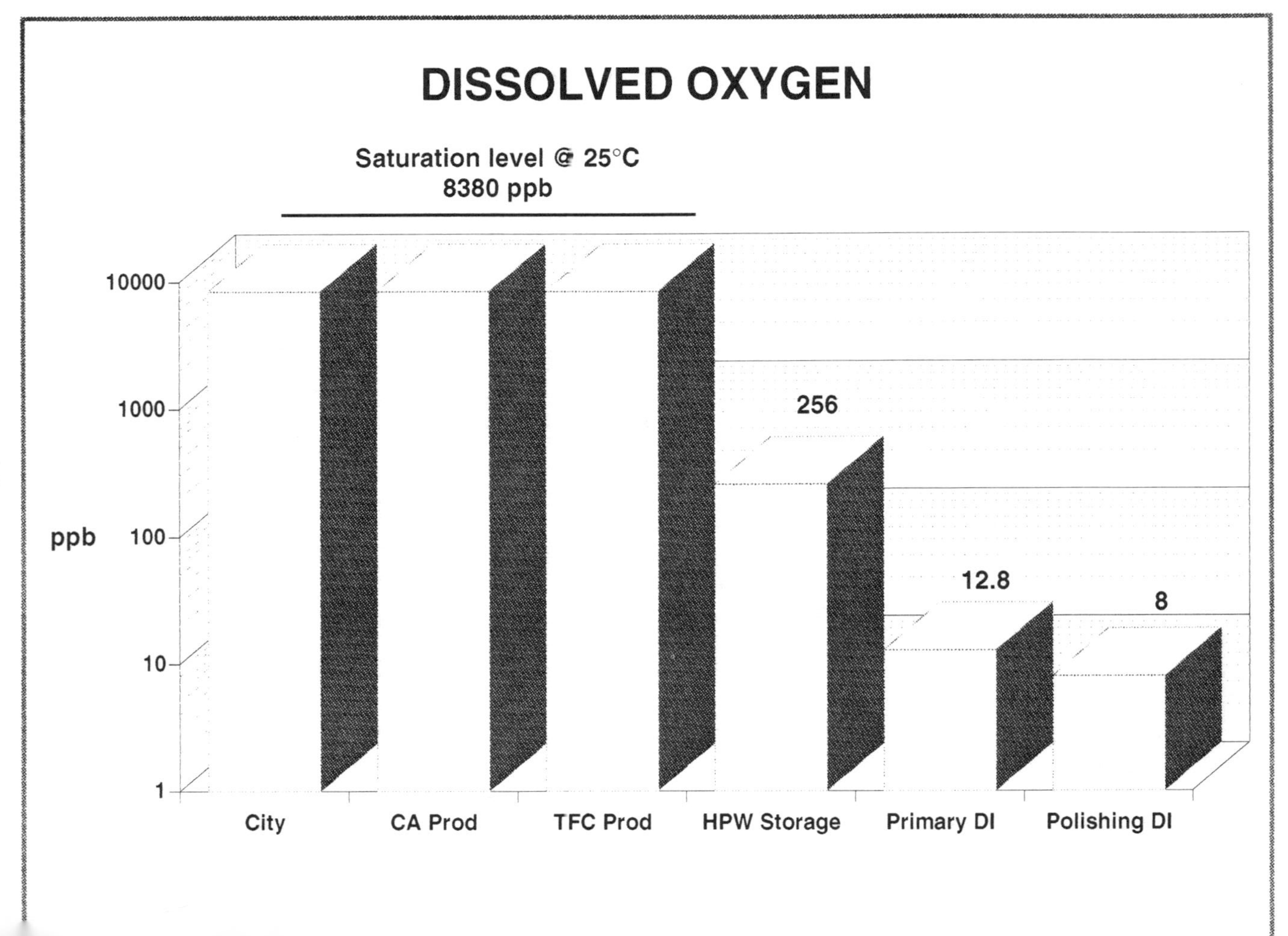

DISSOLVED OXYGEN
Saturation level @ 25°C
8380 ppb
10000
1000
ppb 100
10
1
256
12.8
8
City
CA Prod
TFC Prod
HPW Storage
Primary DI
Polishing DI

SODIUM

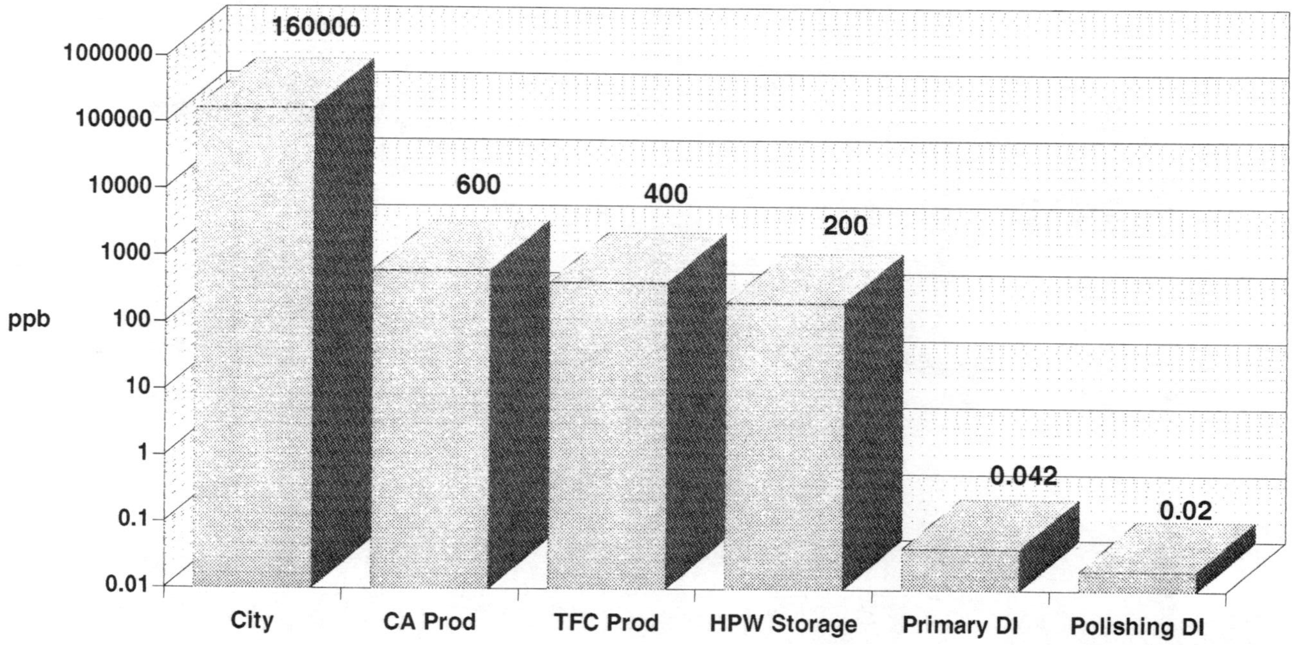

A SYSTEMS APPROACH TO ULTRA HIGH PURITY PROCESS CHEMICALS

Thomas Hackett and Keith Dillenbeck
Ashland Chemical, Inc.
Box 2219
Columbus, Ohio 43216

ABSTRACT

The manufacturing process for integrated circuit devices requires very highly pure and tightly controlled environments that come into contact with the silicon wafer. The complexity of the integrated circuit process has continued to evolve rapidly, as circuit feature sizes fall below the 0.5 μm level and critical silicon dioxide dielectric layers reach below 100Å. During the silicon wafer cleaning process, the purity of the process chemicals, ultrapure water, and wafer drying systems must be carefully optimized. Semiconductor requirements for precess chemical purity have evolved very rapidly relative to ionic and particulate impurity levels. The deleterious effects of impurities within process chemicals are well documented with respect to wafer surface metal deposition, thin SiO_2 dielectric integrity, and silicon surface roughness (1). A systems approach is required in order to control all of the critical parameters relative to the manufacture, distribution and use of high purity process chemicals.

CHEMICAL PURITY ROADMAP

There are about <u>seven</u> process chemicals used to manufacture integrated circuits that are considered critical from a purity standpoint. These critical process chemicals are sulfuric acid, hydrofluoric acid, hydrogen peroxide, ammonium hydroxide, hydrochloric acid, isopropyl alcohol, and buffered oxide etch chemistries. The first five process chemicals are utilized in the RCA wafer cleaning process. The purity of the RCA wafer cleaning chemistries is critical in order to minimize deposition of metallic ions and particles onto the wafer surface during the cleaning process. Some particular impurities have been shown to affect the silicon surface microroughness following the cleaning process. The roughness of the silicon surface has a very dramatic effect on the dielectric integrity of thin SiO_2 capacitors (2). Isopropyl alcohol is a very critical process chemical from a purity standpoint when it is used in the IPA vapor drying process. Impurities in isopropyl alcohol such as moisture, dissolved metallic ions, and non-volatile residue can influence the cleanliness of the wafer surface. Buffered oxide etch chemistries (mixtures of ammonium fluoride and hydrofluoric acid) are used to etch various silicon dioxide layers. Buffered oxide etch chemistries used in submicron integrated circuit processes are also very critical as related to chemical purity.

The purity of semiconductor process chemicals has continued to evolve in order to meet the requirements associated with the increasing complexity of the integrated circuit process. The requirements for advanced chemical purity relate to anion, cation, and particulate impurities at the point where the chemical comes in contact with the wafer surface. These impurities must be measured and certified at the very end of the chemical distribution system, after the chemical has contacted a variety of polymeric surfaces and handling components during the transportation and distribution processes. Table 1 presents a commonly accepted road-map for process chemical purity through the year 1996.

Table 1
Semiconductor Process Chemical Purity Roadmap

YEAR	CATIONS	ANIONS	PARTICLES PER MILLILITER		
			$\geq 0.5\ \mu$m	$\geq 0.2\ \mu$m	$\geq 0.05\ \mu$m
1985	500 ppb	2000 ppb	1000	--	--
1990	10 ppb	500 ppb	100	500	--
1992	1 ppb	100 ppb	20	100	--
1994	100 ppt	50 ppb	10	50	500
1996	10 ppt	1-10 ppb	< 1	10	100

All cation and anion elements are expressed as the individual elemental concentration.

OVERVIEW OF THE CHEMICAL MANUFACTURING PROCESS

The process for manufacturing and distributing very high purity process chemicals involves the integration of several key chemical reaction, purification, and handling steps. Each step of the integrated chemical manufacturing process plays a critical role in increasing or maintaining the purity of the chemical product. This integrated chemical process consists of the following generic steps:

- Chemical Synthesis
- Purification
- Chemical Blending
- Analysis
- Transportation
- Packaging
- Dispensing to the point of use

The chemical synthesis step is the very heart of the integrated chemical manufacturing process. The design of the basic unit operations, selection of the raw materials, and selection of the equipment materials of construction are extremely critical. For example, sulfuric acid is a very critical process chemical used for removing residual organic contaminants and photoresists from the wafer surface. Sulfuric acid is synthesized by burning very pure liquified sulfur in air to form SO_2 gas. The SO_2 is catalytically converted to SO_3 gas using a vanadium pentoxide catalyst. The SO_3 gas is purified via a basic distillation process, then adsorbed into ultrapure water to form 96 weight percent sulfuric acid (H_2SO_4). The detailed design of the unit operation equipment and the selection of the operating conditions can have a significant impact on the chemical purity. The selection of the materials of construction for the gas handling and liquid handling components is extremely critical as well. In the case of sulfuric acid, many of the process components are exposed to 96% sulfuric acid liquid at temperatures in excess of 100°F. It is extremely critical that these process components are constructed of, or lined with a very inert fluoropolymer such as Teflon PTFE. The purity of the ultrapure water used to manufacture sulfuric acid is extremely critical. An ultrapure water system consisting of several stages of deionization, reverse osmosis, and filtration can typically achieve cationic and anionic purity levels well below 0.5 ppb. When all of the individual components and unit operation of the integrated sulfuric acid process are carefully designed and optimized, a high level of chemical purity can be achieved. Table 2 shows the cationic purity levels that can be achieved with an optimized sulfuric acid process. Also shown in Table 2 is the purity of the ultrapure water that is required to manufacture very pure sulfuric acid.

In the detailed design of a chemical manufacturing facility, there are several key elements that significantly influence the purity and consistency of the chemical produced. The design, integration, and operation of these unit operations has changed dramatically in recent years to meet the requirements for semiconductor quality chemical products. It should be noted that many of the criteria for a quality chemical manufacturing and distribution operation are also important criteria for a high purity chemical distribution system located within a semiconductor manufacturing facility. The chemical distribution process can be broken into several major design concepts. These are:

- System Materials of Construction
- Piping Connections
- Chemical Flow Velocity and Pressure
- Chemical Hydraulic Stability
- Design and Materials for Components such as Pumps, Valves, and Regulators

A chemical manufacturing and packaging process designed for sub-one ppb operation must be optimized relative to each of the above concepts.

Table 2
High Purity Sulfuric Acid Process

Parameter	Specification	Typical	Ultrapure Water
Arsenic (As)	1 ppb	< 1 ppb	< 0.5 ppb
Aluminum (Al)	1	0.2	< 0.2
Barium (Ba)	1	0.1	< 0.05
Boron (B)	1	0.6	< 0.15
Cadmium (Cd)	1	< 0.1	< 0.14
Calcium (Ca)	1	0.5	< 0.5
Chromium (Cr)	1	0.1	< 0.2
Cobalt (Co)	1	< 0.05	< 0.1
Copper (Cu)	1	0.6	< 0.2
Gallium (Ga)	1	< 0.05	< 0.1
Germanium (Ge)	1	< 0.1	< 0.2
Gold (Au)	1	< 0.1	< 0.2
Iron (Fe)	1	1.1	< 0.5
Lead (Pb)	1	< 0.1	< 0.1
Lithium (Li)	1	0.3	< 0.1
Magnesium (Mg)	1	0.1	< 0.1
Manganese (Mn)	1	< 0.1	< 0.1
Nickel (Ni)	1	< 0.1	< 0.1
Potassium (K)	1	< 0.1	< 1
Silver (Ag)	1	0.2	< 0.1
Sodium (Na)	1	0.7	< 1
Strontium (Sr)	1	< 0.05	< 0.1
Tin (Sn)	1	0.1	< 0.1
Zinc (Zn)	1	0.2	< 0.5

All cation data are expressed in units of parts per billion (ppb).

PURIFICATION TECHNOLOGY

The integrated manufacturing process for high purity semiconductor process chemicals often includes unit operations for purifying the chemical stream. The purification steps that are typically used for semiconductor process chemicals are filtration, ion exchange, and distillation. Each of these purification steps must always be optimized for the specific chemical being purified. The subject of chemical <u>filtration</u> has been dealt with widely in the literature and hence will not be discussed in this paper. The unit operations of ion exchange and distillation will be discussed in some level of detail in the following sections.

Ion Exchange Technology

The nature of ion exchange materials has been known for a long time. The first observation and published report of the ion exchange phenomenon were of naturally occurring zeolites by H.S. Thompson in 1850 (3). Modern ion exchange technology began in the mid-1940's with the introduction of synthetic polystyrene resins by D'Alelio of General Electric (4). These divinylbenzene crosslinked polystyrene polymers possess exceptional mechanical and chemical stability and greatly improved exchange capacity over previously available materials. Functional groups are attached to the polystyrene resin to convert them to polyelectrolyte compounds. When the polystyrene is sulfonated, a strong cation exchange resin is formed having an SO_3^- group attached to the matrix. A hydrogen ion, the exchangeable ion, is associated with this functional group. The cationic resin will exchange a hydrogen ion for a metallic ion based on natural selectivities and solution concentrations. Anion exchange resins are typically made by chloromethylating the polystyrene resin and reacting the product with an amine. The amine functionality typically has a hydroxide ion as the exchangeable ion.

The major use of ion exchange technology is in ultrapure water treatment. In the past, distilled water was the standard in water purification. Today, the quality of ion exchange water, produced at a fraction of the cost, is as good or better than laboratory distilled water. Ultrapure water may be the most critical chemical in the semiconductor industry. The wafer may come in contact with as much as 500 liters of ultrapure water during manufacture (5). Ultrapure water is blended with other chemicals in standard cleaning operations and used in numerous rinse operations. Contaminants such as cations, organics, microorganisms, and particulates in the ultrapure water have the potential to reduce the useable die on the wafer. Therefore, semiconductor manufacturers exercise extreme care and quality control practices over the production of ultrapure water. Ion exchange units do require maintenance and periodic regeneration of the beds. The specific procedures used for changing and regenerating ion exchange resins can have a profound effect on water purity and consistency.

Ion exchange is only one of a number of unit operations in an ultrapure water system. Other components, such as a reverse osmosis system, may extend the life of the ion exchange bed, but it is only ion exchange that has the ability to remove ions in water to produce ionic quality at the 18.3 Mohm-cm level. Table 3 shows the typical quality of an incoming water supply from a underground aquifer used for process water. The ion exchange process significantly improves the quality from ppm levels to very low ppb levels.

Table 3
The Effect of Ion Exchange Processing on Incoming Process Well Water

Element	Before Ion Exchange	After Ion Exchange
Sodium	4.6 ppm	0.10 ppb
Calcium	39 ppm	0.6 ppb
Magnesium	19 ppm	<0.07 ppb
Chlorides	14 ppm	0.24 ppb
Potassium	1 ppm	0.06 ppb
Total Silicon	7.3 ppm	< 10 ppb
Sulfates	50 ppm	0.32 ppb

Ion exchange is a very versatile technology and is used commercially in many different applications. Ion exchange resins are used as catalysts in chemical reactions, in absorber columns in concentrating precious metals or removing hazardous metals, and in purification of fine chemicals. It is this last category, the purification of chemicals, that is relevant to the semiconductor industry. With the purity roadmap showing the need for 1 ppb process chemicals, ion exchange technology can be used in many instances. An example is the removal of trace metals in hydrochloric acid. In concentrated hydrochloric acid, mercury, platinum, iridium and ferric ions form anionic complexes. A strong basic anionic resin will adsorb these contaminants and elute a purified stream of hydrochloric acid (6). Ion exchange technology is also used to remove metallic impurities from fluorocarbon and hydrocarbon surfactants used in buffered oxide etchants and photoresist developers.

Hydrofluoric acid is a very critical chemical used in semiconductor processing. Since hydrofluoric acid directly contacts the bare silicon surface during a wafer cleaning step, product purity is extremely critical to avoid deposition of ionic and particulate impurities onto the wafer surface. Ion exchange technology has successfully been used to purify dilute hydrofluoric acid solutions (7). Hydrofluoric acid is well suited for ion exchange because HF is a weak acid and ionized H^+ and F^- species are not strongly attracted to the ion exchange resin. However, it is known that cation and anion exchange resins are very rapidly exhausted with full strength 49% HF. The authors have tried unsuccessfully to reduce the metallic content in 49% hydrofluoric acid using an ion exchange process.

Wafer cleaning chemicals are important not only to remove metals and organic contaminants but also particles on the wafer. Since hydrogen peroxide is used in both the RCA SC-1 and SC-2 cleaning steps, its purity is key in overall wafer cleanliness. The presence of metallic impurities within hydrogen peroxide causes the peroxide to catalytically decompose to oxygen and water. This presents serious safety problems. Historically, hydrogen peroxide has been stabilized with stannate compounds which form complexes around metallic ions which cause hydrogen peroxide decomposition. But it is the impurities in hydrogen peroxide which initiate this catalytic decomposition. A very pure and properly handled hydrogen peroxide solution is stable without the addition of any stabilizers. Recent advances in the purity of hydrogen peroxide have reduced metallic impurities to the sub-one ppb levels for individual contaminants. This order-of-magnitude improvement in the purity of hydrogen peroxide is the result of ion exchange technology.

Hydrogen peroxide is similar to ultrapure water in its low degree of ionization and its ability to facilitate the ionization of metallic impurities. These are ideal characteristics for the use of ion exchange to remove ionic impurities in 31 percent H_2O_2. Table 4 shows the purity of hydrogen peroxide before and after treatment with ion exchange resin.

Table 4
Ion Exchange Purification Process for Hydrogen Peroxide

Parameter	Specification	Before Ion Exchange	After Ion Exchange
Arsenic (As)	1 ppb	2 ppb	< 0.05
Antimony (Sb)	1	2	< 0.05
Aluminum (Al), ppb	1	2	0.3
Barium (Ba)	1	2	< 0.05
Boron (B)	1	3	0.4
Cadmium (Cd)	1	1	< 0.05
Calcium (Ca)	1	6	0.3
Chromium (Cr)	1	1	0.25
Cobalt (Co)	1	3	< 0.05
Copper (Cu)	1	2	< 0.05
Gallium (Ga)	1	2	< 0.05
Germanium (Ge)	1	2	< 0.05
Gold (Au)	1	2	< 0.1
Iron (Fe)	1	2	0.2
Lead (Pb)	1	1	< 0.05
Lithium (Li)	1	2	< 0.05
Magnesium (Mg)	1	2	< 0.2
Manganese (Mn)	1	2	< 0.05
Nickel (Ni)	1	1	0.12
Potassium (K)	1	1	< 0.2
Silicon (Si)	1	5	< 2
Silver (Ag)	1	3	< 0.05
Sodium (Na)	1	3	0.4
Tin (Sn)	1	2	< 0.2
Titanium (Ti)	1	1	< 0.05
Zinc (Zn)	1	2	0.15

All cation data are expressed in units of parts per billion (ppb).

Distillation Technology

Distillation is a classical method of purification. Distillation has been used throughout history to enrich alcohol from fermenting mixtures. The basic principle is that the composition of vapors coming from a boiling solution is higher in composition of the more volatile components. Distillation technology came of age with the petroleum and organic chemical industries. Distillation processes are also used to remove ionic impurities from chemical products. In many chemical solutions, cations are generally non-volatile and remain in the liquid phase in the bottom of the distillation column.

The production of semiconductor grade isopropyl alcohol is a prime example of distillative purification. Propylene is contacted with sulfuric acid in an absorber to form propyl sulfate. Water then converts the propyl sulfate to isopropyl alcohol and dilute sulfuric acid, as follows:

1) $$CH_3CH=CH_2 + H_2SO_4 \longrightarrow CH_3CH(OSO_3H)CH_3$$

2) $$CH_3CH(OSO_3H)CH_3 + H_2O \longrightarrow CH_3CH(OH)CH_3 + H_2SO_4$$

A steam distillation process separates the alcohol from the sulfuric acid solution which is recycled to the absorber. The crude IPA enters a second distillation column to concentrate the IPA. Isopropyl alcohol-water forms an azeotrope at 91% IPA and this composition distills overhead. To produce purity in excess of 99.9%, a third distillation column is typically utilized. In this column, a third component is added to entrain the water in a constant boiling ternary mixture. The anhydrous isopropyl alcohol is recovered as the high boiling fraction from the base of the still. Further distillative purification steps are often required to produce ultrapure IPA at the sub-one ppb level. Table 5 presents cationic purity data from several lots of Isopropyl Alcohol that have been purified through a multi-step distillation process.

Although hydrofluoric acid can, in some cases, be purified via ion exchange, distillation is the most robust and efficient purification method. Hydrogen fluoride is produced by the reaction of fluorspar with concentrated sulfuric acid, as follows:

$$CaF_2 + H_2SO_4 \longrightarrow 2HF + CaSO_4$$

The hydrogen fluoride gas is condensed using refrigerated cooling and the resultant liquid typically contains 98.5% HF. The use of distillation increases the concentration to the commercial grade of 99.9% anhydrous HF. This quality of HF is typically not sufficient for high purity applications, and must be purified further to remove metallic species. Typical contaminants in anhydrous HF shown in Table 6 come from the fluorspar ore as well as the materials of construction in the manufacturing process. To produce high quality anhydrous HF which is further absorbed into ultrapure water for semiconductor use requires distillation under controlled conditions using non-contaminating equipment. A major contamination source for high purity anhydrous HF is from entrainment of fine droplets that are carried over into the product condensate. To reduce the effect of mist entrainment, the vapor velocity or rate of distillation can be lowered. By keeping the heat exchanger medium only a few degrees above the boiling point of HF and controlling the evaporation rate based on the available liquid surface area, high purity HF product can be obtained. Another technique to eliminate the entrainment of mist carryover is the placement of high efficiency demister pads in the distillation column. By paying careful attention to eliminate contamination sources, commercial grades of anhydrous HF can be converted to a very high purity form through a relatively simple distillation process, as shown in Table 6. The addition of fluorine gas during the distillation process has proven useful to remove some impurities in the hydrogen fluoride (8).

Table 5

Purity of Several Lots of Isopropyl Alcohol
Using a Distillation Purification Process

Parameter	Lot 1	Lot 2	Lot 3	Lot 4	Lot 5	Lot 6	Average
Al	< 0.3 ppb	0.3	< 0.3	< 0.2	< 0.2	< 0.2	< 0.2
Ca	0.3	0.5	0.2	0.17	< 0.1	0.35	0.27
Cr	< 0.1	< 0.1	< 0.1	< 0.1	< 0.1	< 0.1	< 0.1
Cu	< 0.2	< 0.2	0.4	0.2	< 0.2	< 0.2	< 0.2
Fe	< 0.1	0.2	0.1	< 0.1	< 0.1	< 0.1	< 0.1
K	< 0.06	0.5	0.1	< 0.05	0.06	0.06	0.14
Mg	< 0.05	0.7	< 0.05	< 0.04	< 0.04	< 0.04	< 0.05
Na	0.98	3.0	0.7	0.7	3.2	0.8	1.4
Ni	< 0.3	< 0.3	< 0.3	< 0.2	< 0.2	< 0.2	< 0.2
Pb	< 0.2	< 0.2	< 0.2	< 0.4	< 0.2	< 0.2	< 0.2

All numbers are expressed in units of parts per billion (ppb).

Table 6

Distillation of Anhydrous Hydrogen Fluoride

Element	Before Distillation (ppb)	After Distillation (ppb)
Aluminum	140	3
Barium	6	< 1
Calcium	540	< 1
Chromium	40	< 1
Copper	1000	< 1
Iron	560	3
Magnesium	30	2
Manganese	280	< 1
Nickel	120	< 1
Sodium	300	7
Zinc	70	5

All values are expressed in units of parts per billion (ppb).

Another major area of concern relative to achieving a purity level of 1 part per billion in hydrofluoric acid is secondary contamination. Secondary contamination adds impurities to the product from materials of construction of containers, piping, seals, pumps, valves, and other materials that come in contact with the acid. To control secondary contamination, the selection of materials must be made through the application of experience and extensive testing.

Another technique to achieve very high purity hydrofluoric acid is the distillation of an aqueous solution of hydrogen fluoride. Aqueous solutions of hydrogen fluoride are much safer to handle than anhydrous HF. Anhydrous HF is typically handled in mild steel which adds metal contaminants to the product. Hydrofluoric acid is normally

handled in polyethylene tanks with polypropylene or Teflon process piping. The distillation column is typically made of a fluorinated polymer (e.g., Teflon-PTFE).

The substantial heat of mixing of anhydrous HF and water, 320 Btu/lb anhydrous HF, can be used to help vaporize the 49% HF solution for distillation. This can be used to offset the higher cost of distilling aqueous HF solutions. Table 7 shows the improvement in product quality from the starting material to first distillation. Typically, an order-of-magnitude reduction in impurity levels are seen. To achieve ultrapure quality hydrofluoric acid, a second distillation is performed, as is also shown in Table 7. This second distillation is required because mist from the boiling solution is entrained by the vapor and carried over into the condensate stream. In the subsequent distillation, although this misting cannot totally be avoided, the mist is less contaminating to the final product.

Table 7

Distillation of Aqueous Hydrofluoric Acid

Element	Before Distillation	First Distillation	Second Distillation
Aluminum	70 ppb	19 ppb	<0.1 ppb
Calcium	270	<3	<3
Chromium	20	1.7	<0.1
Copper	500	0.1	0.2
Iron	280	3.9	<2.0
Magnesium	15	1.2	1.1
Manganese	140	4.4	0.2
Nickel	60	5.3	<0.1
Potassium	110	8.1	<3.0
Sodium	150	2.0	0.5
Titanium	No Data	2.4	<1.0

All cation data are expressed in units of parts per billion (ppb).

OPTIMIZATION OF HYDROFLUORIC ACID PURIFICATION

In the beginning months of 1990, the chemical industry in the United States was still struggling to supply process chemicals at the level of purity needed for leading edge semiconductor devices. The objective at that time was to achieve individual cation concentrations of 10 ppb for a variety of chemical products. With current production practices, this level of purity could easily be reached for most metallic contaminants. However, with hydrofluoric acid, the common metals found in the environment such as iron, calcium, sodium, and magnesium frequently failed to meet the 10 ppb concentration target. A higher purity process for hydrofluoric acid was pursued to meet the requirements for the manufacture of megabit semiconductor devices.

Ultra high purity hydrofluoric acid can be obtained in one of two commercial processes. In the first process, anhydrous hydrogen fluoride is purified by distillation and absorbed into ultrapure water as shown in Figure 1. In the second process, aqueous hydrofluoric acid is distilled to obtain high purity product, as shown in Figure 2. The advantage of purifying anhydrous hydrogen fluoride comes from the relative low boiling point of hydrogen fluoride (19.5°C) and subsequently the low energy requirement to distill this material. Techniques to achieve high purity levels are varied, but distillation is the key purification step (8,9). An advantage of the aqueous purification route is that hydrofluoric acid is less hazardous to handle and purification equipment is easier to maintain. Feed to the distillation column is aqueous and purified hydrofluoric acid is distilled overhead while metal impurities are concentrated and removed in the bottoms (see Figure 2).

Figure 1

<u>**ANHYDROUS DISTILLATION SCHEMATIC**</u>

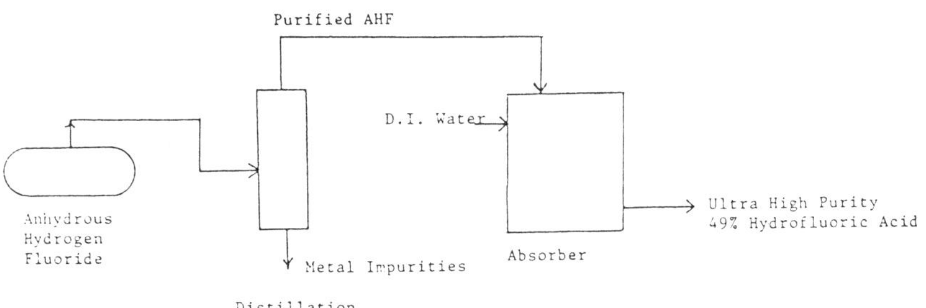

AQUEOUS HYDROFLUORIC ACID PURIFICATION

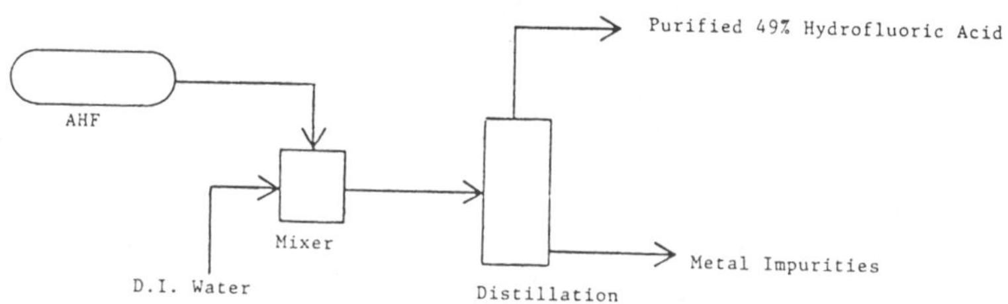

In order to upgrade and optimize an in-house HF purification system, an engineering troubleshooting team was formed. The first step in the optimization process was to determine the independent variables in the process operation that may affect the levels of cationic and anionic impurities in the final product. In a brainstorming session, ideas were organized by using a fishbone type cause-and-effect diagram. In this diagram, general categories influencing the HF quality are drawn as spines attached to a horizontal line leading to our objective of higher hydrofluoric acid purity, as shown Figure 3. These major categories include HF contamination sources such as the environment, human operators, materials of construction, and operating methods. In the brainstorming session, all specific contamination sources were listed under the major categories. For example, if leaks occur in the cooling water system used to control the heat of absorption of anhydrous HF into water, the product will be contaminated. This item is then placed under the major heading of possible contamination from the environment. For the brainstorming session, personnel from various disciplines such as research, manufacturing, maintenance, laboratory, engineering, and management are necessary to ensure the success of the process optimization program.

Figure 3

<u>POSSIBLE INDEPENDENT FACTORS - CATION CONTAMINATION</u>

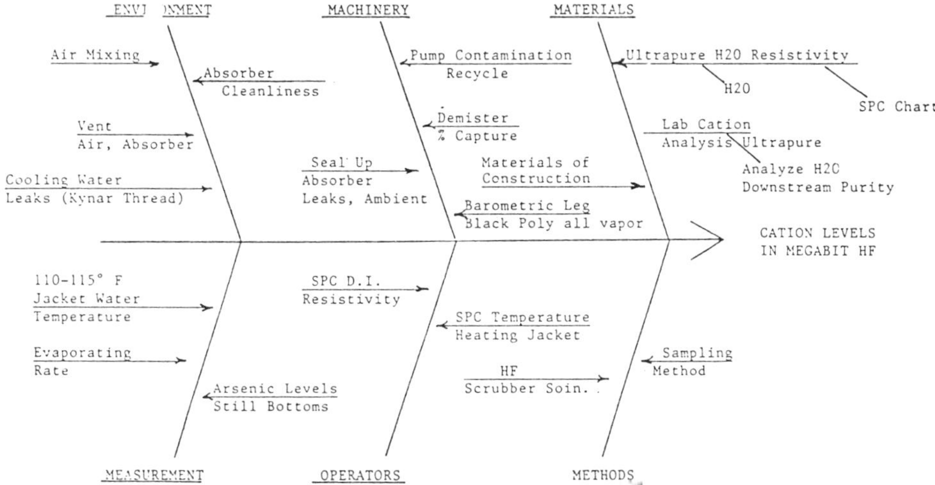

After the fishbone diagram was developed, the possible contamination sources were discussed and prioritized. Action items were assigned both to determine the extent of contamination from various sources and to eliminate known avenues of product contamination. When anhydrous hydrogen fluoride is contacted with ultrapure water, there is an exothermic heat of dilution. The system temperature was controlled by a cooling water system using threaded connections in the HF absorber. The metallic fingerprint of the well water was seen in the hydrofluoric acid, indicating cooling water leaks around the threaded connections. Thorough welding of all the connections eliminated this contamination source. Other environmental contamination sources came from atmospheric air used in agitation within the absorbers, in venting the absorbers, and from leaks around vessel openings for piping and manways. Ultrapure water quality used to absorb the anhydrous HF is also a major factor in the final product purity. The ultrapure water system was upgraded to include a new carbon bed, a change-out of the mixed bed ion exchange resin, and addition of polishing ultrapure beds and reverse osmosis systems. Control charting of ultrapure water quality was instituted on resistivity and cation levels. A significant improvement in aqueous HF purity was experienced as a result of the upgrades made to the ultrapure water system.

268

The materials of construction used in process reactors is critical in the production of high purity chemicals. Catalyst residuals in polyethylene can leach out over a long period of time and contaminate process chemicals. The small molecular structure of hydrofluoric acid allow it to migrate into containment vessels and piping materials, and leach impurities into the acid. Pump seals, shafts and bearings are likely contamination sources. These mechanical parts must be preleached before being put into production. Methods of product sampling and process procedures must also be reviewed to minimize contamination and to achieve ultimate levels of process control. The progress of the system optimization was monitored as each new lot of hydrofluoric acid was produced. Figure 4 shows the progress in both product consistency and purity, for three critical cations (Figures 4a, 4b, and 4c).

Figure 4a

EFFECT OF HF PROCESS IMPROVEMENT - CALCIUM

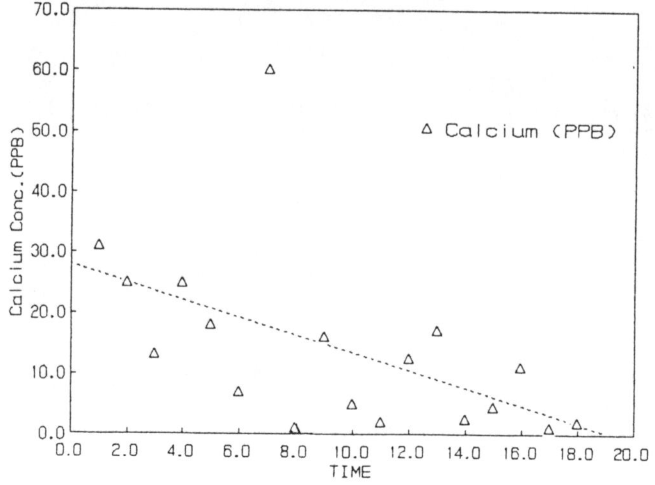

EFFECT OF HF PROCESS IMPROVEMENT - IRON

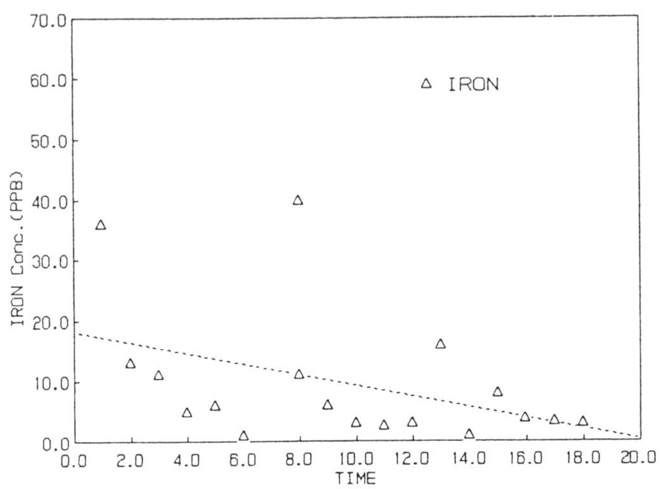

EFFECT OF HF PROCESS IMPROVEMENT - MAGNESIUM

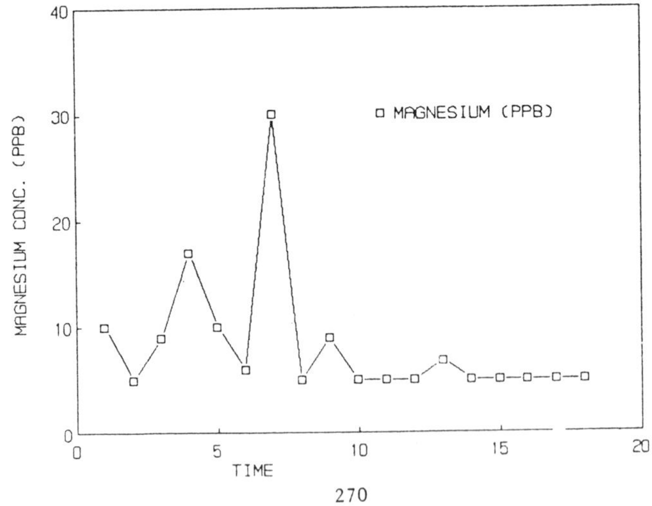

The importance of vapor velocity in HF distillation cannot be overlooked. Very high purity hydrofluoric acid has historically been made in sub-boiling distillation units. Mattinson (10) found that the bursting of bubbles in the distillation process causes tiny droplets. These droplets can be carried over in the distillate which lowers the product purity. The National Bureau of Standards has published its technique for producing ultrapure acids (11). The process of choice was sub-boiling distillation because of two downfalls of fractional distillation. First, still pot liquor can creep up the side walls and eventually contaminate the product. Second, the vigorous boiling process creates small droplet particles that are entrained in the vapor. Although sub-boiling distillation is not amenable to high volume production, the lesson of vapor velocity can be applied to commercial purification of hydrofluoric acid.

In an attempt to optimize the process variables in a distillative aqueous HF purification system, a statistical design was carried out. The feed rate to the distillation column and the reflux ratio (ratio of condensed liquid returned to the column to the distillate product) were varied. At each design condition, the column was operated until steady-state conditions were achieved. Samples were collected and cations were analyzed by ICP-MS and graphite furnace atomic absorption. The concentration of potassium and manganese are the key indicators to examine the effect of process variables. This is because potassium permanganate is typically used to oxidize arsenic, found in the fluorspar raw material, to non-volatile pentavalent compounds (9,12). As a result, the levels of potassium and manganese in the reboiler of the distillation column are high. In typical distillation processes, the reflux ratio affects the degree of separation in each stage. When separating ionic impurities from hydrofluoric acid, theoretical multistage equilibrium analyses are not applicable. Therefore, the effect of reflux ratio must be experimentally determined. In the experimental design, the reflux ratio was varied from zero (no liquid returned to the column), to one (equal amount of distillate and condensed liquid returned to the column). Figure 5 shows that the reflux ratio has very little effect on potassium or manganese levels in the purified hydrofluoric acid product.

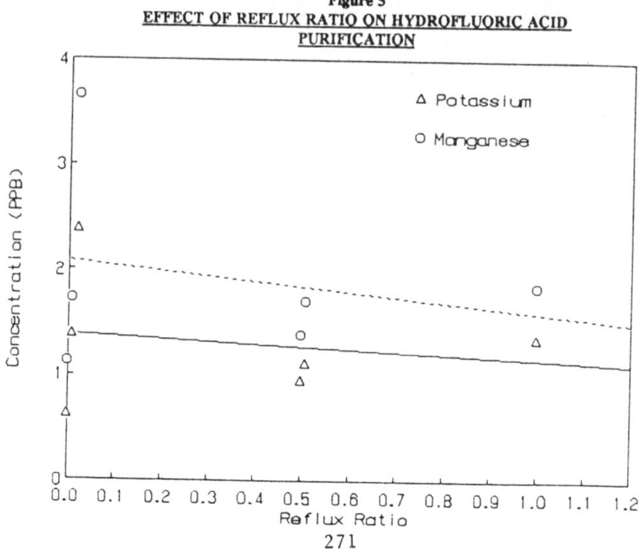

Figure 5
EFFECT OF REFLUX RATIO ON HYDROFLUORIC ACID PURIFICATION

The feed rate to the column was the other independent variable studied in this experimental design. If no bottom product is taken from a distillation system, then at steady state the feed rate is equal to the distillate rate. Figure 6 shows the relationship among the various process streams. In this system the feed rate is directly related to the vapor velocity in the column. As shown in Figure 7, the vapor velocity has a strong effect on the potassium and manganese impurities found in the distillate product. This confirms the lessons learned in sub-boiling distillation and strongly suggests mist entrainment as the mechanism of product contamination.

Figure 6
SCHEMATIC OF PURIFICATION COLUMN SHOWING
RELATIONSHIP BETWEEN VAPOR, LIQUID, AND DISTILLATE RATES

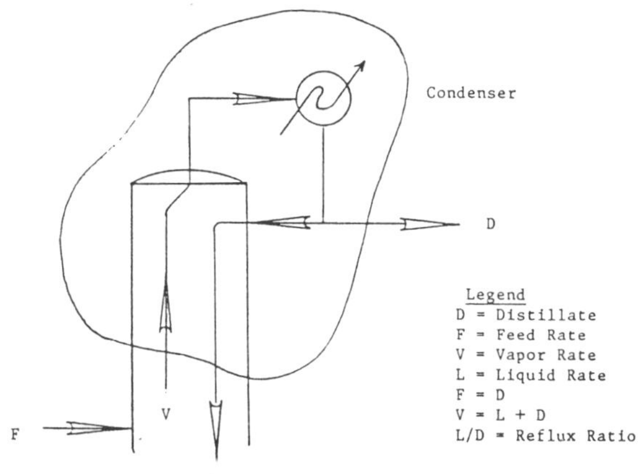

Figure 7
IMPURITY CONCENTRATION AS A FUNCTION OF VAPOR
VELOCITY IN THE DISTILLATON COLUMN

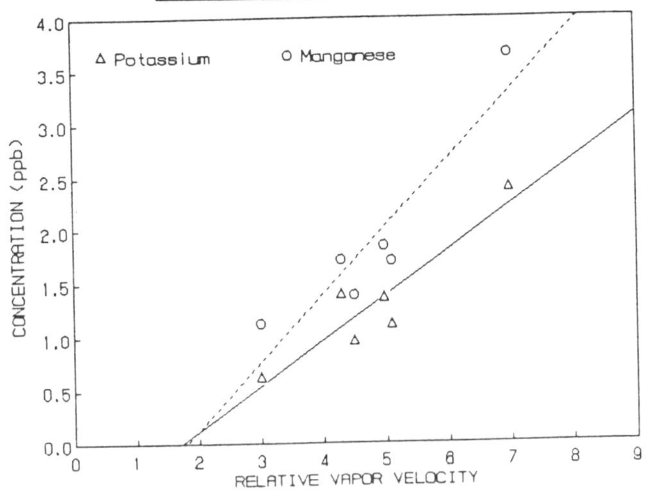

272

PURIFICATION OF AMMONIUM HYDROXIDE

Aqueous ammonium hydroxide (NH_4OH) is a very critical process chemical used in the RCA SC-1 cleaning process. Ammonium hydroxide is produced through the reaction of purified ammonia (NH_3) gas with ultrapure water:

$$NH_{3(g)} + H_2O \rightarrow NH_4OH_{(aq)}$$

Figure 8 shows a simplified schematic of the ammonium hydroxide purification process. In the process, pressurized high purity liquid ammonia is vaporized to form gaseous ammonia. The ammonia vapor is introduced into a Teflon reaction vessel that has been charged with a predetermined amount of ultrapure water. The ammonia gas is absorbed into the ultrapure water through a proprietary sparging device that distributes the gas evenly within the reaction vessel. The reaction vessel is mounted on a high resolution load cell system that automatically controls the flow of reactants to achieve a specific assay level. This ammonium hydroxide is recirculated through a heat exchanger and filtration system to thoroughly mix the chemical product and remove the heat that is generated through the heat of dilution of gaseous ammonia.

Figure 8

HIGH PURITY AMMONIUM HYDROXIDE
MANUFACTURING SYSTEM

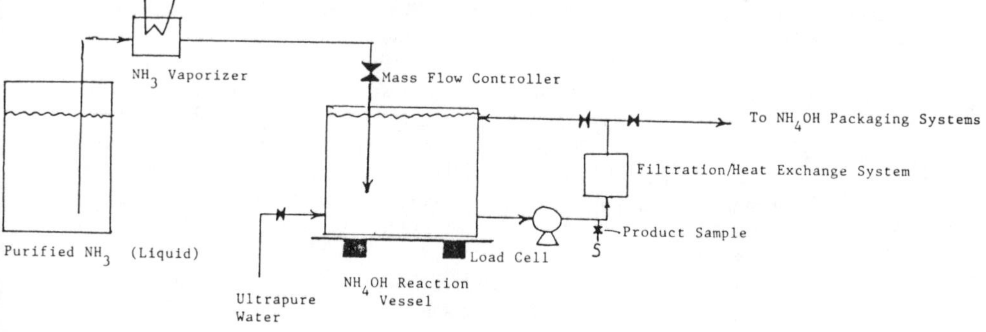

An ammonium hydroxide manufacturing/purification system that is properly designed, fabricated and optimized can produce very high purity product containing less than 0.5 ppb of any individual cationic impurity. There are several critical elements of the ammonium hydroxide process that must be very carefully controlled in order to provide such a high level of purity. These critical process elements are as follows:

- System materials of construction.
- Quality of the ultrapure water system.
- Purity of the liquid ammonia.
- Design of the ammonia gasification system.
- Purity of system components (valves, pumps, etc.).
- System sampling and SPC procedures.
- System preventative maintenance procedures.

Table 8 shows typical purity data from a newly designed ammonium hydroxide purification system. The concentration of all cationic impurities are well below the product specification of one ppb. Also shown in the table is the purity of the ultrapure water used to produce a sub-one ppb purity grade of aqueous ammonium hydroxide.

CHEMICAL PACKAGING TECHNOLOGY

For many high purity chemical products, the **package** represents the most likely source of ionic and particulate contamination. Most containers are not manufactured in cleanroom environments and, when received at the packaging plant, contain particles from many sources. Plastic containers pick up particles from ambient air used in blow molding and in subsequent handling, such as trimming and leak testing. One gallon containers are shipped to packaging plants in cardboard overpacks and some cardboard dust inevitably finds its way into the containers, providing another source of contamination, both chemical and particulate. Any cardboard remaining in the container at the time of fill will cause elevated levels of sodium, calcium, and aluminum. A high purity container manufacturing process avoids contact of the container with cardboard contaminants.

Another factor influencing the cleanliness of a container is static electricity. Plastic containers acquire a static charge during the manufacture of the container. Typical static voltages on HDPE bottles are between 3,000 to 5,000 volts. This electrostatic charge is detrimental in that it increases the collection of ordinary dust, cardboard dust, and polyethylene fine particles in the surrounding air. The electrostatic charge on these containers must be neutralized and the containers washed. Failure to adequately wash containers results in higher particle counts in the chemical (13).

Table 8
High Purity Ammonium Hydroxide - Optimized Process

Parameter	Specification	NH_4OH	Ultrapure Water
Arsenic (As)	1 ppb	< 0.2 ppb	< 0.03 ppb
Aluminum (Al)	1	0.25	< 0.05
Boron (B)	1	0.5	< 0.5
Cadmium (Cd)	1	0.2	< 0.03
Calcium (Ca)	1	< 0.2	< 0.1
Chromium (Cr)	1	< 0.2	< 0.03
Cobalt (Co)	1	< 0.2	< 0.03
Copper (Cu)	1	< 0.2	< 0.03
Gallium (Ga)	1	< 0.2	< 0.03
Germanium (Ge)	1	< 0.2	< 0.03
Gold (Au)	1	< 0.2	< 0.05
Iron (Fe)	1	0.1	0.005
Lithium (Li)	1	< 0.2	< 0.03
Magnesium (Mg)	1	< 0.2	< 0.1
Manganese (Mn)	1	< 0.2	< 0.03
Nickel (Ni)	1	< 0.2	< 0.03
Potassium (K)	1	0.4	< 0.02
Silver (Ag)	1	< 0.2	< 0.03
Sodium (Na)	1	< 2	0.4
Strontium (Sr)	1	< 0.2	< 0.03
Titanium (Ti)	1	0.1	< 0.03
Zinc (Zn)	1	0.2	< 0.1

All data are expressed in units of parts per billion (ppb)

Harder (14) has claimed that plastic containers of perfluoroalkoxy copolymer (PFA), polypropylene, and polyethylene have thousands of particles attached to their surfaces and these particles cannot be removed even by intense cleaning with ultrapure water. Further, these particles shed over time when in contact with semiconductor chemicals. However, this does not have to be the case. By proper container manufacturing practices, one gallon, five gallon, and 55 gallon containers can be manufactured in such a way as to <u>not</u> contribute appreciable levels of particles and ions to the process chemical. The major variable in achieving a non-contaminating container is the air quality of the container manufacturing process.

In a typical blow molding operation, grinders, conveyers, and material handing equipment generate numerous particle sources. Particle counts in excess of one million particles per cubic foot greater than 0.3 microns are not unusual. Container manufacturers are typically not aware of contamination control issues and do not realize the importance of controlling particles in the manufacturing areas. In the blow molding process, molten plastic is extruded into a circular parison. The container molds are clamped around the parison and air is injected to inflate the molten parison forming the container. If compressed air containing a large number of particles is used in this operation, particles in the air will impinge on the inside surface of the container. When the container cools and solidifies, the particles will be attached. These attached particles are difficult to remove by simple washing of the container. Therefore, the air used in blowing containers must be filtered to a high level.

Typical unpurified blow molding air contains between 300,000 and 600,000 particles per cubic foot, ≥ 0.3 μm. Containers blown with such particle-laden air will contaminate semiconductor chemicals by shedding into the chemical over a long period of time. Guaranteeing and predicting particle levels in packaged chemical products has historically been difficult for process chemical suppliers. This has been due in part to poor container manufacturing practices. By controlling the quality of the blow air and the environment around the blow molding operation, particles in packaged semiconductor chemicals can be controlled. It is also important to locate the high purity end of a blow molding machine in at least a Class 10,000 cleanroom in order to keep the containers clean before they are packaged or sealed (15). Plastic containers used in packaging semiconductor chemicals that are produced under cleanroom conditions will mitigate the particle shedding problem (16).

The culmination of packaging technology is the successful filling of containers with high purity semiconductor chemicals with the knowledge that the chemical purity will not degrade with time. That is to say that the container will add neither ionic impurities nor particulates to the process chemical. By optimizing the resin and container manufacturing processes and strictly following cleanroom procedures in packaging high purity chemicals, contamination by the container can be minimized. Table 9 shows the minimal increase in ionic impurities and particulates in 49 percent hydrofluoric acid over a four month period, with three container types. The data in the table demonstrate that one, five, and 55 gallon containers can be fabricated, filled, transported and stored in such a fashion as to minimize the level of impurities contributed by the chemical packaging system.

Table 9

High Purity Chemical Container Technology
Purity of HDPE Containers Filled with 49 % HF

Parameter	One Gallon Bottle		Five Gallon Bottle		55 Gallon Drum	
	1 Wks	4 Mths	1 Wk	4 Mths	1 Wk	4 Mths
Particles ($\geq$0.3 μm)	2	2	3	5	8	15
Aluminum	0.1 ppb	0.5 ppb	< 0.1	< 0.1	0.05	0.08
Calcium	< 0.1	< 0.1	< 0.05	< 0.05	1.2	1.5
Chrome	< 0.1	< 0.1	0.07	0.07	0.03	0.05
Copper	< 0.1	< 0.1	0.02	0.02	0.02	0.02
Iron	< 0.1	0.2	0.05	0.05	0.11	0.14
Nickel	< 0.1	< 0.1	< 0.02	< 0.02	0.03	0.03
Potassium	< 0.1	< 0.1	< 0.1	< 0.1	--	--
Sodium	< 0.1	0.2	0.05	0.07	0.1	0.1
Titanium	< 0.1	0.4	< 0.1	< 0.1	< 0.1	< 0.1
Zinc	< 0.1	< 0.1	0.02	0.08	0.03	0.03

All cation data are expressed in units of parts per billion (ppb).

CHEMICAL MEASUREMENT TECHNOLOGY

As the purity requirements for semiconductor process chemicals continue to evolve, so too must the analytical technology for measuring low levels of impurities. The analytical technology for measuring cationic impurities at the sub-one ppb level can be difficult, time consuming, and expensive. For characterizing chemical processes that produce one ppb chemicals, quantitative analytical methods at the 0.1 to 0.25 ppb range are required. The analytical technology of Inductively Coupled Plasma Mass Spectroscopy (ICP-MS) is effective for measuring most cation impurities at the part per trillion (ppt) level. The ICP-MS technology has been described previously in the literature (17). Some process chemicals produce very significant mass interference signals that make sub-one ppb analysis very difficult. For example, sulfuric acid produces multiple mass interference peaks that make detection at the sub-one ppb level impossible by direct analysis. In sulfuric acid, the aluminum detection limit is 3.1 ppb. By pre-concentrating the sulfuric acid sample in a very clean environment, the detection limit of aluminum can be improved to 0.42 ppb. Table 10 presents typical sulfuric acid detection limit and spike-recovery data for a variety of cationic elements. Shown in the table are the detection limits for a 10% diluted matrix, as well as a 10X concentrated sample. The method spike-recovery values for each cationic element are also listed in the table.

The ICP-MS analytical technique cannot measure low levels of five cationic impurities, regardless of the analytical matrix; this is due to mass interference problems. Calcium, iron, potassium, sodium and zinc require an alternate technique for quantifiable measurement at the sub-one ppb level. Graphite Furnace Atomic Absorption (GFAA) is an analytical technology that is capable of measuring most cationic elements to the ppt level. GFAA technology has operational limitations, in that it detects impurities only one element at a time. Thus, GFAA techniques are very sensitive but are also very time consuming. Table 11 shows the GFAA detection limits and spike recovery levels for the five aforementioned cation elements in 31 percent hydrogen peroxide.

Table 10

ICP-MS Detection Limits and Recovery Values
with Sulfuric Acid

Parameter	Detection Limits Dilute[1]	Concen.[2]	Percent Recovery
Aluminum (Al)	3.1 ppb	0.42 ppb	102 %
Antimony (Sb)	2.5	0.19	112
Arsenic (As)	4.2	0.46	112
Barium (Ba)	1.9	0.32	96
Boron (B)	2.0	0.36	99
Beryllium (Be)	2.0	0.30	99
Bismuth (Bi)	1.7	0.31	94
Cadmium (Cd)	2.9	0.25	98
Chromium (Cr)	2.6	0.74	108
Cobalt (Co)	2.2	0.47	97
Copper (Cu)	4.3	0.56	98
Gallium (Ga)	2.4	0.44	92
Germanium (Ge)	3.1	0.42	98
Gold (Au)	2.7	0.44	79
Lead (Pb)	2.5	0.23	95
Lithium (Li)	2.1	0.36	100
Magnesium (Mg)	3.5	0.52	110
Molybdenum (Mo)	2.9	0.34	100
Manganese (Mn)	2.5	0.36	92
Nickel (Ni)	4.9	0.72	97
Silver (Ag)	4.1	0.41	98
Sodium (Na)	4.2	1.3	120
Strontium (Sr)	3.8	0.26	96
Tantalum (Ta)	1.6	0.22	1006
Tin (Sn)	1.3	0.37	82
Thallium (Th)	2.0	0.34	97
Zirconium (Zr)	2.1	0.53	95

[1]Samples diluted in a 10 % solution prior to analysis.
[2]Samples concentrated ten times in a clean environment prior to analysis.

279

Table 11

**Graphite Furnace Atomic Absorption Analysis
of High Purity Hydrogen Peroxide**

Element	Detection Limit	Percent Recovery at 0.5 PPB Spike
Calcium	0.10 ppb	94 %
Iron	0.10 ppb	108 %
Potassium	0.05 ppb	91 %
Sodium	0.10 ppb	95 %
Zinc	0.05 ppb	101 %

Low levels of anionic impurities in process chemicals are presently measured by colormetric and turbidimetric methods, referenced by the Semiconductor Equipment and Materials Institute (SEMI). These methods are unsatisfactory due to the relatively low sensitivity and the high degree of analytical subjectivity. New analytical technology using high precision ion chromatography is effective for measuring the chloride, nitrate, sulfate, and phosphate ions at the low-ppb level. At the present time, not all process chemical matrices can be measured with ion chromatography, but new analytical in-roads continue to be made in this area. For measuring particles as small as 0.2 microns, high resolution particle counting instruments are utilized. Such instruments have unique laser and optical configurations that enable particle detection to the 0.065 micron level. It is very important that high resolution particle counters be used with sample introduction systems that pressurize the sample prior to analysis. This is necessary in order to avoid detection of small bubbles that arise from typical vacuum sampling systems.

CONCLUSION

A systems approach must be used to produce the ultra high purity process chemicals needed for wafer cleaning and material etching applications for leading-edge semiconductor manufacturing. As integrated circuit process technology becomes more complex, the purity requirements for process chemicals will continue to evolve very rapidly as relates to cation, anionic, particulate and organic impurities. The present requirements for process chemicals are for individual cation impurities to be at or below the one part per billion level. It is very critical that the necessary level of chemical purity be achieved at the point where the chemical contacts the silicon wafer surface. In order to meet the goals for process chemical purity, a systems approach to ultra high purity chemicals is required. Such a program is comprised of:

- Manufacturing Technology
- Purification Technology
- Chemical Packaging Technology
- Chemical Measurement Technology

Meeting future requirements of supplying part per trillion process chemicals to the wafer surface will require advances in each of the above technologies. New purification technologies, such as multi-staged crystallization and improved membrane separations must be explored. Optimization of resins used to package semiconductor chemicals can reduce chemical product contamination from ionic and particulate impurities. Improved resins would also benefit the chemical distribution system which supplies process chemicals to point-of-use. New analytical techniques are required to improve precision and measurement sensitivities. Chemical suppliers are actively pursing these new technologies in a systems approach to meet the needs of the semiconductor industry.

ACKNOWLEDGEMENTS

The authors are indebted to a number of their colleagues from Ashland Chemical who have contributed to the technical work described in this paper. Appreciation is expressed to Morgan Cawthon, Gil Drab, Henry Enenmoh, Rocco Fresoli, Joe Kleinhenz, Tom Marchetto, and Donna Wilkes. The authors would also like to express their appreciation to Mary Hewett for her help in preparing this technical manuscript.

REFERENCES

1. S. Verhaverbeke, M. Meuris, P. Mertens, A. Kelleher, M. Heyns, and R. DeKeersmaecker, "The Effect of Metallic Contamination on Void Formation, Dielectric Breakdown and Hole Trapping in Thermal SiO_2 Layers", presented at the MRS conference, Anaheim, CA, 29 April, 1991.

2. M. Meuris, M. Heyns, W. Kuper, S. Verhaverbeke, "Correlation of Metal Impurity Content of ULSI Chemicals and Defect-Related Breakdown of Gate Oxides", presented at Electrochemical Society Spring Meeting, Washington, D.C., May, 1991.

3. Thompson, H., J. Roy. Agr. Soc. Engl. 11, 68, 1850.

4. D'Alelio, G., U.S. Patent 2,366,007, December 26, 1944.

5. Accomazzo, M., G. Ganzi, and R. Kaiser, "Deionized (DI) Water Filtration Technology", in Handbook of Contamination Control in Microelectronics, Edited by Donald Tolliver, Noyes Publications, 1988.

6. Rohm and Haas Ion Exchange Resins Laboratory Guide, Publication IE-85b, September 1990.

7. Davison, J. and C. Hsu, "The Ultrapurification of Dilute Hydrofluoric Acid by Ion Exchange", 10th International Symposium on Contamination Control, Zurich, Sept. 10-14, 1990.

8. Miki, N., "Process for Purifying Hydrogen Fluoride", U.S. Patent 4,668,497, May 26, 1987.

9. US Patent 4,032,621, "Preparation of Hydrogen Fluoride with Low Levels of Arsenic, Iron, and Sulfite", June 28, 1977, issued to DuPont.

10. Mattinson, J., "Preparation of Hydrofluoric, Hydrochloric, and Nitric Acids at Ultralow Lead Levels", Analytical Chemistry, Vol.44, 1715, (1972).

11. Moody, J., et al., "The Preparation of Ultrapure Acids from Small to Large Scale", 31st International Congress of Pure and Applied Chemistry", Sofia, Bulgaria,July 13-18,1987.

12. US Patent 3,166,379, "Purification of Hydrofluoric Acid", Jan. 19, 1965, issued to UCC.

13.   Personal Communication, N. Miki, Hashimoto Chemical, March 30, 1992.

14.   Harder, N., and M. Wilson, "Chemicals and Supply Systems for ULSI Devices", 9th ICCCS Proceedings 1988.

15.   Hackett, T., and K. Dillenbeck, "Understanding the Particle Shedding Phenomena in Polyethylene Containers for Semiconductor Process Chemicals", Microcontamination 91, San Jose, Oct. 16-18, 1991.

16.   Harder, N., "High Purity Chemical Production: Meeting the Demands of ULSI", Solid State Technology, October, 1990.

17.   S. Tan, T. Chu, and M. Balazs, "Determination of Parts per Trillion Levels of Trace Metals in Bottled and Point-of-Use IC Processing Chemicals by ICP-MS", Proceedings of the SPWCC, Santa Clara, February, 1992.

INTEGRATED APPROACH TO THE REMOVAL OF IMPURITIES FROM GASES AND WATER IN THE MICROELECTRONICS INDUSTRY

Asad M. Haider
Ce Ma
Farhang Shadman

Chemical Engineering Department
University of Arizona
Tucson, Arizona 85721 USA

In ultra-pure systems, the processes for the removal of particles and those for the removal of homogeneous impurities are interdependent and should be integrated. Integrated techniques for control, metrology and analysis of impurity distribution in gas delivery systems are discussed. The interactions of impurity transport mechanisms including convection, dispersion, adsorption/desorption, back-diffusion and permeation are analyzed. A similar approach to the control of organic impurities in water is investigated.

INTRODUCTION

Removal of impurities from gas and water in semiconductor manufacturing involves a number of process steps in series. Interactions of these processes and optimum integration of various steps are important for effective impurity control. This paper summarized research results on examples of these interactions and the advantage of the system approach over the component approach for control of impurities in gases and water.

Impurities in fluids are classified into homogenous (dissolved) and heterogeneous (particles). Removal of particles is accomplished by filtration and that of homogeneous impurities by a variety of purification processes. The technology of filtration is more mature than that of purification and very effective filters for particle removal are available. For purification of gases to low and sub-ppb levels, commercial purifiers consisting of packed beds of getter materials are available [1]. The getters are typically reactive organo-metallic compounds deposited on porous particles or metal alloy getters capable of reacting with impurities. Formation of secondary impurities in purifiers is a problem, particularly in resin-based purifiers [2]. In ultra-pure water systems, particle are removed by filtration and the homogenous impurities are removed by a variety of techniques [3,4]. This study focuses on the organic impurities, primarily removed by filtration, various membrane processes, ion exchange and oxidation. Very little is done to correlate particle removal with the removal of homogeneous impurities and to integrate the two processes.

Removal of impurities at low levels cannot be effectively accomplished by physical processes. Highly irreversible chemical reactions are needed to bring impurity levels down to low and sub-ppb levels. Since such reactions require very effective fluid solid contact and large area, the depth filters have the potential of providing the porosity and intraphase area needed for both filtration and gettering reactions [5]. This presents opportunities for improving the reaction efficiency by using highly porous, low pressure-drop filtration substrates as supports for gettering processes (trapping the impurity completely) or catalytic conversion (converting an impurity into another form which is more easily removable).

An important part of any study on the impurities, particularly in the gas phase, is characterization of their distribution and transport from one point to another [6,7]. These processes make both metrology and control more complicated and should be accounted for in any purification/filtration strategy. Direct measurement of impurity levels is not always possible or practical. In fact, usually the most important impurity concentrations (e.g. concentration at the point of wafer) are also the least measurable ones. Therefore, we depend on predictive techniques to complement measurements for impurity characterization. An objective of this paper is to give examples of a technique for characterizing the primary mode of impurity transport in gases. Similar techniques can be applied to impurities in water.

EXPERIMENTAL SETUP PROCEDURES

The experimental setup for the study of impurity distribution and purification in ultra pure gases consists of a cryogenic gas source, a challenge gas delivery system which can introduce impurities at known levels and known schedules into the test section as schematically shown in Figure 1. The analytical section consists of dedicated oxygen and moisture analyzers, a standard electron impact mass spectrometer with electron impact source and an atmospheric pressure ionization mass spectrometer (APIMS) with sub-ppb sensitivity, and a calibration gas mixing and sample introduction unit.

A reactive filter was developed which consists of a porous high efficiency depth filter made of either ceramic or stainless steel. The gettering sites consisting of reactive metals were deposited on the entire intraphase areas of these filters. The schematic diagram of a simplified version of these reactive filters is shown in Figure 2.

The challenge testing of these purifiers was carried out using two schedules.

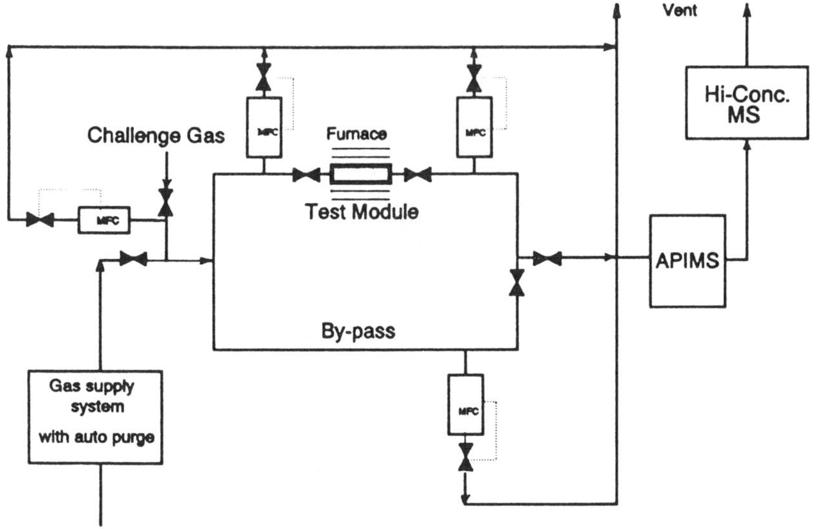

Figure 1. Schematic Diagram of the Test Unit

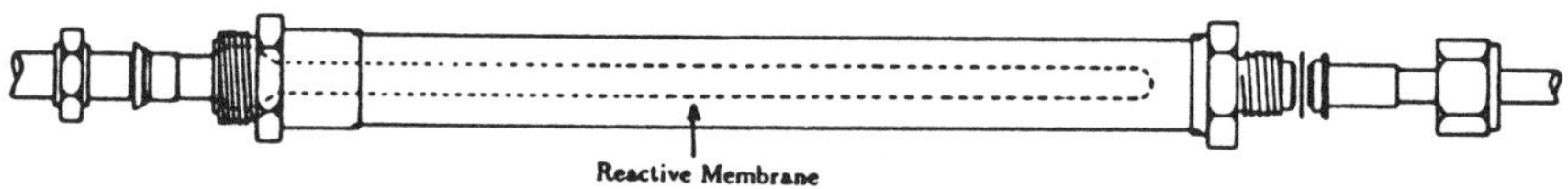

Figure 2. Schematic Diagram of a Single Channel Reactive Filter

286

In the first challenge schedule, impurities were introduced by switching the inlet from ultra-pure gas to a challenge gas containing a known level of those impurities. The second schedule was to introduce a pulse of impurity, representing a surge. In either case, the output was monitored using either APIMS, dedicated analyzers or both.

The tests involving impurity in water were conducted in an ultra-pure water pilot unit consisting of a primary and a polishing loop. The polishing loop had UV, ion exchange and ozonation capabilities; it also allowed installation of filters at various locations. The focus of this study was on the oxidation of known organic impurities (model compounds) using UV, ozone, and UV/ozone combination.

DISCUSSION OF RESULTS

Figures 3 and 4 show the oxygen and moisture removal in pulse impurity injection experiments; dedicated oxygen and moisture analyzers were used in these experiments. Figures 5-6 show the purification results when oxygen and moisture impurities were introduced at a steady challenge level; APIMS was used for trace analysis. Some impurities like hydrocarbons and CO_2 are more difficult to remove. However, in all cases excellent purification down to very low and sub-ppb levels was obtained.

Typically, filters remove particles but due to their high surface area, can adsorb and desorb homogeneous impurities. Packed-bed purifiers remove impurities, but due to various effects, including attrition of the packing, generate particles. The reactive filter is a successful attempt to solve both problems. Moreover, since the active sites are distributed on the support, the heat of reaction is rapidly and effectively dissipated. This prevents hot spot formation which is a major source of secondary impurity generation in the purifiers. Another example of this secondary impurity formation is the evolution of hydrocarbons from the resin-based purifiers [2].

In ultra-pure gas systems, the optimum location of purifiers, and the impurity levels at points of interest, can only be determined by characterizing the transport and transformation of impurities. The impurities in a gas distribution system undergo a number of processes. The most important processes are impurity transport by convection and/or dispersion, adsorption, desorption, back-diffusion and permeation. Presence of these processes complicates impurity monitoring. For example, a typical set up for component evaluation is shown in Figure 7. Because of the above processes, the concentration actually measured, C_M, is not the same as the concentration we are trying to measure, C_I.

To overcome these complications and interaction problems, two approaches can be used: the first technique is to develop a metrology tool that can measure impurity concentrations in situ at any point in the system. This is difficult, if not impossible.

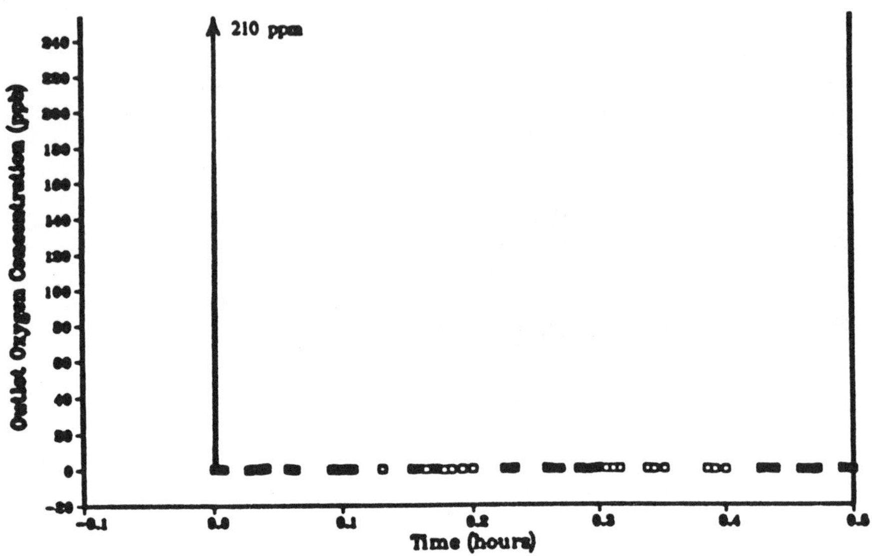

Figure 3. Oxygen Removal by Reactive Filter (Pulse Test)

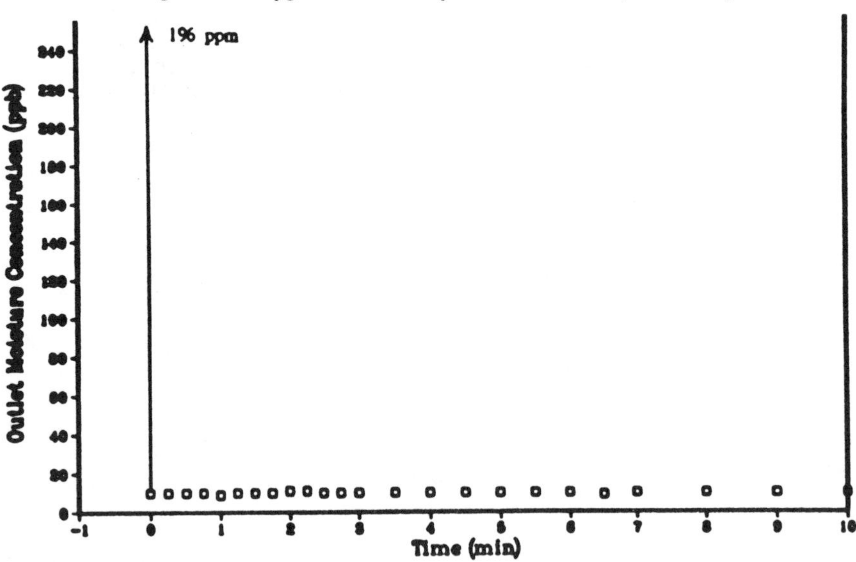

Figure 4. Moisture Removal by Reactive Filter (Pulse Test)

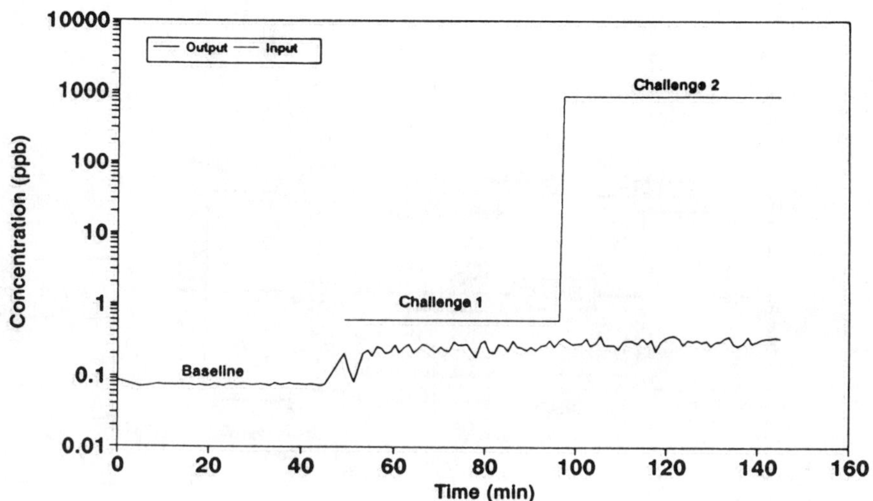

Figure 5. Oxygen Removal by Reactive Filter (Steady Test)

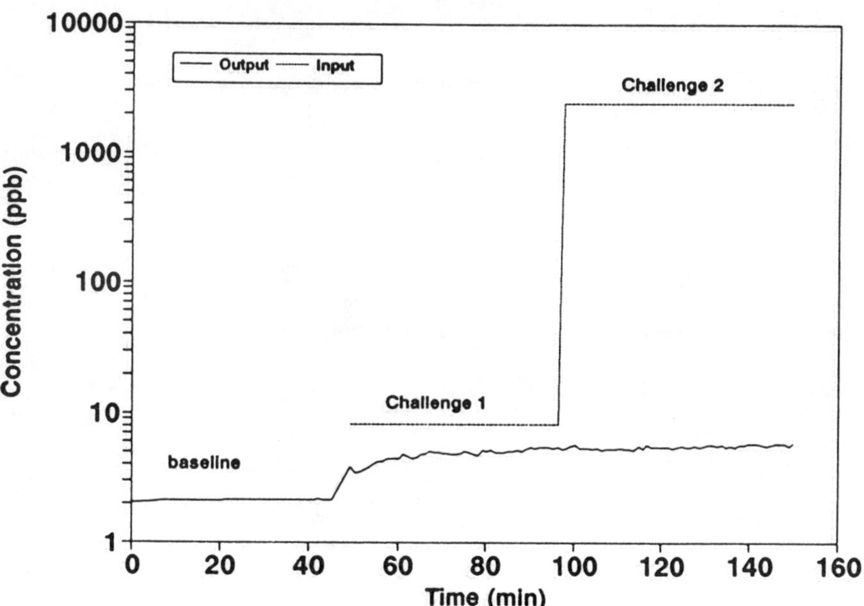

Figure 6. Moisture Removal by Reactive Filter (Steady Test)

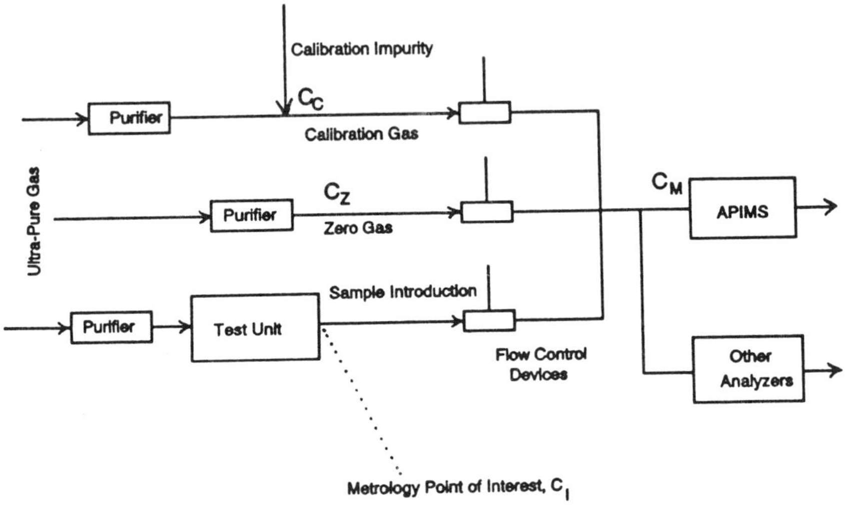

Figure 7. Schematic Diagram of a Typical Trace Analysis Setup

The second method is to develop a technique which allows prediction of impurity concentration at any point if measurements at another point are available. The basic elements of this simulation method are the equations shown below:

$$\frac{\partial C_g}{\partial t} = -U \frac{\partial C_g}{\partial z} + D_z \frac{\partial^2 C_g}{\partial z^2} + S (r_a - r_d)$$

$$\frac{\partial C_s}{\partial t} = r_a - r_d$$

The boundary and the initial conditions are:

$$UC_{g_o} - UC_g - D_z \frac{dC_g}{dz} \qquad at \qquad z-0$$

$$\frac{dc_g}{dz} - 0 \qquad at \qquad z-L$$

$$C_s - C_{so} \qquad at \qquad t-0$$

$$C_g - C_{g_o} \qquad at \qquad t-0$$

Application of the above simulation technique allows various calculations. Following are some examples of utilization of this technique:

1. *Calculation of impurity profiles at various points in a gas distribution system*: This is possible if the impurity profile at another point is known and the flow as well the system geometry are defined. An example of the application is shown in Figure 8 . In this example, impurity is introduced at a certain point as a pulse. The impurity profile at various distances downstream of impurity introduction point are shown.

2. *Development of purge and bake schedules for outgassing and purging:* An example of such results is shown in Figure 9. This graph shows the time needed to achieve certain extent of outgassing (cleaning) in a stainless steel tube. Derivation of such schedules with experiments alone would be very difficult and time consuming. An important finding is that in desorption from most surfaces, the mechanism of outgassing changes significantly when the impurity concentrations drop to very low levels.

3. *Characterization of impurity back-diffusion into a system and the dependence of back-diffusion on the flow rate and system geometry:* Back-diffusion is a critical problem in gas delivery systems. This is particularly important at locations where two different gases are mixed or at the entrance to a tool.

Interaction of particles and homogenous impurities is also important in ultrapure water systems. Traditionally, the design and analysis of ultrapure water systems have focused on the selection and analysis of individual components and processes and not on the interactions between these process steps. In this study, the interaction of dissolved and particulate organic impurities is studied. The importance of combining the particle data with Total Oxidizable Carbon (TOC) data for organic compounds is demonstrated.

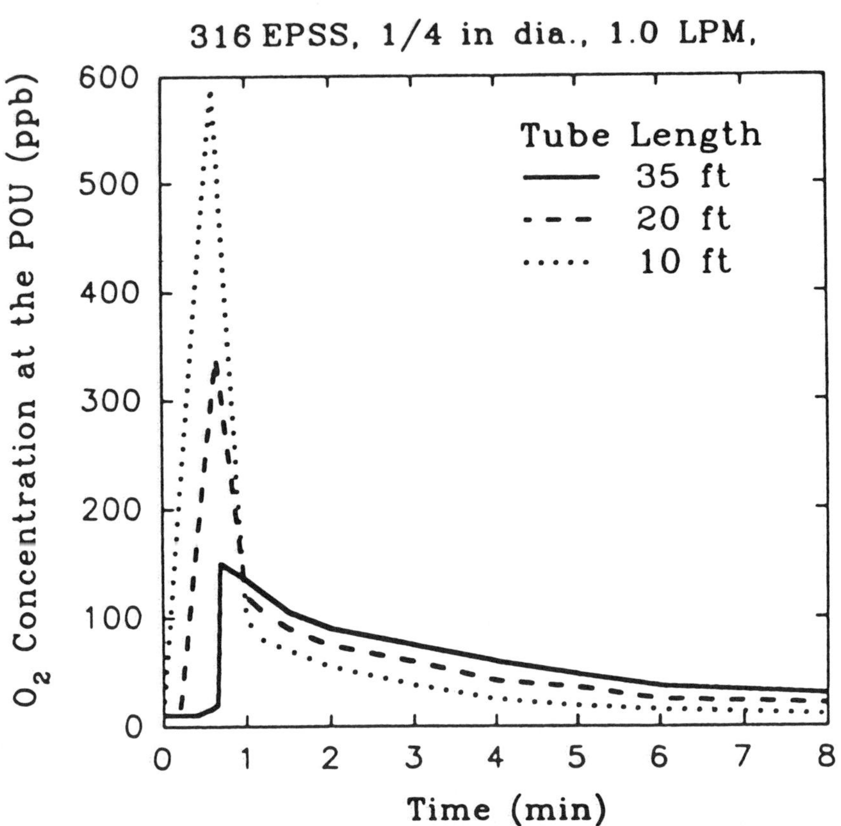

Figure 8. Profiles of Response to a Pulse Impurity Input

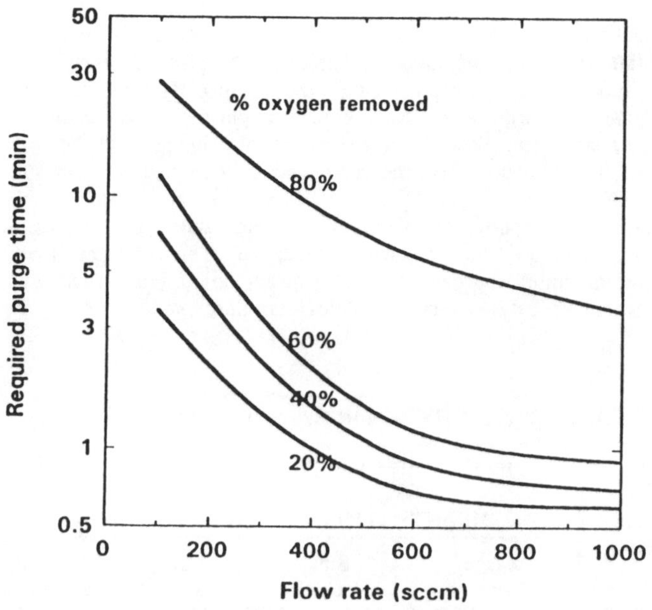

Figure 9. Outgassing Profiles in an Electropolished SS Tubing

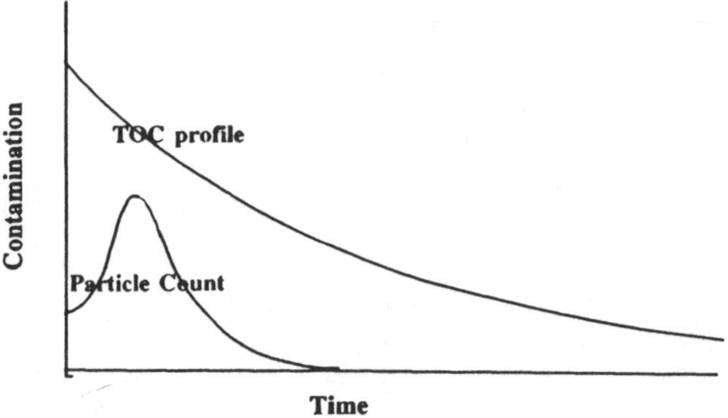

Figure 10. Effect of Oxidation on TOC and Organic Particle Count

Oxidation of organic compounds usually has a straightforward effect on the total TOC but can have a complicating effect on the particle concentration. Test results using model impurity compounds show the presence of two mechanisms (oxidation and fragmentation) during oxidation of organic impurities. The relative significance of these two mechanisms depends on the nature of oxidation and the concentration of impurities. The results of experiments are shown in Table 1. These results indicate that while TOC decreases with any degree of oxidation, the particle number, a very critical parameter for impurity control in semiconductor industry, may increase or decrease depending on the nature and the extent of oxidation. This is illustrated in Figure 10. Therefore, an integrated approach to monitoring and controlling particles and dissolved impurities simultaneously is very critical for ultra-pure water systems.

Table 1. Effect of UV185 and Ozone on Organic Particulates

Oxidizing Agent	HIGH CONCENTRATIONS		LOW CONCENTRATIONS	
	Large Particles	Small Particles	Large Particles	Small Particles
Ozone	O ($\downarrow$)	O ($\downarrow$)	O ($\uparrow\downarrow$)	O ($\downarrow$)
UV185	F ($\downarrow$)	F ($\uparrow$)	O ($\downarrow$)	O ($\downarrow$)
Ozone/ UV185	F ($\downarrow$)	F ($\uparrow$)	O ($\downarrow$)	O ($\downarrow$)

O = Oxidation is the dominant mechanism
F = Fragmentation is the dominant mechanism
$\uparrow$ = Increase in particle count
$\downarrow$ = Decrease in particle count

ACKNOWLEDGEMENT

This work was supported by SEMATECH through Arizona SCOE.

NOMENCLATURE

C_g = Concentration of impurity in the gas phase,
      $C_g = C_g\ (z,t)$

C_s = Concentration of adsorbed impurity
      $C_s = C_s\ (z,t)$

D_z = Dispersion coefficient

r_a = Rate expression for adsorption; $r_a = r_a\ (C_g, C_s, T)$

r_d = Rate expression for desorption; $r_d = r_d\ (C_s, T)$

S = Specific surface area

t = Time

T = Temperature

U = Flow velocity

z = Flow direction

REFERENCES

[1] Carrea, G. and Huling, B., *Proceedings of the 35th Annual Meeting of Institute of Environmental Sciences*, pp. 364-368, 1989

[2] Mulready, J.T., *Microcontamination*, Vol 10, no. 1, pp. 23-28, 1992.

[3] Glaze, W.H , Kang, J.and Chapin, D., *Ozone Science and Engineering*, vol. 7 pp. 47-62 (1985).

[4] Schmitt, S. and Snijders, M.W., *Ultra-pure Water* 7(4), pp. 32-39 (1990).

[5] Haider, A.M, Ma, C. and Shadman, F., *Proceedings of the 38th Annual Meeting of Institute of Environmental Sciences*, pp. 122-126, 1992.

[6] Sugiyama, K. Ohmi, T. Okumura, T. and Nakahara, F., *Microcontamination* Vol.7, no.1, pp. 37-65, 1989.

[7] Haider, A.M., Ma, C. and Shadman, F., *Proceedings of the 38th Annual Meeting of Institute of Environmental Sciences*, pp. 174-179, 1992.

GAS PURITY IN A DYNAMIC DISTRIBUTION SYSTEM

**Gerhard Kasper, Benjamin Jurcik, David Y-E.Li,
James McAndrew and Hwa-Chi Wang**
American Air Liquide
Chicago Research Center
5230 S. East Avenue, Countryside IL 60525 - USA

INTRODUCTION

The influence of gas distributions system on the outlet gas quality has long been recognized as critical. At the same time, this influence exerts itself in a rather complex way which makes predictions and comparisons difficult. On the other hand, a good understanding of this complexity offers opportunities for innovative design.

First of all, a system is dynamic in that any changes of the *input* parameters (gas composition, purity) or operating conditions (flow rate, temperature) will create a time dependent response at the *outlet*. Part of this time dependence is due to purely geometric effects resulting from the system volume and configuration (e.g. dead space) which will affect the arrival of changes at the outlet even in the absence of wall interactions. Of course a system contains surface area with which some gases will interact, causing memory effects or chemical reactions; surfaces can also become sources or sinks of particles. Lastly, one has to deal with potential leaks which are effectively steady state sources of contaminants.

This article attempts to give a general overview of the factors which control the dynamic response of a system and to describe practical test methods for them. The methodology is based on transfer functions (McAndrew et al.,1991) and designed to be flexible enough to handle both inert and reactive gases. The impurities considered are moisture as a typical representative of molecules interacting strongly with surfaces, and particles, also interacting with surfaces, although on an entirely different scale.

The relevance of surface texture and other surface parameters will also be touched briefly.

TRANSFER FUNCTIONS AS GAS SYSTEM DESCRIPTORS AND OPERATING SCENARIOS

The transfer function concept is well established in various engineering disciplines, notably to describe electrical systems. It treats the system as a black box for which the relationship of certain output to input variables is sought.

This concept can be applied to gas systems successfully. As illustrated in Fig.1, one selects an input variable and then tests for the system's effect on a chosen output variable. Typical variables are impurity concentrations such as moisture or particles, or other operating parameters such as flow rate. By defining standard input functions, it is possible to categorize and rank systems, or components of systems, and make general predictions of their behavior in time.

An advantage of using transfer functions is the freedom to express all system performance characteristics in terms of an output parameter which is directly relevant to the operation, e.g. gas purity, rather than direct measurements. The challenge is to devise equipment sensitive enough to perform the task.

EXAMPLES FOR INERT AND REACTIVE GASES

Moisture is one of the major causes for concern in gas distribution systems. It is often the rate determining factor in purging down a system or in upset conditions, where system recovery speed is the issue.

Fig. 2 shows input pulses with a moisture pulse generator capable of generating ppb-level or ppm-level pulses with durations from seconds to minutes. In this figure, the pulse shape is essentially due to the response-time of the electrolytic hygrometer used. Sharper responses have been obtained by using the moisture pulse generator with an APIMS. These pulses have been applied to various types of tubing, filters and other components used in high purity gas systems.

Fig.3 illustrates the response of various tubing samples to moisture pulses of short duration. Such test data describe the uptake and dry-down behavior for moisture under standardized conditions. While these may be somewhat arbitrarily chosen, they have the advantage of representing well defined initial conditions, thus making comparisons of tubing performance independent of packing and other historical factors.

Data are shown to describe the response of tubing to pulses of reactive gas as a means of testing the suitability of the particular component and its surface treatment to non-inert atmospheres. Fig.4 shows a comparison of system responses to a step-up function of ppm-level concentration of a hydride in pure nitrogen. The two tubes were pretreated in different ways. The figure shows that in one case the hydride is strongly retained on the walls (tubing) while there is no surface reaction between the hydride and tubing material in tubing A. From the response curves, one can readily derive the surface reaction rates at both transient and steady status.

MULTI-COMPONENT SYSTEMS

In principle, the transfer function concept is applicable to single components as well as to arbitrarily complex systems without requiring a fundamental change in test methodology.

The effects of leaks, geometry and gas surface interactions on complex, multi-component systems are discussed using simulation as well as experimental data.

PARTICLES

Conceptually, particle deposition is analogous to moisture adsorption and particle reentrainment to moisture desorption. In reality, the deposition and reentrainment rate constants are difficult to obtain because their strong dependence on particle size distribution and operation conditions.

Fortunately, it may not be necessary to consider particle deposition and reentrainment simultaneously in our situations. Diffusional deposition decreases with flow rate while reentrainment increases with flow rate. For example, less than 1% of 0.1-μm particles are lost to the wall of a 1-meter long, 0.25-inch tube at a flow greater than 1 L/min, according to the well-known Gomerly-Kennedy equation. Particle shedding from this tube, on the other hand, is negligible for a flow rate smaller than 1 L/min if the internal surface is electropolished.

Particle deposition is most relevant to particle filters; components downstream of filters see very low particle concentration and thus deposition is not a critical issue. Behavior of particle deposition in filters as a function of particle size is a classic research topic in aerosol science. Predictive equations for filter penetration are generally available if the filter parameters and operation conditions are known.

Particle shedding rate versus time consists of a short term regime with a fairly constant shedding rate, a transition regime with an exponential decay, and a long term regime with a shedding rate approaching 1/t decay (Wen and Kasper, 1989; Jurcik and Wang, 1991). The long term regime of 1/t decay is supported by various experimental evidences from many components (Wen and Kasper, 1989) and can be considered a universal behavior. The short term and transition regimes are difficult to verify due to the need of very high time resolution and the convolution with the start of flow. Direct microscopic observation of particle reentrainment in a turbulent flow cell (Wang et al., 1990) indicates qualitative agreement with the theoretical predictions.

Fig.5 shows the application of a transfer-function test for particles. A step-up in flow rate, which is the chosen input parameter, is applied after each exposure of the test specimen (in this case a section of EP-SS tubing) to corrosive gas.

SURFACE TEXTURE

It is a common misconception that transparency correlates with the surface finishes (average roughness). In fact, it was demonstrated that a surface with low average roughness could be more difficult to clean than a surface with high average roughness (Chesters et al., 1991). It was further shown that fractal roughness, instead of average roughness, was the right parameter for particle cleanability correlation. The fractal roughness concept can be extended down to the nanometer level with the new techniques for surface texture analysis such as scanning tunneling microscopy (STM) or atomic force microscopy (AF). Correlation may be found between moisture interaction and fractal roughness at the near-atomic scale.

REFERENCES

Chesters, S., Wang, H.-C., and Kasper, G. "A Fractal-Based Method for Describing Surface Texture", Solid State Techn., Jan.,73, 1991.

McAndrew,J.J.F.,Brandt,M.D.,Li,Y-E.David,and Kasper,G. "Establishing Moisture Test Methods for Process Gas Distribution" Microcontamination, 9(1) 33-37, 1991.

McAndrew,J.J.F., Brandt,M.D., Kasper,G. and Kimura T., "Moisture Testing of Process Gas Distribution System Components", Microcontamination 91, San Jose 1991, 352-359

Jurcik, B. and Wang, H.-C., "The Modeling of Particle Resuspension in Turbulent Flow"., J. Aerosol Sci., 22, S149-S152, 1991.

Wang, H.-C., Niida, T., Udischas, R., and Wen, H.Y. "Direct Observation of Particle Reentrainment in Turbulent Channel Flow". Proceedings of the Third International Aerosol Conf., p. 681-684, Pergamon.

Wen,H.Y., and Kasper, G. "A Kinetic Model of Particle Reentrainment.", Journal of Aerosol Sci.,20, p.483.

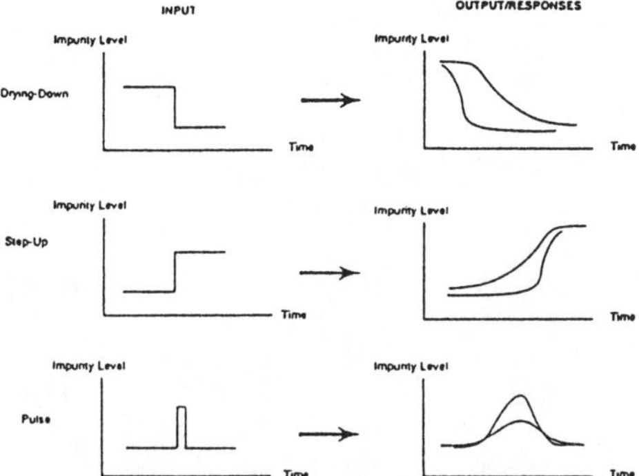

Fig.1.   Typical input functions and their corresponding outputs for hypothetical systems.

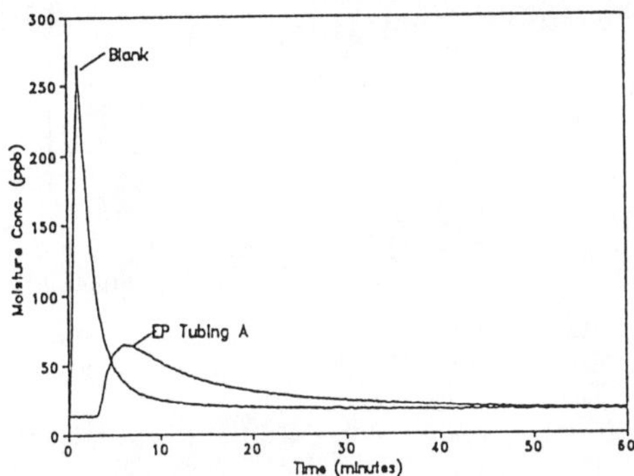

Fig.2.   Input pulses generated by a moisture pulse generator as detected by an electrolytic hygrometer. The effect of introducing a 4m length of electropolished (EP) tubing between generator and detector is illustrated.

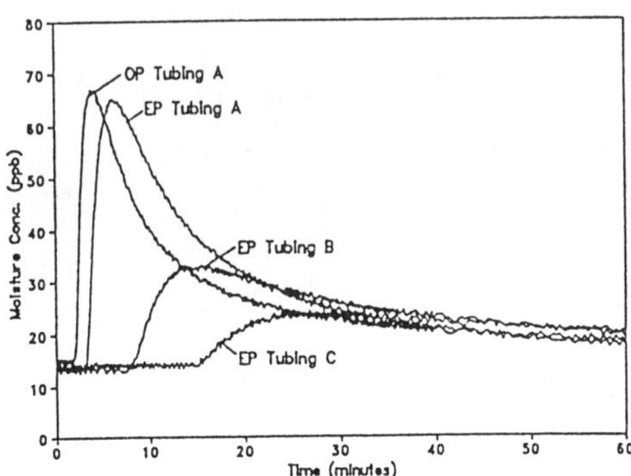

Fig.3.   Response of several different samples of EP and of one sample of oxygen-passivated (OP) tubing to moisture pulses.

299

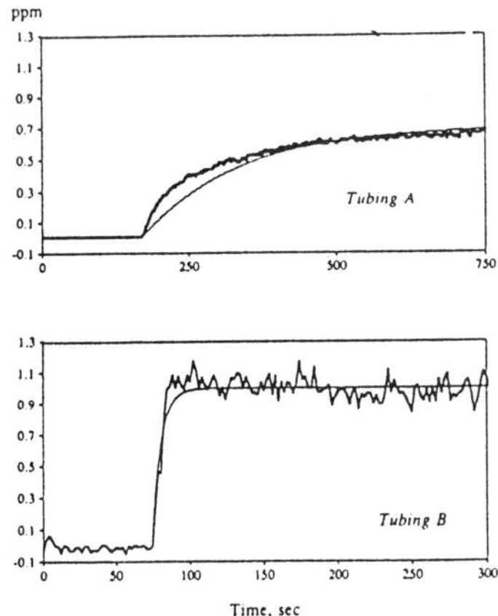

Fig.4. Response of two different tubing samples to a step-up in trace hydride concentration in N_2 carrier gas.

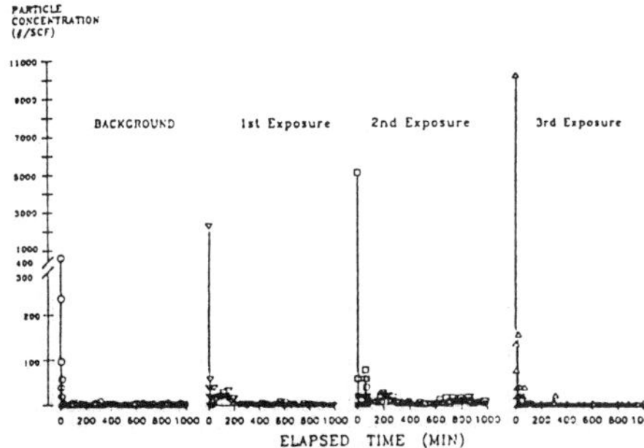

Fig.5. Particles generated in response to a step-up in flow rate applied first to an unexposed sample of EPSS tubing and then to the same sample after each of three consecutive exposures to a corrosive gas.

ANALYTICAL CHALLENGES FOR ULSI PRODUCTION

by

J. V. Martinez de Pinillos, S. N. Ketkar, R. G.

Ridgeway, and D. A. Zatko

Air Products and Chemicals, Inc.
Allentown, Pa 18195

The analyses of chemicals used in ULSI technology
is a very demanding process from the standpoint of
the instruments needed as well as from the quality
of the people required to conduct the analyses.
The sophistication of several techniques are
discussed, including APIMS and moisture analysis
of gases. Data are presented to demonstrate some
significant problems. Challenges associated with
the analyses of reactive chemicals are also
described.

INTRODUCTION

The development of ULSI technology with submicron design
rules requires process chemicals containing very low levels
of impurities (below double digit ppb levels). Instruments
capable of measuring these levels of impurities have only
become available in the last few years and are, in general,
very expensive. Because these instruments are new and have
to perform at or near the limit of detection, they require
performance certification at the low concentration levels.

A high level of expertise is required to accumulate and
interpret the data provided by such instruments.
Difficulties arise from several sources. For example, the
low levels of moisture and oxygen impurities that are
required today are of the same order of magnitude as the
outgassing from: a) the components used to transport the gas
to the analyzers and b) the materials used in the
construction of the analyzers themselves. Also, the ability
to analyze bulk gases on a continuous basis for moisture at
very low levels is hampered by an aging that some detectors
undergo, as they dry-out when exposed to dry gases for
extended periods of time.

ADDITION OF CONTAMINANTS BY THE ANALYTICAL SYSTEM

Current state-of-the-art instruments are used routinely to
measure impurities at the low ppb level. However, a

measurement readout on a sampled chemical (gas or liquid) at
these levels does not guarantee that the quantity of the
impurity indicated is the true value for the source
chemical. For example, a moisture analyzer may be capable
of 1-10 ppb measurement in gases, but if the gas picks up
moisture on the way to the analyzer or while traversing the
analyzer on its way to the sensor, the measurement will not
represent the moisture content of the source gas.
Outgassing of the system components is a common source of
contamination additions.

A calculation can be made to determine the concentration, C
in ppt, of an impurity added to a gas by system outgassing
using the following expression (It will be assumed that the
gas does not contain that impurity. If the gas contains a
concentration of impurity, C_O, the final concentration would
be $C + C_O$):

$$C = (n_{out}/n_T) \times 10^{12} \tag{1}$$

where n_{out} is the number of moles of the impurity produced
by outgassing, n_T is the total number of moles of gas in the
system, and 10^{12} is a factor to express the concentration in
parts-per-trillion. Applying the ideal gas law to the
number of moles in eq. 1, we can see that:

$$n_{out} = P_{out}V/RT \qquad \text{and} \qquad n_T = P_TV/RT \tag{2}$$

where P_{out} is the partial pressure of the outgassing
impurity, P_T is the total gas pressure, V is the total gas
volume, R is the gas constant, and T is the absolute
temperature. V, R, and T are common for both the outgassing
and the total gas. Therefore, eq. (1) becomes:

$$C = (P_{out}/P_T) \times 10^{12} \tag{3}$$

P_{out} can be expressed as R'A/F, where R' is the outgassing
rate, A is the total surface of outgassing material, and F
is the gas flow. Making these substitutions in eq. 3, the
concentration can be expressed as:

$$C = R'A/P_TF \tag{4}$$

Applying the calculation to 500 ft. of 0.5" O.D. 316L
stainless steel line at 90 psig that was outgassed for 75
hrs. under vaccum (the outgassing rate of 316L SS under
these conditions[1] is in SI units 5.73×10^{-8} W/m^2), one can
obvserve (fig.1) the relationship of concentration to flow
rate through the line.

For a flow of 250 scfh, the concentration due to outgassing
will be 246 ppt, and for 10 scfh, it will be 6.1 ppb. The
outgassing from the piping material is a function of
temperature and higher concentrations will be observed, at

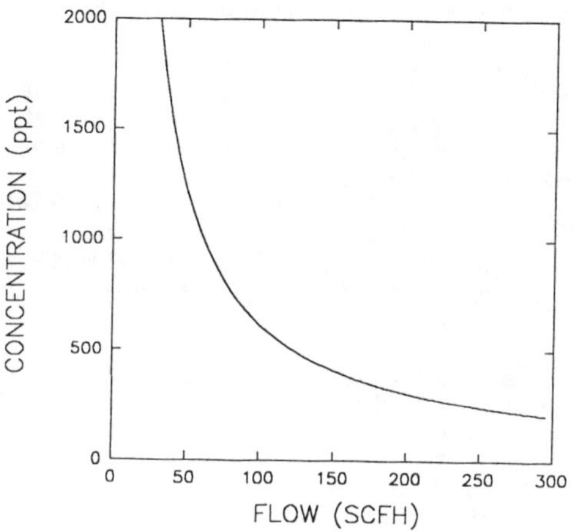

Fig. 1 Concentration of an outgassing impurity in a 0.5"
O.D. x 500'long 316L line at 90 psig after 75 hours of
vacuum.

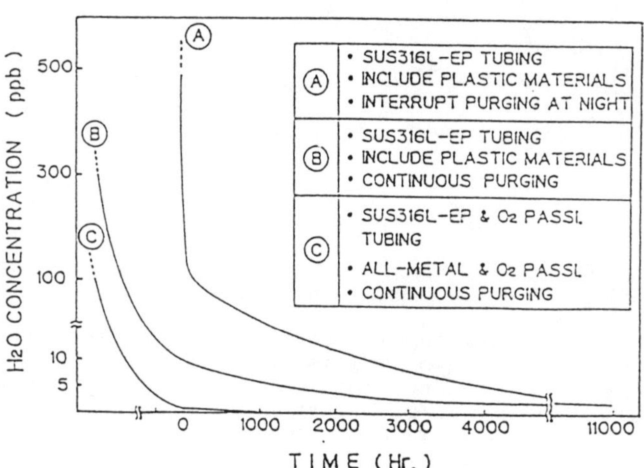

Fig. 2 Outgassing from a 316L stainless steel line subjected
to different conditions[2]

least initially, if the temperature is above room
temperature. At any given temperature, the curve is usually
exponential in time. An example of one of these outgassing
curves obtained at room temperature[2] is presented in Fig. 2.

It can be seen that the moisture concentration starts high
and then decays even though the gas entering this tubing
contains <1 ppb moisture. (The gas used for the measurement
in Fig. 2 was purified to this level). The concentration
read by the analyzer will be that of the gas plus the
moisture contributed by the outgassing. Only when the
outgassing is reduced to levels significantly below that
existing in the source gas will the instrument read-out
correspond to the concentration in the source gas as it
leaves the purifier.

In many facilities where the temperature of the analyzer
environment shifts periodically or the gas tubing/piping is
exposed to diurnal temperature changes, the moisture level
follows these temperature fluctuations indicating shifts in
the outgassing of the tubing/pipes rather than changes in
the moisture concentration of the source gas. Fig. 3 shows
the changes in the moisture content of dry N_2 as a function
of room air temperature.

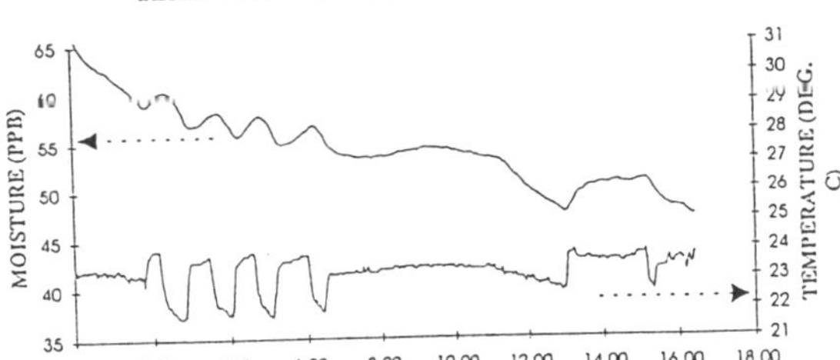

Fig. 3. Moisture concentration and room temperature
fluctuations as a function of time.

These fluctuations are an important factor for semiconductor
fabricators. The changes in moisture concentration due to
ambient temperature fluctuations can have a serious impact
on the quality of the IC's. We should understand the nature
of such fluctuations if we wish to eliminate them.
Thermostating the analyzer may not be sufficient if most of
the impurity changes are due to the temperature fluctuations
in the lines that carry the gases/chemicals from their
sources (tanks or purifiers) to the point of use.

Fig. 4 shows the effect of the temperature variability in
the laboratory as observed with an electrolytic cell and an
APIMS. It can be seen that the highs and lows in the

temperature are followed by the moisture analyzer and the
APIMS. The oscillations in both instruments are of the same
order, 1-2 ppb for temperature oscillations of about 1°C.
In the electrolytic cell, the oscillations are superimposed
on the exponential decay observed when the overall system is
drying down.

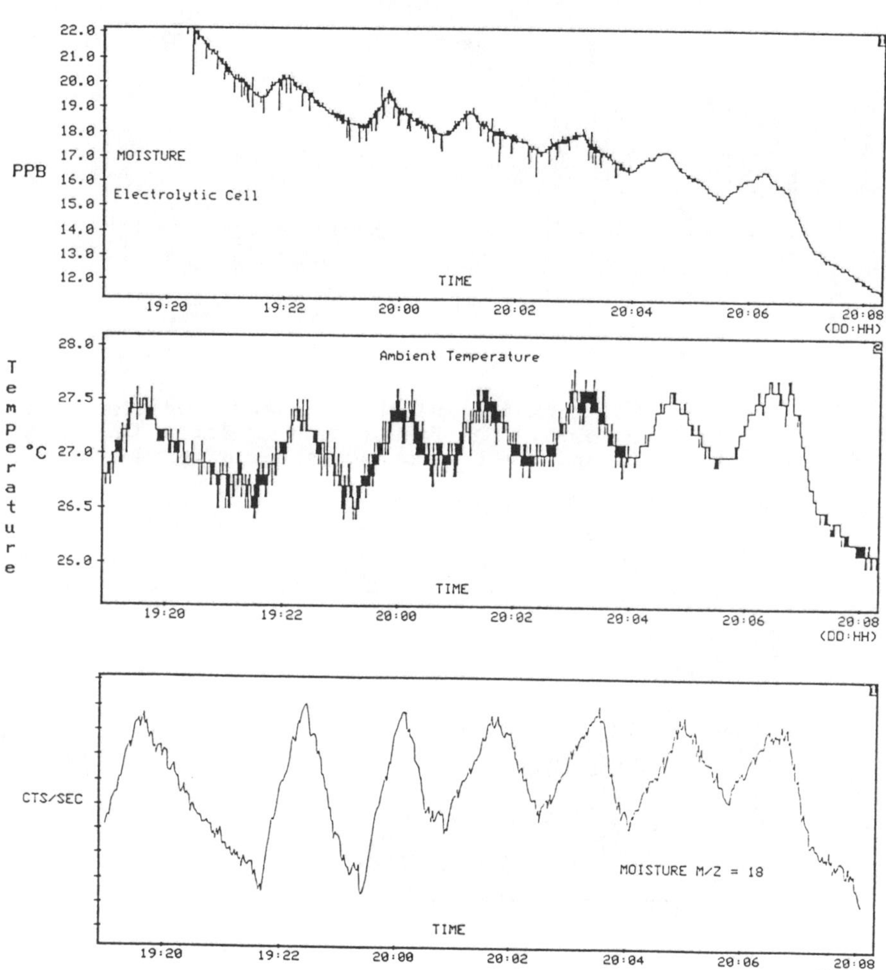

Fig. 4. Moisture concentration variations
observed in an electrolytic moisture analyzer and
an APIMS as the temperature in the room fluctuates.

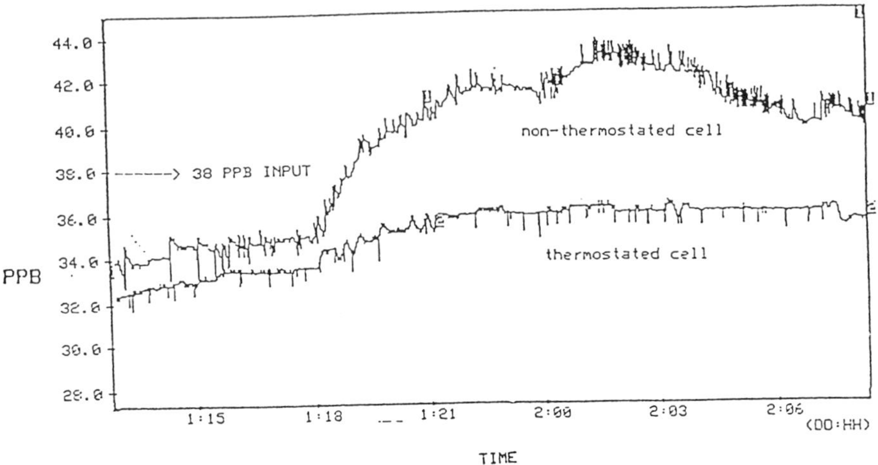

Fig. 5a. Moisture concentration observed by two
electrolytic cells. The top tracing is for a non-
thermostated cell and the bottom one is for a thermostated
cell

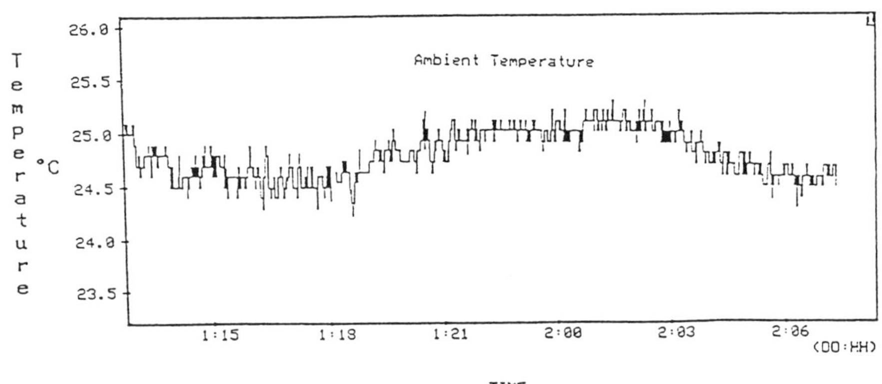

Fig 5b. Temperature variations in the laboratory where
the moisture measurements in fig. 5a were performed.

The oscillations observed in all moisture analyzers due to the temperature fluctuations can be ascribed to two possible causes: the effect of the temperature changes in the tubing carrying the gas to the analyzer and in the electrolytic cell, itself. The latter effect, in turn, is the result of two processes: possible outgassing of the materials of construction of the cell itself and the temperature dependance of the electrolytic process. If the electrolytic cell is thermostated to minimize the effects due to the temperature changes on it, one of the sources of the variability can be removed. Fig. 5a shows this effect by comparing the results obtained when a thermostated and a non-thermostated cells are used to measure the same gaseous stream. The moisture concentration in the gas was maintained at 38 ppb with a moisture generator. Both instruments responded to the moisture intrusion, but the thermostated cell provided a more stable reading than the non-thermostated cell. Moreover, the non-thermostated cell followed the temperature variations in the room as presented in Fig. 5b.

Another problem in instrumentation that impacts low level measurements, centers on the detector functionality. Some detectors operate by detection of surface changes of the sensor. Impurity molecules adsorb on the surface of the detector changing some property that, in turn, changes the signal produced by the detector. For example, thin films of hydrocarbon materials deposited on the surface of detectors may cause significant errors. Such a thin film acts as a diffusion barrier inhibiting the amount of impurity that reaches the sensor surface. The permeation through this layer depends upon the concentration gradient across the layer and, hence, it depends on the concentration of the impurity in the chemical. In UHP chemicals, the concentration of impurities is very low to begin with; therefore, the concentration seen by the detector will be even smaller due to the small driving force. The concentration determined by the analyzer will always lag behind the concentration in the chemical. In such a case, it is important to monitor the detector response with time. The change in signal for a known change in the amount of impurity should be constant with time.

Some moisture analyzers have been found to change their responsiveness when exposed to dry gases for an extended period of time. They depend on the measurement of an electric current through a suitable medium or the measurement of the capacitance of a film. These two properties depend on the moisture content in the sensor. Below certain moisture levels currently under study, the physical characteristics of the medium change, thereby, changing the detector's response to moisture. In all cases, the concentration observed will be lower than the actual concentration present in the gas.

ATMOSPHERIC PRESSURE IONIZATION MASS SPECTROMETER (APIMS)

In recent years, the application of APIMS has provided a
tool capable of very low levels of detection for impurities
in gases. However, accurate analyses at low levels can
still be a challenge. The development of calibration curves
for impurities (especially for atmospheric contaminants)
extending to low ppt levels can be difficult.
Ultra-high-vacuum technologies have been required to obtain
and maintain systems containing low background impurity
levels. One needs to bake out (T > 200°C) the analyzer and
the tubing carrying the gases. Sometimes, this has proven
problematic due to the APIMS' materials of construction or
design constraints.

Cylinder gases containing standards in the ppb region are
difficult to prepare accurately, especially for atmospheric
impurities. Developing calibration standards for low ppt
levels has required dilution systems with ratios 10,000:1 or
higher between the diluent gas (so called zero gas) and the
calibration gas standard. Additionally, obtaining a zero
gas for atmospheric contaminants is very difficult. Most
commercial purifiers yield gases containing a few hundred
ppt of moisture and oxygen. These gases can be used as zero
gases for blending impurities at the high ppb level without
difficulty. One simply ignores their contribution to the
total concentration of the impurities in question. However,
we can not ignore their contribution when their
concentration is of the order of impurities present in the
gas. Zero gases with impurity concentrations <1 ppt are
required and purifier technologies have not reached these
levels, yet.

The accuracy of APIMS measurements depends on the proper
interpretation of the impurity background obtained from
analysis of the zero gas. In many analytical situations, the
background is routinely subtracted from the total
concentration found in the sampled chemical. This may or
may not be applicable for an APIMS analysis. If the
background originates from the tubing, ionization chamber,
and other pieces of equipment, then subtracting the
background from the total signal observed would be
justifiable. However, since this background could be
present only as a contaminant in the purified gas used as a
zero gas and, consequently, not be present during the
analysis of the sampled gas, background subtraction is not
justified. We have decided to report the total
concentration found, the background detected and the
difference between these two figures for any data that we
release.

In APIMS technology, a calibration curve for one impurity
may depend on the concentration of a second impurity. This
is due to the high collision rate present in the ion source.

An initially formed ion may interact with another atom or
molecule of lower ionization potential and transfer its
charge. It can also collide with another atom or molecule
and transfer a proton. If the concentration of the second
impurity is high enough then, the collision frequency
between the two species could be significant. This can be
illustrated by the analysis of CH_4 in N_2. Fig. 6 presents
different calibration curves for CH_4 in N_2 obtained at
m/q=16 and m/q=17 for CH_4^+ and CH_5^+, respectively.
It can be seen that as the CH_4 concentration increases, the
CH_4^+ calibration curve becomes non-linear. The CH_4^+ ions
are possibly reacting with other species in the APIMS source
decreasing their apparent concentration. In this case, the
m/q=17 peak should also be tracked. By adding these two
intensities, an extended region of linearity is obtained by
incorporating the contribution of the following reaction:

$$CH_4^+ + CH_4 \quad ====> \quad CH_5^+ + CH_3$$

$$m/q=16 \qquad\qquad m/q=17$$

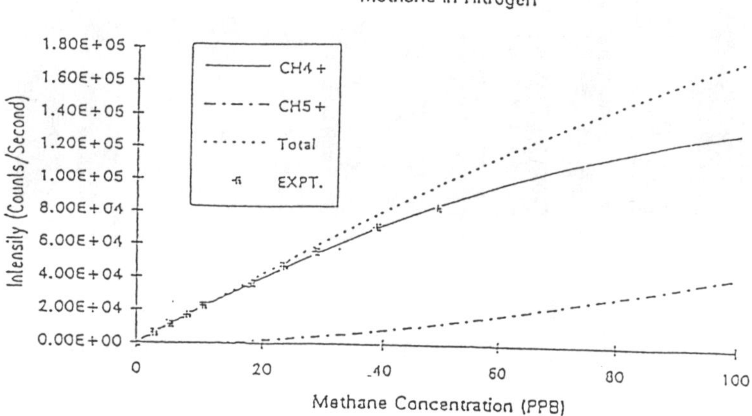

Fig 6. Calibration curve for CH_4 at m/q = 16, 17, and the
sum of the two.

The APIMS can not be used to detect many important
impurities in bulk O_2 because its ionization potential is
lower than that of the impurities. Fig. 7 shows the
ionization potential of common gases in decreasing order.

Any ionized gas can transfer its charge to a gas that
appears to its right, but it can not transfer it to a gas on
its left.

A review of the limitations of the APIMS technology will be
presented at the Microcontamination 92 meeting this fall[3].

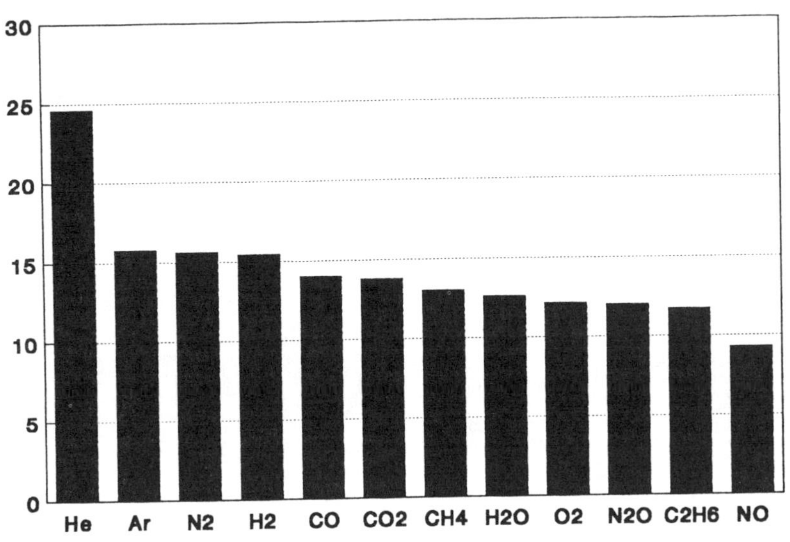

Fig. 7. Ionization potential of different gases.

CALIBRATION:

A challenge that needs to be faced is that of calibration of
the instruments at the low levels of interest. In addition
to the challenges already mentioned, the generation of
standards at low ppb and ppt regions are particularly
trying.

To begin with, calibration curves at the levels of interest
are expected to be linear if the detector is linear.
Detector linearity needs to be established early in the
process.

Additionally, as stated before, there are no reliable
sources of gaseous standards below the ppm region. To
accomplish the required dilution, several schemes with a so
called zero gas have been proposed. The simplest concept
comprises the use of a purifier and two mass flow
controllers (MFC). The purified gas is blended with a
stream coming from a cylinder containing known impurities at
known levels. To avoid further addition of impurities to
the "zero gas", the MFC should be placed upstream of the
purifier to insure that the purifier removes as many

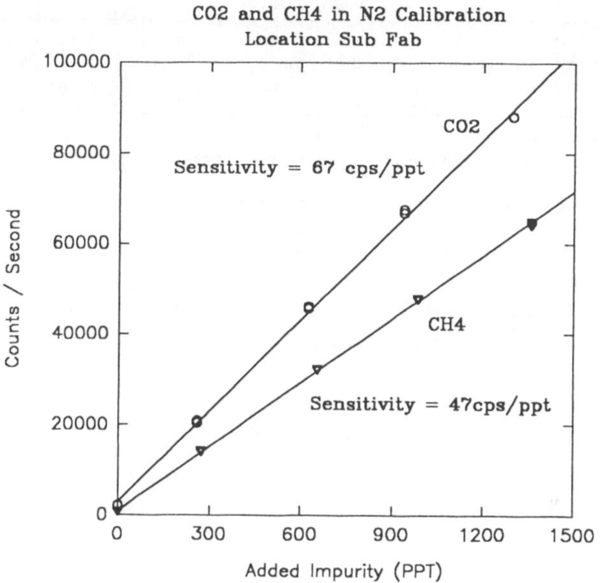

Fig. 8. Calibration curves for an APIMS showing
linearity and near zero or zero intercept.

impurities as possible, including those that could have been
generated by the MFC itself. Most MFCs used for this
purpose are all metal and contain no elastomers. One of the
major problems with flow controller is that due to a zero
drift. If the drift is small most flow controller power
supplies will automatically zero the offset. However, if a
large drift does occur the MFC still maintains its linearity
although the displayed flow value is offset from the "true"
flow value. The problem due to the zero drift can manifest
itself either as a positive offset i.e. the indicated flow
is larger than the "true" flow or as a negative offset i.e.
the displayed flow is smaller than the true flow. Figure 8
displays a calibration curve obtained showing good linearity
and an almost zero intercept. Sometimes the intercept can

311

deviate from zero. If good linearity is still maintained
this is an indication of a background impurity in the
system. This impurity arises due to contributions from the
zero gas itself (i.e. from the purifiers inability to
generate a true zero gas), from impurities outgassing from
the inlet system and from the impurities outgassing from the
instrument itself. The background problem is particularly
evident for the case of moisture since it adsorbs easily
onto surfaces. This is shown in Fig. 9. Figures 10 and 11
show linear calibration curves which do not pass through the
background data point. The data shown in Fig. 10 has a zero
offset which has to be added to the indicated flow to obtain
the " true" flow while for the case of Fig. 11 the zero
offset has to be subtracted from the displayed flow to
obtain the "true" flow. It is the true flow that has to be
used in order to calculate the impurity concentration. In
general, periodic zero point checks have to be performed on
the MFC's to avoid problems due to the zero offset.

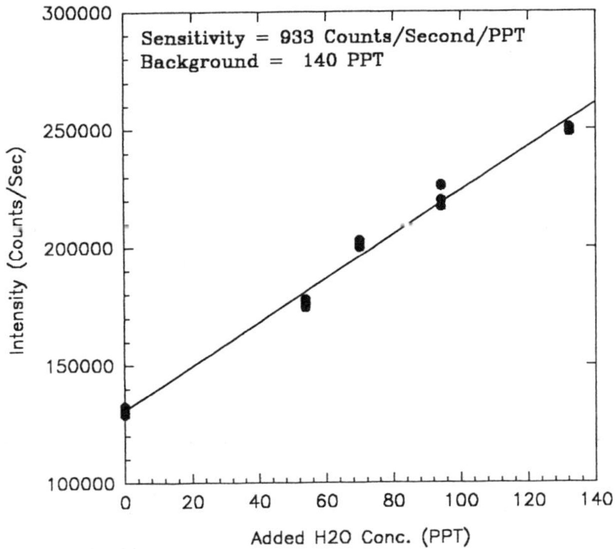

Fig. 9. Calibration curve for APIMS showing linearity but a
non- zero intercept due to background.

The non-linearity observed in Fig. 10 is due to the APIMS
response in this region.

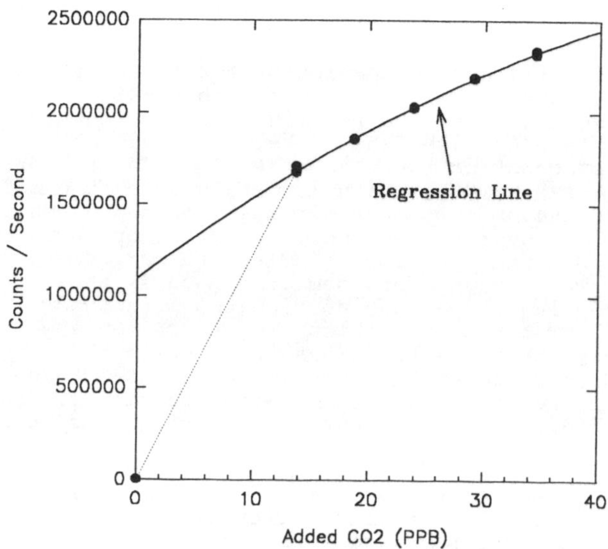

Fig. 10. Non linear calibration curve for APIMS showing a positive intercept.

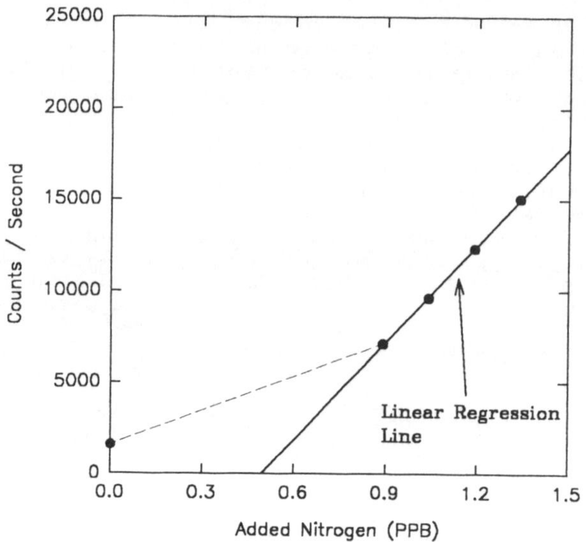

Fig. 11. Calibration curve for APIMS showing linearity and a negative intercept.

313

ANALYSIS OF REACTIVE CHEMICALS

The analysis of reactive chemicals require analytical
systems that are not only equipped with sensitvie detectors
but can also withstand the reactivity of the chemicals being
analyzed. In many cases these chemicals may react with the
materials of construction. Examples include the corrosion
of 316L by gaseous HCl, which is catalyzed by the moisture
adsorbed on system component surfaces, or the deposition of
boron oxide on system components and detector surfaces
resulting from the reaction of BCl_3 with surface adsorbed
moisture. The analysis of chemicals such as these require
special sampling steps such as repeated purge and
evacuations of the sample inlet to reduce the amount of
undesireable chemical reactions occuring in the sampling
system. If such reactions occur, the chemical being
analyzed can become contaminated with reaction products and
the data obtained are no longer representitive of the
chemical's purity.

Instruments used for the analysis of both reactive and
nonreactive chemicals require calibration. Unfortunately,
accurate standards prepared in reactive matrices are not
readily available or standards for reactive impurities are
not available at all. Often instruments are calibrated
using inert matrices and the calibration constants are
applied to data obtained from reactive samples to determine
impurity concentrations. This approach could lead to
inaccurate determinations if unknown matrix effects that
could alter an instruments response to a particular impurity
were present. An analysis demonstrating such an effect is
the determination of N_2F_4 in NF_3 by mass spectrometry.
Table I contains data obtained for N_2, O_2, and N_2F_4 at two
different source pressures. The last column in Table I
contains the ion intensities obtained at lower source
pressure relative to the ion intensities obatined at higher
source pressure. These data indicate that the N_2F_4 signal
is more pressure dependant than that of O_2 and N_2. This was
attributable to the formation of N_2F_4 from NF_3 fragments at
the higher pressure. Calibrations performed in a matrix
such as He would not be expected to show the same pressure
dependance.

Table I

Mass spectrometer signal obtained at two different source
pressures

Impurity	M/Z	Intensity 10^{-4} torr	Intensity 10^{-2} torr	Ratio
O_2	32	52400	142400	0.42
N_2	28	79650	209700	0.42
N_2F_4	104	2651	48100	0.06

COST

The instruments used for low level analysis are, in general,
expensive. Over the last several years, the costs have
generally increased as low level requirements haven been
implemented. Table II lists approximate cost of some
instruments. Unless specified, additional systems to
prepare calibration standards are assumed to be available.

Table II

Approximate Cost of Analytical Instrumentation

Instrument	Cost ($000)
APIMS and Calibration	350-500
Fully Automated Process GC	55
Discrete Sensors	
O_2	30
H_2O	25
Corrosion Resistant MS	150-250

CONCLUSIONS

In conclusion, the analysis of the chemicals that are being
used in ULSI technology require very sophisticated and
expensive instrumentation and highly experienced people.
Suppliers of these chemicals have to commit significant
levels of resources to continue supplying the semiconductor
industry. We are now certifying products at levels where
physical phenomena whose contribution was considered
insignificant have to be seriously considered and their
influence needs to be included as an integral part of the
analysis/calibration being performed.

REFERENCES:

1.- O'Hanlon, J. F., A User's Guide to Vacuum Technology,
John
    Wiley and Sons, NY, 1989, 287.

2.- Kawada, K., Nakamura, M., Ohkura, A., Ohmi, T.,
    Nakahara, F.,   Sugiyama, K., Mizuguchi, Y., and
    Tomari, H., All Metal O_2 Passivation Tubing Technology,
    Ultra-Clean Society, Tokyo, Japan, October 4, 1990, 48.

3.- Ketkar, S. N., Ridgeway, R. G., Scott, Jr., A. D., and
    Martinez de Pinillos, J. V., Microcontamination 92, to
    be presented.

ULTRA TRACE ANALYSIS OF HELIUM BY ATMOSPHERIC PRESSURE IONIZATION MASS SPECTROMETRY

Julio Z. Galarza
University of Arizona
Aerospace and Mechanical Engineering
Tucson, Arizona 85721

Ce Ma[+]
University of Arizona
Chemical Engineering
Tucson, Arizona 85721

Asad Haider
University of Arizona
Chemical Engineering
Tucson, Arizona 85721

Farhang Shadman
University of Arizona
Chemical Engineering
Tucson, Arizona 85721

and

Stephen L. Gilbert[+]
University of Arizona
Electrical and Computer Engineering
Tucson, Arizona 85721

Ultra pure helium gas obtained from liquid helium cooled to 4.2 degrees Kelvin has been investigated by APIMS. Helium has a high ionization potential (24.6 ev) allowing efficient charge transfer (Chemical Ionization) for all impurities. The mass spectra indicate high sensitivity for all impurities, few interferences for hydrocarbon identification, and reveal total impurity concentrations of less than 700 parts per trillion. Ultra pure helium gas can be produced on site by simple phase condensation and re-evaporation yielding a convenient calibration and ultra pure carrier gas.

[+]Contact these authors for experimental details.

INTRODUCTION

Atmospheric Pressure Ionization Mass Spectrometry (APIMS)[1,2,3] has increased analytical sensitivity to trace impurities present in semiconductor processing gases by three to five orders of magnitude. The improved sensitivity directly results from increased ionization efficiency. Ionization by charge transfer upon collision, "soft or chemical ionization", proceeds when an ionized species of higher ionization potential approaches a neutral impurity species of lower ionization potential. Under these conditions an impurity molecule is ionized by a charge transfer reaction. Helium, which has the highest ionization potential, 24.6 ev, can ionize all impurities by charge transfer[4]. Nitrogen and argon, other typical carrier gases, have ionization potentials of only 15.576 and 15.759 ev respectively.

Chemical ionization proceeds very efficiently at low concentrations of impurities in a carrier gas of higher ionization potential at atmospheric pressure due to the large probability a neutral impurity atom will interact with an ionized carrier gas atom. The overall efficiency for this ionization process approaches unity. By comparison, typical ionization efficiencies for electron impact ionization methods approach 1×10^{-3}.

The increase in ionization efficiency yields corresponding improved sensitivity to trace impurities. Analysis and calibration problems arise, however, due to the lack of suitable standards and zero reference gases for these very low concentration levels (ppt).

These problems are currently addressed by linear extrapolation of responses created by analysis of higher concentration standards, or by analyzing diluted standards close to the concentration of the unknown. These approaches are marginally adequate to 0.1 ppb levels.

While several methods for on-site purification of other carrier gases (nitrogen, argon) have met with some success (typically to 1 ppb total impurity levels), there remains a level of uncertainty due to a variety of conflicting claims, and the possibility some impurities are not detected in these carrier gases.

Helium was investigated as an alternative ultra pure carrier gas to provide a known and reproducible level of background impurity concentration, or ZERO GAS, for ppt calibration and measurement. Our objective was to provide a stable and portable standard, easily reproduced on-site, with a well characterized level of background impurities for calibration and carrier gas use.

EXPERIMENTAL

Research grade helium carrier gas was purchased from Air Products and Chemicals, Inc.. Typical impurities detected by gas chromatography are shown in

Table I. A calibration gas mixture containing known amounts of impurities in helium was donated by L'Air Liquide and contents are shown in Table II. One hundred liters of liquid helium refrigerant were obtained from the U.S. Department of Mines, Amarillo, TX, and shipped in a Cryofab 100 liter dewar under atmospheric pressure.

<table>
<tr><td colspan="2">TABLE I</td><td colspan="2">TABLE II</td></tr>
<tr><td colspan="2">Typical impurity levels reported in Research Grade Helium by gas chromatography or FFT spectroscopy</td><td colspan="2">Known concentration of calibration gas species in helium supplied by L'Air Liquide</td></tr>
<tr><td>Impurity</td><td>parts per million</td><td>Impurity</td><td>parts per billion</td></tr>
<tr><td>Neon</td><td>< 0.1</td><td>Neon</td><td>6,300</td></tr>
<tr><td>Oxygen</td><td>< 0.1</td><td>Argon</td><td>7,500</td></tr>
<tr><td>Total Hydrocarbons</td><td>< 0.1</td><td>Nitrogen</td><td>7,500</td></tr>
<tr><td>Water</td><td>< 0.15</td><td>Methane</td><td>7,500</td></tr>
<tr><td>Argon</td><td>< 0.1</td><td>Carbon Dioxide</td><td>7,500</td></tr>
<tr><td>Nitrogen</td><td>< 0.1</td><td>Carbon Monoxide</td><td>7,500</td></tr>
<tr><td>Methane</td><td>< 0.1</td><td>Water</td><td>* 7,630</td></tr>
<tr><td>Carbon Dioxide</td><td>< 0.1</td><td></td><td></td></tr>
<tr><td>Nitrous Oxide</td><td>< 0.1</td><td colspan="2">* water concentration was measured on line with a Meeco Aquamatic+ moisture analyzer</td></tr>
</table>

A VG Instruments Trace+ triple quadrupole APIMS was used in single quadrupole mode for the trace impurity measurements. The APIMS gas delivery system was configured to allow simultaneous measurement of flows and pressures, Fig 1.

Mass flow controllers 1 and 2 (MFC1 & 2) were calibrated to a reference bubble meter with nitrogen gas, and re-calibrated for helium gas. Mass flow controller 2 (MFC2) was fixed at 95.85 cc/minute for all experiments while mass flow controller 1 (MFC1) was varied to provide appropriate helium calibration gas dilution. Dilution ratios were verified by individually measuring the concentration for each of 11 mass peaks at 5 dilution ratios over the range of interest. Linearity and reproducibility were within 1% over this range.

A "U" tube cold trap was constructed of 1/8 inch O.D. 316L stainless steel tubing 8 feet long with an immersed depth in the liquid helium of 4 feet (Fig 1).

The helium carrier gas flow rates through the "U" tube and the APIMS ionization chamber were held constant at 2.41 liters/minute throughout the experiment. Carrier gas variations due to diluent flows (<5 cc/minute) were neglected.

The helium carrier gas and calibration mixture diluent were directly injected into the ionization chamber of the APIMS instrument. A corona discharge was initiated at 950-1,000 volts, significantly lower than the 5,000 volts used for the corresponding nitrogen carrier gas.

Helium carrier gas pressure in the ionization chamber was varied and recorded during analysis to determine maximum sensitivity to each peak and to ascertain the relationship between ionization chamber pressure and sensitivity. A pronounced and sharp pressure dependence was noted, likely caused by increased cluster formation as the ionization chamber pressure was increased. Helium carrier gas flow rate was maintained during these pressure variations by means of a mass flow controller downstream of the APIMS ionization chamber to eliminate possibility of contamination.

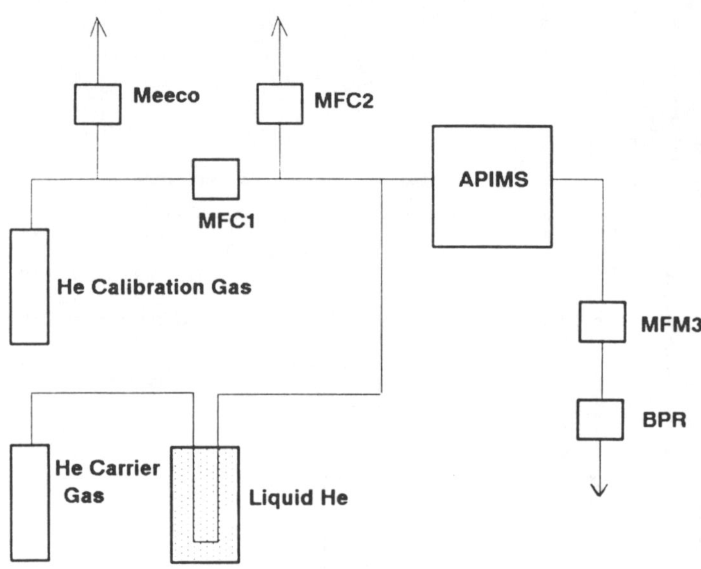

Fig. 1   Carrier Gas and Calibration Gas Flow Schematic

RESULTS AND DISCUSSION

Research Grade Helium APIMS Spectra

The helium corona discharge was initiated at lower applied potential (950-1,000 volts) than normally used for nitrogen carrier gas (5,000 volts) due to the longer meanfree path electrons experience in a helium discharge.

Research grade helium spectra were taken before purification, Fig 2. Unfortunately, the high levels of impurities present, compromise the accuracy of measurement due to significant reduction in the number of helium ions available for impurity ionization. This spectra indicates, however, that several impurity levels approach 100 ppb levels.

Fig. 2   Research Grade Helium APIMS Spectrum

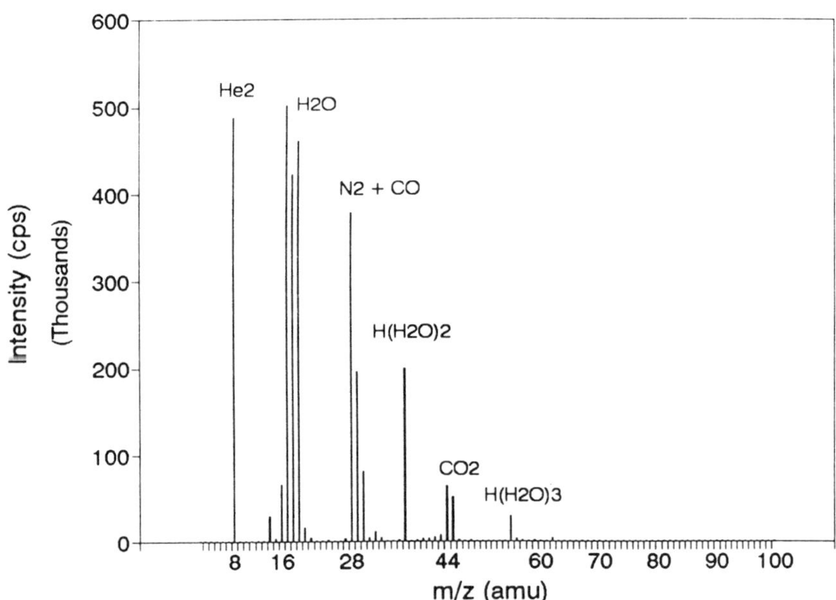

The spectra is complicated by the high levels of impurities present. In particular, the presence of high water levels causes formation of several cluster species, andreduces available helium ions necessary for further impurity ionization events. This lowers overall sensitivity for all impurities.

One should note the nitrogen cluster formation as well. Since this research grade helium feedstock originates as liquid helium, all impurities probably result from transit,container, and filling operations, accounting for the high water and nitrogen levels measured.

Helium Purification

The helium carrier gas was condensed in the "U" tube cooled to 4.2 degrees Kelvin by boiling liquid helium held at atmospheric pressure in the Cryofab dewar. The condensation temperature for the helium carrier gas is raised to 4.56 degrees kelvin as a result of additional pressure, five pounds per square inch above ambient, provided by the supply   pressure regulator. The liquefied helium carrier gas filled   the bottom of the "U" tube and impeded further gas flow. As the supply pressure increases slightly above the static pressure of the liquid helium in the "U" tube, bubbles of helium either pass through the liquefied helium condensing out the trace impurities, or additional helium is condensed on one side of the "U" tube and the latent heat re-vaporizes additional helium gas on the other side. In either case, the trace impurity atoms contact the cold helium liquid for sufficient time to reduce the levels of impurity present in the gas phase.

Calibration For Helium Carrier Gas

Calibration for individual impurities were made by successively diluting the helium standard mixture by ratios of 4.22×10^{-4}, 1.08×10^{-3}, 1.34×10^{-3}, 1.7×10^{-3}, 2.00×10^{-3} with ultra high purity purified helium carrier gas. Mass flow controller 2 (MFC2) was fixed throughout at 94.85 cc/minute and mass flow controller 1 (MFC1) was incremented accordingly to provide the appropriate dilution ratio. Constant flow was maintained through the APIMS ionization cell by the appropriate addition of ultra pure purified helium carrier gas and was indicated by mass flow meter 3 (MFM3). The APIMS ionization chamber backpressure was maintained constant at 5 pounds per square inch above ambient by the backpressure regulator (BPR) to   assure constant and reproducible sensitivity for all species.

Fig. 3   Methane (mass 16) in Helium

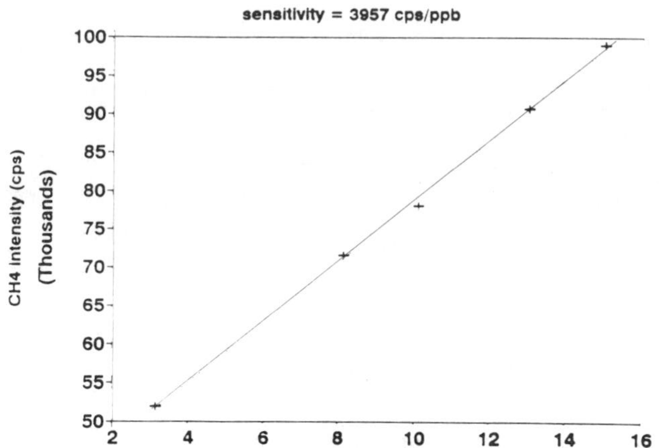

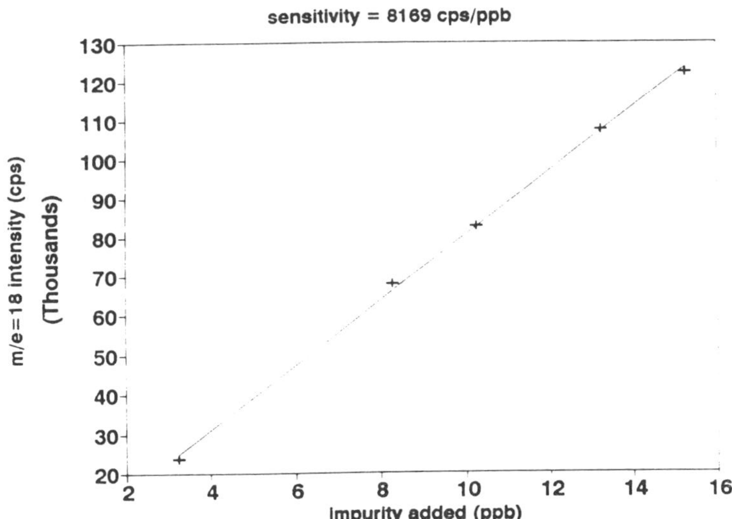

Fig. 4   Water (mass 18) in Helium

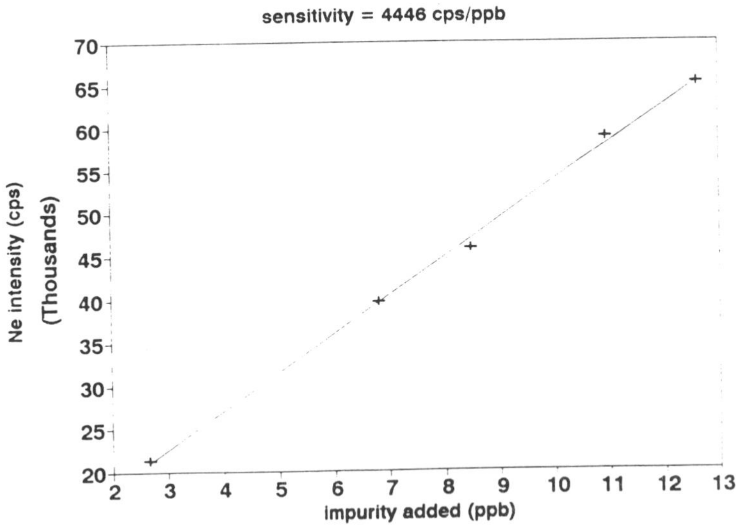

Fig. 5   Neon   (mass 20) in Helium

Fig. 6   Nitrogen/Carbon Monoxide (mass 28) in Helium

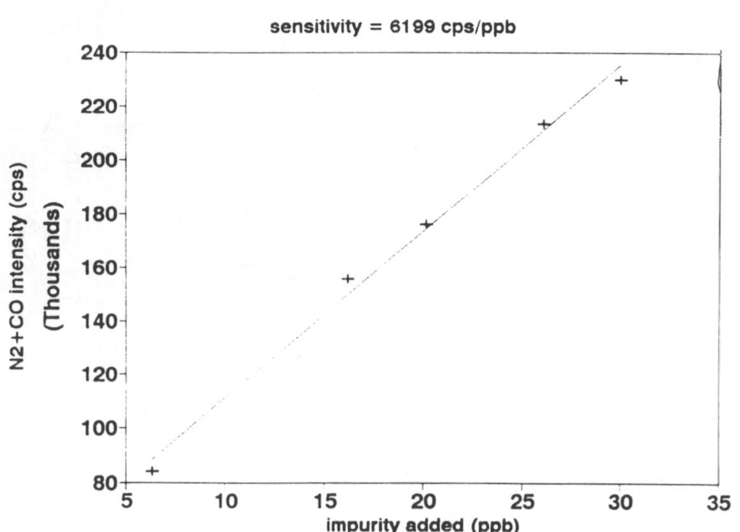

Fig. 7   Argon (mass 40) in Helium

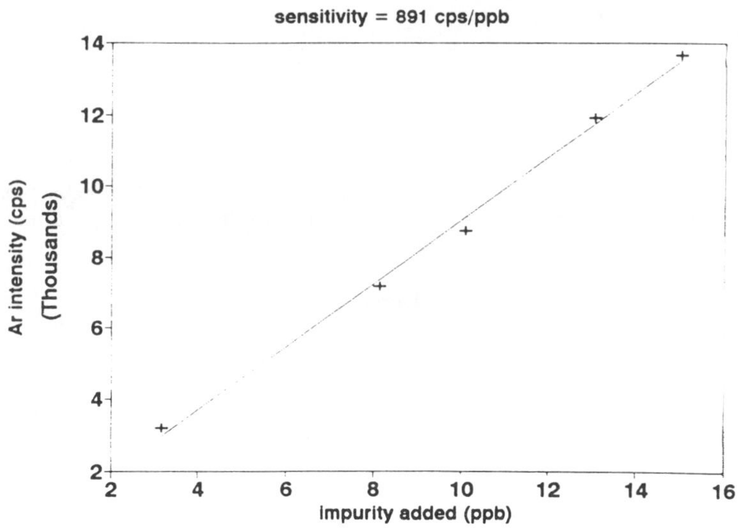

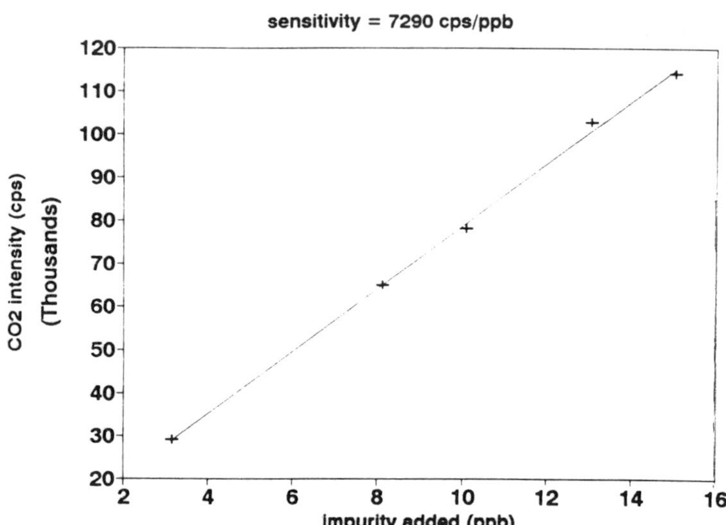

Calibration curves are shown in Figs 3-8 for impurities CH_4, H_2O, Ne, (N_2 + CO), Ar, CO_2. Unfortunately, oxygen was not present in the calibration mixture. This precluded calibration of oxygen levels. Regression analyses indicate good linearity for all measured impurities over this range.

Purified Helium Apims Spectra

After calibration, mass flow controller 1 (MFC 1) was reduced to 10 cc/minute allowing only the ultra pure helium carrier gas be admitted to the APIMS. Equilibrium signal levels were reached in 1-2 minutes for most impurity species, in 10-20 minutes for water. Calibrated sensitivities indicate detected levels of impurities as shown in Table III.

324

TABLE III

Impurity concentrations in helium purified
by condensation and vaporization

Impurity	parts per trillion
Oxygen	not reported
Neon	53
Argon	89
Carbon Dioxide	95
Water	778 **
Nitrogen	2,243 **
Methane	98 *

* Methane peak at mass 16 overlaps with
He_4. The result is given by measuring
CH_3^+ peak at mass 15

** contains background contributions of
instrument. Estimated contribution
comparable with argon and methane
~100ppt

When the helium carrier gas is condensed to liquid, and re-evaporated to gas, the spectra reveals only helium peaks, detecting only the background levels of water and nitrogen normally present in the source. All other peaks disappear to low ppt levels, Fig 9.

Fig. 9   Purified Helium Carrier Gas Spectrum

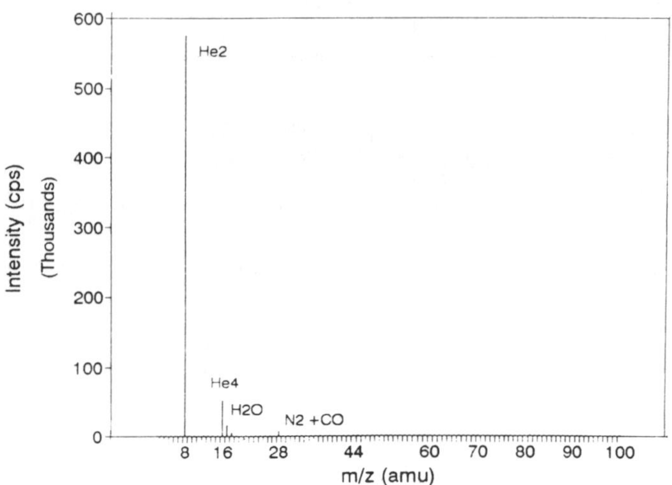

CONCLUSIONS

Purification of helium by condensation to a liquid and re-evaporating to a gas is an efficient and useful method to separate impurity vapor components. All major impurities, water, argon, nitrogen, neon, and carbon dioxide were significantly reduced. Total impurity concentration detected by APIMS in purified gas was under 700 ppt.

The apparatus for producing the ultra pure helium carrier gas is simple and straightforward. This approach to generation of calibration/zero gas standards should find utility in certifying new or recently modified apparatus as well as in determining the hydrocarbon content of specialized samples.

ACKNOWLEDGMENTS

Many thanks for informative discussions with Sam Burton and David Emerson at the U.S. Bureau of Mines, Ken Grosser at Alphagaz and Dr. Catherine Ronge at L'Air Liquide for the donation of helium calibration standards, and Chuck Schneider, Air Products and Chemicals for assistance with helium carrier gas.

REFERENCES

1   D.I. Carroll, I. Dzidic, E.C. Horning, and R.N. Stillwell, Appl. Spectrosc. Rev. 17, 337 (1981).

2   H. Kambara, Y. Ogawa, Y. Mitsui, and I. Kanomata, Anal. Chem. 53, 1500 (1980).

3   E.C. Huang, T. Wachs, J.J. Conboy, and J.D. Henion, Anal. Chem. 62, 713A (1990).

4   C Ma, Steven M. Michael, Mingta Chien, Jianzhong Zhu, and David N. Lubman, Rev. Sci. Instrum. 63 (1), 139 (1992).

HYDROGEN PEROXIDE - A KEY PROCESSING CHEMICAL

R.A. Carpio, J. Baylis, A. Diebold, R. Mariscal,
T. Pinkston, D. Powell, and E. Rosenfeld
SEMATECH, 2706 Montopolis Drive, Austin, TX 78741

ABSTRACT

The analytical methods which are currently used or under development at SEMATECH for evaluating the suitability of hydrogen peroxide for ultra pure processing are outlined. Comparative analytical data for five different types of hydrogen peroxide which were obtained from U.S. suppliers reveal that the stabilizer free grades now on the market are of comparable purity to other processing liquid chemicals from the standpoint of trace metallic and particulate contamination levels. Warranting some concern, however, are the high levels of organic contamination (ppm level). This organic contamination exists partially in the form of mono- and diprotic carboxylic acids. It is shown that the organic contamination level does affect the lifetime of ammonium hydroxide/hydrogen peroxide baths which utilize unstabilized grades of hydrogen peroxide. Hydrogen peroxide stability is linked to the control of processing baths as well as to the storage and distribution of this chemical in its unstabilized form. Considerations for the conversion of a system originally designed for stabilized hydrogen peroxide to one suitable for use with unstabilized peroxide are summarized.

INTRODUCTION

Hydrogen peroxide has occupied an important role in wafer cleaning since the classic paper by Kern on the RCA cleaning process in which hydrogen peroxide is combined with ammonium hydroxide in one step and hydrochloric acid in the second step [1]. Recent modifications of the RCA cleaning process include hydrogen peroxide in combination with HF to prevent metal deposition onto the bare wafer surface [2]. The use of hydrogen peroxide/ sulfuric acid solutions for photoresist stripping and trace organic removal is also a well established application [3]. Hydrogen peroxide is combined with phosphoric acid or ammonium hydroxide for titanium and titanium nitride stripping.

Traditionally, 30 wt% aqueous hydrogen peroxide solutions have been employed in semiconductor processing. Problems which have been encountered with this chemical generally stem its tendency to decompose into water and oxygen. The most common decomposition catalysts are di- and tri-valent metal ions. Stabilizers have been added to hydrogen peroxide solutions to reduce the rate of decomposition. Known stabilizers for hydrogen peroxide include 8-hydroxyquinoline, sodium pyrophosphate, stannic acid, sulfolene, sulfolane, sulfoxide, sulfone, dialklyaminothoxomethyl, thioalklysulfonic acids, aliphatic amines, benzotriazole, nitro-substituted organic compounds such as nitrobenzene sulfonic acids, and thiosulfate. The most commonly encountered stabilizers in semiconductor grade hydrogen peroxide are potassium or sodium stannate, but proprietary stabilizers have also been routinely utilized. Storage of hydrogen peroxide in aluminum containers to minimize decomposition has also led to the introduction of unwanted aluminum impurities. High particulate contamination in hydrogen peroxide has been attributed to the presence of colloidal stabilizers. Moreover, high levels of organic contamination has been reported in hydrogen peroxide [4].

Recent work in the area of in-situ generation of chemicals has been focused upon overcoming some of the contamination problems associated with the use of hydrogen peroxide. It has been shown that, if ozone is used as a replacement for hydrogen peroxide in wafer cleaning, a reduction in metal ion contamination can be achieved [5,6]. An order of magnitude improvement in both particles and metallic impurities can be achieved by the use of the Athens system for the in-situ generation of peroxymonosulfuric acid for use is photoresist stripping when compared to the more the traditional piranha bath made from bottled hydrogen peroxide and sulfuric acid [7,8].

SEMATECH is utilizing stabilizer containing hydrogen peroxide, but the need to switch to the higher purity, stabilizer free grades has been recognized. Aside from purity considerations, the issue of the minimization of surface roughness has emerged as an important processing issue in which hydrogen peroxide plays a competitive role with ammonium hydroxide or HF. Safety regulations which govern the use of processing chemicals have likewise become more stringent.

Trace Metals

Table 1 provides trace metal comparisons between 5 different types of hydrogen peroxide which were obtained from US Suppliers. Four of the five types are current products, while Type A represents a product that was obtained from a vendor approximately 2 years ago and retained in storage under normal room temperature conditions. Also shown are the trace metal levels for ammonium hydroxide. Ammonium hydroxide was selected for this comparison since, from our

Table 1. Trace Metal Levels
Concentrations in ppb (ng/g)

Element	H₂O₂					NH₄OH
	A	B	C	D	E	
Lithium (Li)	<0.03	<0.03	<0.03	<0.03	<0.03	<0.01
Beryllium (Be)	<0.05	<0.05	<0.05	<0.05	<0.05	<0.01
Boron (B)	2.1	<0.8	2.2	3.2	2.4	<0.7
*Sodium (Na)	4.7	2.9	1.3	10	3.7	14
Magnesium (Mg)	1.4	0.6	0.7	1.4	1.8	0.05
*Aluminum (Al)	6.3	2.5	1.4	1.4	31.2	0.8
Potassium (K)	3.9	0.8	2.6	1.9	315	1.2
*Calcium (Ca)	4.3	0.5	1.3	0.6	3.0	3.6
Titanium (Ti)	1.0	0.3	0.3	0.9	0.7	0.9
Vanadium (V)	<0.02	<0.02	<0.02	<0.02	0.04	<0.02
Chromium (Cr)	1.1	0.5	0.5	0.4	4.3	0.1
Manganese (Mn)	0.4	0.04	0.05	0.1	0.5	<0.03
*Iron (Fe)	4.8	0.4	0.4	1.1	7.6	0.8
Nickel (Ni)	0.8	0.2	0.2	0.3	3.9	0.4
Cobalt (Co)	0.04	0.1	0.02	0.02	0.05	<0.03
Copper (Cu)	0.4	0.04	0.05	0.5	0.2	0.8
Zinc (Zn)	5.8	<0.2	1.0	6.5	1.1	1.3
Gallium (Ga)	0.03	0.03	0.03	0.03	0.04	<0.01
Germanium (Ge)	<0.03	<0.03	<0.03	<0.03	<0.03	<0.02
Arsenic (As)	<0.2	<0.2	<0.2	<0.2	<0.2	<0.3
Strontium (Sr)	0.04	<0.01	<0.01	0.03	0.04	<0.01
Zirconium (Zr)	0.02	<0.01	<0.01	0.01	<0.01	<0.04
Niobium (Nb)	<0.01	<0.01	<0.01	<0.01	<0.01	<0.01
Molybdenum (Mo)	0.06	<0.02	<0.02	<0.02	<0.02	<0.1
Silver (Ag)	<0.02	<0.02	<0.02	0.03	<0.02	<0.01
Cadmium (Cd)	0.3	<0.1	<0.1	0.2	<0.1	2.5
Tin (Sn)	<0.03	0.5	<0.03	<0.03	300	0.04
Antimony (Sb)	<0.02	<0.02	<0.02	<0.02	0.1	<0.02
Barium (Ba)	0.1	<0.01	0.07	0.4	0.8	0.04
Tantalum (Ta)	<0.01	<0.01	<0.01	<0.01	<0.01	<0.01
Tungsten (W)	0.08	<0.04	0.07	0.07	0.09	0.03
Platinum (Pt)	<0.04	<0.04	<0.04	<0.04	<0.04	<0.03
Gold (Au)	<0.02	<0.02	<0.02	<0.02	<0.02	<0.1
Thallium (Tl)	<0.01	<0.01	<0.01	<0.01	<0.01	<0.01
Lead (Pb)	0.3	0.05	0.04	0.05	0.2	1.0
Bismuth (Bi)	<0.01	<0.01	<0.01	<0.01	<0.01	<0.01

* Determined by Graphite Furnace Atomic Absorption; all others by ICP-MS

experience, it and HF are the process liquid chemicals with the lowest trace metal contamination levels. It can be inferred from the data that Type E contains a potassium stannate stabilizer. Type E also has the highest aluminum levels. Types B and C are quite close to a 1 ppb specification for all metals, while Types A and D are closest to a 10 ppb specification. The data in Table 1 can be compared to previously published comparisons [9].

Table 2 shows the trace metal levels obtained for Type E hydrogen peroxide over an approximate 6 month monitoring period. The variations in these metals are of interest from the standpoint of insuring processing uniformity. It is clear that the %RSD values are large and show considerable variation. This is partly expected for concentrations at these low levels. On each sampling date, at least two samples were collected from the distribution system in pre-cleaned bottles and then analyzed. The results of the two samples were averaged. Even with modern analytical tools, accurate trace metal analysis is an arduous undertaking which requires meticulous attention to detail.

The data obtained in Tables 1 and 2 were obtained by using inductively coupled mass spectrometry (ICP-MS) and graphite furnace atomic absorption spectroscopy (GFAA). In general, ICP-MS has become the main tool for trace metal analysis due to its low detection limits, multi-element capabilities, and speed. However, mass spectral overlap from atomic and molecular species produced by the plasma gas, impurities in the Ar supply, or from the sample itself is a problem when utilizing a quadrupole mass spectrometer, as was the case for this work (10). For example, $^{38}ArH^+$ will interfere with K; ArO^+ will interfere with Fe; $^{14}N^2$ will interfere with Si; ^{40}Ar and CO_2^+ will interfere with Ca. Due to these interferences or sensitivity considerations, GFAA is utilized for sodium, aluminum, calcium, and trace iron analysis. The graphite furnace technique with the L'vov platform has proven to be popular tool for ultra trace analysis for a decade now (11).

The ICP-MS System used was a Sciex Elan 500 ICP Mass Spectrometer (Perkin-Elmer, Norwalk, CT). The hydrogen peroxide samples were analyzed directly. No special sample preparation was required. It was found that relative to aqueous samples, concentrated hydrogen peroxide, results in only an approximate loss in signal of 5%, and only small departures in the plasma characteristics from optimal conditions. An indium internal standard was utilized to correct for these small variations. A multi-element standard, having elemental concentrations of 20 ug/g, was prepared from single element National Institue of Standards and Technology (NIST) certified standards. For instrument calibration standards having the concentration of 0,1,5, and 10 ng/g were prepared in 1% (v/v) nitric acid. FEP sample bottles were employed for sample containment.

The GFAA work was performed with a Perkin-Elmer 5100PC Atomic Absorption Spectrometer equipped with HGA 500 Zeeman Correction Furnace and an AS-60 Autosampler. A pyrolytically coated L'vov Platform and tube were used for the Al and Fe analysis, while only a pyrolytically coated tube was used for Ca, Na, and K. The hydrogen peroxide samples were analyzed directly. One standard of 10 ppb was used by the autosampler to prepare the other standards.

Recently, attention is being focused upon the use of the new technique of capillary ion analysis (CIA) for trace metal analysis of metals in hydrogen peroxide. This technique has the attributes of speed, multi-component capability, speed, relatively low cost, simplicity, and detection limits in the range of 10-100 ppt. The method developed by Jandik et al. of Millipore Corporation is being employed which utilizes a reduction of peroxide with a platinum catalyst prior to the analysis [12].

Trace Anions and Organic Contaminants

Capillary ion analysis (CIA) procedures are being developed to study anions in hydrogen peroxide. This is a new type of capillary electrophoresis which is used for the separation and detection of both anions and cations [13]. The work described here was performed with a Quanta 4000 Capillary Electropherograph (Waters Chromatography Division of Millipore, Milford, MA) which was configured with a negative power supply and 75 micron I.D. fused silica capillary with a 60 cm total length. The carrier electrolyte consisted of 10 mM of Na_2CrO_4, 68 uM of H_2SO_4, and 0.5 mM of the Waters OFM-BT cationic surfactant which functions as an electroosmotic flow modifier. Indirect UV detection at 254 nm was employed. Figure 1 is the electropherogram of Type A hydrogen peroxide which was diluted 10% v/v prior to the analysis. Ultra pure octane sulfonate was added to serve as the tailing electrolyte during isotachophoretic separation. The peak assignments are indicated in the figure. The acetate and formate peaks are the most prominent peaks. The small unlabelled peak that is adjacent to Peak 3 can probably be attributed to oxalate. Peak 4 is thought to be malonic acid. The peaks which have longer migration times than propionate are probably long chain monoprotic carboxylic acid. In the case of Type E hydrogen peroxide, a stannate peak is present just preceding the phosphate peak, as shown in Figure 2. The largest peak in both Figures 1 and 2 is the acetate peak.

It was found that high concentrations of carboxylic acids, especially acetate, remain after a platinum catalyzed decomposition of hydrogen peroxide. This suggests that some of these organic species may survive severe oxidation processes and may persist in hydrogen peroxide containing solutions. An elevated chloride level was also found after such decompositions which can be explained either on the basis of chloride contamination being released by the catalyst or the presence of halocarbons being present in the hydrogen peroxide which upon oxidation produce soluble halide. This has been shown to occur in electro-oxidative processes of organic compounds [14].

Table 2. Trace Metal Monitoring Data

ELEMENT	MEAN CONC. (ppb)	% RSD	NUMBER OF SAMPLES
Lithium (Li)	0.06	59	5
Boron (B)	15.7	40	10
Sodium (Na)	8.4	69	10
Magnesium (Mg)	3.1	69	10
Aluminum (Al)	25.7	20	10
Potassium (K)	345	2	6
Calcium (Ca)	7.4	72	10
Titanium (Ti)	0.8	52	10
Vanadium (V)	0.04	20	7
Chromium (Cr)	5.0	27	10
Manganese (Mn)	0.4	42	10
Iron (Fe)	7.2	36	10
Nickel (Ni)	3.5	17	10
Cobalt (Co)	0.07	52	5
Copper (Cu)	0.3	101	10
Zinc (Zn)	1.4	32	10
Gallium (Ga)	0.03	52	4
Germanium (Ge)	0.02	25	3
Arsenic (As)	0.05	12	3
Strontium (Sr)	0.04	44	10
Zirconium (Zr)	0.02	113	8
Niobium (Nb)	0.03	---	3
Molybdenum (Mo)	0.83	23	10
Tin (Sn)	238.5	26	10
Antimony (Sb)	0.08	101	9
Barium (Ba)	0.9	44	9
Tantalum (Ta)	0.1	190	5
Tungsten (W)	0.1	59	10
Thallium (Tl)	0.03	15	4
Lead (Pb)	0.1	46	10
Bismuth (Bi)	0.04	---	3

TOTAL   664.9

The inorganic ion anion levels which were obtained by CIA are reported in Table 3. Although the CIA method has excellent sensitivity for fluoride and phosphate, the presence of organic species clearly make their quantitative determination more complicated than for chloride, sulfate, and nitrate. No phosphate stabilizers were found to be present in any of the types which were tested. More details on the capillary electrophoresis investigation of these species will be presented at a later date [15]. Although little attention is normally paid to anionic contamination, anions such as chloride are strong ligands and form complexes with metal ions which are present. It is well known in the field of polarographic analysis of metals that the reduction potentials of metals as well as the mechanism for reduction are dependent upon the type of metal complex present.

Table 3. Trace Inorganic Ion Levels Determined by CIA

TYPE	CHLORIDE	SULFATE	NITRATE
A	91	50	48
B	30	36	22
C	32	42	24
D	27	48	44
E	25	32	121

The total organic contamination levels for these 5 different types of hydrogen peroxide were determined at the Balazs Analytical Laboratory which has developed special analytical methods for process liquids [4]. These data are shown in Table 4, along with the acetate levels which were determined by the CIA method. Figure 3 shows that the acetate levels are directly proportional to the TOC levels for 4 of the 5 types of hydrogen peroxide. The regression equation is:

$$\text{Acetate conc} = -3.55 + 0.99 * (\text{TOC conc})$$

and the correlation coefficient is 0.998

Type C hydrogen peroxide does not conform to this pattern and was omitted from Figure 3. The reason for Type C being an outliner is not clear at this time. However, according to the Balazs Laboratory, this has been the highest value of TOC found for hydrogen peroxide to date, so a new sample from a different bottle has been submitted for verification. This hydrogen peroxide was repackaged prior to shipment to SEMATECH, so the possibility of bottle contamination being a source of organic contamination exists.

Particulate Contamination

The particulate levels below 1 micron for Type A and Type B hydrogen peroxide are displayed graphically in Figure 4. Figure 5 shows the particulate data for ammonium hydroxide for comparison. It appears that the distributions for both chemicals are not exponential over the entire size range. The particulate levels in Type B, unstabilized hydrogen peroxide, are comparable to ammonium hydroxide. Type A, which was the product on the market 2 years ago, has considerably higher particle levels. Many semiconductor facilities will obtain the hydrogen peroxide in bulk and filter the hydrogen peroxide before going into the distribution system and even at point of use, so levels lower than those of the bottled chemicals can be obtained for processing.

These particle counts were obtained with a PMS System CLS-600 Corrosive Liquid Sampler in conjunction with an IMOLV-.2- HF(LD) Integrated Micro-Optical Liquid Volumetric Sensor, and a Micro LPS-16 Laser Spectrometer (Particle Measuring Systems, Inc., Boulder, CO). The CLS-600 is designed to draw a liquid from an unpressurized container. Bubbles are eliminated by the application of pressure. Good reproducibility for hydrogen peroxide was obtained using a compressive pressure of 60 psi. The data reported here represent the means of 4 samples, each 45 ml in volume.

Processing Considerations

Kern noted in his classic paper [1] that the half-life of an SC-1 type bath which consists of 1:1:5 (v/v) H_2O_2-NH_4OH-H_2O is dependent upon the impurity level. This was shown recently by Oshida by adding different iron concentrations to a high purity grade of the Mitsubishi Gas Chemical Company hydrogen peroxide [16]. In this study a comparison of the various grades of hydrogen peroxide was made, using the same ammonium hydroxide, whose composition is shown in Table 1. The appropriate volume of water was preheated to 85° C, and then the hydrogen peroxide and ammonium hydroxide were added and stirred thoroughly before hydrogen peroxide concentration monitoring was commenced. The hydrogen peroxide concentration was established by a visual titration using standardized permanganate. The data that were obtained are plotted in Figure 6. The bath half-life was computed from the data over the range of approximately 10 to 60 minutes where first order reaction kinetics are most applicable. That is, the data were fit to the equation $M = M_o\, e^{-kt}$ where M is the molar concentration and k is the first-order rate constant. The half-life is then computed from k, using the equation $t_{1/2} = 0.693/k$) The experimental half-lives are given in Table 4. For comparison purposes, Kern in 1970 reported a corresponding half-life of approximately 5 minutes for the temperature range of 88-90° C [1], while the data of Meerakker and Straaten in 1990 can be used to compute a half-life of approximately 4.6 min at 85° C [17]. Meerakker and Straaten, however, performed their kinetic study in a flask to which a reflux condenser was attached to prevent excessive NH_3 loss, while in this work the NH_3 is free to escape. The half-lives reported here were determined using a plastic bath container which itself may have served as a heterogeneous catalyst for the hydrogen peroxide decomposition.

It is apparent from Figure 7 which show the data in Table 4 on a relative basis that the half-lives are not solely determined by the iron concentration. It is known that a number of di- and tri-valent ions catalyze the decomposition of peroxide, so the composite metals comparison, consisting of Fe, Zn, Cr, Ni, and the Al concentration, are included. Type E has the highest combined level of catalytic metals , but it also has a stabilizer present. Types B and C have the lowest combined levels. The results for Type D are especially worthy of note, since this type of hydrogen peroxide clearly produces the bath with the longest half-life, even though it's metal content is higher than the levels for Types B and C. The TOC is lower than that of the 3 types of hydrogen peroxide which are stabilizer free. It is known that organic materials can catalyze the decomposition of hydrogen peroxide, so these results lead to the conclusion that the organic contamination in the hydrogen peroxide itself is playing a major role in accelerating the rate of decomposition in the SC-1 type bath in cases were the hydrogen peroxide is stabilizer free. The long bath life for Type C are not consistent with the experimental TOC, just as we have seen earlier for the TOC relationship upon acetate concentration. Unfortunately lacking from this study is a low TOC grade of hydrogen peroxide which is also as low in trace metals as Types B and C. Such a grade of hydrogen peroxide is, however, available.

Table 4. Summary of Lifetime Data

TYPE	ASSAY (wt %)	SC-1 HALF LIFE (min.)	COMPOSITE METALS* (ppb)	TOC (ppm)	ACETATE (ppm)
A	22	24	18.8	12	8.3
B	31.3	51	3.6	8.2	4.8
C	30.9	64	3.5	30	2.9
D	30.9	90	9.7	6.5	2.9
E	32.4	36	48.1	5.7	2.0

* Sum of Fe, Zn, Cr, Ni, and Al concentrations

Megasonic cleaning has become a popular wafer cleaning technique which combines the simultaneous desorption of chemical contaminants with the physical removal of particulate impurities [18]. The use of hydrogen peroxide/ammonium hydroxide/water solutions are widely utilized, but the composition may differ from the standard SC-1 RCA clean and a lower temperature is generally used. Figure 8 shows that the depletion of both H_2O_2 and NH_3 occur rather rapidly from a Megasonic bath which utilizes 1:2:5 (v/v) $NH_4OH:H_2O_2:H_2O$ composition and is operated at 60° C without any wafers being processed. Type E hydrogen peroxide was used for this study. Both the hydrogen peroxide and the ammonium hydroxide concentration must be carefully measured and controlled by appropriate additions of reagents to improve processing uniformity. Studies of the effects of residues and impurities that are introduced during processing on the rate of hydrogen peroxide decomposition are in progress. The control of the ammonium hydroxide and hydrogen peroxide composition in the ammonium hydroxide/hydrogen peroxide/water bath is essential in order to minimize wafer microroughness [19].

Considerations for Using Stabilizer Free Hydrogen Peroxide

A complete system approach is being employed for the hydrogen peroxide delivery system. The hardware selection and installation are critical. The stability of hydrogen peroxide is of great importance from the standpoint of storage and distribution considerations. It is known than hydrogen peroxide can decompose violently when contacted with agents which can function either as homogeneous or heterogeneous catalysts. An appreciable amount of hydrogen peroxide decomposition can lead to a "runaway" reaction with the generation of heat, oxygen, and steam. The most severe safety concern is pressure build up that can lead to rupture of the containment pipes or vessels and which in turn can result in a discharge of peroxide into the facility. Temperature control is an important concern in reducing the rate of peroxide decomposition. This requires that the hydrogen peroxide distribution system be located well away from heat sources such as boilers, motors, or other facility equipment that evolve heat. The issue of trace contaminants has already been discussed, however, that of materials of construction, accidental introduction of foreign material, and back flow from other systems into the hydrogen peroxide or the erroneous introduction of other chemicals into the hydrogen peroxide containment system are worthy of special mention. The cross contamination of a base with the hydrogen peroxide is likely to lead to more serious problems than an acid, due to the pH dependence of the decomposition reaction.

SEMATECH is currently considering converting to unstabilized hydrogen peroxide. All materials of construction of the distribution system which was designed for stabilized hydrogen peroxide had to be evaluated. A listing of compatible and incompatible materials is given in Table 5. The original SEMATECH distribution system is Perfluro Alkoxy, or PFA Teflon, so major material modifications are not required. It was determined that a pressure relief valve needs to be installed between each pair of existing control valves where trapped H_2O_2 can cause a pressure build up. Metal studded air operated valves (AOVs) had to be replaced with nonmetallic AOVs that are constructed of Teflon components only. The elevation of the liquid distribution system also required a pressure relief valve. It was decided to install leak detection equipment in the valve boxes housing the pressure relief valves which will alarm in cases of overpressurization. Any liquid released from the pressure relief system is drained to the acid waste neutralization/collection system.

Table 5. Materials Compatibility Summary

<u>COMPATIBLE OR RECOMMENDED MATERIALS</u>:

Aluminum
* 99.5% minimum purity alloys with Aluminum Association designations 1060, 1260, 5264, 5652, 6063

Stainless Steel
* 304, 304L, 316, 316L

Other Materials
* Chemical glass
* Chemical ceramic
* Polytetrafluoroethylene (PTFE, Teflon®), Teflon®
* Polyethylene (high density, cross linked, unpigmented and UV stabilized), for 50% and lower concentrations of H_2O_2.
* Viton, Kel-F, Tygon

<u>UNACCEPTABLE</u>:

Brass
Copper
Nickel
Iron and mild steel
Bronze
Synthetic rubbers
Polypropylene
Polyvinyl chlorides
* PVC or CPVC

The processing tools which utilize hydrogen peroxide were also reviewed for compatibility. Following the SEMATECH partnering approach, the equipment suppliers were contacted and a dialogue was set up to determine if tool modifications were necessary. It was determined that modifications of plumbing configurations which allowed the liquid chemical to be trapped between valves for extended periods were required. It was also established in one case that a polypropylene beaker dispense system had to be converted to a compatible material. Some of the tools already had the necessary safety features present which provide for the release of material to acid waste in the case of pressure build up.

DISCUSSION

It is clear from the above data that significant improvements have been made in reducing the trace metals and particulate levels in stabilizer free hydrogen peroxide to levels which are comparable to other processing chemicals. These improvements have certainly reduced the hydrogen peroxide decomposition rates, making process control easier in cases where hydrogen peroxide is employed in bath type processing over long terms with intermittent usage and storage. In contrast to the attention that has been paid to trace metal ions in process chemicals, relatively little attention has been paid to the role played by organic impurities and trace anions. This can in part be attributed to the lack of analytical tools for detection low levels of organic contamination, as well as the lack of a complete understanding of the role played by these impurities. The effects of organic contamination is clearly a cause for additional study. The organic contamination is due to the synthetic process which utilizes an auto-oxidation of anthraquinone. The hydrogen peroxide which is formed must be separated from the organic fraction, but residual organics still remain in the ppm level range. These organics are probably straight chained and branched hydrocarbons with a range of molecular weights, polarities, and functional groups, which can be oxidized to various degrees, depending upon the oxidation process. It is also likely that the actual organic fraction used in the synthetic process can vary between batches. The organic species may have beneficial, detrimental or a combination of both properties. From a beneficial standpoint, they might function as chelating agents for trace metals and they may have surfactant properties. Ohmi and associates have in fact shown that aliphatic carboxylic acids and their salts function as surfactants in buffered hydrofluoric acid solutions [20]. On the other hand, some evidence is provided in this investigation that these organic contaminants serve as catalysts for hydrogen peroxide decomposition. Organic contaminants may also be deposited on wafer surfaces and lead to poor film adhesion, nucleation irregularities, and variations in oxide growth rate and quality. Budde and Holzapfol have shown by the use of ion mobility spectroscopy that the organic contamination on a wafer increases by a factor of four as a result of an RCA clean [21]. It should be noted that a great deal of emphasis has been placed upon the measurement and reduction of TOC in process DI water. Now modern semiconductor fabrication facilities utilize DI water with TOC levels (typically, < 5 ppb) which are nearly undetectable by modern analytical instrumentation. Thus, it is quite inconsistent that a chemical as important as hydrogen peroxide which is used in critical wafer cleaning steps has TOC levels in the ppm range.

ACKNOWLEDGEMENTS

Thanks are extended to James Fusco of Particle Measuring System for the use of the PMS Liquid Particle Counting System. Petr Jandik of Waters Chromatography Division of Millipore Corporation is acknowledged for technical and hardware assistance in the CIA area. We are grateful to Genie Kane for her assistance in the manuscript preparation. The anonymous chemical vendors must, likewise, be acknowledged for providing samples and beneficial information.

REFERENCES

[1]   W. Kern and D.A. Poutinen, RCA Rev. **31**, pp. 187-206 (1970).

[2]   T. Imaoka, T. Kezuka, J. Takano, M. Kogure, T. Isagawa, H. Shimada, and T. Ohmi, "The Segregation and Removal of Metallic and Organic Impurities from Interface of Silicon and Liquid Chemicals", in *Chemical Proceedings of the 1992 Semiconductor Pure Water and Chemicals Conference*, Editor M. K. Balazs, 1992, pp. 191-198.

[3]   W. Beck et al. "Method For Stripping Layer of Organic Material" U.S. Patent 3,900,337 Aug. 19, 1975.

[4]   M.J. Camenzind and M.K. Balazs, "Analysis of Organic Impurities in Semiconductor Processing Chemicals," in *Chemical Proceedings of the 1992 Semiconductor Pure Water and Chemicals Conference*, Editor M.K. Balazs, 1992, pp. 143-161.

[5]   R.A. Matthews, "In-Situ Generated Aqueous Chemicals and Their Impact on Chemical Purity," in *Chemical Proceedings of the 1992 Semiconductor Pure Water and Chemicals Conference*, Editor M.K. Balazs, 1992. pp.3-15.

[6]   J.K. Tong, D.C. Grant, and C.A. Peterson "Aqueous Ozone Cleaning of Silicon Wafer" in *Proceedings of The Second International Symposium on Cleaning Technology in Semiconductor Device Manufacturing*, J. Ruzyllo and R. E. Novak, eds., Proc. **92-12** The Electrochemical Society, pp.18-25 (1992).

[7]   C.R. Cleavelin and A.H. Jones, Jr., Solid State Technology, Dec. 1988, pp. 53-56.

[8]   J.B. Davison and J.G. Hoffman "Ultrapure Piranha Solution for ULSI Applications" in *Proceedings of the First International Symposium on Ultra Large Scale Integration Science and Technology*, Vol. **87-11**, S. Broydo and C. M. Osburn,eds., The Electrochemical Society, 798 (1987).

[9]   S. Tan and M.K. Balazs, Semiconductor International, Nov. 1988, pp. 100-102.

[10]   J.W. Oleski, Anal. Chem., **63**, 12A (1991).

[11]   S.R. Koirtyohann and M.L. Kaiser, Anal. Chem. **54**, 1515A (1982).

[12]   E. Fallon, A. Weston, and P. Jandik, "New Approach to the Analysis of Cations In Pure Chemicals" pre-print obtained from P. Jandik of Waters Chromatography Division, Millipore Corp., Milford, Massachusetts.

[13]   P. Jandik, W.R. Jones, A. Weston and P.R. Brown, LC GC 9, 634 (1991).

[14]   T.C. Franklin, J. Darlington, T. Sokouki, and N. Tram, J. Electrochem. Soc. 138, 2285 (1991).

[15]   International Ion Chromatography Symposium to be held September 21-24, 1992 in Linz, Austria. Proceedings to be published in a special issue of J. Chromatography.

[16]   Y. Oshida "High Purity Hydrogen Peroxide for Semiconductor Manufacturing Processes", in *Chemical Proceedings of the 1992 Semiconductor Pure Water and Chemicals Conference*, Editor M.K. Balazs, 1992, pp. 63-82.

[17]   J.E.A.M. van den Meerakker and M.H.M. van der Straaten, J. Electrochem. Soc., 137, 1239 (1990).

[18]   S. Shwartzman and A. Mayer, RCA Review, 46, 81 (1985).

[19]   T. Ohmi, M. Miyashita, M. Itano, T. Imaoka, and I. Kawanabe IEEE Transactions on Electron Devices, 39, 537 (1992).

[20]   H. Kikuyama, N. Miki, J. Takano, and T. Ohmi, Microcontamination, April, 1989 p. 25.

[21]   J.K. Budde and W. Holzapfol, "Measurements of Organic Contaminations from Silicon Surfaces," in *Proceedings of the Institute of Environmental Sciences*, Vol. 1, 1992, pp. 483-488.

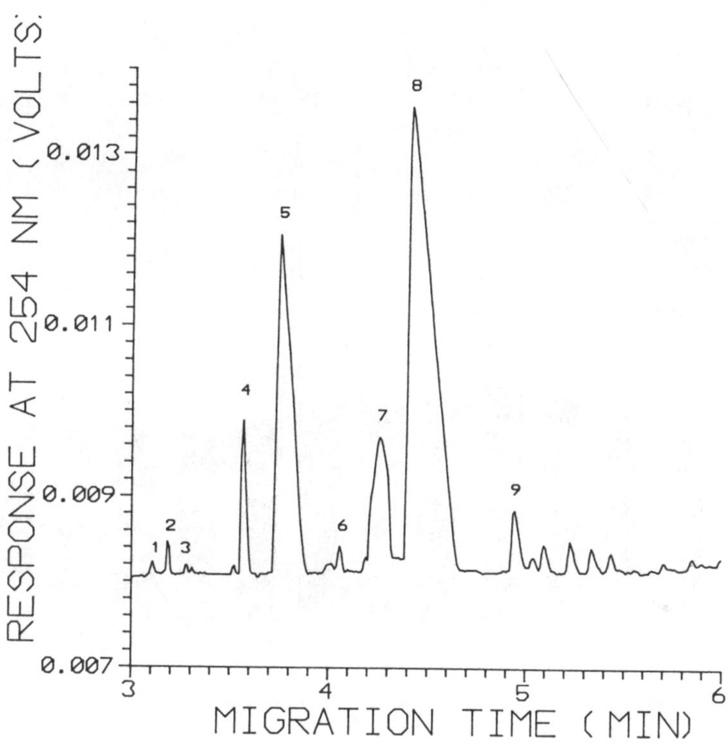

Figure 1. Electropherogram of Type A Hydrogen Peroxide which was diluted 1/10 with Ultra Pure DI Water. Conditions are described in the body of the text. Peak identification: 1, chloride; 2, sulfate; 3, nitrate; 4, unknown; 5, formate; 6, phosphate; 7, carbonate; 8, acetate; and 9, propionate.

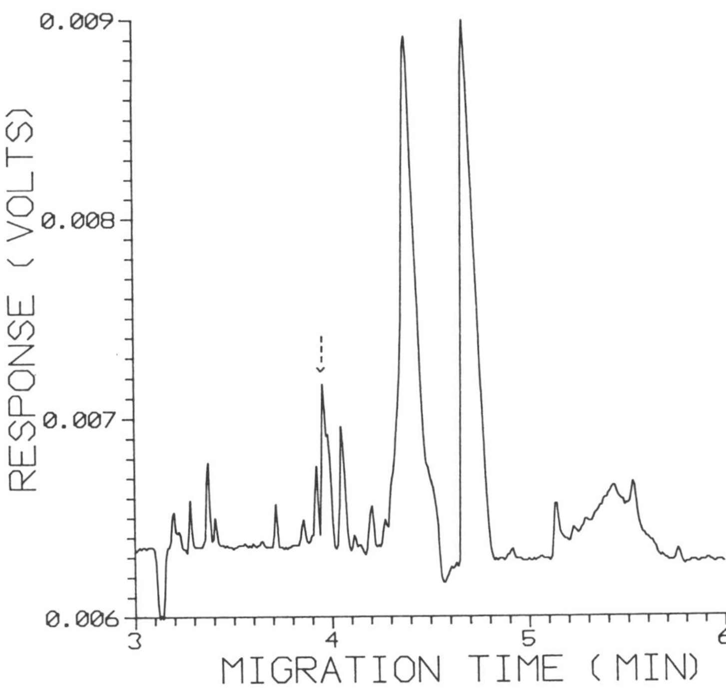

Figure 2. Electropherogram of Type E Hydrogen Peroxide (10% v/v) With Location of Stannate Peak Shown by an Arrow. Same conditions that were used for Figure 1.

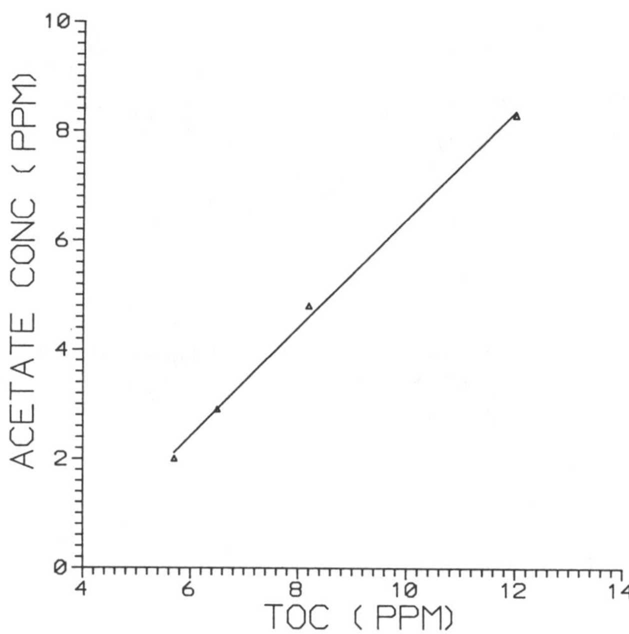

Figure 3. Graph Showing Relationship Between Total Organic Carbon and Acetate Anion Concentrations for Four Different Types of Hydrogen Peroxide.

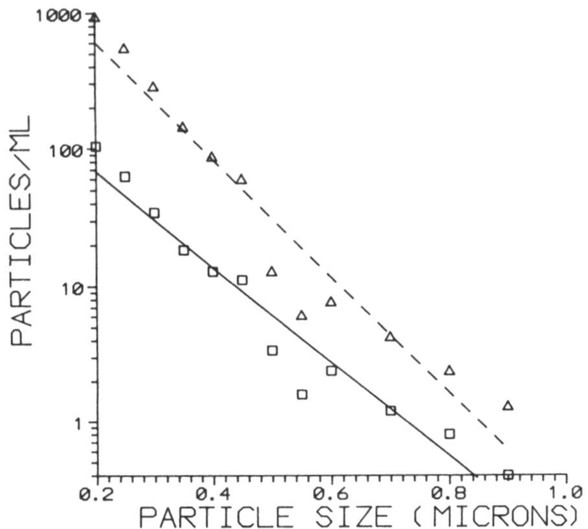

Figure 4. Particle Level Distribution Curves for Two Types of Hydrogen Peroxide. Type A- Dashed Curve. Type B- Solid Curve.

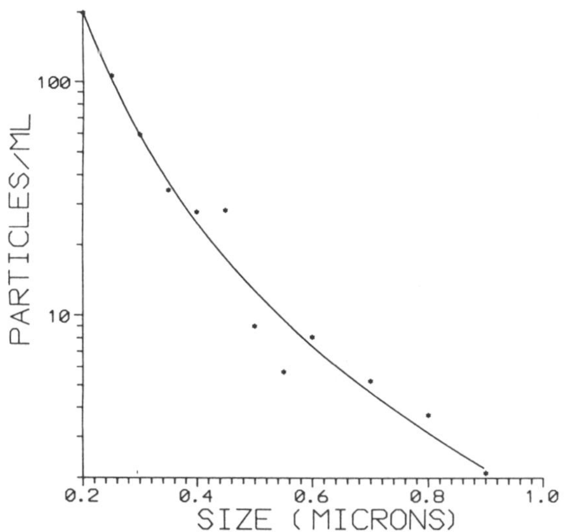

Figure 5. Particle Level Distribution Curve for SEMI Grade NH_4OH (28 % NH_3) Fitted to the Power Curve $Y = 1.59 \, X^{-3.004}$

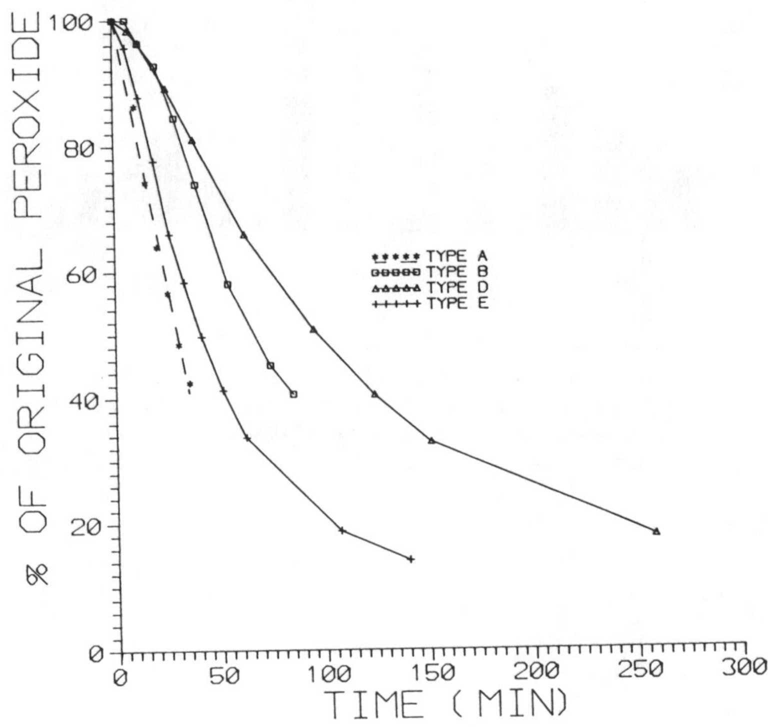

Figure 6. Hydrogen Peroxide Decomposition Rate Curves for 1:1:5 (v/v) NH4OH: H2O2:H20 Bath at 85° C

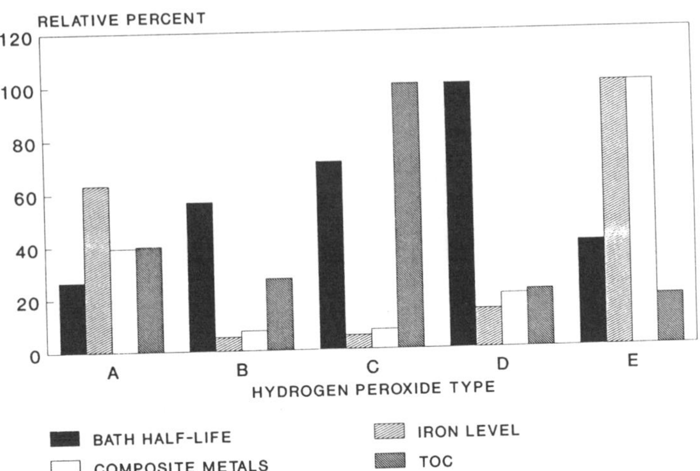

Figure 7. Comparisons on Relative Scale of Hydrogen Peroxide Half-Lives in an SC-1 Type Bath at 85° C, Trace Iron Levels, Major Composite Trace Di- and Tri-Valent Metals, and TOC Levels for Five Different Types of Hydrogen Peroxide

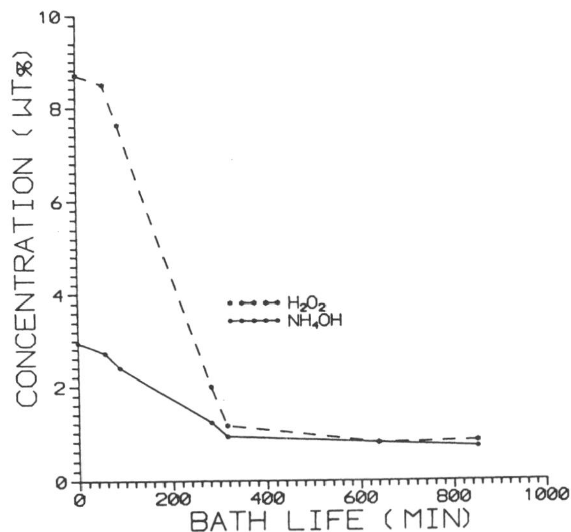

Figure 8. Changes in H_2O_2 and NH_4OH Concentrations for a Megasonically Agitated Bath Consisting of 1:2:5 (v/v) NH_4OH:H_2O_2:H_2O and Ramped from 25 °C to 60°C During First 90 minutes.

STUDY ON AN INFLUENCE OF TOC
IN HYDROGEN PEROXIDE FOR
ADVANCED WET CHEMICAL PROCESSING

K.Kimura, Y.Ogata, F.Tanaka
Santoku Chemical Industries Co. Ltd.
10–2 Ohtake–shinden, Imozawa, Aoba–ku, Sendai, Japan, 989–32

T.Imaoka, J.Takano, T.Isagawa, T.Kezuka, M.Kogure,
T.Futatsuki, and T.Ohmi
Department of Electronics, Faculty of Engineering,
Tohoku University
Aza Aoba, Aramaki, Aoba–ku, Sendai, Japan, 980

An improved high purity hydrogen peroxide, where metallic impurities were removed to the level of ppt concentration, was developed for advanced semiconductor devices manufacturing. After treatment of a Si wafer with this hydrogen peroxide, no metallic impurity was adsorbed onto the Si wafer surface. However a few ppm of organic impurities in the hydrogen peroxide remain as TOC. These organic impurities are caused by for production process of raw hydrogen peroxide and identified as formic acid, acetic acid, and esters. In order to examine the influence of organic impurities on semiconductor devices, we evaluated the electrical characteristics of MOS diodes with gate oxide films, which includes native oxide by the treatment with the improved hydrogen peroxide. In spite of high contents of organic impurities, the experimental results show that the TOC in the hydrogen peroxide did not affect the properties of a 9 nm oxide film.

Keywords: Hydrogen peroxide, TOC, Organic impurities, Metallic impurities, Trace metals, Electrical characteristics, MOS diodes, native oxide.

INTRODUCTION

It is essential to realize ultra clean Si surfaces for device performance and reliability since the manufacturing of semiconductor devices are always carried out on the surfaces and interfaces of Si wafers. Therefore the purity of the chemicals used in the Si wafer cleaning process has become a critical issue.

Especially with hydrogen peroxide, the purest product is required, because respect to hydrogen peroxide solutions are used throughout in RCA cleaning process [1]. Also very recently the use of a mixed solution of diluted hydrofluoric acid and hydrogen peroxide for the final cleaning solution has been evaluated and recommended as an excellent cleaning solution [2].

At present, although the metallic impurities in wet chemicals can be controlled to a level less than 1 ppb, the content of total organic carbon / total oxidizable carbon (TOC) is in general considerably higher, a level of a few ppm [3]. Organic impurities in hydrogen peroxide remain at a level from a few ppm to a few tens of ppm even in the semiconductor–grade. This comes from the fact that almost all hydrogen peroxide makers employ the automatic auto–oxidation method using organic catalysts and extraction solvents.

As an example of the influence of organic impurities upon the Si wafer surface, it is reported that the native oxide cannot be completely removed if some kinds of organic impurities are left on the Si wafer surface [4]. **Figure 1** shows X–ray Photoemission Spectroscopy (XPS) spectra of Si_{2p} on the Si wafer surface. When surfactant molecules remain on the Si wafer surface, native oxide cannot be completely removed after cleaning by a diluted hydrofluoric acid solution. **Figure 2** shows XPS spectra of C_{1s} on the Si wafer surface before cleaning by a diluted hydrofluoric acid solution. A sub–peak of adsorbed surfactant molecules is found on the wafer surface upon which the native oxide cannot be removed by a diluted hydrofluoric acid.

Recently the relationship between increasing the concentration of organic impurities on a Si wafer and a decreasing of the electric breakdown field intensity was reported [5]. Thus, it is clear that organic substances, which easily adsorb on the Si wafer surface and are difficult to remove from it, should be completely eliminated from cleaning chemicals. Also it is important to test the influences of organic matters on the quality of devices fabricated on such surfaces.

We produced a special hydrogen peroxide solution as a cleaning chemical for next generation of semiconductors which is highly purified. Here, we report the impurity content of it and the influence of some organic materials contained in it on the electrical properties of the oxide film on MOS diodes treated by the hydrogen peroxide.

CHEMICAL ANALYSIS OF THE SPECIAL HYDROGEN PEROXIDE

Metallic Impurities

In **Table 1** are shown all impurities except organic impurities contained in the special hydrogen peroxide. The analysis for metals has been carried out mainly using an Inductively Coupled Plasma – Mass Spectrometry (ICP–MS : Seiko SPQ–8000). Some elements for which the ICP–MS is not suitable owing to some interferences, were recorded on a Graphite Furnace Atomic Absorption Spectroscopy (GFAA : Varian Spectra AA–40G) and on an Inductively Coupled Plasma – Atomic Emission Spectrometry (ICP– AES : Jarrell Ash Japan 575 mark–2). No concentration procedures were used for the sample solutions.

We focused our attention on the concentration of metals such as Au, Pt, Ag, Cu of which electronegativities are higher than that of Si and hence are easily adsorbed on the bare Si wafer surface as well as other metals such as Al, Cr, Fe which are oxidized more readily than Si to be preferentially included into the native oxide. All of these elements are removed to a level of ppt or below which is the detection limit of the instruments.

Removal of Silicon as An Impurity

Figure 3 is a typical chart of Energy Dispersive X–ray Spectrometry (EDX), for the particle in a conventional hydrogen peroxide which was captured on a filter for analysis. In general, large silicon peaks are always accompanied by peaks for metals such as Al, Fe, Cu, Ca etc. This observation strongly suggests that silicon as an impurity in a conventional hydrogen peroxide exists as metal silicates. Therefore, it is essential to completely remove the trace silicon to get a practically metal–free hydrogen peroxide.

Now, we have established a way to remove silicon as an impurity to a degree below the instrumental detection limit, 0.5 ppb as shown in **Table 1**. At the same time, other metals seem to be removed together with silicon by the reason described above.

Organic Impurities

Table 2 shows the organic impurities in the special hydrogen peroxide. The TOC analysis was carried out using a TOC meter with a high temperature combustion oxidation / Non–Dispersive Infrared Radiation (NDIR) method, (Simazu TOC–500) optioned with a High Sensitive Catalyst. Nitrogen gas was bubbled into the sample solution in order to exclude carbon dioxide. Decomposition of the hydrogen peroxide was not carried out before injection into the instrument. The content of TOC was 3.0 ppm.

The contents of organic acids in TOC was obtained using an Ion Chromatography (IC) of the type using ion exclusion (Yokokawa IC–100). The hydrogen peroxide in the sample solutions was decomposed by Pt–catalyst before injection to protect the instrumental column. The content as TOC of formic acid was 1.3 ppm and acetic acid was 0.4 ppm. However the data shown in **Table 2** is possibly different from the net contents of acids in the special hydrogen peroxide because the radical decomposition process can be induce some changes in the organic compounds.

Nonvolatile residue was extracted into chloroform and then qualitatively analyzed by Gas Chromatography – Mass Spectroscopy (GC–MS: JOEL JMS–SX102). As shown

TIC mass chromatogram in **Figure 4** a main portion in the nonvolatile residue seems to be esters, which will come from a derivative of anthraquinone as a catalyst and extraction solvents used in manufacturing process of raw hydrogen peroxide.

ADHESION OF METALS ON Si WAFER

Metallic impurity adhesion on the Si wafer surface immersed in the special hydrogen peroxide was analyzed by Total Reflection X–ray Fluorescence Spectroscopy (TRXRF: Technos TREX610), as shown in **Figure 5**. The employed wafers are CZ–n(100) and CZ–p(100), which were immersed in 30% of the special hydrogen peroxide contained in a quartz beaker. The analytical results by TRXRF were compared with that before immersion as references.

As we can see from **Figure 5**, differences could not be found between the samples and the references. No metals were detected in the special hydrogen peroxide adsorbed on the Si wafer surface beyond the concentration of more than 1×10^{10} atoms/cm^2.

ESTIMATION OF ELECTRICAL CHARACTERISTICS OF OXIDE FILM

As discussed above, no problems were found with the metallic impurities contained in the present special hydrogen peroxide for treating Si wafer, however, the amount of TOC is still very large in comparison with that of the metallic impurities. Thus, it is necessary to check whether or not the organic impurities affect the properties of the gate oxide film possibly by adsorbing onto or being occluded into the wafer during the wet process for forming the native oxide film. So we evaluated the electrical characteristics of gate oxide films intentionally including native oxide by treatment of the special hydrogen peroxide.

The experimental set up for the check is shown in **Figure 6**. The Si wafers used were phosphorus doped CZ–n(100) (8–12 ohm•cm). The MOS diodes, Al/SiO$_2$/Si, were prepared by three different ways as follows: the native oxide of the thickness of about 0.7 nm was prepared by immersing the wafers in (1) H$_2$SO$_4$ / H$_2$O$_2$ (SPM), (2) 30% of the special hydrogen peroxide at 85 °C (H$_2$O$_2$), and (3) 2ppm ozone injected ultrapure water (O$_3$), then thermal oxide films of the thickness of about 9 nm with native oxide were formed by dry oxidation at 900 °C on each wafer. As for the reference (control), the native oxide film was removed by diluted hydrofluoric acid just before the thermal oxidation. The thickness of native oxide films were estimated as corresponding to the thermal oxide thickness from the Si$_{2P}$ spectrum by XPS (Surface Science Laboratories: ESCA Model–206).

Any significant differences were not found among those three MOS diodes and the blank run, as tested by the Flat–band Voltage : Vfb (**Figure 7**), the charge to breakdown :Q$_{BD}$ (**Figure 8**), the interface–trap density : Dit(**Figure 9**), the barrier height

(**Figure 10**), and the electric breakdown field intensity :E_{BD} (**Figure 11**) at room temperature, respectively. Especially with the E_{BD}, there were found neither the A–mode nor the B–mode defect at the low field region.

In conclusion, the TOC as impurity in the present special hydrogen peroxide does not affect the devices with the gate oxide film of about 9 nm.

The realization of essentially metal–free hydrogen peroxide is the basis for the remarkable results cited here. Even if the concentration of TOC is much lower, a large deterioration in the electrical characteristics was reported [7] when metallic impurities exist in solutions which contact the Si wafer. When the influence of metallic impurities in hydrogen peroxide was completely eliminated, it was proved the first time that the organic impurities remaining in the hydrogen peroxide in the order of ppm do not affect the device performance.

SUMMARY

A technology to produce ultra purified hydrogen peroxide for advanced wet chemical processing has been developed in an industrial scale, the metallic impurities in which are controlled within a level of ppt. No adsorbed metal elements were detected on Si wafers.

The organic impurities arising from the production process of raw hydrogen peroxide still remain in the newly developed hydrogen peroxide in the amount of a few ppm as TOC, the main components of which have been identified as formic acid, acetic acid, and various kinds of ester. The electrical characteristics of MOS diodes formed on wafers upon which a native oxide had been formed using the special hydrogen peroxide and then a 9 nm gate oxide subsequently grown over top of this oxide by conventional methods were not affected by any of the impurities contained in the special hydrogen peroxide used in these experiments. Therefore the native oxide films which formed by the special hydrogen peroxide treatment are used as a protective film for the Si wafers to make it possible to handle the wafers without any contaminations. And the Si wafers which has the protective film can be proceeded to the next process without removal of this protective film.

However, the present achievement should be estimated when much thinner oxide film, say a thickness of less than 5 nm [8], will have been realized in future. Further efforts to reduce much more all impurities including the TOC should be made in order to realize much more purified chemicals to develop the advanced semiconductor industries.

ACKNOWLEDGMENTS

The experiments were conducted at the Mini Super Clean Room by the Engineering Faculty at the Laboratory for Microelectronics attached to the Electric Communication Laboratory in Tohoku University. We gratefully acknowledge the help from the members. And we exceedingly appreciate Dr. Toshiki Wakabayashi at College of General Education, Tohoku University for precise advice and suggestions on the development of a ultra purified hydrogen peroxide constantly.

REFERENCE

[1]   Werner Kern and David A.Puotien, "Cleaning Solutions Based on Hydrogen Peroxide for use in Semiconductor Technology", RCA Review 31. pp.187–206, June(1970)

[2]   Takashi Imaoka, Takehiko Kezuka, Jun Takano, Masahiko Kogure, Tatsuhiko Isagawa, Hisayuki Shimada, and Tadahiro Ohmi, "The Segregation and Removal of Metallic and Organic Impurities from Interface of Silicon and Liquid Chemicals", Chemical Proceedings, 1992 Semiconductor Pure Water and Chemicals Conference, Santa Clara, pp.162–190, Feb.(1992)

[3]   Mark J. Camenzind and Marjorie K. Balazs, "Analysis of Organic Impurities in Semiconductor Processing Chemicals", Chemical Proceedings 1992, Semiconductor Pure Water and Chemicals Conference, Santa Clara, pp.143–161, Feb.(1992)

[4]   Tatsuhiko Isagawa, Masahiko Kogure, Takashi Imaoka, and Tadahiro Ohmi, "Ozone Added Ultrapure Water Applications for ULSI Advanced Processing", Chemical Proceedings, 1992 Semiconductor Pure Water and Chemicals Conference, Santa Clara, pp.224–249, Feb.(1992)

[5]   Ayako Shimazaki, "Analytical Methods for Wafer Surface Contamination", Extended Abstracts of International Symposium on Semiconductor Manufacturing Technology, Tokyo, pp.157–166, May(1992)

[6]   Tadahiro Ohmi, Takashi Imaoka, Isamu Sugiyama, and Takehiko Kezuka, "Metallic Impurities Segregation on the Interface between Si and Liquid", submitted to JECS

[7]   N. Tsuchiya, M.Tanaka, M. Kageyama, A. Kubota, and Y. Matsusita, "Nondestructive Evaluation of Trace Metals and Application to Defect Generation", Extended Abstracts of the 22nd Conference on Solid State Devices and Materials, Sendai, pp.1131–1134, (1990)

[8]   Tadahiro Ohmi, Mizuho Morita, Akinobu Teramoto, Kouji Makihara, and K.S.Tseng, "Very thin oxide film on a silicon surface by ultraclean oxidation", Appl. Phys. Lett., 60(17), pp.2126–2128, April(1992)

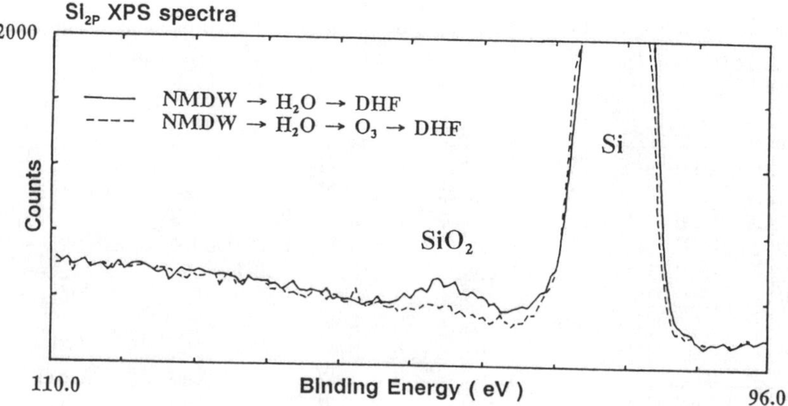

Fig.1. Influence of residual surfactant molecule on Si surface for the removal of native oxide by Si$_{2P}$ XPS spectrum.

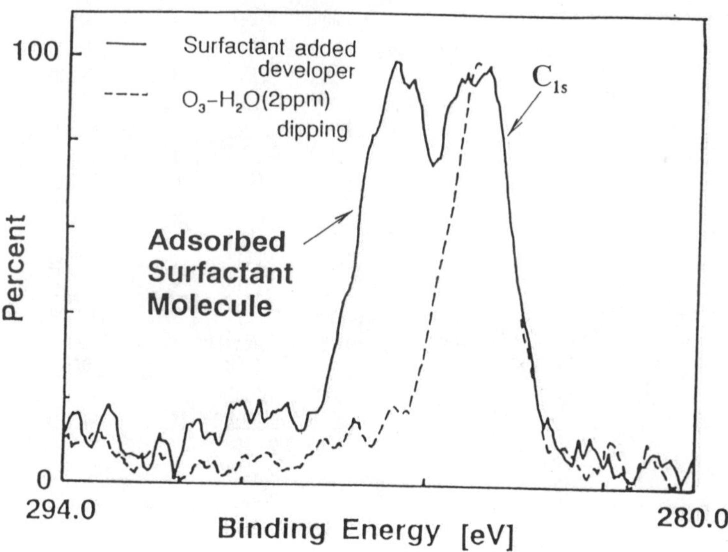

Fig.2. C$_{1s}$ XPS spectrum of surfactant adsorbed wafer.

Table 1.   Analysis data of metallic impurities in the special hydrogen peroxide.

<u>Sample No. SS-911127</u>

<u>1kg HDPE Container</u>

<u>Item</u>		<u>D.L.</u>	<u>Value</u>	<u>Equipment</u>	<u>Item</u>		<u>D.L</u>	<u>Value</u>	<u>Equipment</u>
Assay	(wt%)	—	31.6	—	L i	(ppb)	0.0003	N.D.	ICP-MS
Residue	(ppm)	0.2	N.D	—	B e	(ppb)	0.0006	N.D.	ICP-MS
Free Acid	(ppm)	0.1	2	—	B	(ppb)	0.004	0.100	ICP-MS
p H	(at20℃)	—	3.9	—	N a	(ppb)	0.0003	0.0012	ICP-MS
C l$^-$	(ppb)	1.0	N.D	I.C	M g	(ppb)	0.0003	N.D.	ICP-MS
N O$_2^-$	(ppb)	2.2	N.D	I.C	A l	(ppb)	0.0003	0.0027	ICP-MS
P O$_4^{3-}$	(ppb)	7.0	N.D	I.C	S i	(ppb)	0.476	N.D.	GFAA
N O$_3^-$	(ppb)	2.9	3.0	I.C	P	(ppb)	0.1352	N.D.	ICP-MS
S O$_4^{2-}$	(ppb)	3.2	4.3	I.C	K	(ppb)	0.017	N.D.	GFAA
N H$_4^+$	(ppb)	1.0	N.D	I.C	C a	(ppb)	0.09	N.D.	ICP-AE
					T i	(ppb)	0.0024	N.D.	ICP-MS
					V	(ppb)	0.0004	N.D.	ICP-MS
<u>Particles(count/ml)</u>					C r	(ppb)	0.0027	0.0119	ICP-MS
≧0.3μm	9.5			Particle	M n	(ppb)	0.0009	N.D.	ICP-MS
≧0.5μm	0.7			Counter	F e	(ppb)	0.045	N.D.	GFAA
					C o	(ppb)	0.0005	N.D.	ICP-MS
					N i	(ppb)	0.0006	N.D.	ICP-MS
					C u	(ppb)	0.059	N.D.	GFAA
					Z n	(ppb)	0.0033	N.D.	ICP-MS
					G a	(ppb)	0.0006	N.D.	ICP-MS
					G e	(ppb)	0.0012	N.D.	ICP-MS
					A s	(ppb)	0.0020	N.D.	ICP-MS
					S r	(ppb)	0.0003	N.D.	ICP-MS
					Z r	(ppb)	0.0008	0.0026	ICP-MS
Equipment					N b	(ppb)	0.0004	0.0032	ICP-MS
I.C　:Dionex 2010i					M o	(ppb)	0.0012	N.D.	ICP-MS
ICP-MS:Seiko SPQ-8000					P d	(ppb)	0.0012	N.D.	ICP-MS
GFAA　:Varian Spectra AA-40G					A g	(ppb)	0.0082	N.D.	ICP-MS
ICP-AE:Nippon Jarrell Ash ICAP-575MarkⅡ					C d	(ppb)	0.0013	N.D.	ICP-MS
Particle Counter : Rion KL-20K					I n	(ppb)	0.0004	0.0061	ICP-MS
					S n	(ppb)	0.0021	N.D.	ICP-MS
					S b	(ppb)	0.0015	N.D.	ICP-MS
					B a	(ppb)	0.0006	N.D.	ICP-MS
					T a	(ppb)	0.0004	0.0058	ICP-MS
					P t	(ppb)	0.0013	N.D.	ICP-MS
					A u	(ppb)	0.0020	N.D.	ICP-MS
					H g	(ppb)	0.0047	0.0147	ICP-MS
					T l	(ppb)	0.0005	N.D.	ICP-MS
					P b	(ppb)	0.0009	N.D.	ICP-MS
					B i	(ppb)	0.0010	0.0050	ICP-MS

Table 2.   Analysis data of organic impurities in the special hydrogen peroxide.

I t e m	Value
T O C [1]	3.0 ppm
Formic Acid (asTOC) [2]	1.3 ppm
Acetic Acid (asTOC) [3]	0.4 ppm

[1]   Measured without decomposition
[2,3] Measured with decomposition
[2]   The content as formic acid is 5ppm
[3]   The content as acetic acid is 1ppm

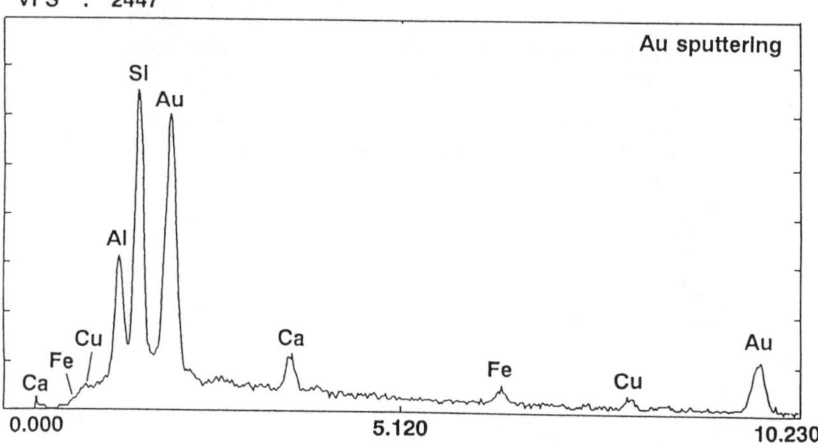

Fig.3. X-ray spectrum by EDX for the particle in the conventional hydrogen peroxide.

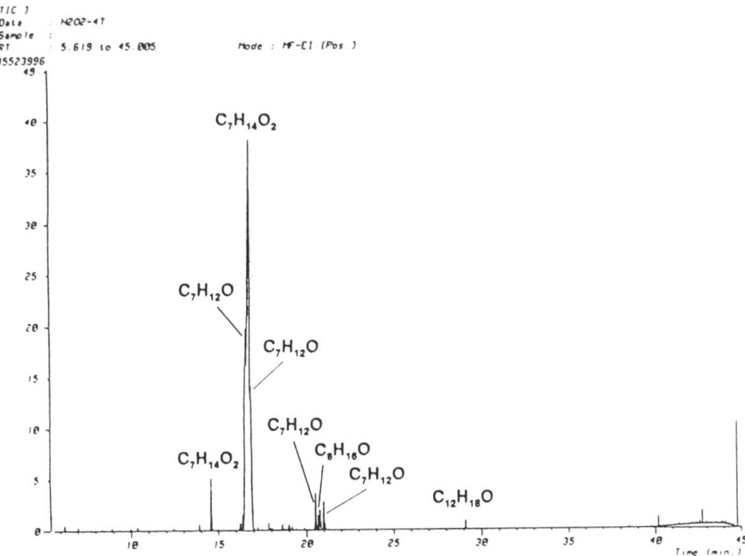

Fig.4. Mass chromatogram by GC–MS for the residue after evaporation in the special hydrogen peroxide after chloroform extraction.

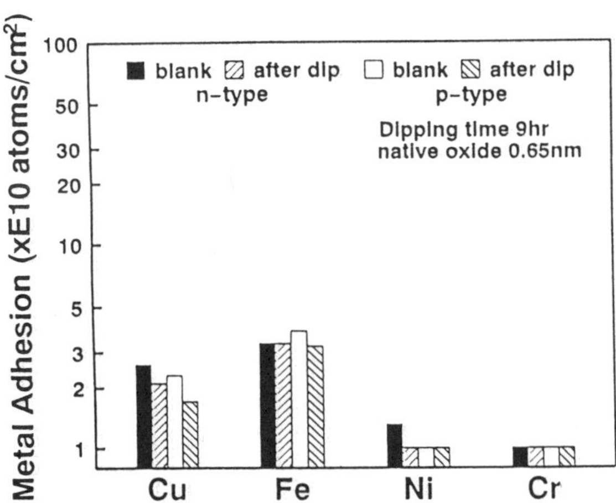

Fig.5. Metal adhesion onto the wafer surface in the special hydrogen peroxide by TRXRF.

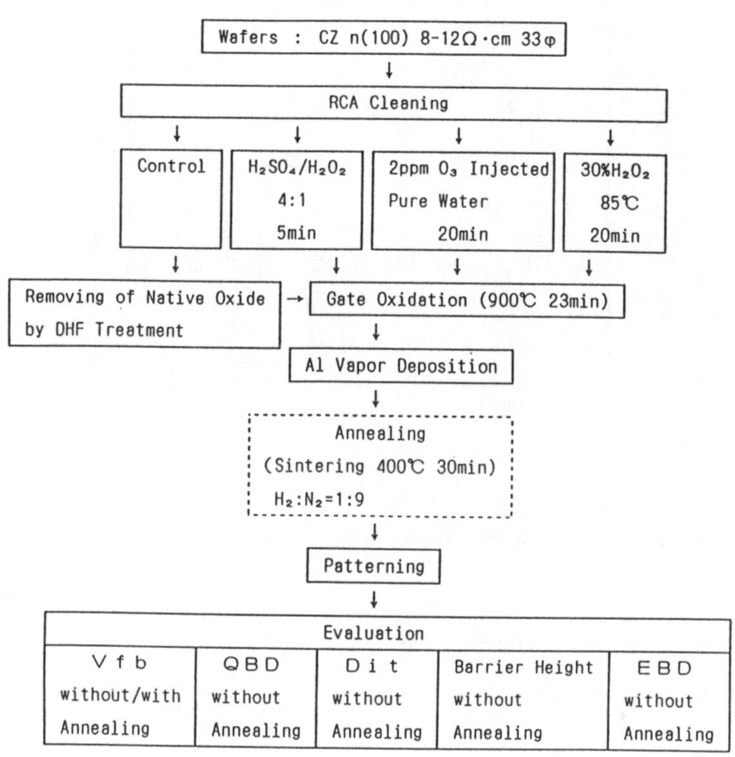

Thickness of films unit : nm

Treatment	Native Oxide	Gate Oxide
Control	—	9.6
H_2SO_4/H_2O_2 4:1 5min	0.776	8.4
2ppm O_3 injected Pure Water 20min	0.736	9.3
85℃ H_2O_2 20min	0.776	9.2

※The thickness of the native oxide films were estimated as corresponding to the thermal oxide thickness using Si_{2p} spectrum of XPS.

Fig.6. Experimental method for estimate of electrical characteristics of MOS diodes.

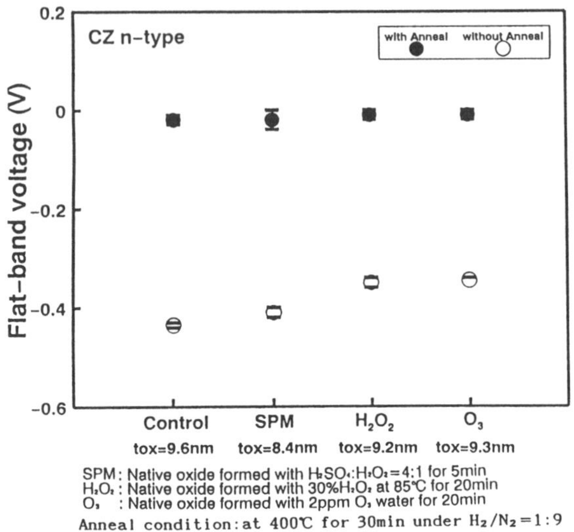

Fig.7. Flat–band voltage at room temperature for MOS diodes.

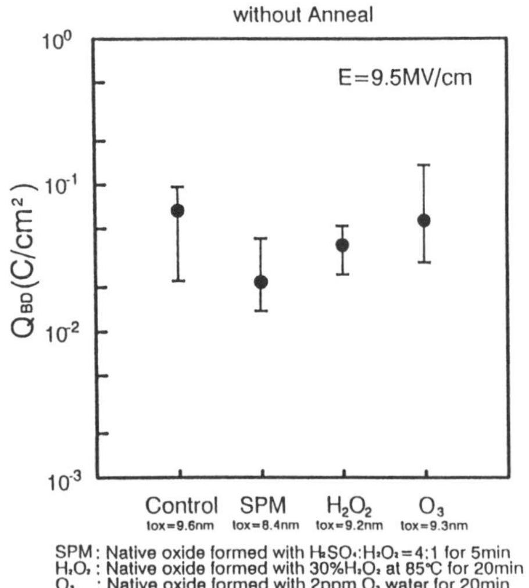

Fig.8. Charge to breakdown at room temperature for MOS diodes without annealing (measurement area : $1.6 \times 10^{-4} \text{cm}^2$).

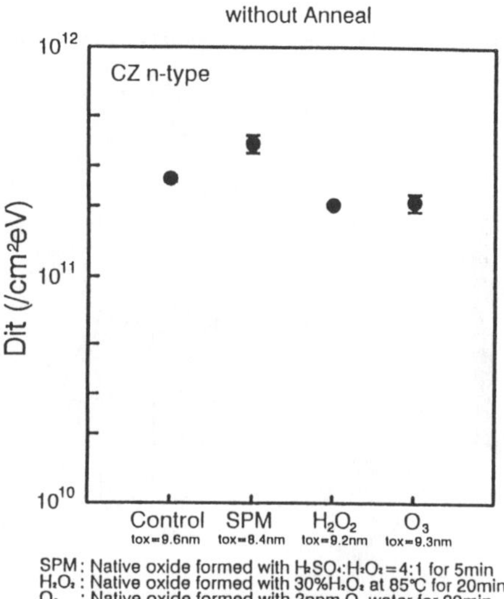

Fig.9. Interface–trap density at room temperature for MOS diodes without annealing.

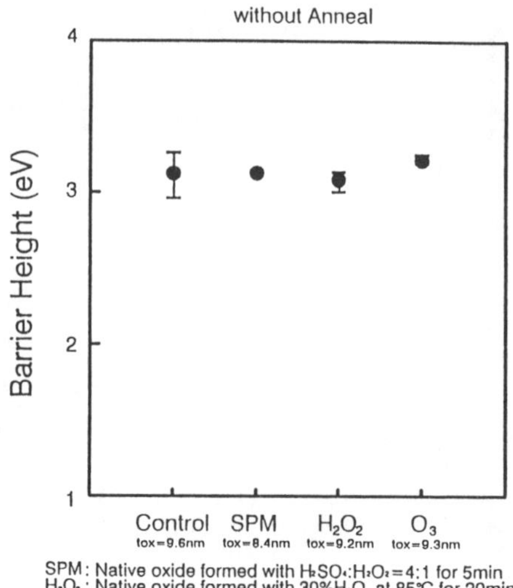

Fig.10. Barrier heights at room temperature for MOS diodes without annealing.

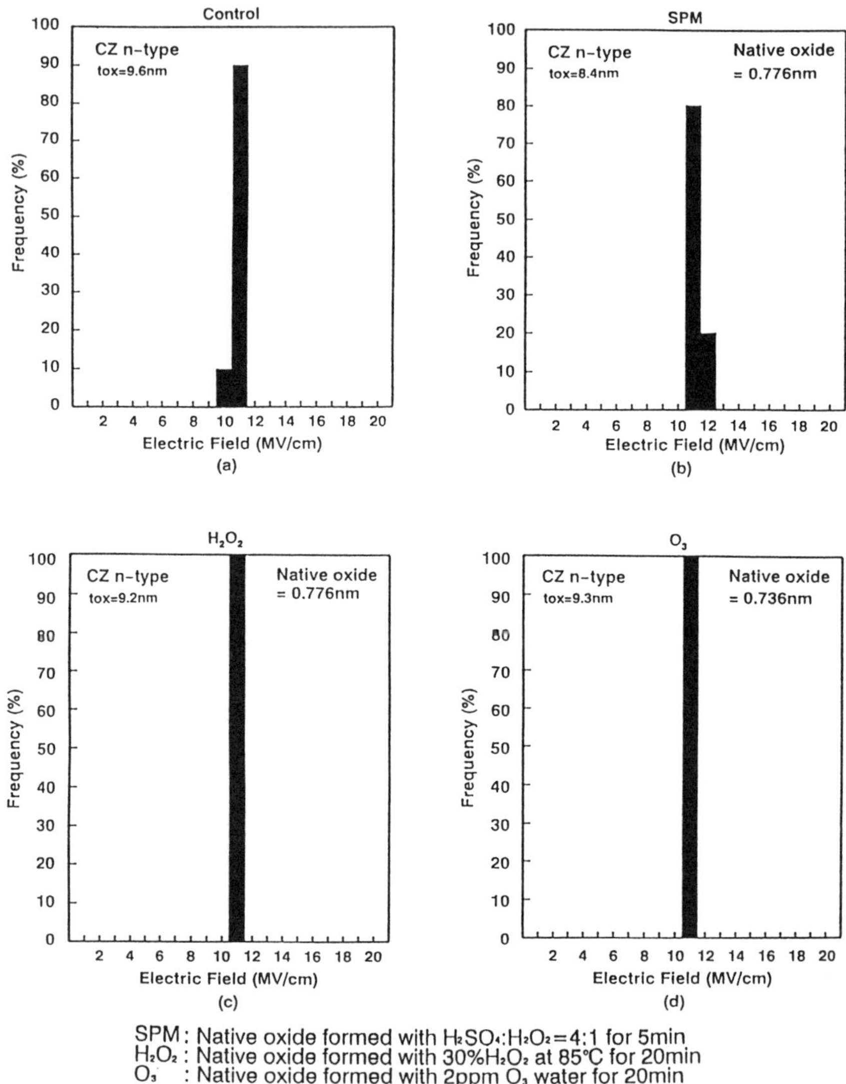

SPM : Native oxide formed with H_2SO_4:H_2O_2=4:1 for 5min
H_2O_2 : Native oxide formed with 30%H_2O_2 at 85°C for 20min
O_3 : Native oxide formed with 2ppm O_3 water for 20min

Fig.11. Electric breakdown histograms at room temperature for MOS diodes (measurement area : 1.6×10^{-4}cm^2, judgment current : 1×10^{-4}A).

STUDY ON ADHESION AND REMOVAL
OF METALLIC IMPURITIES ON PFA SURFACE

Katsuhide Ohtani
Department of Electronics, Faculty of Engineering,
Tohoku University, Sendai 980, Japan

Kiyohiko Ihara
Chemical Division, Daikin Industries, Ltd.,
Nishi–Hitotsuya, Settu 566, Japan

and
Tadahiro Ohmi
Department of Electronics, Faculty of Engineering,
Tohoku University, Sendai 980, Japan

The adhesion and removal behavior of metallic impurities on PFA sheet surface was examined with the TRXRF analysis. It was found that metals such as Fe, Cu were rather strongly adhered, and that they could be removed to some extent by the SPM or the FPM cleaning. Also it was discovered that the surface metallic concentration of PFA sheets didn't increase when they were dipped into UPW or some chemicals contaminated with metals, so it is assumed that metallic impurities don't adhere to PFA surface from liquids. Moreover the metallic transfer tests showed that a cleaned PFA sheet didn't contaminate Si wafer but the un–cleaned sheet extremely polluted the wafer with metallic impurities. Therefore it has been confirmed that when PFA molded parts such as wafer carriers and tubes are used in the wet process of the semiconductor manufacturing, they need to be cleaned by the SPM or the FPM cleaning.

INTRODUCTION

Fluoro–resin is widely used in the semiconductor manufacturing process since its heat– and chemical–resistant natures are superior to other plastic materials. In the fluoro–resin. PFA (Perfluoroalkoxy resin) is the most suitable for use in the semiconductor field as well as PTFE (Polytetrafluoroethylene), because their properties are especially excellent due to its molecular structure of strong C–F covalent bonds. Table 1 shows their molecular structures and some properties of these resins. PFA is easier to mold than PTFE. and is indispensable to the wet process of the semiconductor manufacturing as a

Keywords– PFA. Metallic Impurities, TRXRF, Surface Analysis, RCA Cleaning

material for wafer carriers, chemical tanks or tubes in various piping systems.

From the view point of wafer contamination, however, this resin still has some problems to be solved. These problems become more serious as the trend of high density and integration in LSIs increases. Examples of this are the elution of fluoride ion generated by the hydrolysis of unstable polymer end groups [1], the chemical carryover [2], the elution of particles caused by the oligomer (low molecular weight polymer) which is contained in the PFA resin or by being scrubbed with Si wafers in the case of wafer carriers, the susceptibility to electricity charge because of its highest insulation property, the elution of metallic impurities which is observed for other materials, and so on [3]. These contamination will have a great impact on the yield in the manufacturing of sub-micron devices.

The introduction of the Total Reflection X-ray Fluorescence Spectroscopy (TRXRF) [4] has made it possible to observe the trace metallic contamination on the surface of Si wafers, then the segregation and removal mechanism of metallic impurities has been cleared [5]–[8]. In addition, it has been evident that these have a fatal effect on the device characteristics [4]. For example, the metallic contamination of 10^{10}atoms/cm^2 level such as Fe and Cu can give rise to the oxide breakdown and the decrease of minority carrier lifetime [9]. On the other hand, the behavior of metallic impurities on PFA surface has never been examined, although only the metallic elusion from PFA materials has been reported [10]. The PFA molded parts which is represented by wafer carriers have many opportunities to meet Si wafers through Ultra Pure Water (UPW) or chemicals at the wet cleaning process. Therefore, it is likely that the metallic impurities eluted from the inside or released from the surface of them can pollute the cleaned wafers.

Table 1. Molecular structure and some properties of Fluoro-resin.

	PFA	PTFE
Molecular structure	$-(CF_2-CF_2)_m -(CF_2-CF)_n-$ $\quad\qquad\qquad\qquad\quad \mid$ $\qquad\qquad\qquad\qquad OC_3F_7$ m:n=99.6 ~ 96.0:0.4 ~ 4.0	$-(CF_2-CF_2)_n-$
Melt viscosity at 380°C (poise)	$(2 \sim 25) \times 10^4$	$10^{11} \sim 10^{13}$
Melting point (°C)	302 ~ 310	327
Maximum temperature for continuous use (°C)	260	260
Volume resistivity (Ω–cm)	$> 10^{18}$	$> 10^{18}$
Chemical resistance	excellent	excellent

Thus it has become more important to understand the adhesion and removal phenomena of metallic impurities on PFA surface in order to realize the completely cleaned wafer surface, as the density and integration in LSIs are getting higher. In this study, we reported the metallic adhesion in UPW or some chemicals, the removal effect of metallic impurities on the surface of PFA sheets by some chemical cleanings, and also the metallic contamination for Si wafers from PFA sheets with the TRXRF analysis.

EXPERIMENTS

Flat PFA sheets (thickness 1mm, surface roughness (Ra 200~400Å) which were smoothed by compression molding and cut into 33mmφ circles, were prepared in this study. Phosphorus doped n-type CZ(100) Si wafers which were also cut into 33mmφ circles were prepared in the metallic transfer tests, and these were cleaned by the SPM and the FPM cleaning before use. Beakers and carriers made of PFA, were used in all the experiments of cleaning or dipping, and were also cleaned fully many times by those two cleanings. Then the volume of chemicals used was 800ml. High pure chemicals of the semiconductor grade were used through in these experiments. To dry the sheets or wafers, the high purity nitrogen gas blow was employed. And all the experiments were carried out in the clean bench.

The standard concentrations of chemicals and the cleaning methods in this study are listed below.

Hydrogen Peroxide (H_2O_2) : 30wt%
Sulfuric acid (H_2SO_4) : 97wt%
Hydrofluoric acid (HF) : 50wt%
Hydrochloric acid (HCl) : 36wt%
Nitric acid (HNO_3) : 70wt%
Ammonium Fluoride aq.solution (NH_4F) : 40wt%

the SPM cleaning; H_2SO_4:H_2O_2=4:1 (volume ratio), 10min
the FPM cleaning; HF:H_2O_2:H_2O=0.5:10:89.5 (wt%), 10min
the HPM cleaning; HCl:H_2O_2:H_2O=1:1:6 (volume ratio), 90°C, 10min
All cleanings were followed by rinsing with UPW for 10min.

The analysis of metallic impurities were measured with the TRXRF (Technos, TREX610). The incident X-ray was excited by the tungsten target with the condition of 30kV-200mA, and its glancing angle is 0.05° (except for the Angle Scan). The metallic concentration per unit area on the PFA surface was calibrated by using the value of the Si wafer standard sample, since it was difficult to make a PFA standard sample.

RESULTS AND DISCUSSION

The TRXRF Analysis of the PFA Surface

Figure 1 shows a typical fluorescence spectrum of the PFA sheet just after rinsing with UPW. Strong peaks of Fe, Cu and some weak peaks such as Cr, Ni except the incident X-ray W-Lβ1 arising from the surface roughness appeared. By the way, false peaks are rarely observed in the TRXRF analysis such as some resists on Si wafers. Then several PFA sheets of different contamination levels were measured by the TRXRF and the relationship between the X-ray intensities of Fe-Kα and that of Fe-Kβ was examined. There is a clear proportional relation as it is shown in Figure 2, so it is recognized that these peaks express true metallic contamination.

Figure 1 also indicates that metallic impurities on the PFA surface are mainly Fe, Cu. And Ca, Cr, Mn, Ni are minors. Therefore all the experiments below were evaluated by paying attention to only these six metals. Figure 3 shows the metallic contamination levels of several PFA sheets just after rinsing with UPW. These contamination levels were different individually, but the total of these six metals were within the range of $1\sim4\times10^{13}$ atoms/cm^2. This contamination levels are much higher than that of Si wafers. These impurities probably come from any metallic materials used in the molding process of PFA parts, since the metallic concentrations on the PFA surface calculated by the metallic analysis of the ashed PFA pellets are less than 10^9 atoms/cm^2 [10].

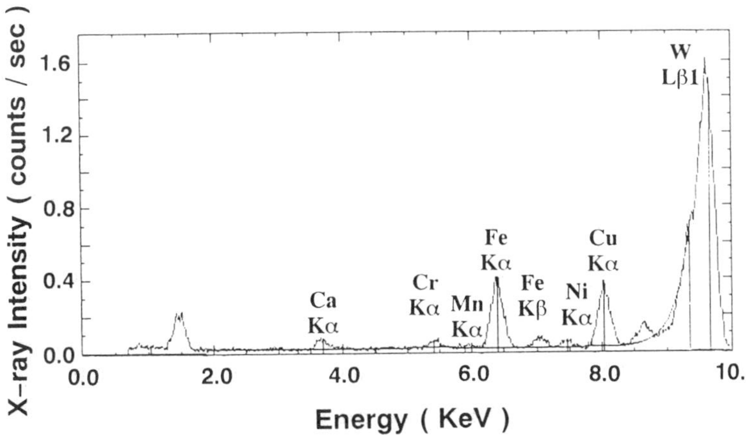

Fig.1.   A fluorescence spectrum of a PFA sheet just after rinsing with UPW by the TRXRF analysis.

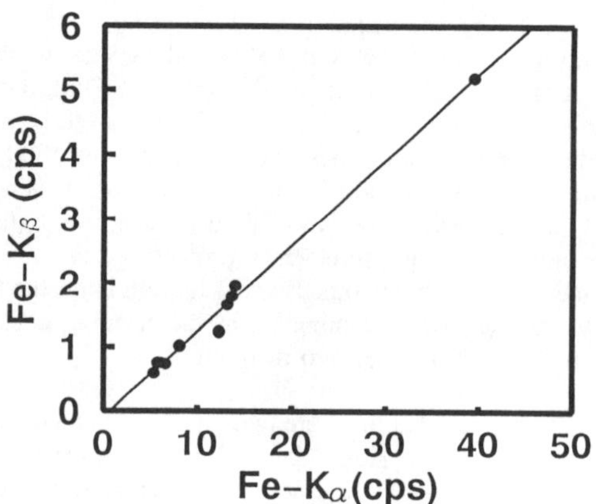

Fig.2. The relationship between the X–ray intensities of Fe–Kα and Fe–Kβ for some PFA sheets.

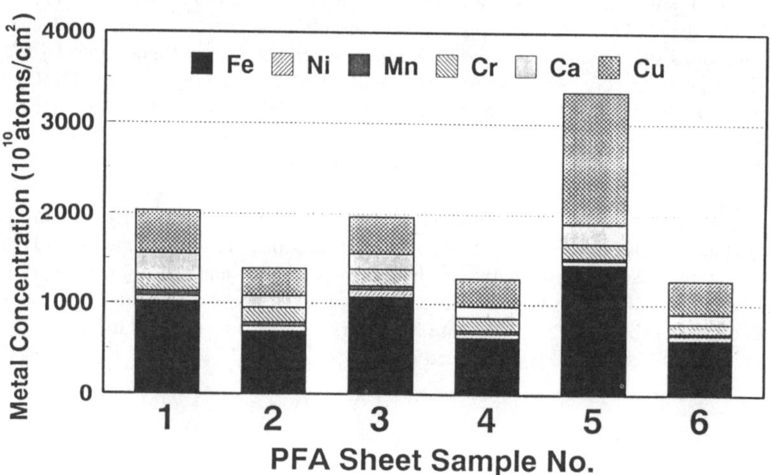

Fig.3. The metallic concentration on several PFA sheets just after rinsing with UPW.

Removal of Metallic Impurities on PFA Sheets

To remove metallic impurities from PFA sheets, some chemical cleanings were tried, which are normally used for cleaning of Si wafers, what is called the RCA cleaning [11]. The cleaning effect of PFA sheets by the SPM, the FPM, and the HPM cleaning is shown in Figure 4, 5, and 6 respectively. It was found that these three cleaning methods were able to reduce metallic impurities. The SPM and the FPM cleanings were more effective to reduce the contamination by half by only one cleaning. Also it was cleared that the contamination levels became almost the same (6 metals total about 1×10^{13} atoms/cm^2) after cleaning. Moreover, unexpectedly, this level never decreased any longer with more cleanings. On the other hand, Figure 6 indicated that the HPM cleaning gradually reduced the metallic contamination as the cleaning times increased, but that it could never reach the level of other two methods.

These results suggest that there are two types of the metallic adhesion on the PFA surface. Namely, one is the physical adhesion onto the surface, and another is firmly buried in the polymer network. The former can be removed by these chemical cleanings but the latter can never be eliminated.

Metallic Adhesion onto the PFA Surface

Next, some metallic adhesion tests were tried with the PFA sheets cleaned by the SPM and the FPM methods. Figure 7, 8, and 9 indicate the variation of metallic concentrations on the surface on the PFA sheets when these were dipped into UPW, HF, and HNO_3, respectively, which were intentionally contaminated by 1ppm Fe, Ni, Cu each. It was found that the level of each metallic impurity didn't increase after dipping even for 4days in all cases. In other words, metals didn't adhere onto the surface of PFA sheets. Moreover, no metal showed the preferential adhesion.

In the case of Si wafers, it has been well known that the surface contamination with metallic impurities becomes more severe as the concentration of metals in UPW or chemicals increases. For example, it had been reported that the dipping into UPW polluted by 0.1ppm Cu caused the Cu contamination of about 1×10^{14} atoms/cm^2 level on the Si surface [8]. The difference from Si wafers can be explained by the fact that PFA has no dangling bond and generates no oxide.

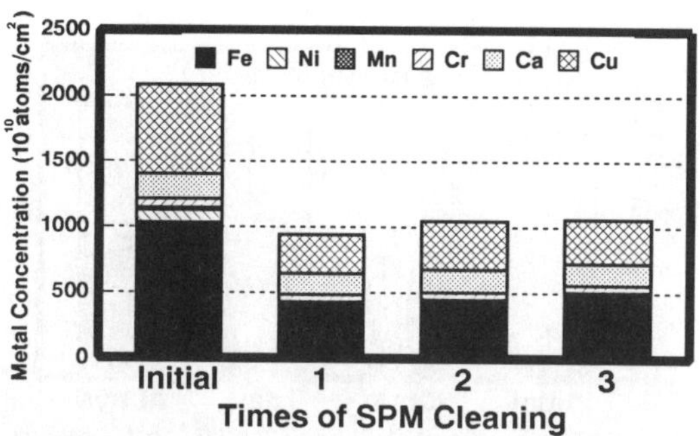

Fig.4.   The metallic concentration on a PFA sheet before and after the SPM cleaning ($H_2SO_4 : H_2O_2$).

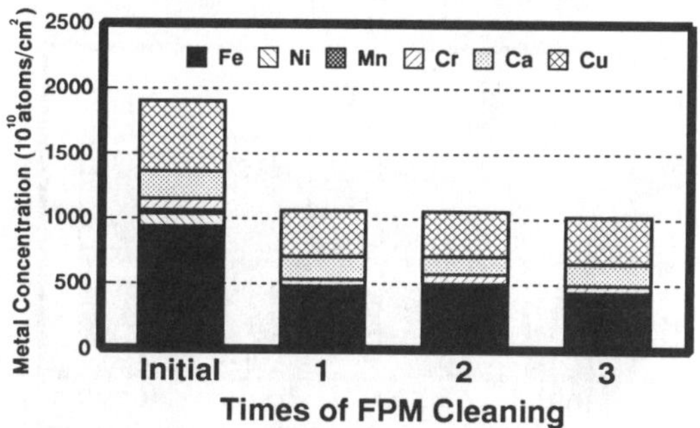

Fig.5.   The metallic concentration on a PFA sheet before and after the FPM cleaning ($HF : H_2O_2 : H_2O$).

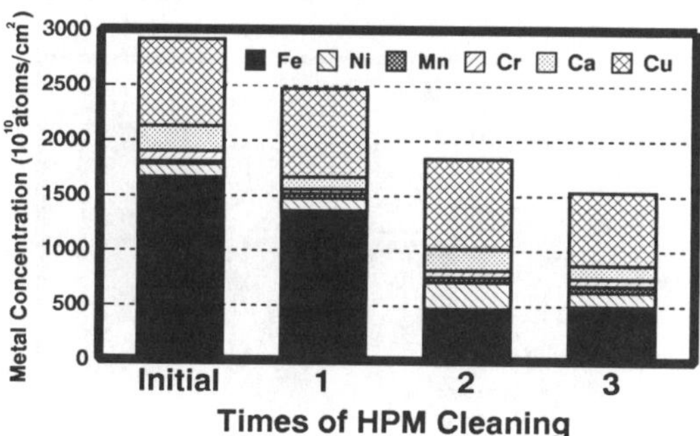

Fig.6.   The metallic concentration on a PFA sheet before and after the HPM cleaning ($HCl : H_2O_2 : H_2O$).

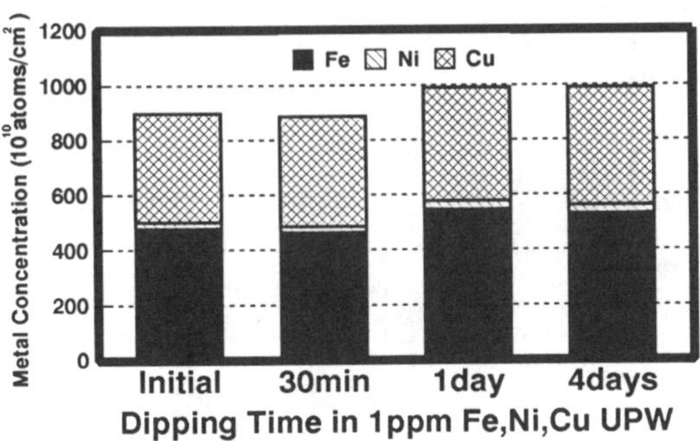

Fig.7.   The variation of metallic concentration on a PFA sheet when dipped into UPW contaminated by 1ppm Fe, Ni and Cu each.

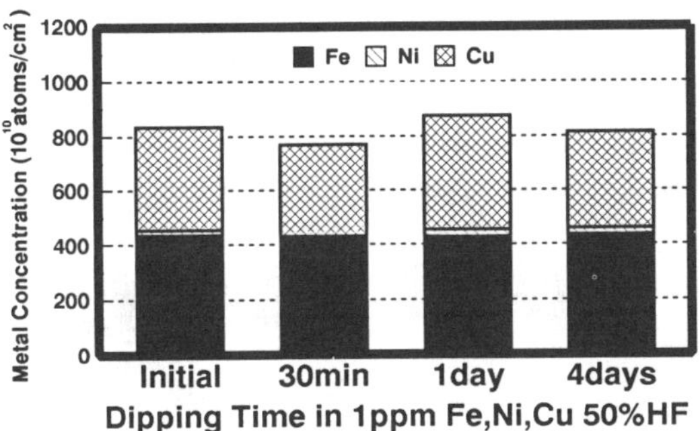

Fig.8.   The variation of metallic concentration on a PFA sheet when dipped into 50%HF contaminated by 1ppm Fe, Ni and Cu each.

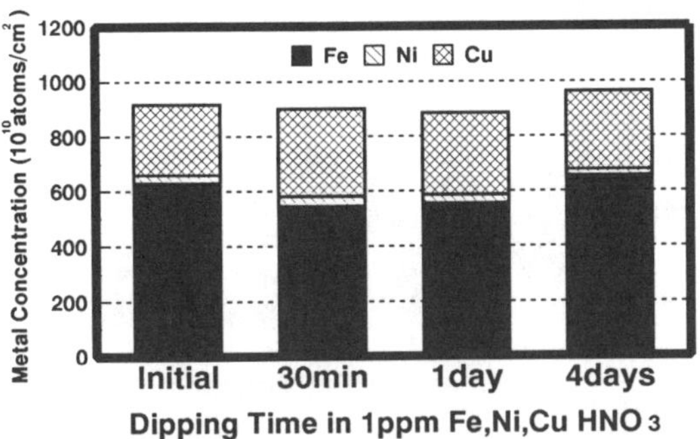

Fig.9.   The variation of metallic concentration on a PFA sheet when dipped into HNO_3 contaminated by 1ppm Fe, Ni and Cu each.

The Metallic Depth Analysis on the PFA Surface

In the TRXRF analysis, the depth profiles of metallic impurities of some thin films can be measured. As the glancing angle of the incident X-ray is altered, the incident depth into the substrate is changed from several nm to about 1μm (called the Angle Scan) [12]. As to Si wafers, the X-ray fluorescence intensities of metallic impurities which exist only on the surface of the wafer have peaks around 0.18° that is the critical angle of total reflection proper to silicon, while the metal impurity profiles from the surface to the inside at the same concentration increase like the intensity profile of the substrate Si as the glancing angle becomes larger [13].

The result of the Angle Scan for a PFA sheet cleaned by the SPM and the FPM is showed in Figure 10. Though the critical angle of PFA is calculated about 0.17°, no peak of metallic X-ray intensity was observed around this angle. Therefore another PFA sheet was scrubbed with the ECB (electro-chemical buffing)-treated SUS316 (chemical composition Fe:Cr:Ni:Mn:Cr=65:18:13:2:2) plate to contaminate only the top surface intentionally. Figure 11 shows the result of this sheet. The X-ray intensities of Fe and Ni considered to exist on the surface at high concentration didn't have peaks around the PFA's critical angle but increased extensively above about 0.1°. If we assume that the X-ray intensities of metals existing only on the top surface of PFA sheets give such profiles, from Figure 10 we can say that the metallic impurities on the cleaned PFA sheets exists widely from the top surface to the inside.

The Metallic Pollution on Si Wafers from PFA Sheets

Finally, the influence of metallic impurities contained in PFA sheets on Si wafers was examined. Some Si wafers were left to touch the PFA sheets face to face before and after the SPM cleaning, and then they were dipped into UPW intact. Figure 12 shows the variation of metallic contamination on the surface of the Si wafers in this test. It was found that the wafer touched the cleaned PFA sheet was never contaminated but one touched the non-cleaned (only UPW rinse) PFA sheet was extremely polluted with Fe, Cu, and so on. The variation of metallic concentration on wafers is indicated in Figure 13 when they were touched the cleaned PFA sheets and were dipped in HF or NH$_4$F aq. solution in the same way. Though the concentration levels had some unevenness, it is clear that the metallic concentration did not increase. Also Figure 14 indicates that the cleaned PFA sheets did not contaminate the wafers in the SPM or HPM solutions. At the end, the direct metallic transfer test was carried out. The variation of metallic concentration on the Si wafers is shown in Figure 15 when they were left to touch face to face directly the PFA sheets cleaned by the SPM method, and then were pressed by some load weights. The contact time was 1 minute, the wafers did not suffered any metallic contamination in this case.

These results suggests that the metallic impurities on the surface of cleaned PFA sheets could transfer to Si wafers but that the metals of about 1×10^{13} atoms/cm^2 on the cleaned PFA surface were tightly built in the resin network and could never be eluted into UPW or chemicals.

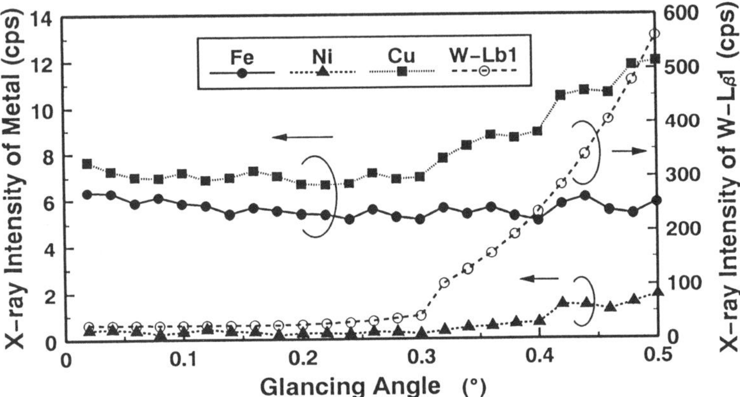

Fig.10. The depth profiles of metallic impurities on a cleaned PFA sheet by the Angle Scan.

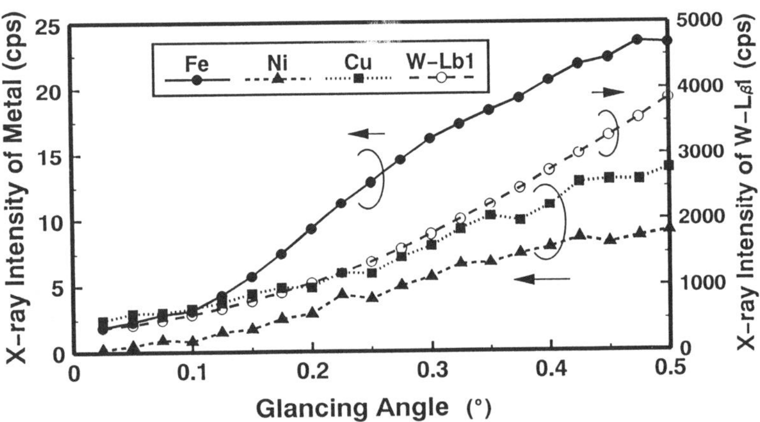

Fig.11. The depth profiles of metallic impurities on a PFA sheet whose top surface was contaminated by being scrubbed with the ECB–treated SUS316L plate.

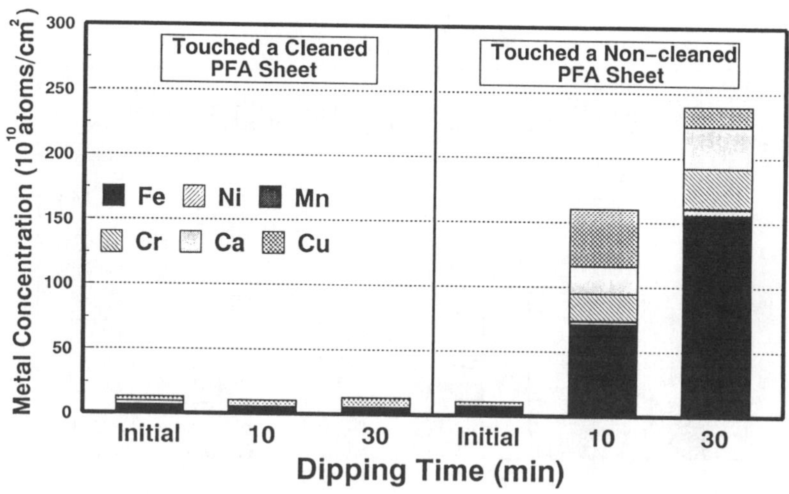

Fig.12. The influence of a cleaned or a non-cleaned PFA sheet on Si wafers. The variation of metallic concentration on Si wafers when touched the PFA sheets before or after cleaning by the SPM and dipped into UPW.

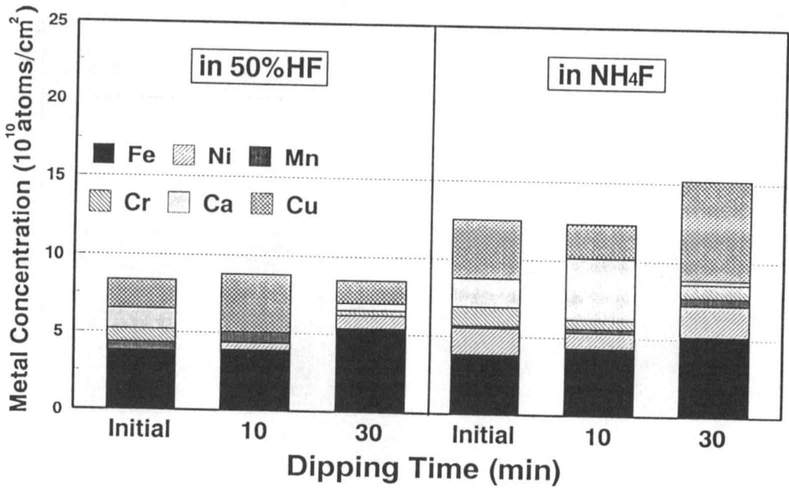

Fig.13. The variation of metallic concentration on Si wafers when touched the cleaned PFA sheets and dipped into 50%HF or NH$_4$F.

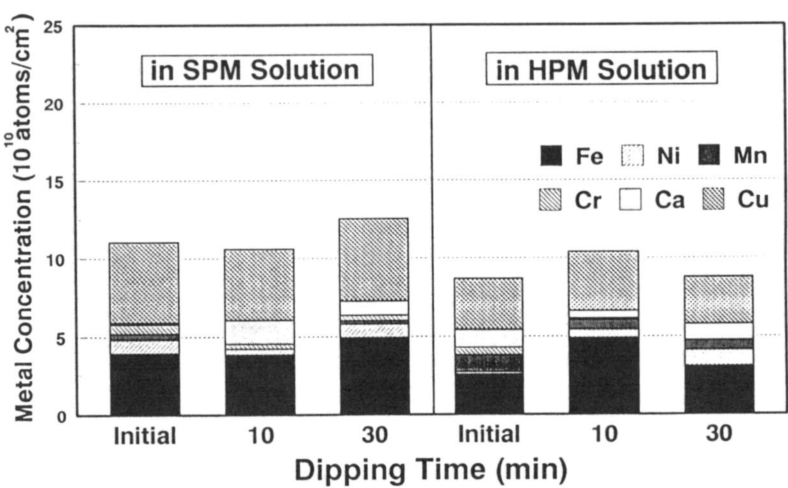

Fig.14.   The variation of metallic concentration on Si wafers when touched the cleaned PFA sheets and dipped into the SPM or the HPM solution.

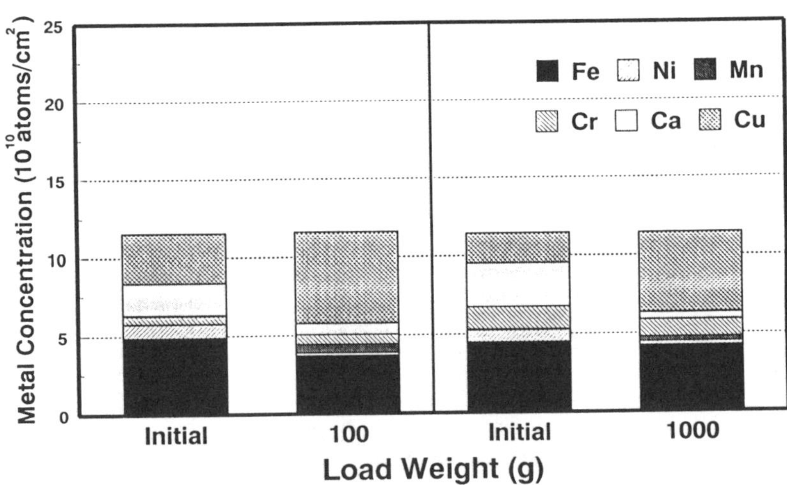

Fig.15.   The variation of metallic concentration on Si wafers when directly touched the cleaned PFA sheet and pressed by some weights in the clean bench.

SUMMARY

The adhesion and removal behavior of metallic impurities on the surface of PFA sheets was examined with the TRXRF analysis. It was found that metallic impurities such as Fe, Cu were rather strong adhered onto the PFA surface. We considered that there were two types of the metallic adhesion on the surface ; one is a physical adhesion onto the top surface, and another is a firm buried in adsorption into the polymer network. The former could be removed by some chemical cleanings such as the SPM cleaning which are generally used for Si wafer cleaning, but the latter could never be eliminated. The depth analysis of metallic impurities on the cleaned PFA surface made it clear that those trapped in the polymer network existed widely from the top surface to the inside of PFA sheets.

Also it was discovered that the surface metallic concentration of PFA sheets didn't increase when they were dipped into UPW or some chemicals intentionally contaminated with metals. This suggests that metallic impurities do not adhere to the PFA surface. Moreover the metallic transfer tests showed that a cleaned PFA sheet did not contaminate the Si wafer, but that a non-cleaned sheet extremely polluted a wafer with some metallic impurities in UPW or chemicals. We therefore conclude that when PFA molded parts such as wafer carriers and tubes are used in the wet process of the semiconductor manufacturing, they need to be cleaned by the SPM or the FPM cleaning methods.

ACKNOWLEDGMENTS

The research reported in this article was conduced in Mini Super Cleanroom at Faculty of Engineering and in Laboratory for Microelectronics attached to the Electric Communication Laboratory in Tohoku University.

REFERENCES

[1]   K.Ihara, K.Ohtani and H.Aomi, "High-quality PFA for Semiconductor", Proceedings of the 37th Annual Technical Meeting, Institute of Environmental Science, San Diego, pp.802–807, May 1991.

[2]   J.Goodman, "Extractable Fluorides and Chemical Carryover from PFA Products", Ultra Clean Technology, Vol.2, pp.465–473, 1990.

[3]   S.Koyata, Monthly Semiconductor World, pp.121–127, March 1992.

[4]   N.Tsutiya, M.Tanaka, M.Kageyama, A.Kubota and Y.Matsushita, "Nondestructive Evaluation of Trace Metals and Application to Defect Generation", Extended Abstract of the 22nd(1990 International) Conference on Solid State Devices and Materials, pp.1131–1134, Sendai, 1990.

[5]   T.Imaoka, I.Sugiyama, T.Kezuka and T.Ohmi, "Metallic Impurities Segregations on the Interface between Silicon and Liquid", Proceedings of 11th Workshop on ULSI Ultra Clean Technology, Tokyo, pp.39–80, June 1991.

[6]   T.Imaoka, T.Kezuka, J.Takano, I.Sugiyama and T.Ohmi, "Metallic Impurities Segregation and Removal at the Interface of Silicon Wafer Surface and Liquid Chemicals", Proceedings of IES 38th Annual Technical Meeting, Vol.1, pp.466–474, May 1992.

[7]   M.Meuris, M.Heysn, S.Verhaverbeke, P.Mertens and A.Philipossian, "Metallic Contamination and Silicon Surface Roughening in Wet Preoxidation Cleaning Treatment", Proceedings of Microcontamination '91 Conference and Exposition, San Jose, pp.658–665, October 1991.

[8]   Y.Shimanuki, Monthly Semiconductor World, pp.116–120, March 1992.

[9]   T.Shimono, M.Morita, Y.Muramatsu and M.Tsuji, "Device Degradation by Metallic Contamination, and Evaluation and Cleaning Metallic Contaminants", Proceedings of 8th Workshop on ULSI Ultra Clean Technology, Tokyo, pp.59–68, December 1990.

[10]   "PFA Immersion Test", Proceedings of 6th Workshop on ULSI Ultra Clean Technology, Tokyo, pp.165–222, July 1990.

[11]   W.Kern and D.A.Puotien, "Cleaning Solution Based on Hydrogen Peroxide for Use in Silicon Semiconductor Technology, RCA Review 31, pp.187–205, June 1970.

[12]   A.Iida, "Semiconductor Surface Characterization by Total Reflection X–Ray Fluorescence Analysis", Extended Abstract of the 21st Conference on Solid State Devices and Materials, pp.501–504, Tokyo, 1989.

[13]   Y.Matsushita, N.Tsuchiya, Nikkei Microdevices, No.64, pp.99–106, October 1990.

SYNTHESIS OF PURE FLUORITE
FROM SPENT FLUORIDE CHEMICALS

I. The Reaction Mechanism of Binary System HF–H_2O and Calcite in a Column Reactor

N.Miki, M.Maeno, T.Fukudome, M.Miyashita and T.Ohmi[*]

Hashimoto Chemical Corporation, Kaizan–cho Sakai 590, Osaka, Japan,

[*] Department of Electronics, Faculty of Engineering,
Tohoku University, Sendai 980, Japan

ABSTRACT

Scientific study is carried out to optimize the LSI manufacturing wet process in order to keep balance between industrial activities and environmental conservation through efficient recycling use of resources and materials without waste. This study aims at establishing a stoichiometric synthesis method of Fluorite (CaF_2) using Calcite ($CaCO_3$) from spent fluoride chemicals in a column reactor under crystallogenic and kinetic investigation. The fluorine concentration in effluent can reduce as low as 5ppm which is below than the solubility of crystal CaF_2. This special decreasing effect of fluorine concentration is discussed concerning mechanism of " solid to solid formation " and surface activity of formed fluorite. The conversion rate from calcite into fluorite is elucidated by analysis of break–through curve. Purity of synthesized fluorite is higher than 99.5% and pure effluent from column can be recycled into a purewater system.

INTRODUCTION

" Wet process " employing chemicals and ultrapure water continuously plays an important role in LSI manufacturing process. Large volume of chemicals and ultrapure water is used at present. The cleanliness level of spent chemicals and ultrapure water is so high that they are hardly considered as wastes. However spent chemicals are not effectively processed to non–harmful substances at present. This is because this processing step has not been scientifically studied.

Hydrofluoric acid and buffer hydrogen fluoride (hereunder referred to as " HF " and " BHF ") are essential in the LSI wet process. At the same time, they are hardest to be processed : in order to suppress fluorine below the legally–set level (15ppm or 8ppm), a lot of $Ca(OH)_2$ is added to include fluorine, which results the generation of large volume of sludge (industrial waste) containing fluorine. Cost of chemicals used to process sludge constitutes another big problem. Besides sludge requires large space for processing. Especially, dispose of sludge make worse the environment.

This study aims at establishing a closed system for wet chemical processing through a recycling use of fluorine and water without adding any chemicals and discarding any waste materials.

SCIENTIFIC APPROACH

Authors recognize that a scientific approach to realization of non–disposal process in LSI manufacturing plant is to intend the regeneration of used chemical materials and the minimization of additive chemical materials. We propose three principles [1][2].

Resource Circulating Purification : A resource circulating purification system can be constituted with the regeneration of pure fluorite from spent fluoride chemicals in wet etching process, in which HF is generated from natural fluorspar, purified and synthesized to pure fluoride chemicals for wet etching. Regenerated fluorite after processing serves as more valuable resource of HF generation instead of fluorspar.

Stoichiometric Synthesis of Fluorite : For the completion of such system, a condition of stoichiometric reaction, i.e, the equivalent reaction of calcite and fluorine in spent chemicals, is essentially required, which is different from the conventional chemical waste processing.

Our investigation concentrated in mechanism of " solid to solid formation of fluorite from calcite grain ".

Effluent System Separation in Wet Processing : In order to implement an effective regeneration system, the effluent from HF process and BHF process must be separated for recycling use of fluorine and purewater. The effluent from HF process is large volume with diluted concentration less than 0.5% of HF and that of BHF process is small volume with high concentration more than 30% of NH_4F containing additional HF. We have confirmed essential difference in the reaction mechanism of calcite between binary system $HF-H_2O$ and triple system $NH_4F-HF-H_2O$.

EXPERIMENTAL

Crystallogenic Mechanism of " Solid to Solid Formation " and Reaction Kinetics of Calcite and $HF-H_2O$:

The gradual transition of calcite into fluorite in $HF-H_2O$ system proceeds under a mechanism of " Solid to Solid Formation " in inner grain of calcite without changing original form and grain size. A difference of density between calcite (2.711) and fluorite (3.181) makes inner surface, thus the fibrous aggregates of fluorite have grown in calcite grain. The X–Ray diffraction patterns of formed fluorite indicate somewhat poor crystallinity comparing to natural crystal of fluorspar (CaF_2) as shown in Fig.1. Purity of used calcite is 99.5% including MgO 0.3%, SiO_2 0.1%, Fe_2O_3 and Al_2O_3 less than 0.1%. As basic information, we examined a conversion rate of calcite grain with mean diameter of 0.30mm in $HF-H_2O$ as shown in Fig.2. Under the stoichiometric condition at mol ratio $[CaCO_3]$ / 2[HF], i.e, the mixing ratio is $CaCO_3$ 10g / HF 4g at the HF concentration of 5,000ppm, the conversion rate is indicated by a gradual decrease of the

HF concentration. Under the condition of mol ratio $[CaCO_3]$ / 10[HF], i.e, the mixing ratio is $CaCO_3$ 10g / HF 20g at the HF concentration of 6,250 ppm, it is expected theoretically that the HF concentration equilibrates at the level of 5,000 ppm after perfect conversion from calcite to fluorite consuming 2mol [HF]. The HF concentration, surprisingly, approaches to a level of 4mol [HF] consumption at initial stage and goes back to a level of 2mol [HF] consumption after 10hrs. The equilibrium concentration of HF after mixing of calcite and $HF-H_2O$ at the HF concentration ranging from 25ppm to 10,000ppm is examined under the condition of same mixing volume which nearly corresponded the volume ratio of calcite and aperture in filled column. Figure 3 demonstrates a characteristic decrease of the fluorine concentration having a minimum which there in the range around the initial HF concentration of 1,000 ~ 5,000ppm, and the fluorine concentration equilibrates rapidly in a few minutes. The fluorine concentration is determined by a fluorine ion meter after filtration and by a conductivity cell compensating solubility of calcite in H_2O.

Behavior of Column Reaction of Calcite and $HF-H_2O$:

Figure 4 shows a column reaction system with a flow controller and a conductivity measuring unit. A volume ratio of calcite grains (0.3mm) and aperture in filled column is 0.96. Feeding of $HF-H_2O$ is designed as downflow in order to keep constancy of a fresh column zone from variation of a reaction zone due to generation of CO_2 gas and others. Behavior of column reaction was studied as changing a flow velocity and a concentration of HF. The fluorine concentration in effluent and its enhancement so-called break-through curve are observed. Figure 5 shows some break-through curves where the feeding HF concentration is 5,000ppm and the space velocity is varied at 1 and 4. It is elucidated that the column reaction converts calcite into fluorite, rapidly and perfectly, without leaking fluorine at the concentration as low as 5ppm. Chemical analysis has found that the conversion from calcite into fluorite is 100% after beak-through, and the purity of fluorite discharged from column is over 99.5%. Dissolved SiO_2 in spent HF passes through column, thus synthesized fluorite is free from the SiO_2 inclusion. A study of column reaction is extended in NH_4F-HF H_2O system as shown in Fig.6. Its break-through curve indicates poor conversion ability compared with $HF-H_2O$ system.

DISCUSSION

Author first observed a special behavior in the reaction of calcite and $HF-H_2O$ as shown in Fig.2. In the initial reaction stage, 1mol $[CaCO_3]$ consumes 4mol [HF], which corresponds a formation of $CaF_2 \cdot 2HF$, nevertheless the HF concentration is very low level, less than 1%. Past studies only recognized a formation of $CaF_2 \cdot 2HF$ in the high HF concentration, for example, exceeding 55% [3][4]. Moreover, a transitional normalizing phenomenon from $CaF_2 \cdot 2HF$ structure to CaF_2 structure is observed. Such special mechanisms may be attributable to " Solid to Solid Formation " of fluoride in inner surface of calcite. Activity of calcite surface depends on a special inner surface structure.

In the static reaction of calcite and $HF-H_2O$, author also recognized a special decreasing effect of fluorine concentration less than the solubility of crystal CaF_2, which is well known as 8ppm of fluorine, as shown in Fig.3. Equilibrium concentration of fluorine decreases as low as 5ppm depending critically on the initial HF concentration ranging from 1,000ppm to 5,000ppm. A column reaction makes exactly a reproduction of decreasing effect as same as a static reaction, for example, the fluorine concentration in effluent from column passing through $HF-H_2O$ at the HF concentration of 5,000ppm is just 5ppm as shown in Fig.6. Such low fluorine concentration is noticeable because an ordinary treatment of fluorine ion with excess hydrated lime cannot reduce fluorine concentration less than 8ppm. It is considered that the special decreasing effect depend on a strong activity of fluorite surface from " Solid to Solid Formation ".

Author elucidated a conversion rate on column reaction. A conversion rate R of calcite grain on column reaction can be estimated with W/SV where W(l) is width of two inflection points of break-through curve as shown in Fig.5 and SV is space velocity (l/hr). The R for calcite grain with mean diameter 0.3mm in $HF-H_2O$ at the HF concentration of 5,000ppm at 25°C is found 2hr as the result of examination shown in Fig.5. Behavior of column reaction of calcite and $NH_4F-HF-H_2O$ is remarkably different from $HF-H_2O$ system as shown in Fig.6, in spite of the same fluorine concentration. The R of $NH_4F-HF-H_2O$ system is 10hrs as the result of W/S.V analysis and the fluorine concentration in effluent is constantly 50 ~ 70ppm which is much higher than that of $HF-H_2O$ system. Difference of such behavior is attributable to the reaction products as shown following equation 1) and 2).

$$2HF + CaCO_3 \quad = CaF_2 + H_2O + CO_2 \qquad 1)$$
$$2NH_4F + CaCO_3 = CaF_2 + (NH_4)_2CO_3 \qquad 2)$$

The reaction 1) is a simple neutralization and well proceed, however, the reaction 2) is difficult course because generated alkali $(NH_4)_2CO_3$ heavily retards the conversion. Synthesis of fluorite in the $NH_4F-HF-H_2O$ system is discussed in separated paper, in which another solution will be presented. In the column reaction of calcite and $HF-H_2O$, byproducts are only H_2O and CO_2. The purity of effluent from column including only minor fluorine and silica is sufficiently higher than that of natural water. Thus, column system serves as a water recirculation system. The separated treatment of effluent from HF process with large volume of water is effective way to organize a water saving closed system.

CONCLUSION

Synthesis study of fluorite with column reaction of calcite in $HF-H_2O$ system has been studied through fundamental research of kinetics and mechanism. Calcite grains can be convert rapidly into fluorite with the purity higher than 99.5% without losing fluorine in effluent at the concentration as low as 5ppm. Spent hydrofluoric acid is perfectly regenerated into fluorite in column reaction, and pure effluent can be recycled to a purewater system. A closed wet process for HF etching has been established with the calcite column system without any consumption of chemicals and disposal materials.

REFERENCES

[1] N.Miki, M.Maeno, T.Fukudome and T.Ohmi, Semiconductor Purewater and Chemicals Conference, Chemical Proceedings, pp.24–43, Santa Clara, Feb (1992)
[2] N.Miki, M.Maeno, T.Fukudome, M.Miyashita, K.Tsuru and T.Ohmi, 17th Work Shop on ULSI Ultra Clean Technology, Ultra Clean Society, Proceeding, pp.9–104, Tokyo, March (1992)
[3] Ikrami, D.D. and Nikolaev, N.S., Zh. Neorg. Khim., 15(9), 2538–41 (1970)
[4] Ikrami, D.D., Izv. Akad. Nauk Tadzh. SSR, Otd. Fiz.–Mat. Geol.– Khim. Nauk, (3) 40–6 (1971)

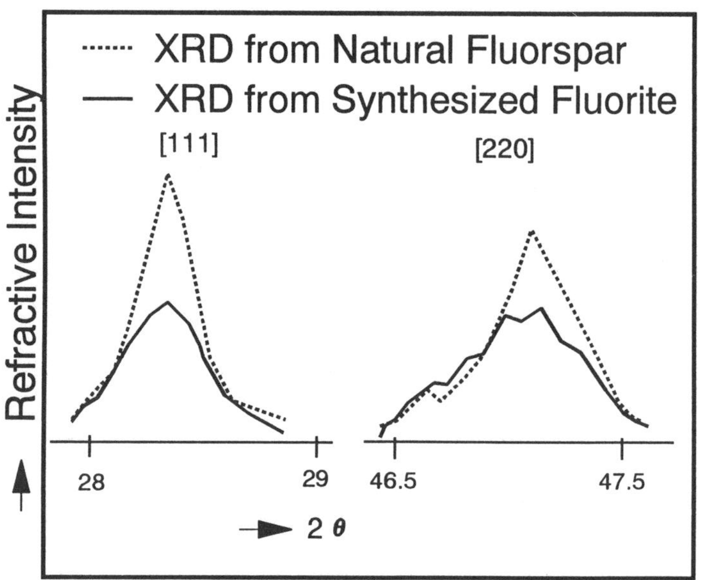

Fig.1. Crystallinity of Synthesized Fluorite

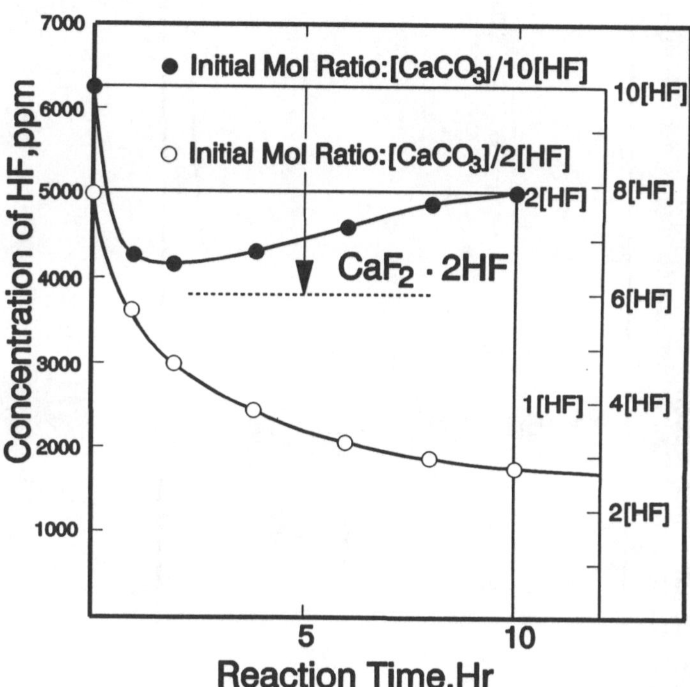

Static reaction at 25℃ under the mixing ratio
● $CaCO_3$ 10g/HF 20g(Conc.6250ppm)
○ $CaCO_3$ 10g/HF 4g(Conc.5000ppm)
Calcite : Grain Size 0.3mm

Fig.2. Reaction Behavior of Calcite and HF-H_2O
Converting into Fluorite

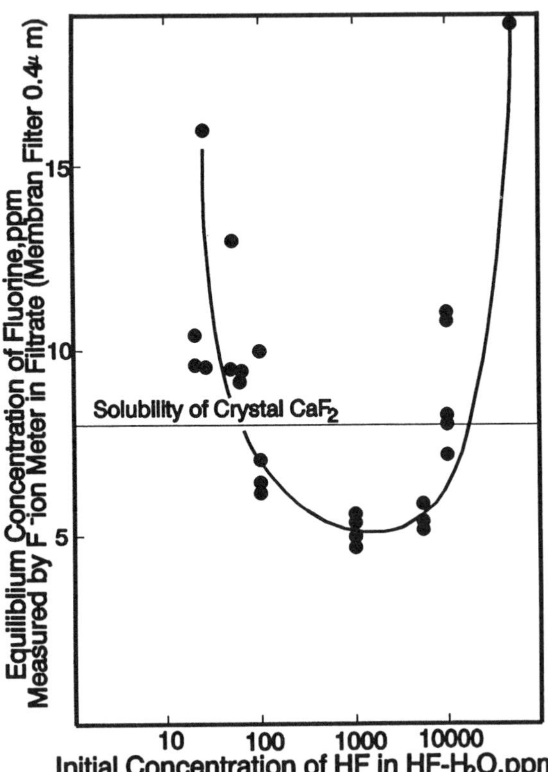

Static reaction at 25°C under the mixing ratio
CaCO₃500g/HF-H₂O 200g (each initial concentration)
Calcite : Grain size 0.3mm

Fig.3. Equilibrium Concentration of Fluorine in Reaction of Calcite and HF-H₂O

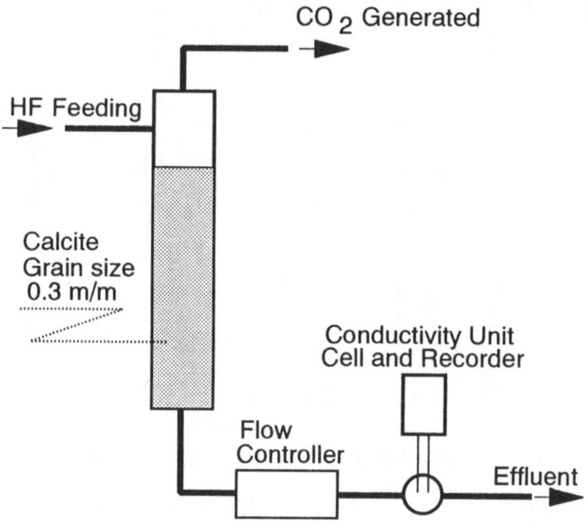

Fig.4 Schematic Flow of Column Reaction System

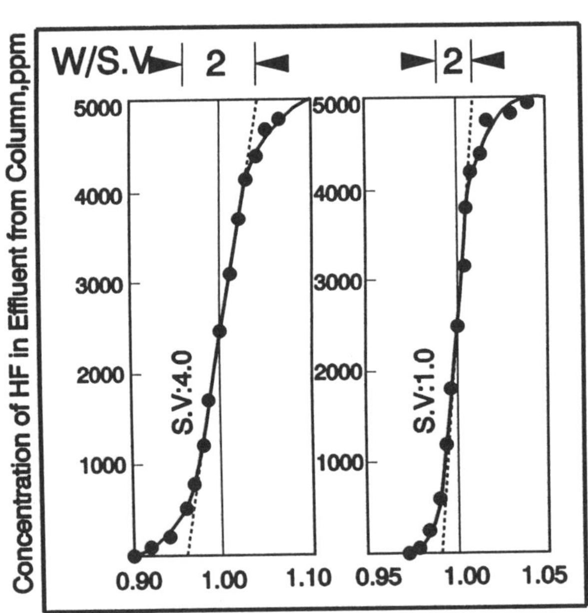

Column   :Calcite(grain size 0.3mm) 1320g/l
HF-H_2O :HF concentration,5000ppm
                 Equivalent Volume,105   l
W/S.V   :Width of break-through curve (l)
                ─────────────────────────────────
                       space velocity (l/hr)
Temperature :25 °C

Fig.5. Synthesis of Fluorite by a Column Reaction
       of Calcite and HF-H_2O

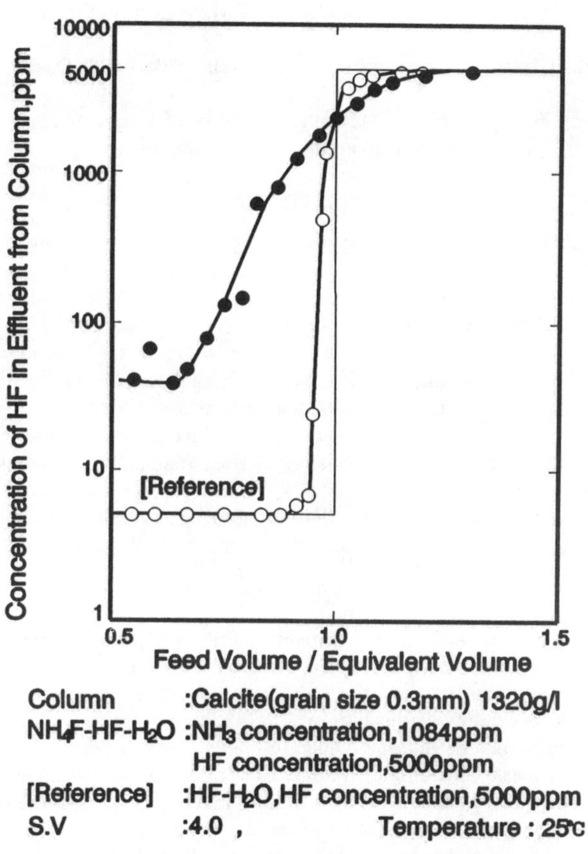

Column :Calcite(grain size 0.3mm) 1320g/l
NH_4F-HF-H_2O :NH_3 concentration,1084ppm
HF concentration,5000ppm
[Reference] :HF-H_2O,HF concentration,5000ppm
S.V :4.0 , Temperature : 25°c

Fig.6. Synthesis of Fluorite by a Column Reaction of Calcite and NH_4F-HF-H_2O

SYNTHESIS OF PURE FLUORITE
FROM SPENT FLUORIDE CHEMICALS

II. The Enhanced reaction of Ternary System $NH_4F-HF-H_2O$ and Calcite in an Evacuating Reactor

M.Maeno, N.Miki, T.Fukudome, M.Miyashita and T.Ohmi[*]

Hashimoto Chemical Corporation, Kaizan-cho Sakai 590, Osaka, Japan,

[*] Department of Electronics, Faculty of Engineering,
Tohoku University, Sendai 980, Japan

ABSTRACT

A synthesis method of Fluorite (CaF_2) by converting Calcite $(CaCO_3)$ in spent buffered hydrogen fluoride $(NH_4F-HF-H_2O)$ is studied. Kinetics and mechanism of reaction of calcite and $NH_4F-HF-H_2O$, especially difficulty to convert of coarse calcite into fluorite in high NH_4F composition, are made clear. In order to enhance the reaction, it is essential to remove generated NH_3 and CO_2 from the reaction system under an evacuating condition at elevated temperature. A perfect stoichiometric synthesis of fluorite from $NH_4F-HF-H_2O$ has been established by evacuating synthesis system consisting of a heating reactor and an ejector. The concentration of fluorine in $NH_4F-HF-H_2O$ can be reduced from 10,000ppm to 15ppm level. Synthesized fluorite keeps original size and shape of calcite without including silica from spent chemicals.

INTRODUCTION

Hydrofluoric acid and buffer hydrogen fluoride (hereunder referred to as "HF" and "BHF") are essential in the LSI wet process. At the same time, they are hardest to be processed to be discarded : in order to suppress fluorine concentration in chemical waste to be discarded below the legally-set level (15ppm or 8ppm), a lot of $CaCO_3$ and $Ca(OH)_2$ is added to include fluorine, which results in the generation of large volume of sludge (industrial waste) containing fluorine. Cost of chemicals used to process sludge constitutes another big problem. Besides sludge requires large space for processing. Especially, discarding sludge does harm to the environment.

Past studies and technologies for treatment of fluorine contained in waste water concentrated on purify the effluent usually employing excess hydrated lime, chemical composition $Ca(OH)_2$, and following chemical coagulation treatment. There are many reasons not to employ hydrated lime for synthetic regeneration of Fluorite as discussed later. Authors recognize that natural mineral Calcite, chemical composition $CaCO_3$, is optimal raw material for regeneration and reuse of Fluorite, which is abundant resource. The first kinetic study for the replacement of Calcite to Fluorite in ammonium fluoride (NH_4F) solution was reported by N.S.Baer and S.Z.Lewin in 1970 [1]. W.Augustin et al studied the synthesis reaction of Fluorite with coarse calcite from NH_4F solution and developed a Calcite–fixed bed technology [2]–[5]. D.Simonsson studied kinetics of Calcite–fixed bed process for application to effluent from so–called ammonium–uranyl–carbonate process to convert of UF_6 to UO_2 [6][7]. V.A.Kuznetsov reported a good filterability of precipitated fluorite using Calcite from fluoride solution in manufacturing process of phosphate fertilizer [8]. A.J.Cipolla et al presented an industrial regeneration technology of fluorite with hydrated lime from waste water in fluoride chemicals manufacturing plant [9]. In order to complete the optimized technology for synthetic regeneration of Fluorite applicable to LSI wet process, further study has to be pursued under a concept of closed processing.

EXPERIMENTAL

Kinetics for Reaction of Calcite and NH_4F–HF–H_2O : Wet etching processes in semiconductor manufacturing plant mainly consist of two lines in terms of the chemical composition of etchants. One of the wet processes is cleaning stage of wafer or device surface using mixed acid–based HF or diluted HF while the other process is pattern etching stage using BHF with several compositions. All the compositions of etchants containing fluorine are displayed as ternary system HF–NH_4F–H_2O as shown in Fig.1. The chemical composition of BHF is usually determined by the mixture ratio of $NH_4F40\%$ and $HF50\%$, ranging from 5:1 to 30:1. This ternary system can be characterized with the ratio of NH_4F mol and total mol of NH_4F plus HF ranging from 0 to 1, in which typical mol ratios of 0, 0.5 and 1.0 correspond to the chemical formula HF, NH_4HF_2 and NH_4F respectively. The chemical reactions of typical formula solution with calcium carbonate are expressed as follows.

$$2HF \quad + CaCO_3 \; = \; CaF_2 \; + \; H_2O \; + \; CO_2 \qquad 1)$$
$$NH_4HF_2 \; + CaCO_3 \; = \; CaF_2 \; + \; NH_4HCO_3 \qquad 2)$$
$$2NH_4F \; + CaCO_3 \; = \; CaF_2 \; + \; (NH_4)_2CO_3 \qquad 3)$$

The reaction 1) is a simple neutralization and well proceeds with sufficient conversion from $CaCO_3$ to CaF_2, with no complication remaining. However, reactions 2) and 3) are difficult courses because the generation of alkali heavily retards the conversion. The kinetics of stoichiometric reaction of calcite in ternary system HF–NH_4F–H_2O was investigated first. Figure 2 shows the degree of silica inclusion in regenerated calcium fluoride, compared with the source materials of calcite and calcium hydroxide. Apparently, high inclusion , that is more than 90%, is caused in calcium hydroxide while the reaction with calcite inhibits the inclusion of silica colloid. As the result, pure fluorite including negresibly small silica can be obtained from calcite. Taking other process parameters such as the handling character of precipitate into consideration, calcite is better material for advanced treating processes because the grain size which closely relates to the size of regenerated fluorite can be determined as required. Figure 3 shows the settling velocity of the regenerated fluorite from calcite and calcium hydroxide.

Almost all of fluorite regenerated in the reaction with calcite settles within 15 minutes, however, fluorite obtained from calcium hydroxide shows poor settling characteristic.

In order to achieve the synthesis of pure fluorite without the inclusion of excess calcite, the high conversion from calcite to fluorite under the stoichiometric condition is required. The reaction mechanisms relating the chemical composition of etchant and the reaction kinetics relating the grain size of calcite were studied to reveal the condition which can realize the high conversion. Figure 4 shows the relation between conversion and composition in the stoichiometric condition, in which pure calcium carbonate in calcite and fluorine contained in etchant is exactly equivalent to the ratio of Ca : 2F. In this examination, fine powder of calcite is used and the reaction process reaches equilibrium in 0.5 or 1hr at 70°C or 30°C respectively. The more NH_4F ratio increases, the more conversion decreases, and especially typical NH_4F medium heavily retards the conversion. In order to keep the conversion higher than 97%, the mol ratio of less than 0.8 and the reaction temperature of higher than 70°C are required even for fine calcite. One of the parameters for conversion is alkalinity, i.e, pH value of reacted solution, which is closely relates to the chemical composition of etchants. For example, pH value exceeds 9.0 after reaction of typical etchant having mol ratio 0.8. Another important parameter of the conversion is the grain size of calcite. Figure 5 shows the effect of grain size indicated in specific surface area ranging from $40cm^2/g$ of coarse grain to $6400cm^2/g$ of fine powder. The time dependence of conversion remarkably increases along with the increase of grain size and the size effect increases in accordance with the increase of concentration. However, in the case of the reaction temperature of 70°C, the conversion even for large grain calcite also reaches up to 99% after 8 hours of reaction. **Behavior of Evacuating Reaction of Calcite and NH_4F–HF–H_2O :** Figure 6 shows an evacuating reactor–closed type with live–steam heater, air bubbler and materials feeder for calcite and liquid. After stoichiometric feeding of materials with the ratio of $[CaCO_3]$ / $2[F]$, the reactor is heated to 60°C and reaction is continuously proceeded under evacuation by using an ejector via a mist separator. Reacted materials are drained from reactor with stirrer after finish of reaction. NH_4F–HF–H_2O with the concentration of NH_4F 15.35% and HF 1.70% which is a typical composition drained from an etching process with BHF, is treated in this evacuating reactor. Figure 7 shows a reaction proceeding curve indicating the residual fluorine concentration which directly shows the conversion from calcite into fluorite due to the stoichiometric feeding without loss of HF. Grain size of calcite used for the reaction is 0.05 ~ 0.15mm. In the case of ordinary heating reactor, fluorine concentration in the reactor cannot reach the level less than 100ppm. However, evacuating reaction enhances the conversion compared with a normal heating reaction, especially, suppressing the residual fluorine concentration as low as 15ppm level. It is recognized that fluorite synthesized from calcite keeps size and shape of original calcite grain, the same as that synthesized from HF–H_2O.

DISCUSSION

Reaction of calcite with NH_4F–HF–H_2O is essentially difficult course due to the generation of $(NH_4)_2CO_3$ and NH_4HCO_3 especially when the NH_4F concentration and grain size of calcite increase. An elevation of reaction temperature is required for decomposition of such alkali compounds to NH_3 and CO_2. Volume ratio of gas/liquid is caluculated 93 for NH_3 and 56 for CO_2 at normal temperature and pressure for liquid with the concentration of NH_4F 15.35% and HF 1.70%. Major decomposition proceeds reaction, but, residual alkalinity retards tenaciously the complete conversion of calcite in core part. An evacuation method is the most effective to promote the residual reaction expelling minor ammonia. Fluorine is not missing from a reactor to an ejector through a mist separator, because the vapor pressure of NH_4F is negresibly small in alkali phase. A treatment of scrubbed ammonia with the biotechnology is easy due to absence of fluorine ion, in which fluorine ion poison microorganisms. Grain size of synthesized fluorite which keeps original size and shape of calcite is practically valuable for handling and reuse.

CONCLUSION

Synthesis of fluorite with evacuating reaction of calcite and NH_4F–HF–H_2O has been studied through fundamental research of kinetics and mechanism. Calcite grains can be converted perfectly into fluorite, reducing the concentration of residual fluorine in effluent as low as 15ppm. A closed wet process for BHF etching in LSI manufacturing has been established by introducing the enhanced evacuating reaction system with calcite without consuming of any chemicals and disposal materials. Separation of the treatment process for BHF system contributes effectively to the treatment of HF system without being disturbed by NH_4F.

REFERENCES

[1] N.S.Baer and S.Z.Lewin, The American Mineralogist, Vol.55, March–April, (1970)
[2] W.Augustyn, M.Dziegielewska, A.Kossuth, and Z.Librant, Journal of fluorine Chemistry, 12, 281–292, (1978)
[3] W.Augustyn, Mater, Ogolnopol, Symp, Zwlazki Fluorowe, Meeting Date, 47–51, (1980)
[4] W.Augustyn, Chem, Stosow., 30(3), 353–60, (1986)
[5] W.Augustyn and D.Krystyna, Przem, Chem.,68(4), 153–5, (1989)
[6] Daniel Simonsson, Ind. Eng. Chem. Process Des. Dev., Vol.18, No.2, 288, (1979)
[7] Per Ekdunge and Daniel Simonsson, J.Chem. Tech. Biotechnol. 34A, 1–9, (1984)
[8] Kuznetsov, V.A., Novikov, A.A., Zaitsev, V.A., Rodin, V.i., Mishin, N.I. and Makarov, S.V., Khim. Prom-st. (Moscow), (2), 90–3, (1983)
[9] Anthony J.Cipolla, Edward J.Shields and Constance P.Wickersham, Chemical Processing, 48(14), 126–128, (1985)

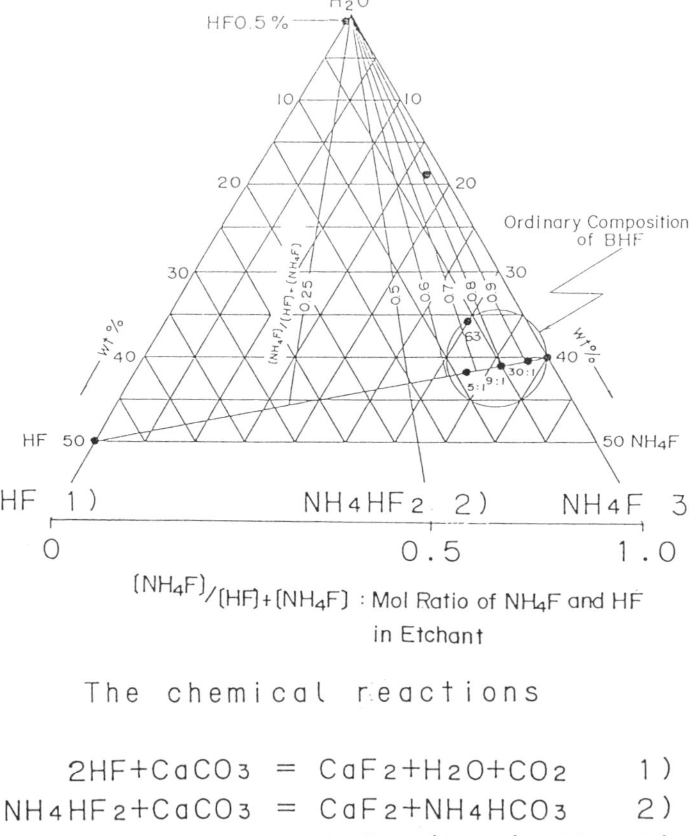

$$\frac{[NH_4F]}{[HF]+[NH_4F]} : \text{Mol Ratio of } NH_4F \text{ and } HF$$
in Etchant

The chemical reactions

$$2HF + CaCO_3 = CaF_2 + H_2O + CO_2 \qquad 1)$$
$$NH_4HF_2 + CaCO_3 = CaF_2 + NH_4HCO_3 \qquad 2)$$
$$2NH_4F + CaCO_3 = CaF_2 + (NH_4)_2CO_3 \qquad 3)$$

Fig. 1. Ternary System NH_4F-HF-H_2O

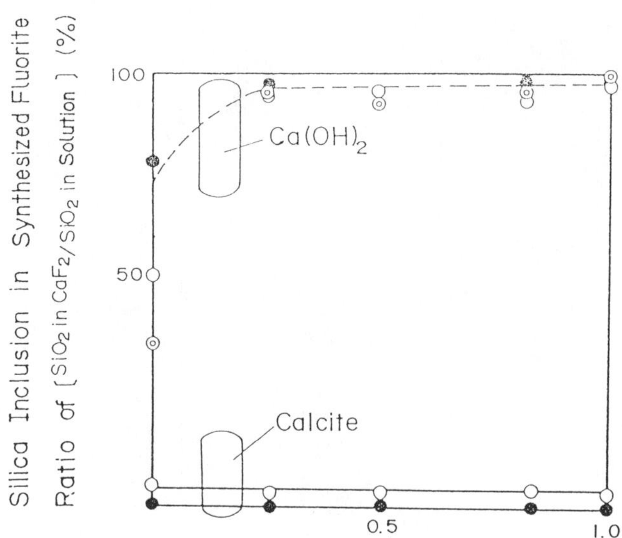

Fig. 2. Silica Inclusion into Fluorite
in the Reaction with $CaCO_3$
or $Ca(OH)_2$

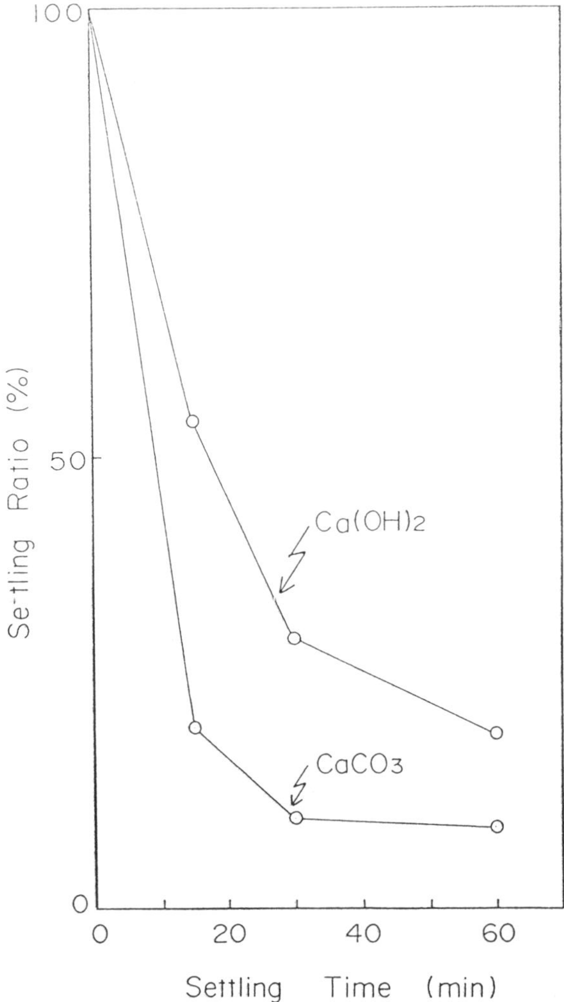

Fig.3   Comparison of Settling Velocity
the Fluorite in the Reaction with
CaCO3 or Ca(OH)2

392

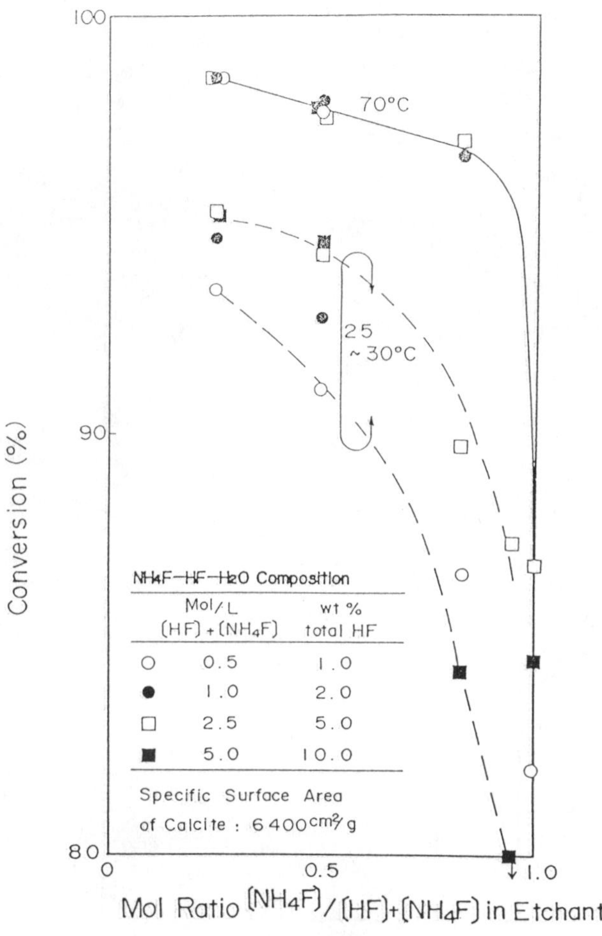

Fig.4. Conversion of Calcite into Fluorite in the Stoichiometric Reaction as a Function of NH4F—HF—H2O Composition and Temperature

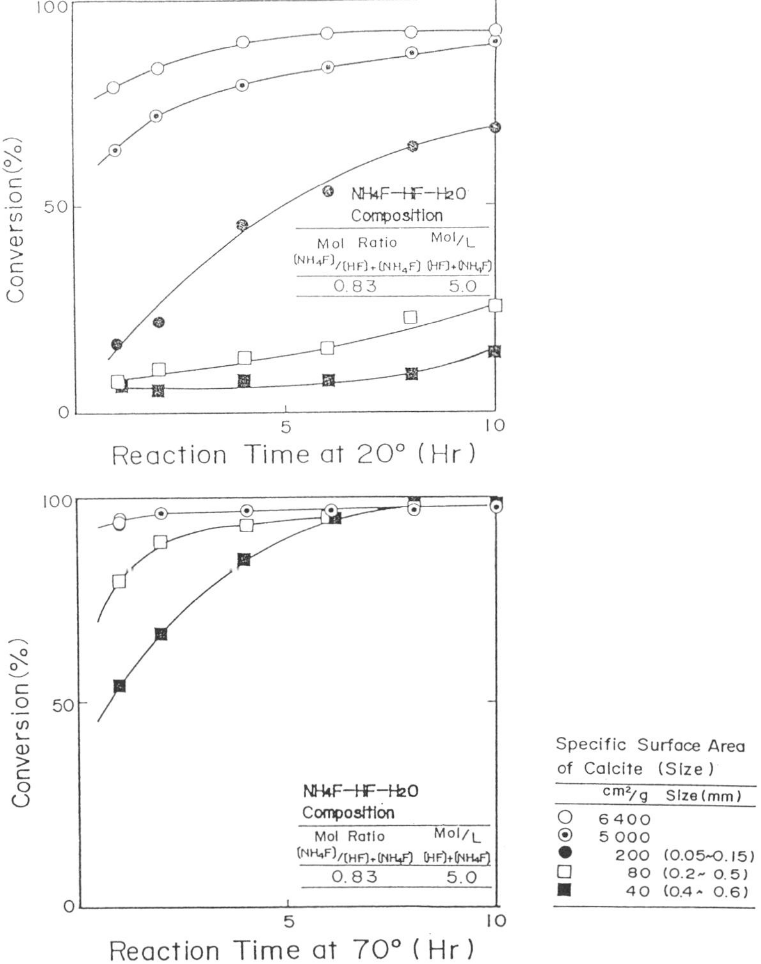

Fig.5.   Conversion of Calcite into Fluorite
in the Stoichiometric Reaction
Relating to the Surface Area of Calcite

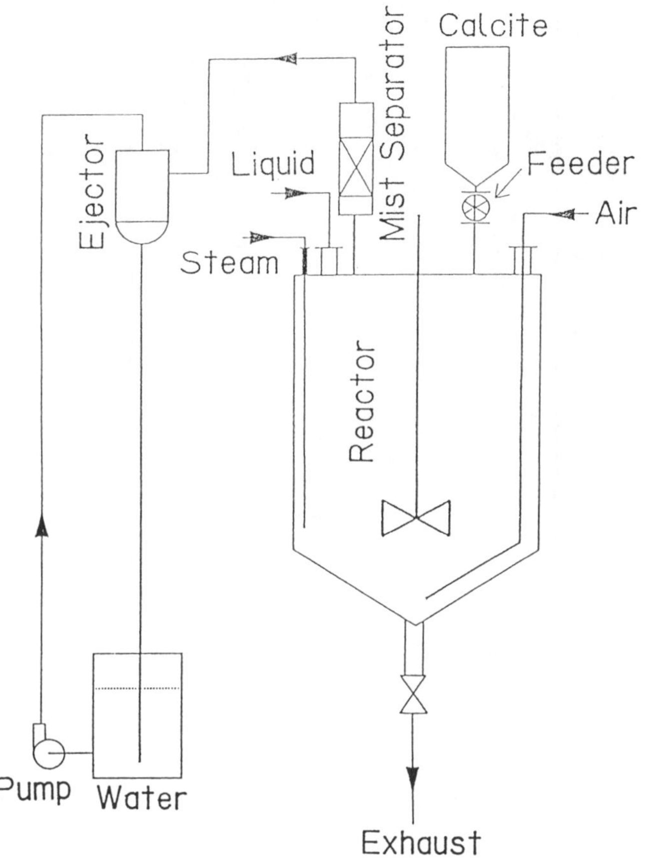

Fig.6. Evacuating Synthesis System

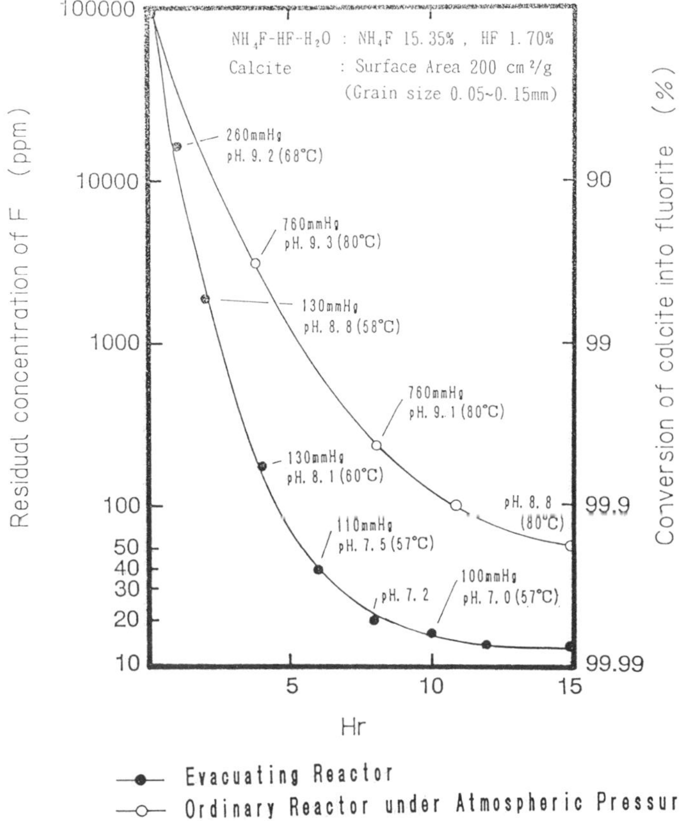

Fig. 7 Enhanced Stoichiometric Reaction of Coase
Calcite and NH4F—HF—H2O in Evacuating Reactor

DEFECT REDUCTION IN DRAM MANUFACTURE USING ION EXCHANGE-PURIFIED HF

Vikram Doshi and Lindsey Hall
Texas Instruments, DMOS IV and COD, Dallas, Texas 75265

and

John Davison
Athens Corp, Oceanside, CA 92056

An ion exchange purification system was used to purify dilute hydrofluoric acid for use in silicon wafer cleaning operations. Impurity levels of 34 elements in the treated HF were routinely below 1 ppb. The ion exchange purified hydrofluoric acid performed significantly better in refresh characterization and full bar gate oxide integrity split lot tests than the high purity imported HF then being used in the fab. A 5% improvement in multiprobe yield was observed after the ion exchanged hydrofluoric acid was placed into production in a MOS memory fab making 1 Megabit DRAMS.

INTRODUCTION

Hydrofluoric acid (HF) purity has been demonstrated to have an impact on wafer fab yields, although this has not been well documented in the past. In late 1989, TI's DMOS IV wafer fab in Dallas, Texas which manufactures 1 Megabit DRAMS, experienced severe problems with the wet clean process which was found to be related to dilute HF. These problems were corrected by switching to high purity HF from Japan, and implementing some process modifications. Additional modifications in 1990 further improved the wet clean process. Ultimately, a dilute HF-last process was adopted based, in part, on the fact that the high purity HF was the second cleanest chemical in the process after DI water.

Also in 1989, the first hydrofluoric acid reprocessor, or HFR, was successfully tested at DMOS IV [1]. The HFR used filters and ion exchangers to remove particulate and ionic impurities from dilute HF. Developed under a co-sponsorship arrangement with TI, the first HFR was designed to recycle and repurify the hydrofluoric acid solutions used to clean furnace tubes and other quartz parts. Tests at TI and Athens soon revealed that ion exchange technology has the ability to produce ultrapure HF solutions [2].

Consequently, in late 1990, another HFR was installed by TI's Chemical Operations Department (COD) for the purpose of manufacturing high purity dilute HF solutions at low cost. The HFR proved capable of making high purity dilute HF solutions reliably and consistently. The next step was to qualify the HF for use in the wet clean process. The qualification took place in DMOS IV since they had a demonstrated need for high purity dilute HF. The ion exchange (IE) purified HF from the HFR was compared directly to the high purity HF from Japan then in use in a series of split lot tests carried out simultaneously in identical wet process hoods. No changes to the equipment, and only minor changes to the chemical distribution systems, were required. Thus, the only variable in the tests was the HF. The results of these tests indicate with a high degree of certainty that the IE purified HF gives cleaner wafer surfaces, fewer defects, and higher yields.

Description of the HF Purification System at the DCC

The theory and operation of the HFR have been described in detail elsewhere [1-3]. Suffice it to say that the HFR uses state-of-the-art membrane filter technology to remove insoluble contaminants, and ion exchange technology to remove soluble ones.

The purification system at TI's Dallas Chemical Center is shown in Figure 1. 49% HF is purchased in bulk, stored in a 10,000 gallon PTFE-lined pressure vessel, and transported in PFA piping to the weigh tank. The weigh tank is suspended from a strain gauge and the chemicals are weighed in this tank before being discharged to the mix tank for blending and filtration. The 49% HF is filtered to 0.1 microns immediately before the weigh tank. The weigh and mix tanks are high density polyethylene. The materials of construction downstream from the mix tank are precleaned and leached PVDF and PFA. The process begins when 49% HF is diluted with ultrapure DI water to 6% in the weigh tank. The dilute HF is then sent to the mix tank where it is blended by recirculating through a 0.1 micron filter for 2 hours. This material is then pumped through a 0.1 micron filter; a cation exchange column filled with a strong acid cation resin; an anion exchange column filled with a weak base anion resin; and 0.1 and 0.05 micron filters in series. The 6% HF then enters the product tank from which an assay sample is taken in a class 10 cleanroom. The HF in the product tank is further diluted to its final concentration depending on the need. The product tank sits on a weigh pad and the dilution is achieved by adding the appropriate weight of ultrapure DI water. The product HF is blended by recirculating through a 0.05 micron filter for one hour. A sample is taken in the class 10 cleanroom and analyzed for trace impurities. The HF is then packaged in either 200 gallon PFA- or PTFE-lined pressure vessels, or 1 gallon high density polyethylene bottles. A final analysis of the packaged material is conducted before delivery to the fab.

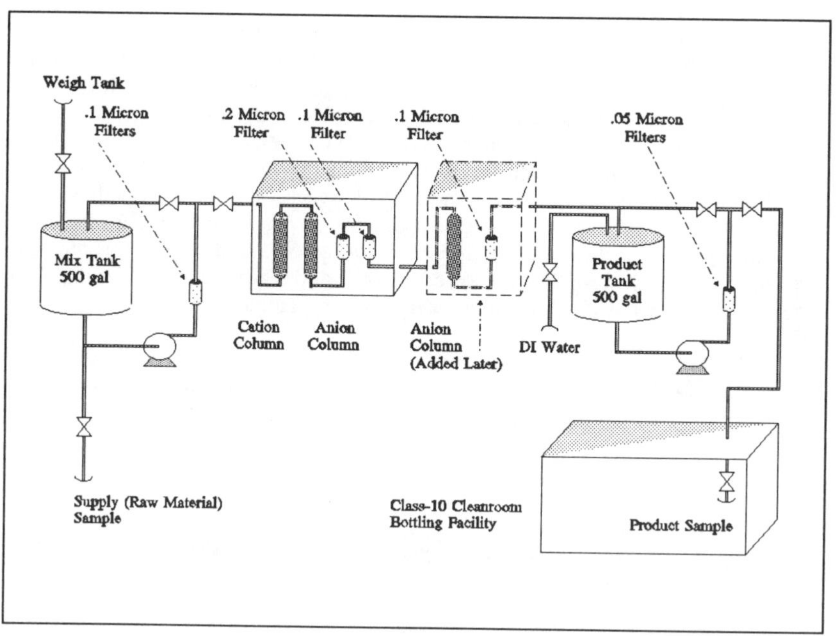

Figure 1. The HF purification system at the Dallas Chemical Center.

The HFR has proven to be an extremely reliable piece of equipment. The unit was placed in production service in early 1991 for the DMOS IV evaluations. The machine uptime was 95.3% with 100% chemical availability during this four month period. The unit has operated with > 95% uptime and 100% chemical availability since then. Other than the need for routine regeneration of the ion exchange columns, machine maintenance has been minimal.

Purity of IE Purified HF

All analyses were performed by the Texas Instruments Analytical Service Laboratories managed by the COD. Aluminum, arsenic, calcium, iron, potassium and sodium were analyzed by GFAA. All other metals and boron were analyzed by ICP/MS. Sample introduction areas for both instruments were under laminar flow HEPA filters to minimize sample contamination.

Contaminant levels in the IE purified and imported HF are compared in Table 1. There is data for IE purified HF that was purified with one and two anion columns. (The two anion column purification system will be discussed later.) The data for the imported HF and the 1 anion column IE HF are averages from the time of the qualification tests (early 1991). The data was handled as follows: non-detected values were taken as zero and detected values were used as-is. The average and standard deviation were calculated for each element. If all values were below detection, then the detection limit is simply given. While no universally accepted method exists for comparing actual values with not-detected values, it is felt that data obtained from the same instruments at the same time can be compared in this manner. Of those metals detected, it appears that the IE purified acid was consistently lower in metals, i.e. lower average and/or standard deviation, than the imported HF except for potassium. Zinc was equivalent in both acids.

Table 1. Comparison of ionic impurity levels (in ppb) in imported and IE purified 4.9% HF. Data for IE purified HF with 2 anion columns was obtained in 1992, after the 1991 HF qualification tests. Avg = average; S.D. = standard deviation.

Element	Imported HF		IE W/1 Anion Column		IE W/2 Anion Columns		
	Avg.	S.D.	Avg.	S.D.	Avg.	S.D.	Best
Silver	< 1.0		< 1.0		< 1.0		< 1.0
Aluminum	< 1.0		< 1.0		0.4	0.3	< 0.1
Gold	< 1.0		< 1.0		< 1.0		< 1.0
Barium	< 1.0		< 1.0		< 1.0		< 1.0
Beryllium	< 1.0		< 1.0		< 1.0		< 1.0
Bismuth	< 1.0		< 1.0		< 1.0		< 1.0
Boron	< 1.0		< 1.0		<1.0		< 1.0
Calcium	0.6	1.6	0.4	0.3	< 0.1		< 0.1
Cadmium	< 1.0		< 1.0		< 1.0		< 1.0
Cobalt	< 1.0		< 1.0		< 1.0		< 1.0
Chromium	< 1.0		< 1.0		< 1.0		< 1.0
Copper	0.2	0.5	< 1.0		< 1.0		< 1.0
Gallium	< 1.0		< 1.0		< 1.0		< 1.0
Iron	2.3	2.8	2.2	1.4	0.4	0.2	< 0.1
Potassium	0.2	0.3	0.4	0.7	0.1	0.1	< 0.1
Lanthanum	< 1.0		< 1.0		< 1.0		< 1.0
Magnesium	0.1	0.4	< 1.0		< 1.0		< 1.0
Manganese	< 1.0		< 1.0		< 1.0		< 1.0
Molybdenum	< 1.0		< 1.0		< 1.0		< 1.0
Sodium	0.2	0.8	0.2	0.2	0.2	0.2	< 0.1
Nickel	< 1.0		< 1.0		< 1.0		< 1.0
Lead	< 1.0		< 1.0		< 1.0		< 1.0

Element	Imported HF		IE W/1 Anion Column		IE W/2 Anion Columns		
	Avg.	S.D.	Avg.	S.D.	Avg.	S.D.	Best
Palladium	< 1.0		< 1.0		< 1.0		< 1.0
Platinum	< 1.0		< 1.0		< 1.0		< 1.0
Antimony	< 1.0		< 1.0		< 1.0		< 1.0
Tin	< 1.0		< 1.0		< 1.0		< 1.0
Strontium	< 1.0		< 1.0		< 1.0		< 1.0
Tantalum	< 1.0		< 1.0		< 1.0		< 1.0
Vanadium	< 1.0		< 1.0		< 1.0		< 1.0
Tungsten	< 1.0		< 1.0		< 1.0		< 1.0
Zinc	0.2	0.5	0.2	0.6	< 1.0		< 1.0
Zirconium	< 1.0		< 1.0		< 1.0		< 1.0

Elemental impurities were the main contaminants of interest early in the program. The anions sulfate, nitrate, phosphate, chloride, and fluosilicate were determined to SEMI levels with no difference seen between the two acids. Particle levels were not readily comparable because of sampling difficulties encountered due to differences in packaging of the two acids.

Addition of a Second Anion Column

A second anion column and filter were added in late 1991 after the qualification tests described in this article. The same anion resin was used in both anion columns. The second anion column was added for two reasons: first, to increase the production capacity of the system by increasing the volume of HF that could be purified in between ion exchange resin regenerations. And second, to enable the system to successfully deal with the periodically high levels of iron in the 49% HF. Iron exists as an anionic fluoride complex in HF, so it is removed by the anion resin. The second column provided an additional benefit of decreasing the elemental contaminants and their variation, see Table 1. After installation of the second anion column, sulfate, chloride, nitrate and phosphate levels were determined to be less than 100 ppb. Sulfate was measured by ion chromatography. Chloride, nitrate and phosphate were measured using modified SEMI methods. Fluosilicate, also determined by a modified SEMI method, was found to be less than 300 ppb. A large number of production runs have had all elements at levels below detection.

Unfortunately, none of the electrical tests were repeated after the addition of the second anion column to see if the improved HF purity further enhanced device performance. However, a positive impact on performance might be expected.

EXPERIMENTAL

Liquid Particle Counts

The difficulties encountered in making particle comparisons between the imported and IE purified HF were mentioned previously. However, it was possible to compare the particle levels in the IE purified HF with those in the dilute HF solutions prepared by the COD standard method, see Table 2. This study was conducted with 4.9% HF after the installation of the second anion column. The IE purified HF was prepared as described above. The standard HF was prepared from domestic 49% HF using the same weigh and mix tanks, and the same class 10 cleanroom for sampling and packaging. The only difference between the two preparations was that the standard HF was not purified through the HFR. As shown in Table 2, the average particle counts for the IE purified HF was consistently lower than the standard HF, although not by a significant amount. However, the variation in particle counts between the two acids was significantly lower for the IE purified HF. The data suggests that the ion exchange resin is not a significant particle source, or that any resin-generated particles are effectively controlled by the filters in the system. Obviously, the particle levels in the chemicals used in critical process steps are always a matter of concern.

Table 2. Comparison of the particle levels (counts/ml) in dilute standard and IE purified HF solutions prepared at TI's COD.
Avg. = average; S.D. = standard deviation.

	Standard 4.9% HF		IE Purified 4.9% HF	
	Avg.	S.D.	Avg.	S.D.
< 0.2 microns	61.8	70.2	54.5	34.9
< 0.3 microns	15.7	18.1	13.9	9.9
< 0.5 microns	2.4	3.4	1.7	1.5

Device Testing

Preliminary testing to determine the suitability of IE purified HF for wafer cleans consisted of bath analysis for metals, etch rate consistency, particle levels in the HF bath, and particles added to a non-patterned wafer. These preliminary tests showed no significant differences between the two acids.

The next tests were electrical measurements consisting of minority carrier lifetimes, GOI breakdown voltages, stress CVs, refresh times and multiprobe yields. The wafers used were 150 mm P-type. All HF cleans were performed in recirculating filter baths equipped with 0.1 micron filters and operated at 25 $\pm$ 1°C. Split lot tests were performed in identical baths at the same time. The lots were split only for the cleaning sequence using the two acids. In all cases, wafers were rinsed after HF immersion for at least 10 minutes in a DI cascade rinse to constant conductivity, and then dried in an IPA vapor dryer.

Two to four wafers were split between the imported HF and the IE purified HF. The tests were repeated five or six times. After this initial testing, three split lots (48 wafers/lot) were then committed, and the split was done at critical gate cleans. These lots were then processed normally through multiprobe. Following favorable results at multiprobe, twenty full lots were then processed through the new acid. After favorable results from the full lot qualification, the IE purified HF was released to production.

RESULTS AND DISCUSSION

Electrical Tests

Minority Carrier Lifetime (MCL) - MCL is related to many device properties. In DRAM devices, it limits the time between the refresh cycles that update the data. Transition metals, particularly copper and gold, are the contaminants that most affect MCL [4].

After HF cleaning and drying, the split lot test wafers were placed in an oxidation furnace together. MCLs on the oxidized wafers were determined by the microwave reflectance technique using a LEO GILKEN "Wafer Tau" Model LTA-130A apparatus. In most of the comparisons, the IE purified HF was found to be as good or better than the imported HF. Perhaps it is not too surprising that no significant difference between the two acids was seen in this test, since MCL tests are often dominated by impurities present in the Czochralski crystal from which the wafers originate.

Capacitance-Voltage (CV) Analysis - This test used six groups of 4 wafers. Each group of wafers was split, cleaned, dried, and oxidized as before. Polysilicon caps were then deposited on top of the oxide to make test capacitors. The capacitance of each MOS test structure was measured as a function of the applied voltage. The CV test is sensitive to conditions at the Si/SiO_2 interface. Among other things, CV tests can provide information regarding mobile ion concentrations in the oxide or at the interface [5]. Mobile ions like sodium and potassium can come from the cleaning process, or from the oxidation furnace [6]. As in the previous test, no significant differences were observed between the IE purified HF and the imported HF. Again, this result may not be too surprising since mobile ions, especially sodium, are not known to deposit in HF cleaning steps [7,8]. Regardless, the IE purified HF had a lower level of sodium, on average, than the imported HF.

Full Bar Gate Oxide Integrity (FBGOI) Tests - FBGOI tests all the "bars," or chips, on a wafer. The performance of each gate on each bar is tested in order to determine the yield of good bars per wafer. The initial split lot tests were favorable, so the test was repeated with 20 full lots of wafers. The results of the FBGOI tests are shown in Figure 2. The data indicates a significantly higher yield of good devices in the IE purified HF split. FBGOI tests are sensitive to both particulate and ionic contaminants, but we believe the observed FBGOI yield difference is due primarily to the lower levels of metals in the IE purified HF.

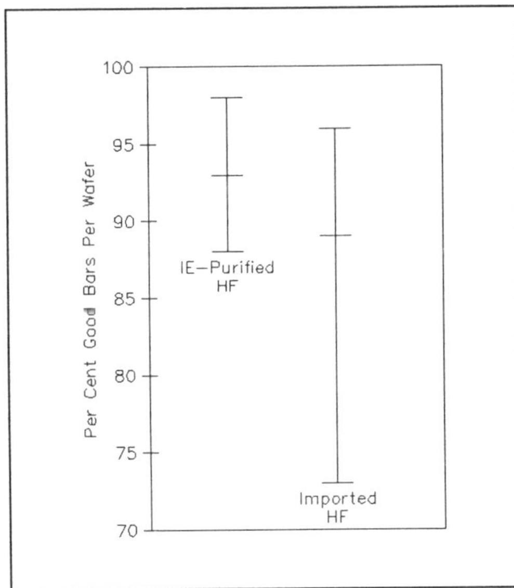

Figure 2. FBGOI test results comparing ion exchange purified HF to the fab standard (imported) HF.

Refresh Characterization - Refresh characterization tests for current leakage. A cell will lose its value if there is a leakage path for electrons. The test is performed by writing "ones" to the entire array, then a 120 millisecond pause occurs during which time the array is not refreshed. After the pause, the entire array is read. The number of cells retaining the "one" value is then determined. The IE purified HF demonstrated significantly fewer refresh failures as compared to the imported HF. This is believed to be related to the lower metallic content of the IE purified HF.

Multiprobe Yield - The ultimate test for any semiconductor device is multiprobe yield. The impact seen in the refresh characterization and FBGOI tests was also seen at multiprobe. A yield increase of 5% was observed for those split lots processed through the IE purified HF as compared to the imported HF. The yield difference was seen as significant with 90% confidence. Furthermore, multiprobe yield was compared for each hood before and after it was converted from imported to IE purified HF. All lots that experienced problems during other parts of the process were removed from this statistical analysis. This additional verification also showed a 5% yield increase with a 95% confidence level in one hood, and a 99.9% confidence level in another.

The very low levels of calcium, copper, magnesium and other metals in the IE purified HF may be responsible for the yield improvement. Calcium, for example, once thought to be benign [6], has recently been linked to degradation of gate oxide integrity [9,10]. Copper is notorious for its tendency to plate on silicon surfaces exposed in HF [11-14]. Copper can diffuse into silicon, even at room temperature [15] and, if present in sufficient quantity, degrade minority carrier lifetime [4]. In addition, high temperature operations can cause copper to precipitate as a silicide, forming various crystallographic defects [16].

CONCLUSIONS

Ion exchange technology has demonstrated the ability to further reduce the already low levels of ionic impurities in dilute VLSI-grade HF. The need for such extremely low levels of contamination in dilute HF has been advocated by SEMA-TECH [17] and others [12,13]. This work demonstrates a substantial yield improvement in a high volume DRAM manufacturing facility from the use of IE purified HF. To the best of our knowledge, this is the first report of a quantified yield improvement directly linked to a change in a single chemical. Results of the split lot FBGOI and refresh characterization tests suggest that the yield improvement is likely due to the low ionic impurity levels in the HF. In the year since switching to the IE purified HF, defect levels have remained down, and not a single problem related to HF has been experienced. Furthermore, in over a year of operation in a production environment, the ion exchange purification system has proven itself to be reliable and cost effective.

ACKNOWLEDGMENTS

The authors wish to thank the technical staff of DMOS-IV, COD and Athens for their assistance with the data collection.

REFERENCES

[1] J. Davison, C. Hsu, E. Trautmann and H. Lee, in *Cleaning Technology in Semiconductor Device Manufacturing*, J. Ruzyllo and R. Novak, eds., Electrochem. Soc. Proc. **90-9**, 83 (1990).

[2] J. Davison and C. Hsu, *Proceedings of the 10th International Symposium on Contamination Control* in Swiss Contamination Control **3**, 246 (1990).

[3] J. Davison, Solid State Technology, in press.

[4] J. Atsumi, S. Ohtsuka, S. Munehira and K. Kajiyama, in *Cleaning Technology in Semiconductor Device Manufacturing*, J. Ruzyllo and R. Novak, eds., Electrochem. Soc. Proc. **90-9**, 59 (1990).

[5] C. Osburn and S. Raider, J. Electrochem. Soc., **120**, 1369 (1973).

[6] S. Banerjee, in *Microcontamination Conference Proceedings 1991*, 621 (1991).

[7] L. Zazzera and J. Moulder, J. Electrochem. Soc., **1236(2)**, 484 (1989).

[8] R. Poliak, R. Matthews, P. Gupta, M. Frost and B. Triplett, in *Microcontamination Conference Proceedings 1991*, 511 (1991).

[9] S. Verhaverbeke, M. Meuris, P. Mertens, A.Kelleher, M. Heyns, R. DeKeersmaecker, M. Murrell and C. Sofield, in *Cleaning Technology in Semiconductor Device Manufacturing*, J. Ruzyllo and R. Novak, eds., Electrochem. Soc. Proc. **92-12**, 187 (1992).

[10] P. Mertens, M. Meuris, S. Verhaverbeke, M. Heyns, A. Schnegg, D. Graf and A. Philipossian, in *1992 Proceedings of the Institute of Environmental Sciences*, 475 (1992).

[11] W. Kern, RCA Review, 234, June (1970).

[12] F. Kern, M. Itano, I. Kawanabe, M. Miyashita, R. Rosenberg and T. Ohmi, presented at the *37th Annual Technical Meeting of the Institute of Environmental Sciences*, May (1991).

[13] T. Imaoka, T. Kezuka, J. Takano, I. Sugiyama and T. Ohmi, in *1992 Proceedings of the Institute of Environmental Sciences*, 466 (1992).

[14] R. Gruver, R. Gaylord, B. Bilyou and K. Albaugh, *ibid.*, 460 (1992).

[15] P. Jones, Y. Zhang, J. Liu, J.-Z.. Yuan, C. Ortiz, B. Baufeld, H. Bakhru, J. Corbett and S. Pearton, in *Defect Control in Semiconductors*, K. Sumino, ed., Elsevier Science Publishers B.V., 317 (1990).

[16] M. Hourai, S. Sadamitsu, K. Murakami, T. Shigematsu and N. Fujino, in Defect Control in Semiconductors, K. Sumino, ed., Elsevier Science Publishers B.V., 305 (1990).

[17] SEMATECH SEMASPEC #90120407A-STD. (1991).

CLEANING PROCESS FOR REMOVING
OF OXIDE ETCH RESIDUE

Jae Jeong Kim, Eun Gu Lee, Woo Shik Kim
Min Sung Choi, Hee Cook Lee, Chang Soo Kim

Semiconductor Research Laboratory,
GoldStar Electron Company, LTD.
Woomyun-dong, Seocho-gu, Seoul 137-140, KOREA

Cleaning process of an oxide etch residue on a silicon substrate which retarded further thermal oxidation and increased junction leakage was studied. The residue was found to be C, F-based polymer (CF_x-polymer) and compound with Si-C and Si-O bonds (SiO_yC_z). SiO_yC_z, in which the ratio of y to z was monotonously changed with depth, was positioned at the underneath of CF_x-polymer. Since wet cleaning only was insufficient to remove the residue, two step cleaning was applied: dry and wet cleaning. The most effective cleaning process of dry and wet cleanings was silicon light etch, followed by $NH_4OH-H_2O_2$ mixture and aqueous HF dip. It was observed that two step cleaning recovered minority carrier life time to the extent of the initial state and reduced junction leakage nearly to the wet oxide etching.

INTRODUCTION

In a dynamic random access memory (DRAM) fabrication process, a silicon substrate is exposed by an oxide dry etching and overetching process for an electrical contact. The oxide etch chemistries and etch mechanisms, fluorocarbon-based reactive ion etch or magnetic enhanced reactive ion etch, inevitably lead to the deposition of fluorocarbon polymer on the contact area and cause Si substrate damaged. After the oxide etching, it was found that an oxide growth was retarded on the exposed silicon substrate in the subsequent thermal oxidation step.

Papers have reported etch residue composition and post treatment[1-7]. Up to now what is revealed in an oxide etch residue is that it consists of C, F-containing film and silicon carbide. There has been no report on the influence of an oxide etch residue on device and process. Little study has been done on how to get rid of the residue. Detailed analysis of residues and cleaning methods have not been studied sufficiently enough to be applied for DRAM fabrication.

Since the structure and composition of an etch residue practically depend on an etching gas chemistry and composition, work to identify the etch residue for each etching gas should be done. This paper deals with the layer structure and chemical composition of an oxide dry etch residue on a silicon substrate when $CHF_3/CF_4/Ar$ plasma was used, and cleaning process to remove the residue.

408

EXPERIMENT

Using SC1 (NH4OH–H2O2–H2O) and HF solution, bare p-type Si (100) wafers with resistivity of 9-12 ohmcm were precleaned and then CVD oxide of 2000 Å thickness was deposited. The oxide layer was etched and overetched with $CHF_3/CF_4/Ar$ plasma. The residual layer was examined using XPS and SIMS. Cleaning steps, either wet or dry-wet cleanings, were applied to removing the oxide etch residue. Three wet cleaning processes were used: SPM (H_2SO_4–H_2O_2), SC1, and HF. Dry cleanings included O_2 plasma ashing, ultraviolet/O_3 ashing, silicon light etch (CF_4/O_2 plasma), and silicon etch (HBr/Cl_2 plasma). For electrical evaluation, minority carrier life time and junction leakage were measured. Figure 1 shows the experimental procedure.

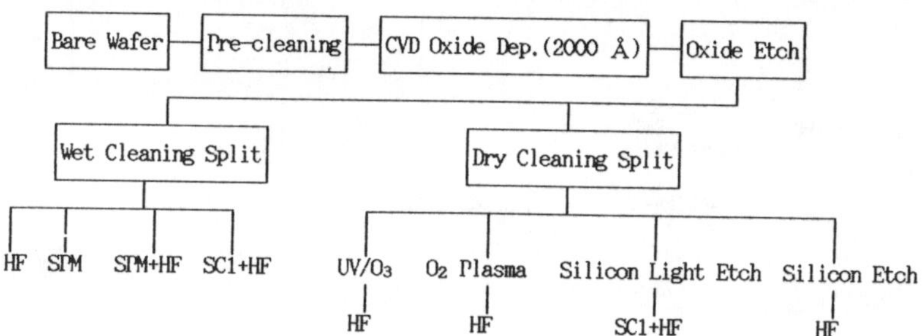

Figure 1. Experimental Scheme

RESULTS

Etch Residue

As shown in Figure 2, main residual elements by oxide etch were silicon, carbon, oxygen, and fluorine. Major bonding states of Si_{2p} were Si–Si (99.5 eV), Si–C (100.2 eV), and Si–O (102.5 eV). Si–O bond state was different from that of thermal oxide of which typical binding energy is 103.5 eV. In C_{1s} peak, C–Si (283.4 eV), C=O (284.9 eV), C–F (287.4 eV), C–F_2 (290.0 eV), and C–F_3 (292.5 eV) were observed. The last three bonding states were attributed to fluorocarbon polymer. O_{1s} peak mainly came from oxygen content in residual layer and adsorbed molecular oxygen from atmosphere. F_{1s} peak (686 eV) was hard to identify the fluorine bond states in detail due to the highest electronegativity.

As appearing in Figure 3, the C–F_x bond states disappeared by 3 KeV Ar^+ sputtering for 25 sec, and Si–O and Si–C bonds were gone by additional 195 sec sputtering, which revealed that the top surface of the residue was covered with a CF_x-polymer and a layer of Si–O and Si–C bonds formed the underneath layer. The position of the CF_x-polymer at the top surface was also confirmed by Figure 4 (b) which indicated that C_{1s} peaks had relatively less dependency on photoelectron take-off angles, compared to Si_{2P} peaks.

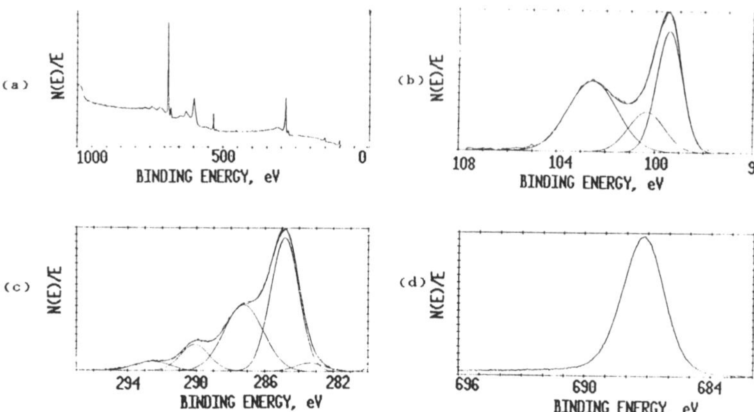

Figure 2. XPS Spectra of the oxide etch residue: (a) wide scan
(b) Si_{2P} peak   (c) C_{1s} peak   (d) F_{1s} peak

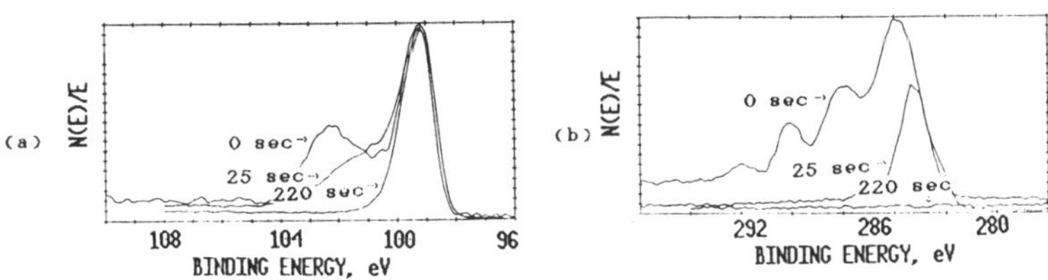

Figure 3. XPS data with sputtering: (a) Si_{2P} (b) C_{1s}

The Si-O and Si-C bonds seemed to form one compound of SiO_yC_z. The reason Si-O
and Si-C bonds were not interpreted as SiO_2 and SiC, respectively is that the Si-O
bond of the etch residue was chemically inert in HF acid, which is a typical wet etch
chemical of oxide, so that there was no change at all in Si_{2P} peaks after HF
treatment irrespective of dip time. Si-O bond did not seemed to represent simply the
existence of silicon dioxide. Instead, Si-O and Si-C formed one compound like
SiO_yC_z.

The quantitative explanation of y and z in SiO_yC_z is as follows: First, with
increasing the photoelectron take-off angle, the area ratio of Si-C to Si-O
increased, which indicated that carbon content was increased and oxygen content was
decreased with depth.(Figure 4 (a)) Table 1 summarizes the dependence of the area
ratio of Si-C bond to Si-O bond on photoelectron take-off angles. Second, the
maximum concentrations of oxygen and carbon were detected at 3.2 min and 3.6 min,

410

Table 1. Change of area ratio depending on photoelectron take-off angles

Take-Off Angle	Chem. Bond	Si-C (100.2eV) (cps)	Si-O(102.5eV) (cps)	RATIO $\left(\dfrac{Si-C}{Si-O}\right)$
20°		30	73	0.411
30°		80	150	0.533
45°		140	250	0.560
60°		306	347	0.862

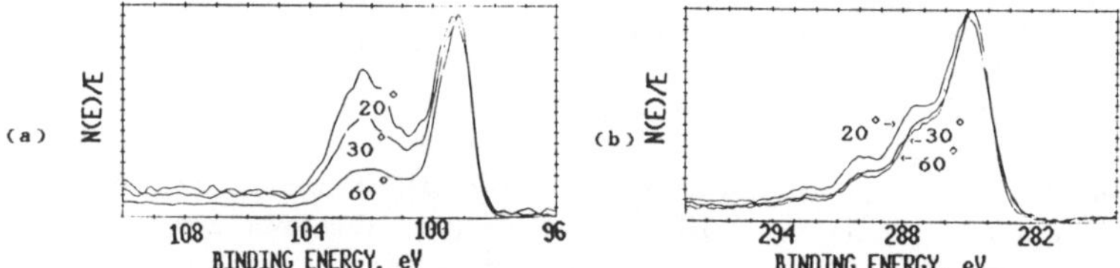

Figure 4.   XPS spectra with photoelectron take-off angle   (a) Si_{2p} (b) C_{1s}.

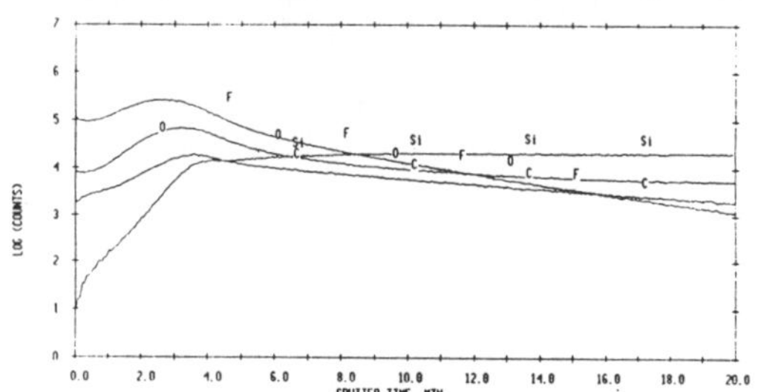

Figure 5. Depth profile of C, F, and O in SIMS

Therefore, the ratio of y to z in SiO_yC_z layer was monotonously changing with depth: y/z was maximum at the interface of CF_x-polymer and SiO_yC_z, and minimum at the interface of SiO_yC_z and silicon substrate.

Considering that no Si-C bond was detected and the residue was simply CF_x-polymer/SiO_2 in case of underetched wafer with 300 Å thick remaining oxide, it was thought that Si-C bond was originated from a chemical reaction of carbon-containing plasma with damaged silicon substrate during the oxide overetching process.   Si-O bond seemed to form at the vacancy of Si sites by the diffusion of oxygen in the ambient which penetrated into the damaged silicon substrate through the CF_x-polymer.

Cleaning

To remove the oxide etch residue, three wet cleaning methods were applied: SPM, SC1, HF, and their combination. Aqueous HF solution was inactive in removing both the CF$_x$-polymer and SiO$_y$C$_z$, while SPM could get rid of the CF$_x$-polymer even though small amount. Among the wet chemicals, the most effective one in removing the etch residue was SC1. But the major portion of the residue still remained on the wafer. Any wet cleaning was revealed to be ineffective for removing the oxide etch residue.

Since wet cleaning only was ineffective in removing the oxide etch residue, two step cleaning was proposed: dry and wet cleaning. Four dry cleaning methods, O$_2$ remote plasma, ultraviolet/O$_3$ ashing, silicon etch (HBr/Cl$_2$), and silicon light etch (CF$_4$/O$_2$ remote plasma) were tried. SC1 and HF were mainly used for the following wet cleaning process.

It was found that O$_2$ remote plasma and UV/O$_3$, followed by HF dip, had nearly same effectiveness on removing the residue. These dry cleaning methods got rid of carbon polymer but left the SiO$_y$C$_z$ unremoved.

When silicon light etch was applied, the residual layer was changed to CF$_x$-polymer/ SiO$_2$, that is, Si-C bond vanished. In narrow scanning of C$_{1s}$ peak, only a small amount of the CF$_x$-polymer was detected. Si-O bond state was observed in Si$_{2p}$ peak uniquely and the thickness of oxide was around 20Å. CF$_x$ layer seemed to be removed through CO, CO$_2$, COF$_2$ etc., SiO$_y$C$_z$ layer by forming SiF$_4$, CO and CO$_2$, and silicon substrate by way of SiO$_2$ and subsequently CO$_2$ and SiF$_4$. The silicon light etch residue, CF$_x$-polymer and SiO$_2$, was completely cleaned by SC1 and HF dip.

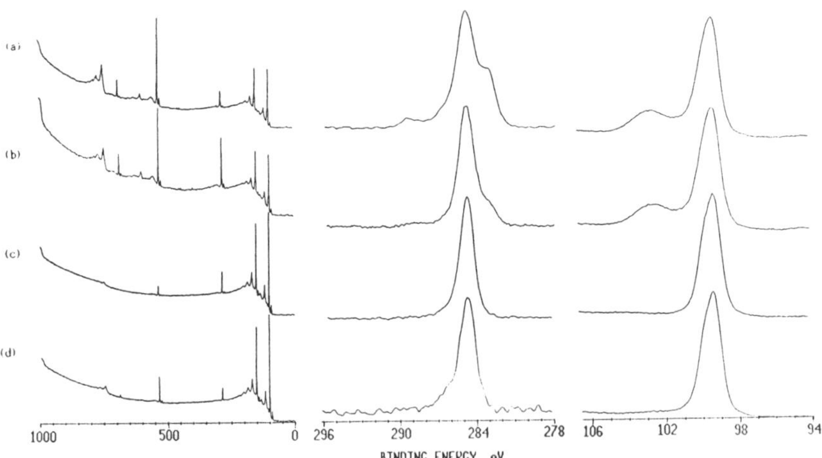

Figure 6. XPS spectra after dry-wet cleanings: (a) O$_2$ plasma + HF (b) UV/O$_3$ + HF (c) silicon light etch + SC1 + HF (d) silicon etch + HF

The XPS spectrum of wafer at this stage was exactly the same as that of an initial wafer which was just dipped into HF acid.

Silicon etch with HBr and Cl_2 plasma left small quantity of CF_x-polymer, Cl and Br. HF cleaning after silicon etch cleaned the fluorocarbon residue by lift-off. HBr/Cl_2 plasma removes the oxide etch residue by lift-off of the silicon substrate by forming highly volatile $SiBr_4$ and $SiCl_4$.

Oxide etch residue was completely removed by silicon light etch and silicon etch, accompanied by wet cleaning. Silicon etch had two weak points, compared to silicon light etch: one was it caused silicon substrate damaged due to its removal mechanism of lift-off, the other was that there was a possibility to etch the silicon substrate too excessively because it was hard to control the etching depth in silicon etch (silicon etching rate = 90 Å/sec). Among the dry and wet cleanings, the most effective cleaning process was silicon light etch(CF_4/O_2), followed by $NH_4OH-H_2O_2$ mixture and HF dip. No retardation of thermal oxidation was found when the silicon surface was completely cleaned by silicon light etch with SC1 and HF dip, which was mentioned in the next section.

Effect of Cleaning

The effect of silicon light etch was evaluated in three ways: monitoring oxide growth thickness, minority carrier life time, and junction leakage. As mentioned before, the etch residue was found to block further thermal oxidation. Table 2 shows the oxide thickness grown by thermal oxidation at 850 ℃ for 30 min by changing post-treatments of oxide etch.

Table 2. Oxide thickness depending on post-treatments

treatment	as-etched	no cleaning	SPM	RCA	HF	silicon light etch + SC1 + HF
$T_{ox}.(Å)$	0	0	2	1	0	71

When the etch residue remained on the wafer in the case of as-etched, no cleaning and only wet cleaning treatment, oxide was not grown on the subsequent thermal oxidation step, while the same thickness of oxide as that of a monitoring wafer was deposited on the dry and wet cleaned wafer.

Variation of minority carrier life time with process is appearing in Figure 7. Wet cleanings could not recover the damaged surface of silicon but dry and wet cleanings recovered the damage to increase the minority carrier life time. The silicon substrate with silicon light etching and subsequent wet cleaning was almost recovered from the plasma etch damage. Compared to initial minority carrier life time, silicon lightly etched wafer recovered minority carrier life time most effectively but silicon etch, ultraviolet/O_3, and O_2 plasma did not recover the etch damage that much.

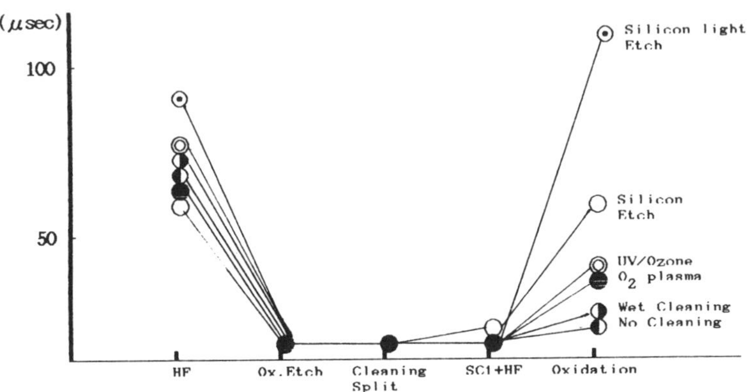

Figure 7. Dependance of minority carrier life time
on the post treatment of etch process

Electrically, the effect of silicon light etch, SC1 and subsequent aqueous HF cleaning was examined by measuring cell junction leakage for 512 K cell, depending on the cleaning split. When the cleaning was applied to gate side-wall dry etch, cell junction leakage was around 4.9nA at 3.3V, while it was 6.4nA without cleaning and 4.1nA for gate side-wall wet etching.(Table 3)

Table 3. Dependence of junction leakage on cleaning methods

Etch	Cleaning Method	Junction Leakage (nA at 3.3V)
Dry + Wet etch	SC1 + HF	4.1
10% overetch	SC1 + HF	6.4
30% overetch	SC1 + HF	60.0
30% overetch	Silicon light etch + SC1 + HF	4.9

CONCLUSION

The composition of oxide etch residue was found to be CF_x-polymer and SiO_yC_z. The oxide dry etch residue of CF_x-polymer and SiO_yC_z was not effectively removed by wet cleaning only, while it was cleaned up by silicon light etch and subsequent wet cleaning: NH_4OH-H_2O_2 mixture and then aqueous HF dip. The dry and wet cleaning recovered the silicon substrate damage and reduced the cell junction leakage nealy to that of oxide wet etch process.

Acknowledgement

The authors would like to thank Ik Nyun Kim and Seon Mee Kim for carrying out experiments and surface analysis.

References

1. G. S. Oehrlein, J. Appl. Phys., Vol. 59, 3053 (1986)
2. H. P. Struck et al., J. Electrochem. Soc., 135, 2876 (1988)
3. G. S. Oehrlein et al., J. Electrochem. Soc., 133, 1002 (1986)
4. K. W. Kwon, H. H. Park, S. M. Lee, S. J. Kang, O. J. Kwon, B. W. Kim, Y. K. Sung, J. Korean Vacuum Soc., 1, 145(1992)
5. R. D'agostino et al., Thin Solid Films, 143, 163(1986)
6. G. J. Coyle Jr. and G. S. Oehrlein, Appl. Phys. Lett., 47 (6), 604 (1985)
7. G. S. Oehrlein, R. M. Tromp, J. C. Tsang, Y. H. Lee and E. J. Petrillo, J. Electrochem. Soc., 132, 1441 (1985)

THE FORMATION TECHNOLOGY OF CHROME OXIDE PASSIVATED SURFACE FILM

A. Ohki, Y. Nakagawa, M. Nakamura, K. Kawada,
S. Miyoshi, T. Watanabe, S. Takahashi,
M. S. K. Chen and T. Ohmi
Department of Electronics, Faculty of Engineering
Tohoku University, Sendai Japan 980

A near 100% chrome oxide passivation technology has been developed on the metal surface suitable for use in the high purity gas delivery systems and process chambers. The optimal thermal treating condition starting with an electropolished SUS316L stainless steel material was found by using a gas mixture of 10% hydrogen, 1 ppm oxygen and the balanced argon at 500 °C for one hour. The results of a 5-day HCl gas (containing 1.4 ppm moisture) corrosion test at 5 Kg./cm^2 and 100 °C showed no sign of corrosion on the passivated chrome oxide surface. The chemical stability tests for silane and diborane specialty gas thermal decomposition on this surface also showed a remarkable non-catalytic activity as compared with the conventional surfaces.

Keywords: Passivation. Chrome oxide. SUS316L. Electropolishing. HCl. Corrosion. Silane. Diborane. Thermal decomposition. Metal filter.

1. INTRODUCTION

As the semiconductor devices become more highly integrated, the substrate feature dimensions has steadily advanced toward sub-half micron scales and miniaturization. It is just now becoming possible to achieve a complete uniformity and reproducibility no only on a single wafer but also from wafer to wafer reliably in the fabrication such sub-half micron ULSI devices [1,2].

Presently in the two core semiconductor processes, Chemical Vapor Deposition (CVD) & Reactive Ion Etching (RIE), a variety of very reactive and corrosive specialty gases are used [3]. Among these gases, the halogen gases are well known to corrode the gas delivery system due either to the presence of the residual moisture in the supply gas or to the interaction with the adsorbed moisture on the surface of the system. The corrosion often leads to metal contamination which deteriorates the device performance, yield and reproducibility as well as causes serious interruption of plant operations in order to replace the gas delivery lines [4]. Therefore, it was the objective of this study to research and develop an effective surface passivation technology for metal surfaces to improve their corrosion resistance to these specialty gases, to reduce the surface moisture adsorption and to lower the adsorption energy for easy removal of these impurities from the surfaces.

SUS316L stainless steel has been the main material used in today's gas delivery system. Currently the most popular metal surface passivation method has been the wet process by using chemicals such as nitric acid [5]. However this wet passivation method cannot achieve a complete chrome oxide passivated surface film and therefore cannot attain the goal of the complete corrosion resistance, even though it has improved the corrosion resistance considerably as compared to the untreated surface. Furthermore, because the wet processing often leaves chemical residues on the surface, the passivated surface is not entirely satisfactory for use in the ultra high purity specialty gas services. For this reason, we looked for an alternative dry technique to replace the wet process by using a dry gas mixture of hydrogen, oxygen and argon to simultaneously reduce the iron oxide and oxidize the chromium selectively to achieve a near 100% chromium oxide (Cr_2O_3). passivated surface film, which has a superior corrosion resistance to halogen gases, a low tendency to adsorb moisture and a high chemical stability. Because of the dry gas processing in this new technique, the formed chrome oxide surface contains no residual moisture and is smooth and dense in the texture.

2. EXPERIMENTAL

Electropolished SUS316L stainless steel surfaces were exposed to a gas mixture of ultra high purity hydrogen, oxygen and argon to selectively reduce the iron oxide and oxidize the chromium to achieve a near 100% chrome oxide (Cr_2O_3) passivated surface film [6]. Figure 1 shows a schematic of the experimental setup for passivating the SUS316L stainless surface. It consists of a reaction furnace and a gas supply system. The reaction furnace is made of a 100 mm long 1/2" electropolished SUS316L tube equipped with a 300 W sheath heater and a PID system to control the temperature to within 1 °C. All plastic materials in contact with the gas were removed and the gas supply system were made of dead space-free, all-metal valves and mass flow controllers. The dilution system can mix argon with hydrogen in any proportion to achieve a gas mixture with 1 ppb or less impurities [7,8]. The impurity was measured by using APIMS. The samples for chrome oxide passivation were electropolished SUS316L material with a size of 7x20 mm and a thickness of 0.5 mm. The sample composition is shown in Table 1.

The passivation process involves first placing the samples into the furnace and then purging the furnace at room temperature with an ultra purity argon gas containing 50 ppt moisture at 1 l/min for one hour to remove the adsorbed moisture from the

TABLE 1

SUS316L Composition

(wt %)

C	Si	Mn	P	S	Ni	Cr	Mo	O ppm	H ppm	N ppm
0.006	0.36	0.65	0.024	0.001	12.22	16.57	2.09	54	3	93

sample surface. Then at the same argon flow of 1 l/min, the temperature was raised from room temperature to 500 °C. in one hour and held it there for two hour baking. After complete removal of the moisture and other impurities at 500 °C, the argon gas was then switched to a mixture of argon, hydrogen and oxygen to carry out the surface oxygen passivation. The gas mixture consists of hydrogen from 2 to 10% and oxygen from 100 ppb to 100 ppm in order to find out the optimal composition. The passivation time ranged from 30 minutes to 5 hours. The resulting passivated film were examined by XPS for surface and depth profile composition evaluation.

The near 100% chrome oxide passivated surfaces were evaluated for their corrosion resistance and chemical stability with respect to very reactive and corrosive specialty gases. The corrosion tests were conducted by sealing in 1/4" 300 mm SUS 316L stainless tubes with a HCl gas containing 1.5 ppm moisture and a moisture-free HCl gas at 5 Kg/cm^2 and 100 °C for 5 days. The moisture-free HCl experiment was conducted by passing the moist HCl gas through an all-metal filter installed ahead of the sample tube [9]. These filters were first purged and baked simultaneously with an ultra high purity argon gas at 400 °C to remove the moisture and then cooled to room temperature. After 5 days, HCl gas was replaced by an argon gas purge at 500 l/min for 12 hours. Samples were examined by SEM. Both the chrome oxide passivated surface and the electropolished surface were evaluated.

For chemical stability evaluation, 1/4"x190 mm SUS 316L stainless sample tubes were first purged with argon gas at 450 °C for 10 hours to remove the adsorbed moisture and then cooled to 25 °C. A silane gas diluted with argon at 100 ppm and 5 cc/min was then flowed through the tube for 10 hours, Then the temperature was

418

raised at 0.1 °C/min to 500 °C to evaluate the silane decomposition behaviors. Silane concentrations were monitored by gas chromatography. Four different surfaces including the chrome oxide passivated surface, electropolished surface, iron oxide passivated surface and Hastelloy X surface were compared [10]. To further evaluate the chemical stability at low temperature and large gas contact surface area, silane and diborane gas decomposition behaviors were observed on metal filters with and without passivation.

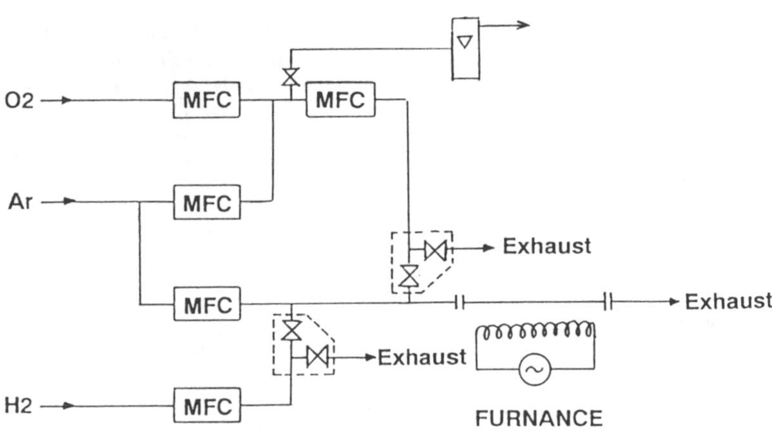

Fig.1   Schematic of chrome oxide passivation experimental setup

3. RESULT AND DISCUSSION

3.1 Chrome oxide passivation technology

Figure 2 shows the XPS results of an electropolished SUS316L surface and a 100% oxygen-passivated electropolished surface at 400°C . The vertical axis shows the elemental composition and the horizontal axis the etching time in the depth direction. The etching speed was 70 A°/min. In the pure oxygen atmosphere, the passivated surface has about a surface layer of 60 A° iron oxide, under which oxides of iron and chromium co-exist. From this XPS depth profile, we could understand why the direct oxygen passivation could not achieve a chrome oxide surface with superior corrosion resistance.

By comparing the values of free energy of formation listed in Table 2, we observe that the chrome oxide has a relative larger negative value than other oxides, indicating its stability. Therefore, we proceeded to investigate the chrome oxide passivation of the electropolished SUS316L stainless surface under the reducing atmosphere.

TABLE 2

STANDARD HEAT OF FORMATION ($\Delta H_f°$) AND GIBBS FREE ENERGY OF FORMATION ($\Delta G_f°$) FOR VARIOUS METAL OXIDES AND FLUORIDES

Oxides	PdO	CuO	MnO_2	SnO_2	MoO_2	Fe_2O_3	NiO	Cr_2O_3	SiO_2	Ta_2O_3
$\Delta H_f°$ kJ/mol	-85	-157	-385	-581	-520	-824	-464	-1140	-911	-2040
$\Delta G_f°$ kJ/mol		-130	-363	-520	-465	-742		-1058	-857	-1911

Fluorides	FeF_2	NiF_2	CrF_3
$\Delta H_f°$, kJ/mol		-651	-1159
$\Delta G_f°$, kJ/mol		-604	-1088

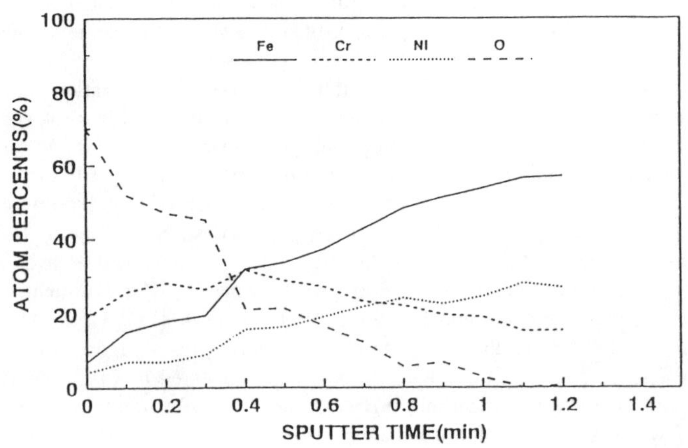

(a) Electropolished surface

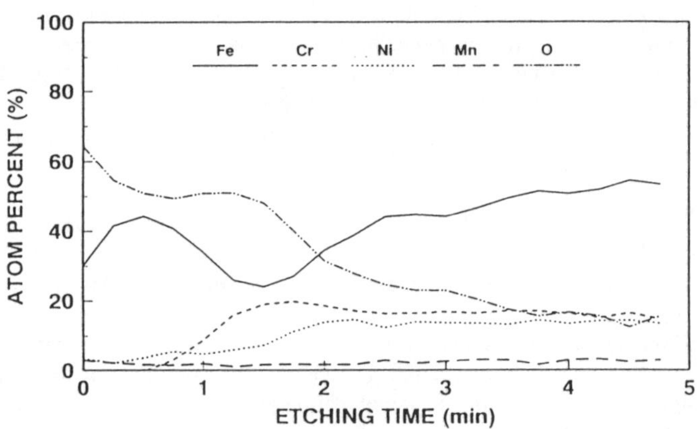

(b) Iron oxide passivated surface

Fig.2   **XPS depth profiles of electropolished and iron oxide passivated surfaces in 100% O_2 at 400°C**

Figure 3 shows the XPS depth profiles for surfaces thermally treated with 1 to 10% H_2 in argon at 500 °C for two hours. The vertical axis shows the element composition and the horizontal axis the etching time. the etch speed was 70 $A°$/min. This figure shows that the profiles are nearly independent of hydrogen concentration and the surface contains more and less a constant amount of chrome oxide. Figure 4 shows the XPS spectral profiles of chrome and iron on the upper surface layers for the samples before and after the passivation by 10% H_2 in argon at 500 °C for two hours. The vertical axis shows the photoelectron intensity and the horizontal axis the binding energy of the elements. First let's examine the iron binding peak positions in Figure 4(a) before and after the thermal treatment. Before the treatment, we observed a peak at 709.8 eV (Fe_2O_3) appeared to the left of the 706.6 eV (Fe) peak After treatment, the 709.8 eV peak (Fe_2O_3) disappeared and left with 706.6 eV peak (Fe) only. This indicates a reduction in Fe_2O_3 by the passivation. Similarly, from Figure 4(b) of chromium binding energy peak position, we saw the peak appeared at 577.0 eV (Cr_2O_3) both before and after the treatment and this peak increased after the treatment. From these results, we might think of a mechanism that the iron oxide is first reduced by the hydrogen and the released oxygen then oxidizes the atomic chromium in the bulk phase.

Furthermore, in order to increase the chrome oxide layer and its proportion, a small amount of oxygen was added to the 10% hydrogen-argon mixture in the passivation process. Four level of oxygen, 0.1, 1, 10 and 100 ppm, were tried. The condition was at 500 °C and one hour. The XPS results are shown in Figure 5 With the addition of oxygen, not only the chrome oxide layer was increased but also the iron disappeared from the surface for cases of 1 and 10 ppm, resulting a near 100% Cr_2O_3 film formation. By comparing 1 and 100 ppm oxygen addition cases, the 1 ppm one shows a less iron content at the surface. The 100 ppm oxygen case shows a stronger oxidation power began to enrich iron oxide again in the chrome oxide layer.

Next, we investigated the effect of the treating time. Figure 6 shows XPS profiles for 0.5, 1, 2 and 5 hours at 10% H_2, 1 ppm O2 in argon at 500 °C. From this figure, we observed that for time over 2 hours, the iron content has increased.

From the above results, we understand that if oxygen is greater than 10 ppm and the time is over 2 hours in the passivation process, the iron oxidation begins to appear again to compete with chrome oxidation and a proper chrome oxide layer cannot be formed [11]. Therefore, in order to realize a complete chrome oxide passivated surface film, It is important to control the oxygen and other impurities at the ppm level and it is necessary to carry out the process in a complete closed system by using the ultra clean technology.

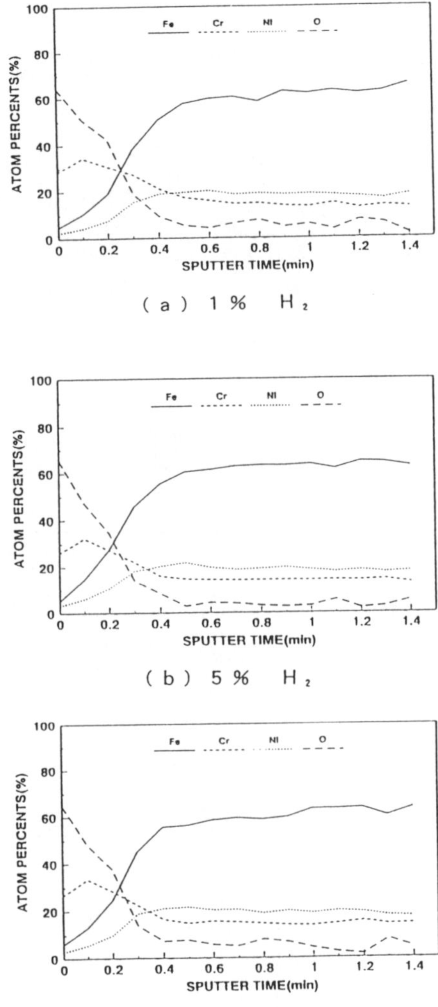

Fig.3   XPS depth profiles of thermally passivated surfaces in H_2–Ar gas mixture at 500°C, 2 hours – effect of % H_2

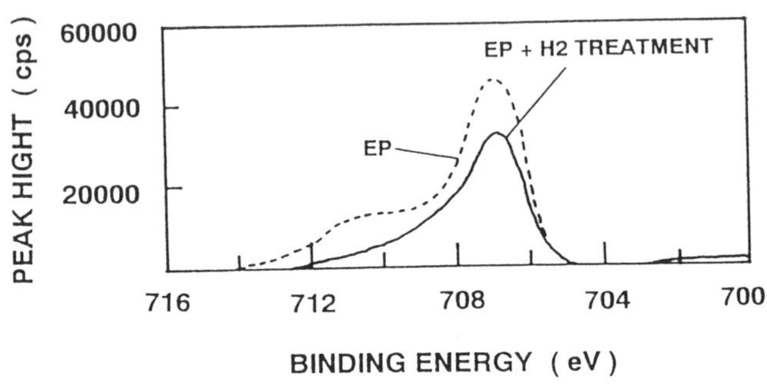

(a) F e   2 p   3 / 2

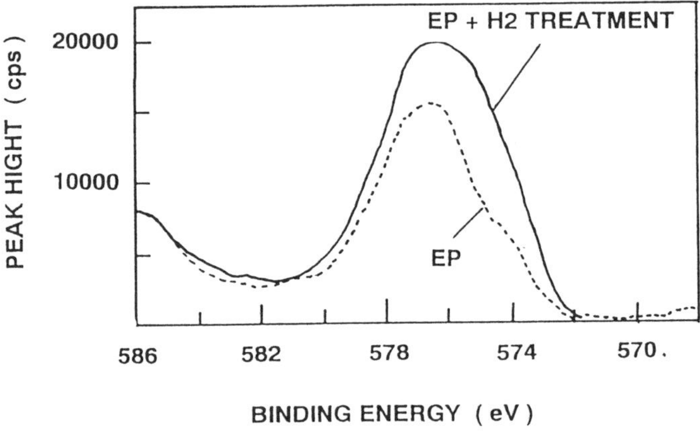

(b) C r   2 p

Fig.4   XPS spectrums of Fe and Cr on thermally passivated surfaces in 10% H_2 – Ar gas mixture at 500°C , 2 hours

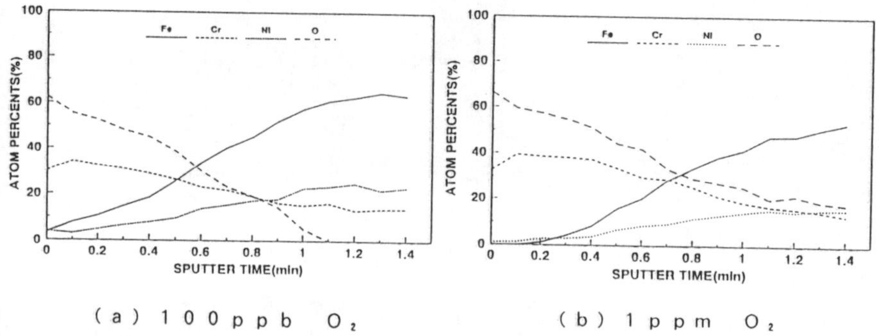

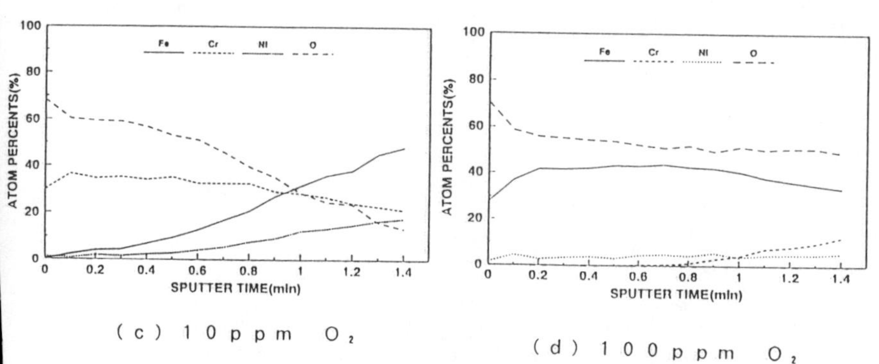

Fig.5 **XPS depth profiles of chrome oxide passivated surfaces – effect of trace O addition in 10 % H_2–Ar gas mixture at 500°C , 1 hour**

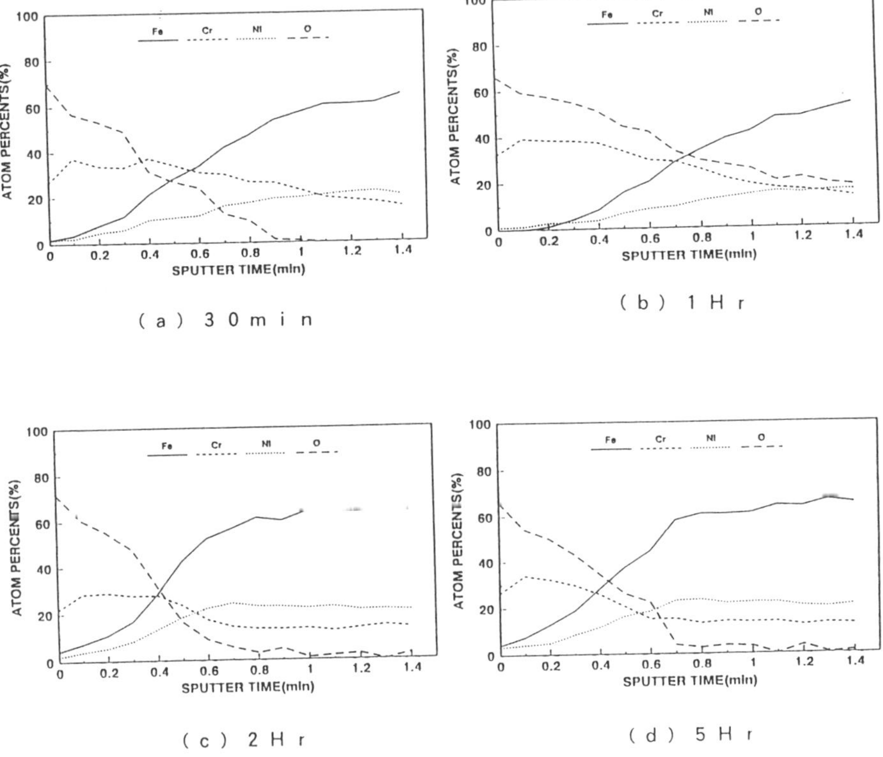

Fig.6 XPS depth profiles of chrome oxide passivated surfaces in $10\%H_2$–1ppmO_2–Ar gas mixture at 500°C – effect of time.

3.2 Anti-corrosion performance of chrome oxide passivated film

Corrosion resistance tests was conducted for the passivated chrome oxide film. Two kinds of samples were used: the passivated surface under 10% H_2, 1 ppm O_2 in argon at 500°C for one hour as obtained from the above passivation technique and the electropolished SUS316L stainless steel surface. Both the HCl gas containing 1.5 ppm moisture and the dry HCl gas obtained by passing the gas through a metal filter to were used. These gases were sealed into 1/4" 300 mm long SUS316L tubes at 5 Kg/cm^2 at 100°C. for 5 days. The exposed tube inner surfaces were then inspected by SEM. The results were shown in Figure 7 and 8. Figure 7 shows the SEM results with 1.4 ppm moisture and Figure 8 the results without the moisture. From Figure 7 we observed a localized pitting corrosion had occurred on the electropolished surface while the chrome oxide passivated surface was not affected. From the SEM images of Figure 8 we observed that both surfaces were not corroded. These results clearly indicate that corrosion would take place only in the presence of moisture. In other words, the corrosion may be considered to be set off as the moisture, either adsorbed on the surface or existed in the bulk phase, begins to dissociate the absorbed HCl into protons and chloride ions attacking the metal.

3.3 Chemical stability of the passivated surface

So far in the field of catalysis, various metals and metal oxides have been studied for their catalytic activities for their practical uses [12]. Table 3 shows some examples for their catalytic properties. Stainless steel is a metal alloy, made of iron, chromium, nickel as major constituents, with some added trace components such as manganese and molybdenum. We can imagine that depending on the metal composition, its surface possesses many catalytic activities   Presently, various stainless steels are the key material used in the gas delivery system and the process chambers. Their extreme stability is required for use in contact with the very reactive specialty gases. For this reason, we focused our chrome oxide passivation studies on this material for its chemical stability in specialty gases.

First of all, the chemical stability of various surfaces with respect to silane gas will be described. To evaluate chemical stability, 1/4"x 190 mm SUS 316L stainless sample tubes were first purged with argon gas at 450 °C for 10 hours to remove the adsorbed moisture and then cooled to 25 °C. A silane gas diluted with argon at 100 ppm and 5 cc/min was then flowed through the tube for 10 hours, Then the temperature was raised at 0.1 °C/min to 500 °C to evaluate the silane decomposition behaviors. Silane concentration and the decomposed product hydrogen gas were monitored as a function of time by gas chromatography . Four different surfaces including the chrome oxide passivated surface, electropolished surface, iron oxide passivated surface and Hastelloy X surface were compared.

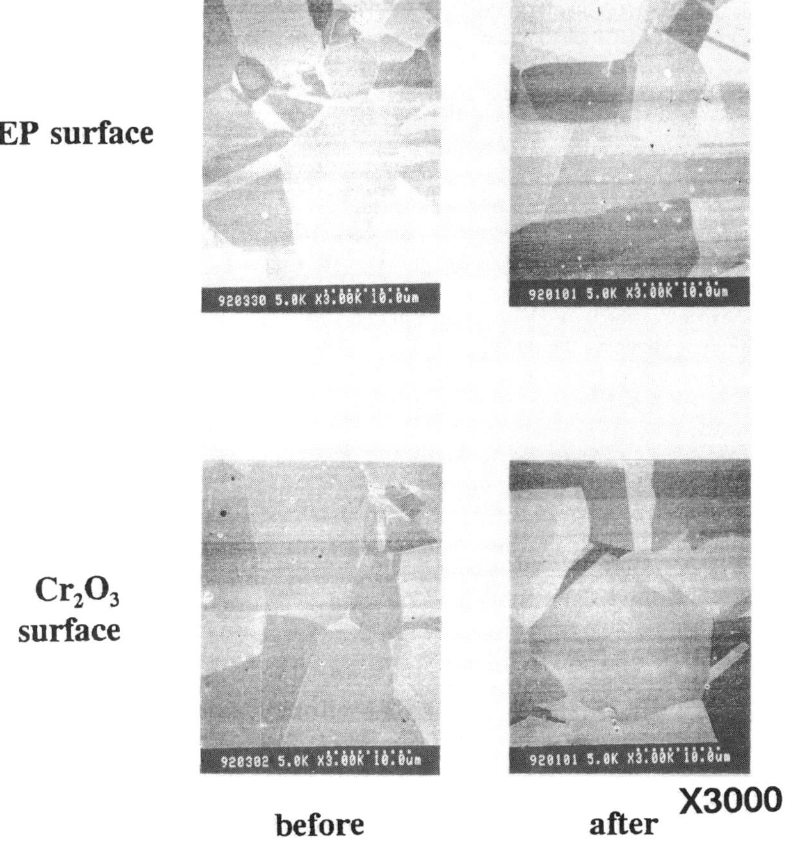

Fig.8   SEM images of electropolished and chrome oxide passivated surface – before and after exposure to dry HCl gas at 100°C for 5 days

428

EP surface

Cr$_2$O$_3$ surface

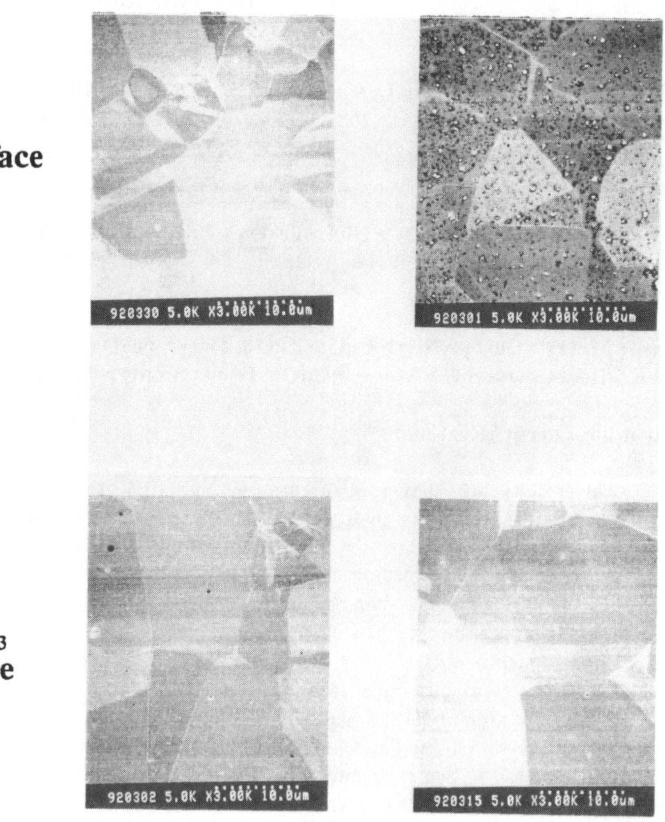

Fig.7 SEM images of electropolished and chrome oxide passivated surfaces – before and after exposure to HCl gas with about 1.5 ppm H$_2$O at 100°C for 5 days

TABLE 3

ORDER OF METAL OXIDE CATALYTIC ACTIVITIES

Reaction	Order of Catalytic Activities
O_2 isotope exchange	$Co_3O_4 > MnO_2 > NiO > CuO > Fe_2O_3 > ZnO > Cr_2O_3 > V_2O_5 > TiO_2$ E 16 22 24 26 33 40 42 46 (39)
H_2 oxidation	$CoO_3 > CuO > MnO_2 > NiO > Fe_2O_3 > ZnO > Cr_2O_3 > V_2O_5 > TiO_2$ E 11 13 14 14 15 24 18 18
Co oxidation	$MnO_2 > CoO_3 > NiO > CuO > ZnO > TiO_2 > Fe_2O_3 > V_2O_5 \sim Cr_2O_3$ StoneCoO > NiO > MnO_2 > CuO > Fe_2O_3 > ZnO > TiO_2 > Cr_2O_3 > V_2O_5

E: Activation Energy (Kcal/mol

The results are shown in Figures 9 and 10. Figure 9 is for silane concentration and Figure 10 for hydrogen concentration. The horizontal axis shows both the time and the temperature. the vertical axis shows the concentrations of either silane or hydrogen. For the chrome oxide and iron oxide passivated surfaces, silane started to decompose around 360 °C as indicated by the reduction of silane concentration and appearance of hydrogen. As the temperature is further increased, more decomposition proceeds till 460 °C where the reaction is completed. On the other hand, on the electropolished surface, silane started to decompose at around 310° C, a peak to occur at around 360 °C and completed at 450 ° C. in an apparent two-step decomposition process. On the electropolished Hastelloy X surface, silane also started to decomposed at 310 °C, which is lower than that of chrome oxide, a peak at around 360 °C and completed at 430 °C, which is the lowest of the four surfaces tested. Moreover, at the end of silane decomposition, all four surfaces yielded the same hydrogen concentration at 140 ppm. This means that for each part of silane decomposed, 1.4 parts of hydrogen was formed. If we consider the complete reaction as $SiH_4 \text{---}> Si + 2 H_2$, then SiH, SiH_2 and SiH_3 may still exist on the surface. From these results, we see that except for the chemically stable passivated chrome oxide and iron oxide surfaces, some catalytic activities exist on other surfaces for they lowered the starting silane decomposition temperatures.

430

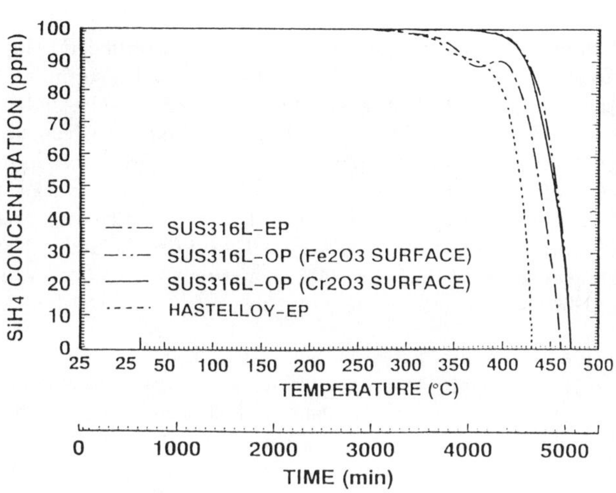

Fig.9   Thermal decomposition of 100% SiH$_4$ in Ar on various surface (SiH$_4$)

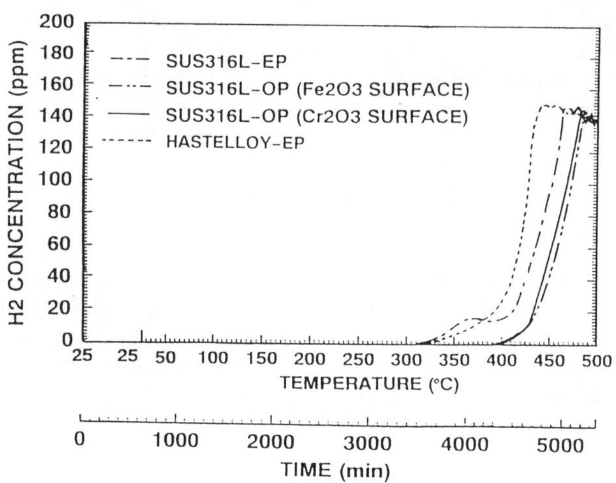

Fig.10   Thermal decomposition of 100% SiH$_4$ in Ar on various surfaces (H$_2$)

We further evaluated a small amount of silane decomposition on filters with large surface area to observe the catalytic activities on the passivated surfaces. We did this at low (room) temperature. We first removed the adsorbed moisture by baking the stainless filter in ultra pure argon at 450 °C for 10 hours. After it was cooled to 25 ° C, argon gas was switched to a gas of 500 ppm of silane in argon at 5 cc/min for room temperature thermal decomposition evaluation. Both silane and hydrogen concentration were measured by gas chromatography. The chrome oxide passivated surface and two other untreated metal filters were tested this way. Two different kinds of all-metal filters obtained from two separate companies, A & B were used. A filter was made of long single elemental fibers while B filter made of short single elemental fibers.

The results are shown in figures 11 and 12. Figure 11 shows the silane concentration and Figure 12 hydrogen concentration. The horizontal axis shows the exposed time. The vertical axis shows the silane or hydrogen concentration. From Figure 11, we observed that in the case of untreated A filter (O), silane was not observed for the first 50 minutes. After that, silane concentration gradually rose to 450 ppm over a period of 5 hours and never quite returned to the 500 ppm initial level. On the other hand, the chrome oxide passivated A filter (●), silane appeared after 30 minutes, rose to reach 480 ppm in 5 hours and could not quite return to the 500 ppm initial level. In comparison, in the case of B filter, either treated (▲) or untreated (Δ), the silane concentration recovered much faster. In both A and B filters, the chrome oxide passivated surface showed faster recovery of silane concentration Figure 12 shows the corresponding hydrogen concentration as the result of the silane decomposition. The chrome oxide passivated B filters showed much less hydrogen generation than filter A. From the above results, we saw that the chrome oxide passivated filters did suppress silane decomposition at room temperature but complete suppression was not obtained. Comparing B filter to A filter, B filter surface appeared to have more chrome oxide passivation than A filter but further improvement to achieve a complete chrome oxide passivation is still necessary.

Similar to the silane experiments, the diborane thermal decomposition behavior was observed on A filter. Two pretreatments were conducted: one with no heat treatment and with only argon gas purging for 12 hours and the other with argon baking at 120 C. Filter as received and filter with the pretreatment were evaluated. Diborane concentration was 104 ppm at a flow rate of 200 cc/min. Diborane concentration was monitored by IR.

The results were shown in Figure 13 and 14. Figure 13 shows the results with the pretreatment of argon purge only. Figure 14 shows the results with pretreatment with argon purge and baking at 120 °C for one hour. The horizontal indicated the exposed time and the vertical axis the diborane concentration. These two figures show

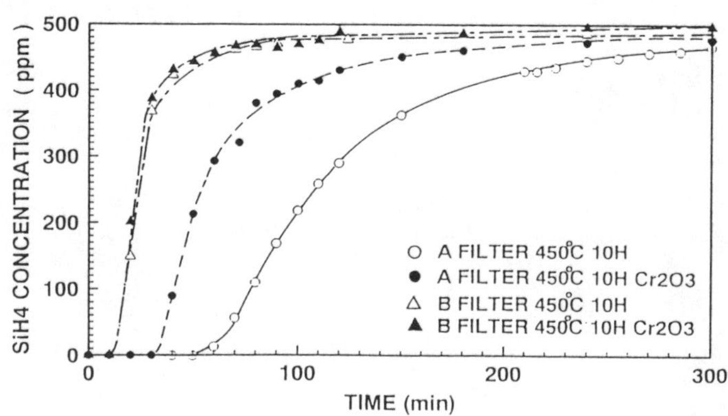

Fig.11   Thermal decomposition of 500 ppm SiH$_4$ in Ar on various all–metal filter surfaces (SiH$_4$)

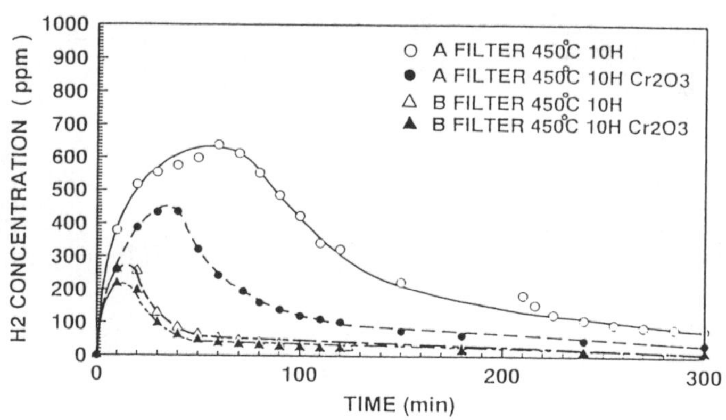

Fig.12   Thermal decomposition of 500 ppm SiH$_4$ in Ar on various all–metal filter surfaces (H$_2$)

433

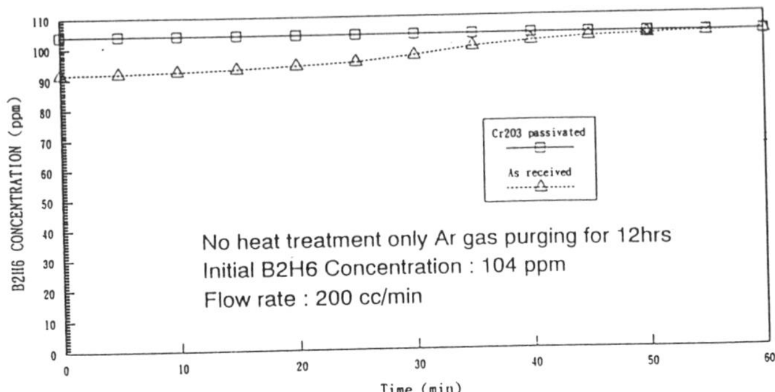

Fig. 13 Thermal decomposition 100 ppm B_2H_6 in Ar on Ar-purged only all metal filter at room temperature

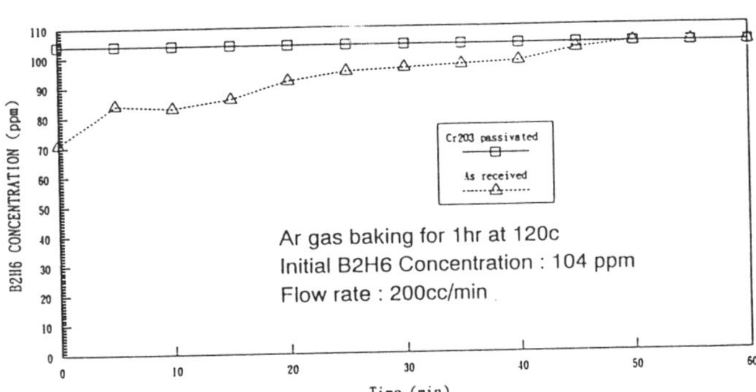

Fig.14 Thermal decomposition of 100 ppm B_2H_6 in Ar on Ar-baked all metal filter at 120°C

that for the as received filter with argon purged only the diborane started out at 92 ppm and gradually returned to the inlet concentration after 50 minutes while the argon baked filter started out at 72 ppm and gradually returned to the initial concentration in 50 minutes. On the other hand, the chrome oxide passivated filter shows no diborane decomposition in either pretreated and un- pretreated cases.

Presently, polymerization reaction of diborane appears to cause a flow problem for the mass flow controller and the pressure let-down valves in the gas delivery system. From the results of Figure 13 and 14, we think that this problem may be mitigated if we passivate the gas contact surface in these components to stabilize it with the chrome oxide.

4. CONCLUSION

The target properties of the metal surface in the gas delivery system and in the process chambers in contact with the specialty gases include no corrosion and no reaction with the gas, low tendency to adsorb moisture and other impurities and low activation energy for the adsorbed impurities to be removed quickly from the surface once adsorbed. We believed we had developed a viable passivation technology for producing a near 100 % chrome oxide on the metal surface that satisfies these targets.

Corrosion is strongly influenced by the presence of moisture. To prevent corrosion in the present special gas system is not possible because they contain 0.1 to 1 ppm moisture. Therefore, under the current situation, it is necessary to have a technology to passivate the surface to prevent corrosion at 1 ppm moisture level. Based on the results presented in this study, we believed we had developed a near 100% chrome oxide passivated surface that showed no sign of corrosion under a very severe condition of exposing it to a HCl gas containing 1.4 ppm of moisture for 5 days at 5 kg/cm^2 at 100 °C. Furthermore, the chrome oxide passivated surface appears to have a superior chemical stability and non-catalytic activity toward specialty gases such silane and diborane. It is well suited for use in the gas delivery system

With this newly developed chrome oxide passivation technology, the process equipment and the gas delivery system are expected to have a much longer service life. This in turn should help reach the goal of having un-interrupted semiconductor plant operations and minimize the ill-effects on the production yields as the devices become much more highly integrated in the future.

5. REFERENCES

1. Tadahiro Ohmi, "An Approach to Automated IC Manufacturing," Ultra Clean Technology, pp 107-124, Realize Inc., Tokyo, 1990.

2. Tadahiro Ohmi, "Ultra Clean Processing," Microelectronic Engineering, Vol. 10, pp 163-176, 1991.

3. Tadahiro Ohmi, Masakazu Nakamura, Atsushi Ohki, Koji Kawada, Keiji Hirao and Tuyoshi Watanabe, "Thermal Decomposition of SiH_4 and Si_2H_6 and the Influence of Residual Moisture and Oxygen", Proceedings, 15th Workshop on ULSI Ultra Clean Technology, Tokyo, pp. 125-133, November, 1991.

4. Tadahiro Ohmi, Tadashi Shibata and Mizuho Morita, "Break Through for Scientific Semiconductor Manufacturing in 2001", ISSMT'92, Tokyo, May 14-15, 1992.

5. Kazuhiko Sugiyama, Tadahiro Ohmi, Takeshi Okura and Fumio Nakahara, "Electropolished, Moisture-free Piping Surface Essential for Ultra Pure Gas System," Microcontamination, Vol. 7, No. 1, pp 37-65, January, 1989.

6. Masakazu Nakamura, Tadahiro Ohmi, Koji Kawada, "All Metal and Oxygen Passivation Tubing Technology for Ultra Clean Gas Delivery System," Proceedings, 37th Annual Technical Meeting, "Technical Solutions Through Technical Corporation," San Diego, pp 605-613, May 1991.

7. Kazuhiko Sugiyama, Masakazu Nakamura, Atunobu Ohkura, Yasumitsu Mizuguchi and Tadahiro Ohmi, "Calibration of APIMS and Application for Process Evaluation," Ultra Clean Technology, Vol. 1, No. 1, pp 23-33, 1990.

8. Yasumitsu Mizuguchi, Kazuhiko Sugiyama, Fumio Nakahara and Tadahiro Ohmi. "Precision Gas Dilution System for Semiconductor Gas Delivery System," Electronics Information & Communication Society Technical Research Report (Silicon material Device Research Committee), No. 89-2, pp 7-12, April, 1989 (in Japanese)

9. Kenich Sado, Masakazu Nakamura, Tadahiro Ohmi, "The Applications of All Metal Filters in Semiconductor Manufacturing," Special Edition on Metals, The Breakthrough Technology - the Bases and Applications - Part 2, (Agune Co.) pp 104-109, Aprial, 1991. (in Japanese)

10. Tadahiro Ohmi, " All Metal and Oxygen Passivated Ultra Clean Gas Delivery System for ULSI Manufacturing," Program and Abstract Book, Gas Separation International Austin, pp 46-47, April, 1991.

11. Carl Wagner, "Theoretical Analysis of the Diffusion Processes Determining The Oxidation Rate of Alloys," Journal of the Electrochemical Society, Vol. 99, No. 10, pp 369-379, October, 1952.

12. Ohara and Ono, General Chemistry, 34 Catalyst Design, 1981. (in Japanese)

ULTRA CLEAN WELDING TECHNOLOGY WITHOUT ACCOMPANYING CORROSION

Shinji Miyoshi, Masakazu Nakamura, Atushi Ohki,
Koji Kawada, Tsuyoshi Watanabe, Shinji Takahashi,
Michael S. K. Chen and Tadahiro Ohmi

Department of Electronics, Faculty of Engineering
Tohoku University, Sendai, Japan 980

Metal fume generated during the welding often contributes to the metal contamination in the process chamber, especially when exposed to the halogen gases. In this study, we showed the metal fume can be reduced substantially by reducing the welding heat input and by increasing the rotation speed. We also demonstrated the effectiveness of removing the metal fume from the surface by performing ultra pure water wash.

Keywords: Welding. Metal fume. Corrosion. Ultra pure water. HCl.

1. INTRODUCTION

In the ULSI Manufacturing, it is essential to keep the silicon wafer surface ultra clean in order to realize low temperature and high selectivity processing [1]. An ultra clean wafer surface means we have to remove particles, organics, metal impurities, native oxide, surface microroughness and other adsorbed impurities from the surface. With respect to the metal contamination in particular, we conducted a metal contamination study in the gas delivery system as part of the overall in process system. The main reason for the metal contamination is due to the corrosion metal products generated as a result of using semiconductor specialty gases, especially the halogen compounds contacting the metal surface. In the two key processes, Chemical Vapor Deposition (CVD) and Reactive Ion Etching (RIE), a variety of very reactive and corrosive gases are used [2]. If there is a residual moisture left in the halogen gases such as HF, HCl and HBr, the moisture will adhere to the surface of the gas delivery system or the process chamber. The adsorbed moisture will then absorb and dissociate these halogen molecules to produce reactive ionic species such as H^+, F^-, Cl^- and Br^- even at room temperature [3]. These ions corrode the metal surface to generate particles, metal impurities or other reaction products which often cause serious deterioration of the process performance. The explanation for this is that the binding energies of gaseous HF, HCl and HBr are 5.8 eV, 4.4 eV and 3.8 eV respectively, which are overwhelmingly greater than 0.026 eV of the room temperature thermal energy of molecules. Therefore they should not dissociate in the gas phase at room temperature. However, when they are dissolved in the water, their binding energies, derived from the electrostatic attractive forces, are reduced by nearly two orders of magnitude due to the large dielectric constant of water ($\varepsilon=81$). As the result, they ionize easily even at room temperature.

In order to improve the corrosion resistance, we have developed a technology to form a corrosion resistant Cr_2O_3 passivated film on the surface of electropolished SUS316L. As compared to the conventional stainless steel surface, the Cr_2O_3 passivated surface exhibits a markedly improved corrosion resistance, chemical stability and non-catalytic property toward specialty gases at room temperature. However, while the Cr_2O_3 passivation has achieved the surface chemical stability and the corrosion resistance, there is still a problem in the area of metal welding, which is a necessary procedure in assembling the all whole gas piping system together. Because the metal must be melted at the weld, metal fume is inevitably generated. This metal fume consists mainly of those high vapor pressure metals such as Mn which would form deposits in the vicinity of weld points. This deposited Mn containing metal fume near the weld point is responsible for the observed low corrosion resistance when it is contacted with the moist specialty gases. We tried to remove the metal fume by washing the welded area with ultra high purity water to improve the corrosion resistance and we also tried to increase the welding speed in order to reduce the heat input to lower the metal fume deposits. Presently, we are tackling the problem of how to confine the metal fume within the weld bead by applying a magnetic field during the welding. We will study the effects of the magnetic strength and direction on the fume behaviors and study the possibility of Cr_2O_3 passivation on the weld itself in order to attain a complete corrosion resistance at the weld.

2. EXPERIMENTAL

We will now describe the sample materials used in the present experiments and their preparation methods. Samples were prepared by taking 1/4" electropolished SUS316L tubes (twice vacuum-melted materials: VOD+VAR) and welding them together with an external coupling-type automatic, electric arc welding machine (Astro Arc Co.) [4]. Figure 1 shows a schematic flow diagram of the welding test system. The back seal gas and arc gas are made of a mixture of H_2 and Ar gases. The reason for the hydrogen addition is to reduce the heat input for the arc as well as to reduce the CrC endothermically to form CH_4 and other volatile gases in order to reduce the corrosion taking place at the CrC grain boundary. Ceramic filter was used to remove particles from the arc gas and the back seal gas. The internal pressure at the weld point was measured and controlled by using a differential pressure gauge. To ensure the surface smoothness and cleanliness, the electric current was controlled to within 0.1 A and the speed of tungsten electrode to within 0.1 sec. The samples of the weld surface and the surface near the weld were evaluated for the metal composition depth profile by using XPS (Shimadzu Co.). Metal fume compositions especially near the weld points were measured at 3 mm, 5 mm, 10 mm and 15 mm locations both upstream and downstream from the weld.

Next, we will describe how we evaluated the effectiveness of metal fume removal from the downstream of welds with ultra pure water. The welded samples were washed with 80 °C ultra pure water (with a resistivity of 18.25 MΩ-cm at 25 °C) at 250 cc/min for 6 hours to remove the deposited metal fume. The composition depth profile of downstream weld surfaces as well as the weld points were measured by XPS.

We will describe how we measured the metals generated from the corrosion of the welds. Figure 2 shows the experimental setup. To prevent the metal contamination from the supply gas, a 0.1 μm filter was installed at the inlet of the sample tube. The samples are prepared by using 1/4" x 40 cm long electropolished SUS 316L tubes previously mentioned and 22 evenly spaced welds were made along the tube length by the arc welding technique. The 22 welds were then exposed to a HCl gas and the corroded metals were then stripped by Ar gas and deposited on the downstream 5" wafer surface, which was cleaned by using a dilute HF + H_2O_2 solution to remove the native oxide film. The captured metals were measured by using TRXRF (Total Reflection X-ray Florescence, made by TECHNO Ltd.). The experiment began by purging the sample tube with an ultra clean Ar gas (50 ppt moisture). Then the sample tube was sealed in the HCl gas containing 1.4 ppm moisture at 1.5 kg/cm^2 for 15 hours at room temperature. Thereafter, the sealed HCl gas was discharged to the downstream outlet end where all the stripped metals were captured by the native oxide free Si wafer. To ensure all the metal was captured after HCl gas flow stopped, the sample tube was further purged with ultra high purity Ar at 500 cc/min for 6 hours. The silicon wafer was then removed from the chamber under the fan filter unit in the clean room and the metal composition was examined by TRXRF. To determine the stripped metals from the HCl exposed weld surface, we evaluated both the samples of 40 cm, 22 welds and the samples which were washed with 80 °C ultra pure water (with a resistivity of 18.25 MΩ-cm at 25 °C) at 50 cc/min for 6 hours. Furthermore, the weld point metal surface was inspected for changes by using SEM.

Next, we evaluated the effectiveness of using ultra pure water for to remove deposited metal fume from downstream of weld point with a resistivity measurement meter. Figure 3 shows the experimental setup. Samples were made of 1/4" x 40 cm long electropolished SUS316L tube, which has 9 evenly spaced 4 cm apart electric arced welds. All-metal valves used in the experimental line were highly integrated, extremely low dead space valves. Thus, it is possible to achieve a good flow control

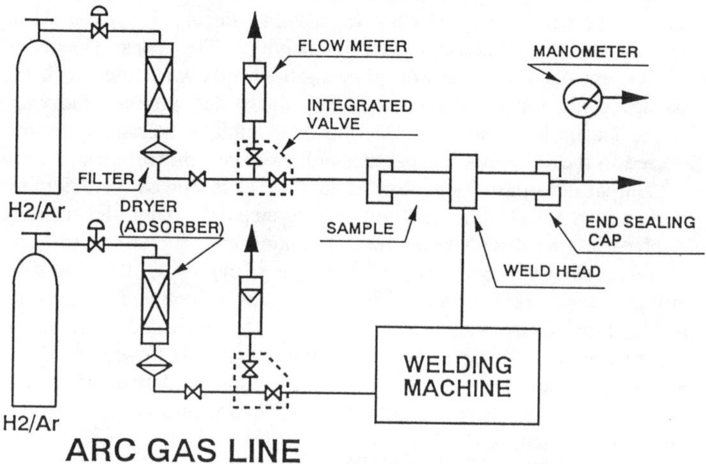

Fig.1 Schematic flow diagram of welding test system.

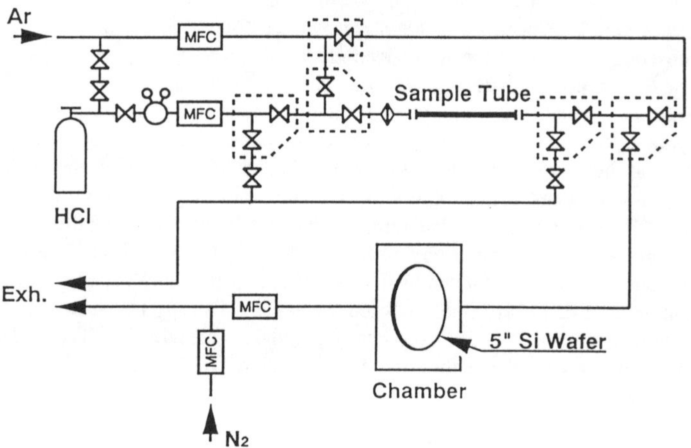

Fig.2 Schematic flow diagram of metal corrosion test system.

by operating these air-actuated multiple valves. The flow meter located at the inlet line allowed us to control freely the ultra pure water supply to the sample line and the by-pass line. A resistivity meter (Denki Kagaku Instruments, Ltd.) located at the end of the line was used to monitor the water resistivity. The measurement methods were shown in Figure 4. First, the well placed sample tube was filled with the ultra pure water at 4.5 l/min for 10 seconds and then the sealed sample tube was allowed to dissolve the metals for 5 minutes. During the 5 minute interval, the ultra pure water was directed to the valve line for background resistivity measurement. Next, the valve was shut and an ultra pure water flow at 15 l/min was flowed for 7.5 minutes to flush out the dissolved metals for resistivity measurements. After the measurements, the sample tube was rinsed with water for 7.5 minute intervals at set flowrates of 15, 45 and 70 l/min. Considering a total of 20 minute as one cycle, the total dissolved metal measurement in one sample was repeated until the detection limit was reached. To evaluate the flow dependence, three flow rates at 15, 45 and 70 l/min were used but the sampling times were all the same at 7.5 minutes for all flowrates. Within this 7.5 minute interval, the measured resistivity reflecting the dissolved metals decreased instantaneously from the baseline value of 18.25 MΩ with time and then returned to the baseline value and stabilized.

Next, we will describe the evaluation on the relation between the metal fume and the welding thermal energy input. In these experiments, the objective was to reduce the metal fume by studying the possibility of reducing heat input and welding time with the welding machine. The welding machine we used was capable to increase the motor rotation speed up to 19.8 rpm with a fine adjustment of 0.1 rpm. Thus it was possible to increase the weld head electrode revolution speed in a multiple of the current base speed. First, we had to make sure of the smoothness and cleanliness on the weld joint. For this purpose, we studied the conditions for the weld bead width and smoothness of the inside piping surface by conducting high speed one revolution welding at both 15 rpm and 19.8 rpm. We determined that the welding surface condition was indeed comparable with the conventional two revolution welding. Then we took the electric arc welded samples and nearby surface samples for composition depth profile measurements by using XPS for 15 rpm and 19.8 rpm welding evaluation. In addition, for the evaluation of the relationship between the welding time and the amount of metal fume, we used the same resistivity measurement techniques as previously described, that is, by preparing samples from 1/4" x 40 cm electropolished SUS316L tubes with 9 evenly spaced, 4 cm apart welds which was rinsed with ultra pure water.

3. RESULT AND DISCUSSION

Figure 5 shows the metal fume generated in the vicinity of the weld point as measured by XPS: (a) for the depth profile of the metal composition for the sample

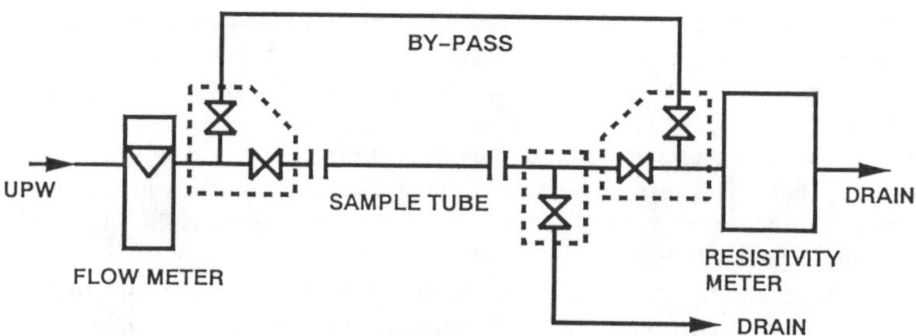

Fig.3   Schematic diagram of cleaning effect for welding points
by ultra pure water rinsing.

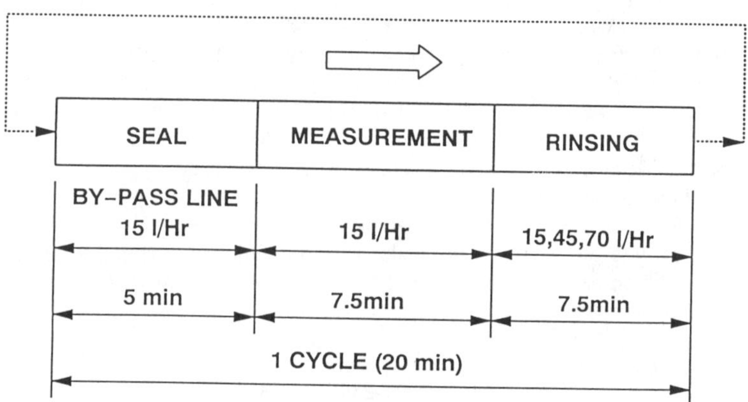

Fig.4   Measurement method of UPW rinsing.

located 5 mm upstream against the back sealed gas flow and (b) for the sample located 5 mm downstream. The vertical axis shows the metal position percentage and horizontal axis shows the sputtering time. Sputtering etching was about 60 °A/min. Time zero on the horizontal axis indicates the results on the upper most surface. From (a), we observed that at the upstream location, the Mn composition profile was fairly constant from the upper most surface into the depth. (b) indicates that at the downstream location, the Cr and Ni contents were lower while the Mn was higher at the upper most surface and decreased in the direction of the depth and tapered off gradually. These XPS results indicated the Mn profile at upstream location was about the same as the base material while the Mn was exceptionally high at 5 mm downstream location. We can understand this by noting that because Mn has the highest vapor pressure among the stainless steel constituents, its vapor created during the welding was carried with metal fume by the back seal gas and deposited on the downstream location. Next we washed off the metal fume from the surface with 80 °C ultra pure water (with a resistivity of 18.25 MΩ-cm at 25 °C) at 250 cc/min for 6 hours. The washed XPS results are shown in Figure 6.: (a) for the downstream 3 mm location and (b) for the downstream 5 mm location. Both (a) and (b) show the Mn was reduced at the upper most surface layers in the depth direction. From this experiment, we understand that the deposited metal fume can be removed by warm water wash.

Next we evaluated the corrosion resistance and the mounts of corrosion product metals from the exposure to HCl gas containing 1.4 ppm moisture for the downstream samples with and without ultra pure water wash under the identical welding conditions. The metals collected on the silicon wafer were analyzed with TRXRF. Figure 7 shows the results: (a) for the sample without water wash and (b) for the sample with water wash. From (a) we determined Fe = 1.8 x 10^{14} atoms/cm^2, Cr = 4.9 x 10^{13} atoms/cm^2, Ni = 2.8 x 10^{13} atoms/cm^2 and Mn = 7.8 x 10^{12} atoms/cm^2. From (b) of the washed sample, we observed metals were below the TRXRF detection limit of 1.0 x 10^{10} atoms/cm^2. The analyzed metals here were corrosion products. Figure 8 shows the SEM images of these two samples. (a) for the sample without the water wash and (b) with 80 °C ultra pure water at 250 cc/min for 6 hours. We observed chloride formation on the sample without the wash, indicative of HCl corrosion while the washed surface was essentially corrosion free with markedly reduced corrosion.

The metal fume removal efficiency with ultra pure water wash was evaluated at 15, 45, 70 l/min to see their flow rate dependence. As a baseline reference, Figure 9 shows the resistivity changes of the washed samples without welds for the samples prepared from 1/4" x 400 mm electropolished tube material as received. Figure 10 shows the resistivity changes for the samples containing metal fume from 9 welds. The vertical axis show the metal resistivity values calculated as NaCl equivalent [μg] and the horizontal axis shows the number of rinse cycles. These experiments were carried out by using the conventional 2 revolution welding technique. From these results, we

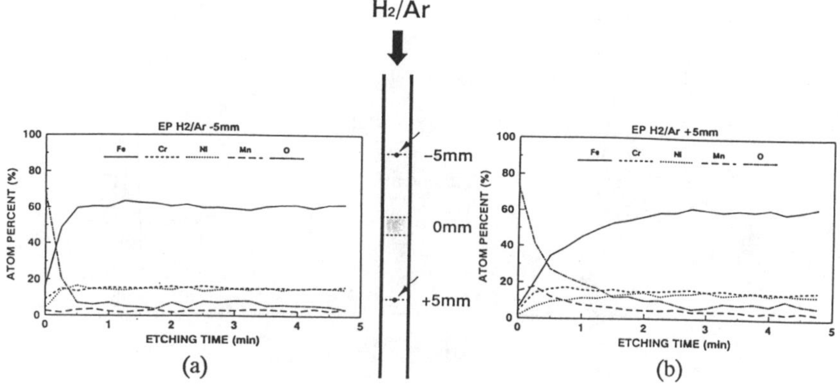

Fig.5   The depth profile of welding point vicinity by XPS.
(a)5mm upstream location from the welding point.
(b)5mm downstream location from the welding point.

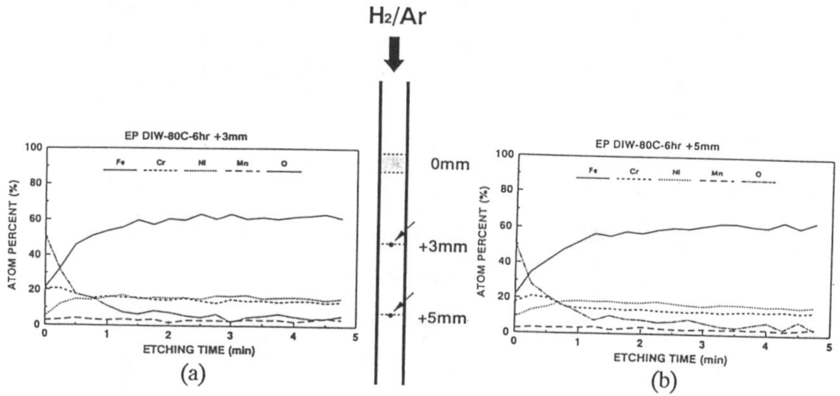

Fig.6   The depth profile of downstream location from the welding point
after hot UPW rinsing by XPS.
(a)3mm downstream location from the welding point.
(b)5mm downstream location from the welding point.

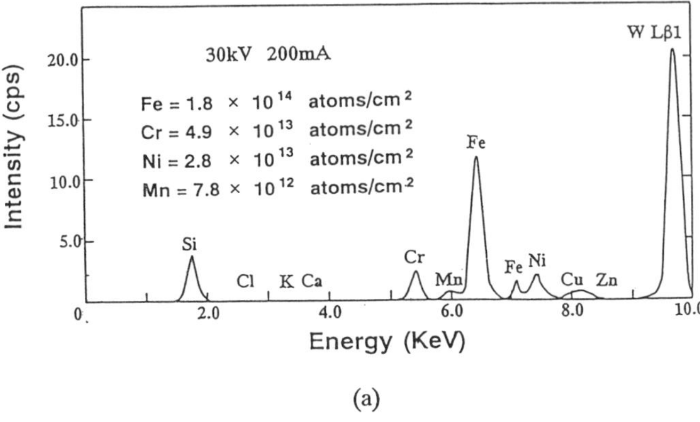

(a)

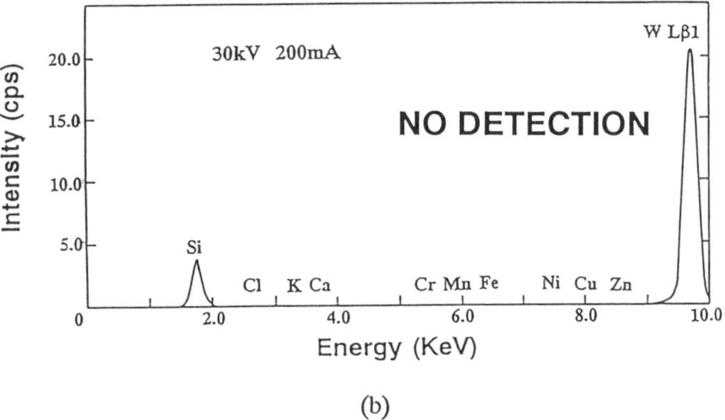

(b)

Fig.7   The fluorescence spectrum of stripped metals after scaled
with the HCl gas by TRXRF.
(a)Without Hot UPW rinsing.   (b)With Hot UPW rinsing.

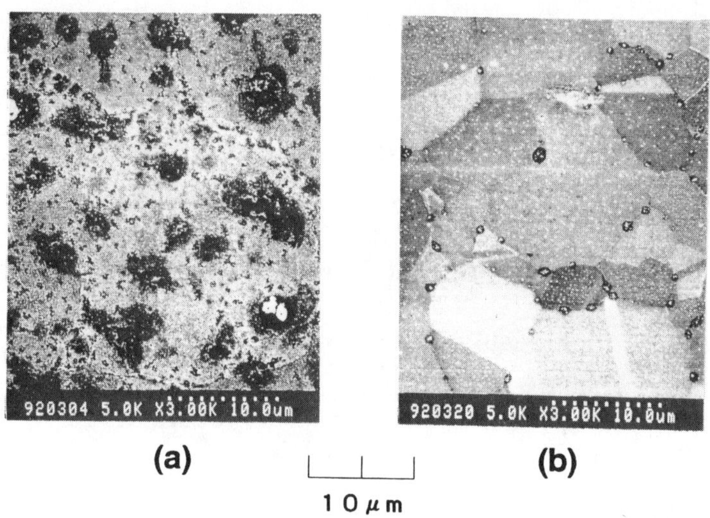

(a)   (b)

1 0 μ m

Fig.8   SEM images of downstream surface from the welding point after sealed with the HCl gas.
(a)Without Hot UPW rinsing.   (b)With Hot UPW rinsing.

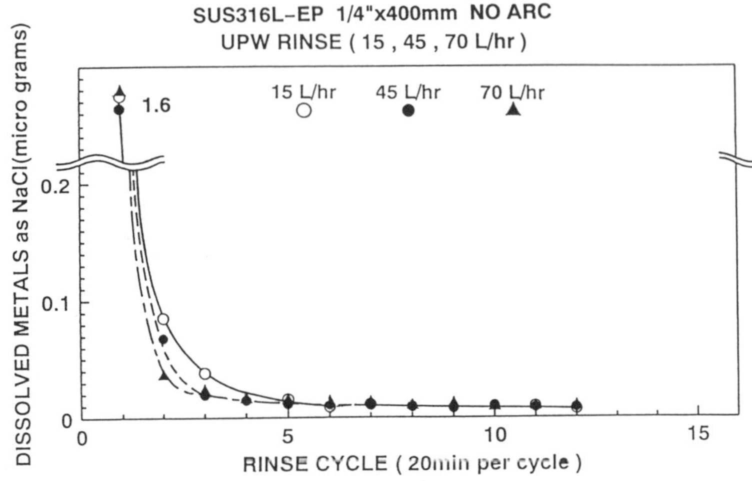

Fig.9   Frow rate dependence of metal fume removal effectiveness by UPW rinsing, as received EP tube.

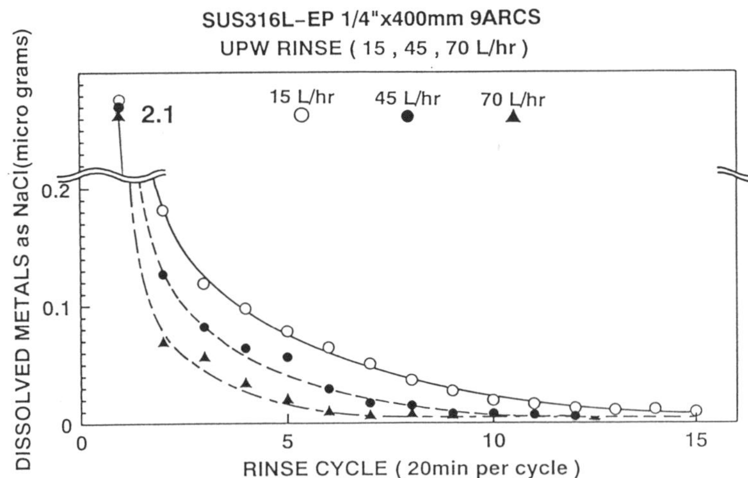

Fig.10 Flow rate dependence of metal fume removal effectiveness by UPW rinsing, welded EP tube.

understand that the number of rinse cycles to remove the metal fume from downstream location of the weld point can be reduced if we increase the water flow rates to reach the background level.

Next, we evaluated the interrelationship between the ultra pure water rinse and the welding heat input and the amount of the generated metal fume. Table 1 compares the results of the conventional 2 revolution at 7.5 rpm with the one revolution at 15 rpm and 19.8 rpm. We observed that the heat input for the high speed one revolution at 15 rpm was about 1/2 and for high speed one revolution at 19.8 rpm was 2/5 of the conventional one. Also the welding time of the high speed revolution at 19.8 rpm was about 2/5 of the conventional one. The right hand column shows the dissolved metal values (NaCl equivalent) from the rinse water resistivity measurements. These data will be explained later.

Next, we contrast the effects of the conventional 2 revolution welding speed with the high speed one revolution on the surface Mn amounts in the vicinity of weld bead in Figure 11 from XPS. The vertical axis shows the Mn atomic percentage and the horizontal axis shows the locations. 0 mm indicates the weld point. (-) indicates the upstream and (+) the downstream locations from the weld bead. As we can see that the upstream Mn content is about the same as in the base metal material but the downstream Mn content is substantially higher. We also note that the high speed one revolution at either 15 rpm or 19.8 rpm generated less amount of Mn than the conventional two revolution at 7.5 rpm. In other words, reduced heat input and the welding time can effectively suppress the Mn generation from the welding.

We further investigated the effect of the heat input at the same welding speed (15 rpm) on the generated metal fume for 1, 1.5 and 2 revolution three cases. Samples were made of 1/4" x 40 cm long electropolished SUS316L tube with 9 evenly spaced 4" apart electric arced welds. Metals were determined by the resistivity in the rinse water at 15 l/hr and we ignored the problems related to the surface smoothness. Figure 12 shows the results. For each weld, we calculated the metals of 0.04 μg for 1 revolution, 0.11 μg for 1.5 revolution and 0.53 μg for 2 revolutions. This clearly indicates that the reduction in heat input can suppress the metal fume generation. The data are plotted in Figure 13. The vertical axis shows the log of metal values (NaCl equivalent) and the horizontal axis shows the revolution in liner scale. The graph suggests a exponential relationship with the number of welding revolution.

The right hand column in Table 1 for the metal results of the conventional 2 revolution at 7.5 rpm and the one revolution at 15 rpm and 19.8 rpm from the continuous resistivity measurement of rinse water at 15 l/hr on sample tube with 9 welds is further plotted in Figure 14. We observed that the metal can be removed much faster for the case of one revolution 19.8 rpm than the other two cases. The

Table 1   Comparison of heat input and welding time in each welding condition.

	HEAT INPUT [J]	WELDING TIME [sec]	DISSOLVED METAL in UPW as NaCl [μg]
7.5RPM × 2ROT.	2830	18	0.14
15RPM × 1ROT.	1296	4.8	0.04
19.8RPM × 1ROT.	1159	3.7	0.004

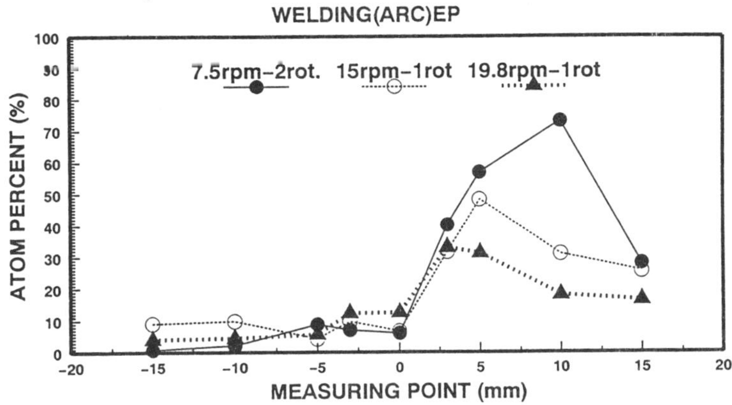

Fig.11   Mn content of the upper most surface layer in vicinity of bead under difference welding conditions.

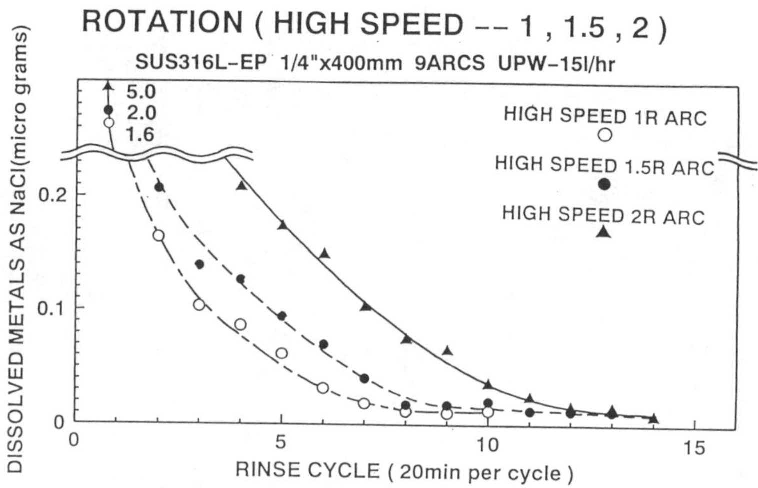

Fig.12   Metal fume removal effect of UPW rinsing with the difference of heat input.

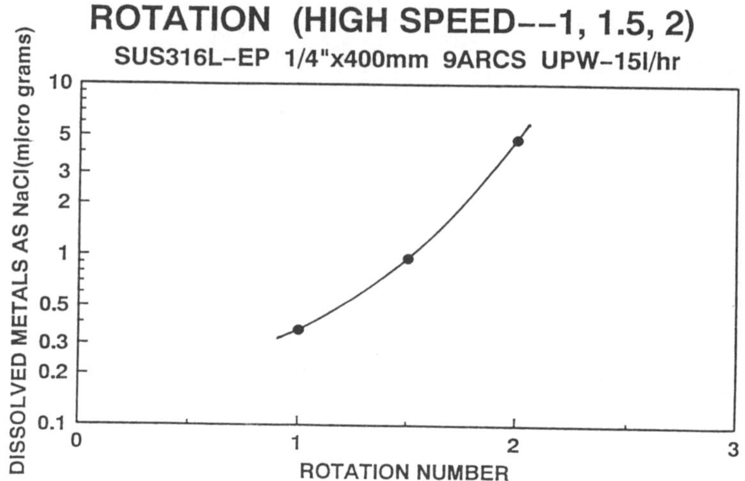

Fig.13   The effect of welding conditions on metal fume removal by UPW rinsing.

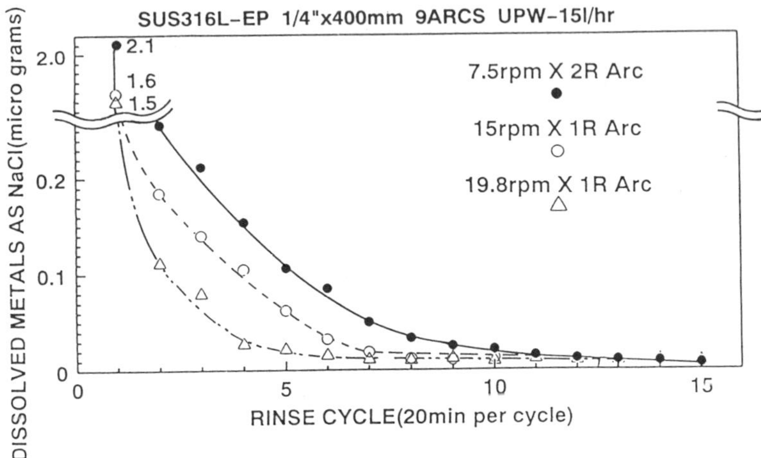

Fig.14   Relation between heat input and the amount of metal fume generation.

amount of metals were calculated as follows: 0.14 μg for the conventional 2 revolution at 7.5 rpm, 0.04 μg for the high speed one revolution at 15 rpm and 0.004 μg for the high speed one revolution at 19.8 rpm. In other words, the metal amount is the lowest and the removal rate is the fastest for the high speed one revolution at 19.8 rpm.

4. CONCLUSION

We have developed a Cr_2O_3 passivation technology on the stainless steel surface and achieved a markedly improved corrosion resistance on the gas piping surface. Since the welding is a necessary procedure in assembling the whole system together, we need to improve the corrosion resistance in the welding area as well.. We now understand that in the vicinity of the weld point, the deposited metal fume with Mn as a major constituent generated during the welding is the major cause for the reduced corrosion resistance. We elucidated in this study the metal fume composition and its generation mechanism, and demonstrated the metal removal efficiency by using the ultra pure water wash to improve the corrosion resistance. Although the corrosion resistance of the washed weld surface is still not as good as the Cr_2O_3 passivated surface, we showed that it is possible to wash the weld surface to reduce the metals such as Fe, Cr, Ni and Mn by 4-5 orders of magnitude. We further showed that it was possible to reduce the metal fume generation by reducing the welding speed to reduce the heat input to the weld. While it is important to study the techniques to remove the metal fume and to reduce fume generation further, it is more urgent to develop a completely metal-fumeless welding technology. We are currently exploring the possibility of adding a magnetic field to the welding process and study the possibility of passivating the weld surface with Cr_2O_3 to achieve a complete corrosion resistance. For an early recovery of the huge investment in the super clean fab system for the ULSI manufacturing, it is essential to maintain an ultra cleanliness in the process ambiance and to extend the useful life of the process system. We believe the welding technology as part of the overall metal surface treatment is very important to satisfy that desire.

5. ACKNOWLEDGMENT

The authors would like to express their sincere thanks to Osaka Sanso Kogyo Ltd., Motoyama Eng. Works Ltd. and Kobe Steel Ltd. for their supports. This study was carried out in the Mini Super Clean Room of the Facaulty of Engineering, Tohoku University.

6. REFERENCES

1. Tadahiro Ohmi, "Importance of ULSI Clean Technology in Advanced ULSI Processing," Symposium on Advanced Science and Technology of Silicon Materials, Kona (Hawaii), pp. 315-326, November 1991.

2. Tadahiro Ohmi, Masakazu Nakamura, Atushi Ohki, Koji Kawada, Keiji Hirao and Tsuyoshi Watanabe, "Thermal Decomposition on SiH_4 and Si_2H_6 and the Influence of Residual Moisture and Oxygen," Proceedings, 15th Workshop on ULSI Ultra Clean Technology, Tokyo, pp. 125-33, November 1991.

3. Masakazu Nakamura, Tadahiro Ohmi, Kazuhiko Sugiyama, Yasumitsu Mizuguchi, Atsunobu Ohkura and Koji Kawada, "All Metal and O_2 Passivation Ultra Clean Gas Delivery System for Submicron ULSI," Extended Abstracts, 178 th Electrochemical Society Meeting, Seattle, Abstract No. 433, pp. 633-634, October 1990.

4. Yasumitsu Mizuguchi, Kazuhiko Sugiyama and Tadahiro Ohmi, "Welding Technology for Passivated Tubing Systems," Microcontamination 89 Conference Proceedings, Anaheim, pp. 570-64, October 1989.

THERMAL DECOMPOSITION CHARACTERISTICS OF SiH$_4$

Shinji Takahashi, Tsuyoshi Watanabe,
Shinji Miyoshi, Atsushi Ohki,
Koji Kawada, Masakazu Nakamura,
Michael S. K. Chen and Tadahiro Ohmi

Department of Electronics, Faculty of Engineering,
Tohoku University, Sendai, Japan 980

In this study, we evaluated the thermal decomposition characteristics of SiH$_4$ with respect to different dilution rates, diluent gases and different surfaces on which thermal decomposition reactions take place. We observed that even when we used the same gas, the initial decomposition temperatures, the decomposition rates, the decomposition behaviors and the activation energies were radically different.

Keywords: Silane. Thermal decomposition. Kinetics. Activation Energy. Polysilicon. Cr$_2$O$_3$. Fe$_2$O$_3$. Passivation. Filter. Hastelloy. Semiconductor. CVD. ULSI.

1. INTRODUCTION

In fabricating micro devices such as TFT for ULSI and LCD, thin-film forming process, photolithography and pattern etching are repeatedly employed. While every process is very important, the high performance thin film technology appears to be the most critical to achieve a good pattern modification in the future advanced microfabrication. Thin film formation has traditionally been based on the key process, Chemical Vapor Deposition (CVD) in which a variety of specialty gases such as SiH$_4$ are used. However, much is still not well understood about the properties of these very reactive specialty gases. It is not too exaggerating to say that in the current CVD process, many important parameters such as gas concentration, diluent gas, flow rate, film forming time, temperature as well as chamber pressure or high frequency power input etc. are by and large empirically determined [1]. For high yield production of deep sub-micron devices, it is essential to remove impurities completely from these specialty gases in the process ambiance and to gain a complete control over these process parameters [2]. For these reasons, it is necessary to study their physical and chemical properties and reaction behaviors of these specialty gases on various surfaces.

Studyies of silane decomposition went as far back as in the 1930s when T. R. Hogness et al studied the extents of silane thermal decomposition by measuring the pressure changes in a static system [3]. Gas chromatography was used to measure the gas composition of the reactor at the end of reaction. Their studies had revealed many valuable observations but the results were obtained in the gas delivery system using parts not quite adequate by the present standard. Besides, the specialty gas purity was not as high. In recent years, although there were many reports on the reaction mechanisms about these specialty gases, these studies are primarily concerned with the problems of predicting film growth rates in the present process.

We had previously evaluated the SiH_4 thermal decomposition characteristics in a completely closed system within the limits of outgassing from the internal metal surface and external leaks [4]. We observed that even when we used the same gas, the initial decomposition temperatures, decomposition rates, decomposition behaviors and activation energies were radically different with different dilution rates and diluent gases. Furthermore, we observed that the gas contact metal surfaces had profound effects on the decomposition characteristics. On this later point, we made further investigations into the SiH_4 decomposition behaviors, by focusing on the materials used in the gas delivery system and the process chamber,i.e., on various surface-modified stainless steel reactor tubes and all-metal filters so that we could clearly delineate the different surface catalytic effects on SiH_4 decomposition behaviors. The following is our report.

2. EXPERIMENT

2.1 Experimental setup

Figure 1. shows a schematic flow diagram of the experiment. The system is composed of a precision gas dilution system, a thermal decomposition tube reactor and a high sensitivity gas chromatography (GC) [5]. Diluted SiH_4 gas at a chosen dilution rate near atmospheric pressure was fed to the reactor and the extent of thermal decomposition was measured by GC.

In the gas dilution line, all of plastic materials that tend to trap the moisture were replaced by all-metal, dead space-free, highly integrated parts [6]. All parts used in the system are made of SUS316L and their internal surfaces were electropolished and mirror-finished. All-metal C rings were used in the fittings with an external leak rate of $< 2 \times 10^{-11}$ Torr-l/sec. The use of this ultra clean system made it possible to study the true thermal decomposition characteristics of specialty gases, free from the effects of gas impurities, outgassing from the metal surface and the external leaks. The mass flow controller (MFC) used in the precision dilution system can achieve precise dilution rate and flow control. In order to match the desired residence time with the time for the gas to flow through the reactor tube, it was necessary to control the reactor temperature distribution and SiH_4 flow rate precisely.

The reactor was made of 1" OD x 40 cm long electropolished SUS316L tube with an effective reactor volume of 169 cc. The reactor external was wrapped with alumina foils, sheath heaters and insulation. Temperature was measured with a thermal couple inserted in center of reactor and controlled to within 2°C by using several PID controllers along the reactor tube length.

Downstream of the reactor, a high sensitivity GC equipped with TCD (Thermal Conductivity Detector) was used to measure the specialty gas concentration. This high sensitivity GC was made of all-metal parts. All internal gas contact area was electropolished and mirrored finished to avoid trapping any residual impurities. 1/16" tubings were used to minimize the sample gas contact area and the outgassing effect.

2.2 Experimental Procedures

By using the above-mentioned experimental setup, we evaluated the SiH_4 thermal decomposition characteristics with respect to different dilution rates and diluent gases. Ar, N_2 and H_2 were used as diluent gases to prepare the SiH_4 concentration ranging from 0.01% to 10%. Gas was introduced to the reactor at about one atmosphere. The temperature and flow rate were varied as parameters to determine the amount of SiH_4 decomposition. The reactor tube was prepared by depositing a polysilicon surface film by SiH_4 thermal decomposition. To make sure we had the same poly-Si surface characteristics, we frequently returned to the same experimental run condition to measure the amount of SiH_4 decomposition at 0.01% SiH_4/Ar at 400 °C. Ultra high purity SiH_4 with a moisture level of <6 ppb [7] was used. The diluent gas Ar had a moisture about 50 ppt [8]. H_2 gas was purified by passing the supply gas through a Pd membrane unit [9]. The experimental results was approximately analyzed by using the first order kinetic equations shown in Table 1. In the data treatment, we took into account the effects of inlet temperature ramp and the volume expansion due to reaction on gas flow variations. First-order kinetic rate constants were computed for each diluent gas, dilution rate and temperature. Activation energies were then determined from the Arrhenius plots. These results allowed us to clearly see the differences in the activation energy for different diluent gases, different dilution rates and different initial concentrations.

Next, we evaluated the SiH_4 thermal decomposition characteristics on various stainless steel surfaces used in the gas delivery system and the process chamber. 4 kinds of 1/4" x 190 mm tubes were used: electropolished SUS316L surface, oxygen

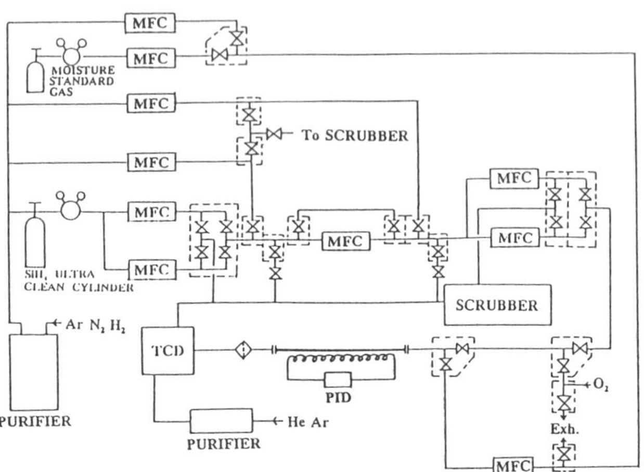

Fig–1　Schematic diagram of thermal décomposition evaluation system.

Table–1　SiH_4 reaction rate equation.

$$-\frac{d[SiH_4]}{dt} = k \cdot [SiH_4]$$

$$t_0 \Rightarrow [SiH_4]_0 \quad \cdots \quad \text{Initial}$$

$$t \Rightarrow [SiH_4]_t$$

$$[SiH_4]_t = [SiH_4]_0 - x$$

$$x \quad \cdots \quad \text{Decomposed } SiH_4 \text{ Concentration}$$

$$-\int_{[SiH_4]_0}^{[SiH_4]_t} \frac{1}{[SiH_4]} d[SiH_4] = \int_0^t k \cdot dt$$

$$\text{Decomposition Rate} = \frac{[SiH_4]_t - [x]}{[SiH_4]_0} = e^{-kt}$$

$$k = A\exp\left(-\frac{E_a}{RT}\right) \qquad k = \text{Rate Constant}$$

passivated Fe_2O_3-rich surface [10], passivated Cr_2O_3-rich surface [11] and electropolished Hastelloy surface. These reactor tubes were wrapped with the aluminum foil and sheath heaters. First, we passed an ultra high purity Ar gas through the tube and baked it at 400 °C for 10 hours to remove all the adsorbed moisture and to activate the surface. After cooling down to 25 °C, a 100 ppm SiH_4 flow at 5 cc/min was passed through the tube for 10 hours. The temperature was then raised to 500 °C at 0.1 °C/min and the gas at the reactor exit was monitored continuously by GC to measure the SiH_4 and H_2 concentrations.

Next, we investigated the SiH_4 thermal decomposition characteristics on all-metal filters which have relatively large effective areas at room temperature. Two kinds of filters were employed. Filter A was made of single gland long fiber elements with an effective area of 0.8 m^2 and filter B short fibers with an effective area of 0.66 m^2. We evaluated 4 samples : two were the filters as received and the other two were Cr_2O_3 passivated on the surface. I this experiment, we wrapped the samples with alumina and sheath heater. First we baked them at 450°C for 10 hours and then cooled to 25 °C. A gas flow of 500 ppm SiH_4 was then passed through the filter sample at 5 cc/min. Gas from the downstream was analyzed by GC for SiH_4 and H_2 concentrations as a function of time.

3. RESULT AND DISCUSSION

3.1 Effect of different surfaces on the SiH_4 thermal decomposition characteristics

First, we will describe the effects of different surfaces on SiH_4 thermal decomposition characteristics. The thermal decomposition was carried out in an 1" electropolished SUS316L tubes as described previously. The reactor internal surface with Fe_2O_3 as a major constituent was oxygen passivated, hydrogen reduced (400 °C x 1 hour) and Ar annealed (400 °C x 1 hour). This passivation resulted in a surface with a 65% Cr_2O_3 depth of about 60 °A. Polysilicon was deposited on this surface by thermally decomposing 100% SiH_4 at 70 cc/min at 450 °C for 4 times. Each time we measured the amount of 0.01% SiH_4/Ar thermal decomposition at 400 °C. The results are shown in Figure 2. The vertical axis shows the amount of decomposition and the horizontal axis shows the residence time. The "initial evaluation" data shows the rate of the original Cr_2O_3 surface without polysilicon deposition. We observed that each time we deposited more polysilicon on the surface, the rate of 0.01% SiH_4/Ar thermal decomposition increased. The surface was stabilized at the fourth time deposition (with a total deposition time was 2.5 hours). Thereafter even with further polysilicon deposition, the rate of 0.01% SiH_4/Ar would not change. It was on this stabilized surface in the reactor that we conducted all the subsequent decomposition runs with different diluent gases and different dilution rates. As mentioned before, we also periodically returned to the base run to confirm the same polysilicon surface condition.

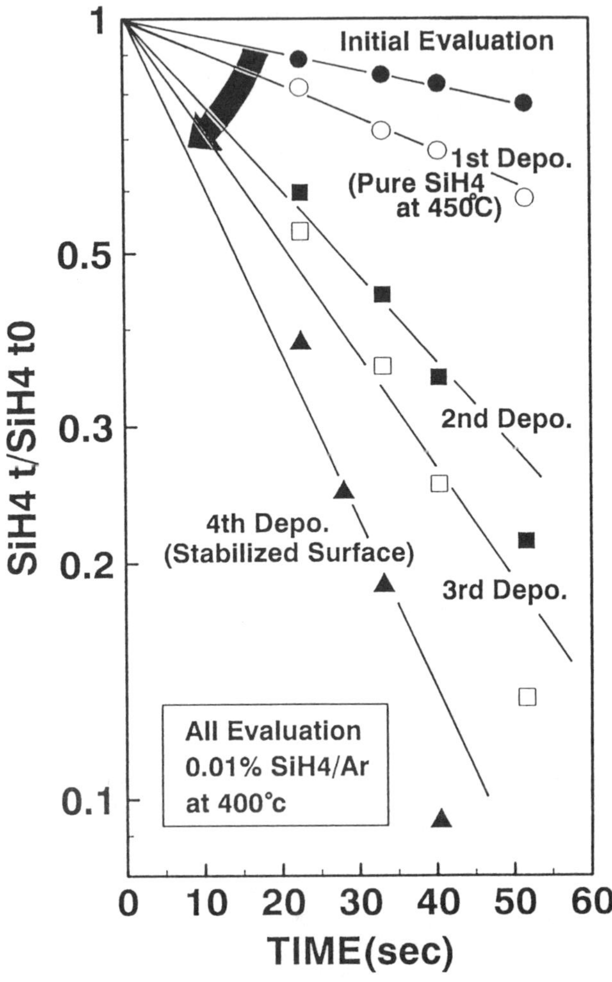

Fig-2 Thermal decomposition characteristics with surface condition transition.

The results shown in Figure 2 are very important because they have a strong implication on the thermal decomposition on the wafer surface and the process chamber surface. The effects of metal surfaces inside the process chamber and SiH_4 decomposition will be discussed later.

3.2 The effects of different diluent gases and different dilution rates on SiH_4 thermal decomposition characteristics

Figure 3(a) shows the decomposition rate of 0.01% SiH_4/Ar balance gas as a function of reaction time. The vertical axis shows the rate and the horizontal axis the time (sec). Figure 3(b) shows the corresponding activation energy for 0.01% SIH_4/Ar. The vertical axis shows the decomposition rate constants and the horizontal axis show the reciprocal temperature. For 0.01% SIH_4/Ar, we measured the rate on every 10 °C interval from 380 °C to 420 °C. Similarly, Figure 4(a) and 4(b) show the results for 0.1% SiH_4/Ar, Figure 5(a) and 5(b) for 0.6% SiH_4/Ar and Figure 6(a) and 6(b) for 10.0 % SiH_4/Ar. These results show that as the SiH_4 concentration increases, both the time needed to decompose to the same fractional degree and the activation energy increase. Figures 7, 8, 9 and 10 (a) and (b) show similar results for 0.01%, 1.0% and 10.0% SiH_4 using N_2 as the diluent gas. Comparing the results of using N_2 with H_2 as the diluent gas, we observed that their activation energies are about the same but the rates with Ar are faster. Therefore, for the same degree of fractional conversion, it takes more time for SiH_4 to decompose in N_2 than in Ar. In other words, SiH_4 will decompose more in Ar than in N_2 for the same reaction time. Even in the case of N_2 as the diluent gas, it takes more time to decompose and the activation energy also increases with SiH_4 concentration. The reason for this is that in both Ar and N_2 diluent gas cases, SiH_4 decomposition is an exothermic reaction (enthalpy of formation is 34.3 KJ/mol), the reaction rate would normally increases as the temperature increases. However, in reality, as the SiH_4 concentration increases, more and more SiH_4 would adsorb onto the surface and gradually saturate the surface. As a result, the surface reaction would proceed more slowly and generate more H-containing intermediate reaction products and suppress the overall decomposition rate.

Next, we will describe the thermal decomposition characteristics in the H_2 diluent gas. Figures 11, 12, 13 and 14 (a) and (b) show results for 0.01%, 0.1%, 1% and 10% SiH_4 initial concentrations respectively. Comparing Figure 11,12 13 and 14 (a) of H_2 with the corresponding cases with Ar and N_2 as diluent gases, we observed that H_2 suppresses the reaction significantly. It took more than one order of magnitude in time to reach the same fractional decomposition level, and the starting decomposition temperature was also higher. At long reaction time, we also observed that the deviation from the first order behavior became more significant. With respect to the

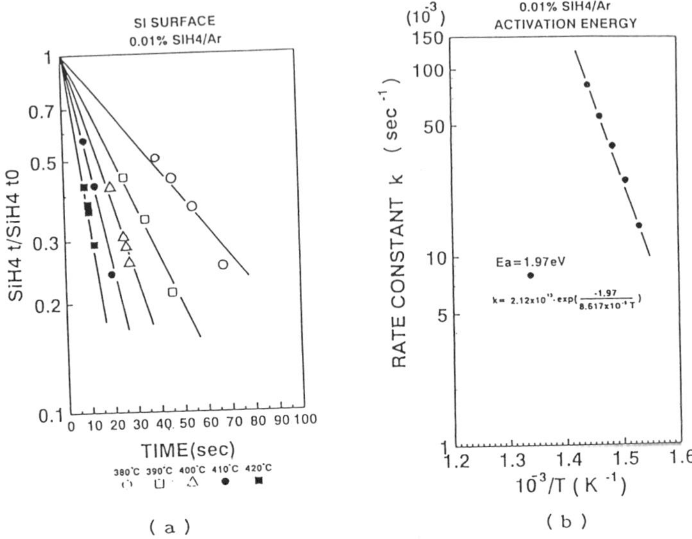

Fig-3 Thermal decomposition characteristic with Ar dilution.

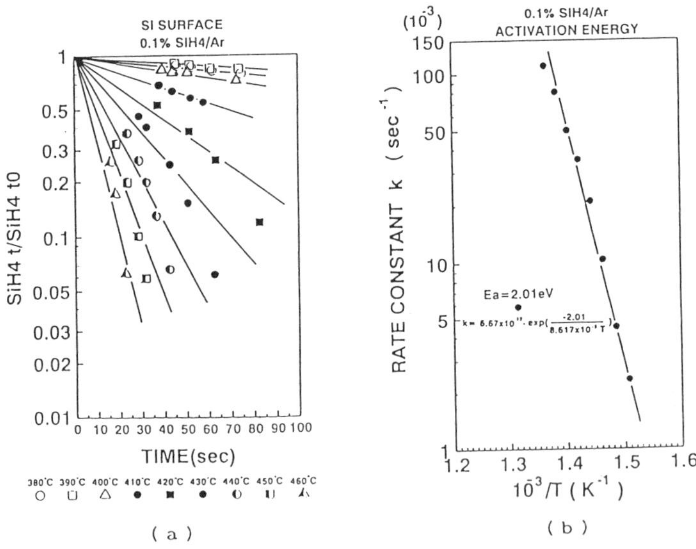

Fig-4 Thermal decomposition characteristic with Ar dilution.

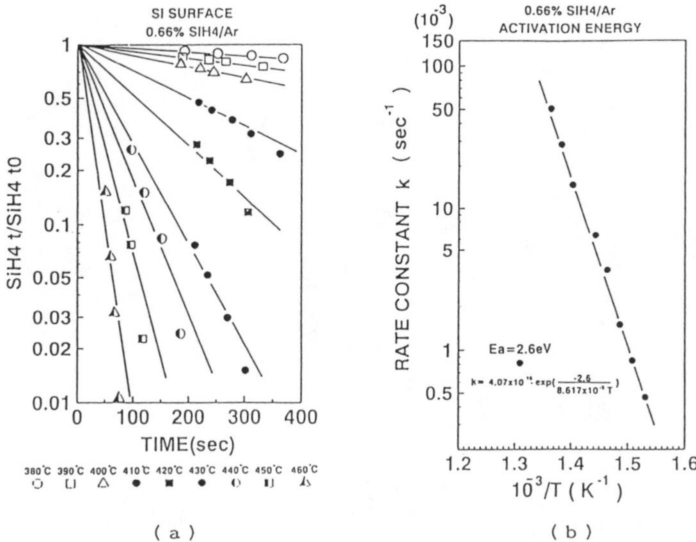

Fig–5 Thermal decomposition characteristic with Ar dilution.

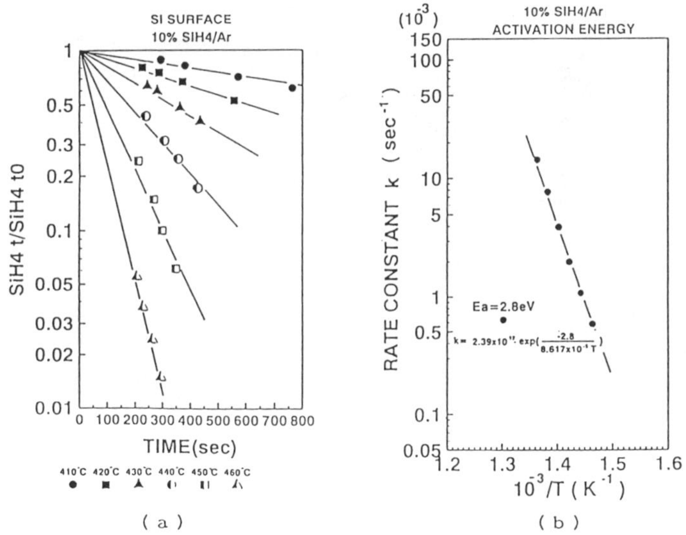

Fig–6 Thermal decomposition characteristic with Ar dilution.

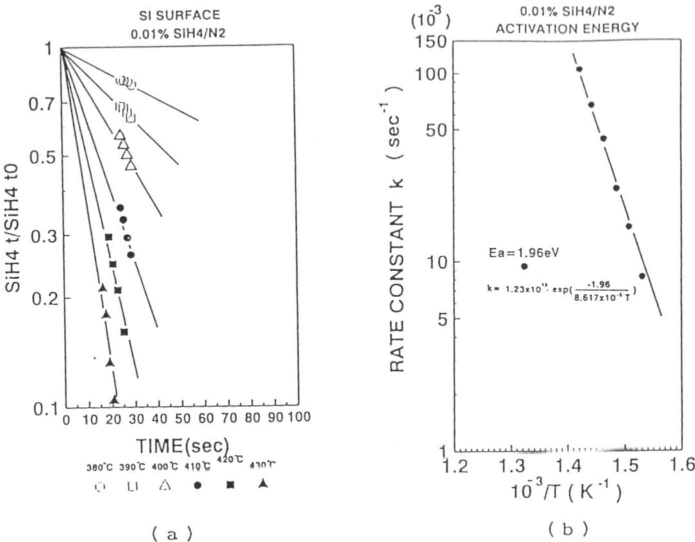

Fig–7 Thermal decomposition characteristic with N_2 dilution.

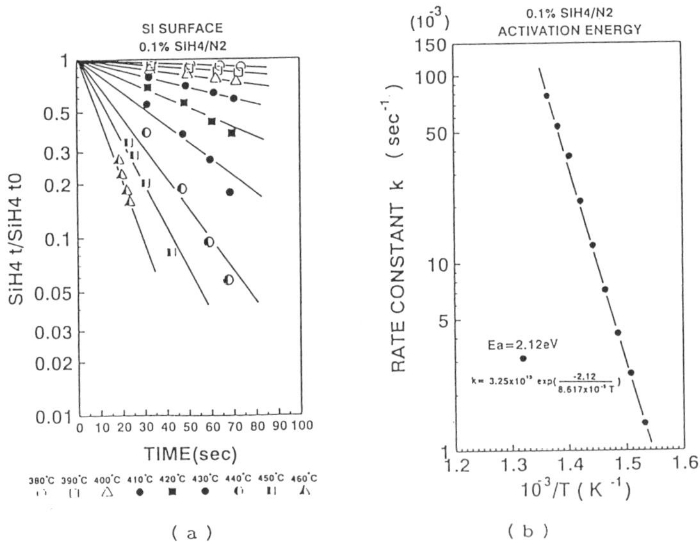

Fig–8 Thermal decomposition characteristic with N_2 dilution.

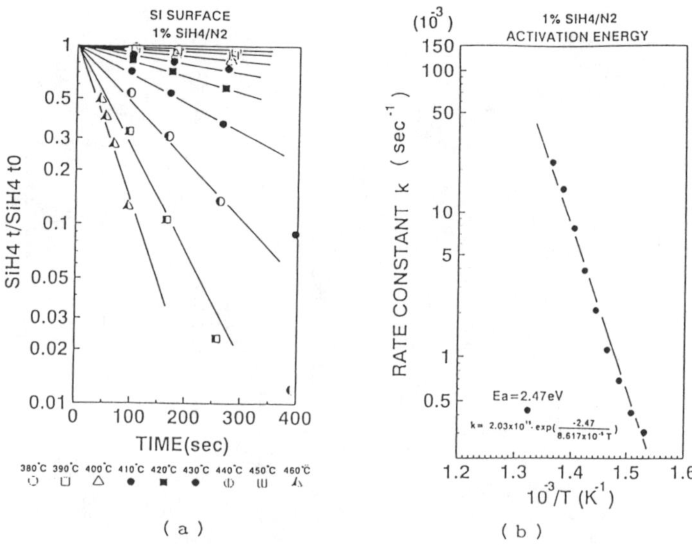

Fig-9   Thermal decomposition characteristic with N_2 dilution.

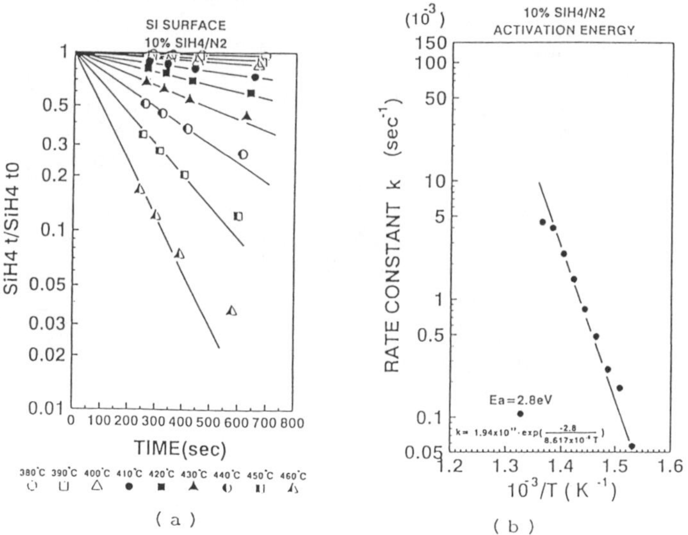

Fig-10   Thermal decomposition characteristic with N_2 dilution.

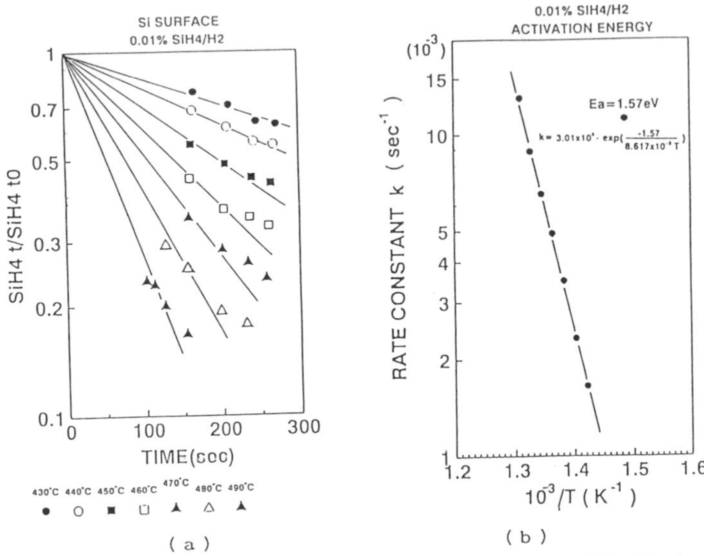

Fig-11 Thermal decomposition characteristic with H_2 dilution.

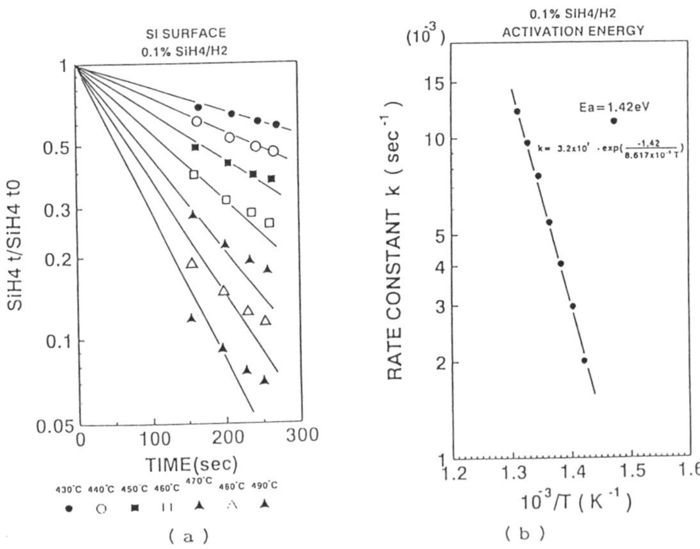

Fig-12 Thermal decomposition characteristic with H_2 dilution.

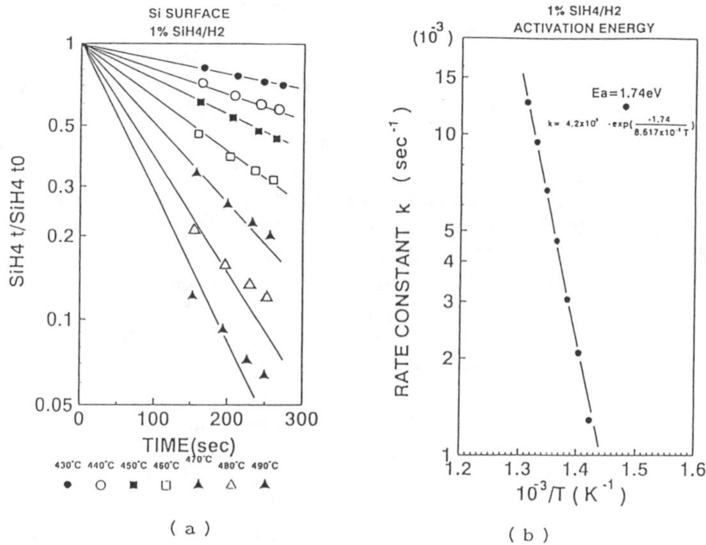

Fig–13 Thermal decomposition characteristic with H_2 dilution.

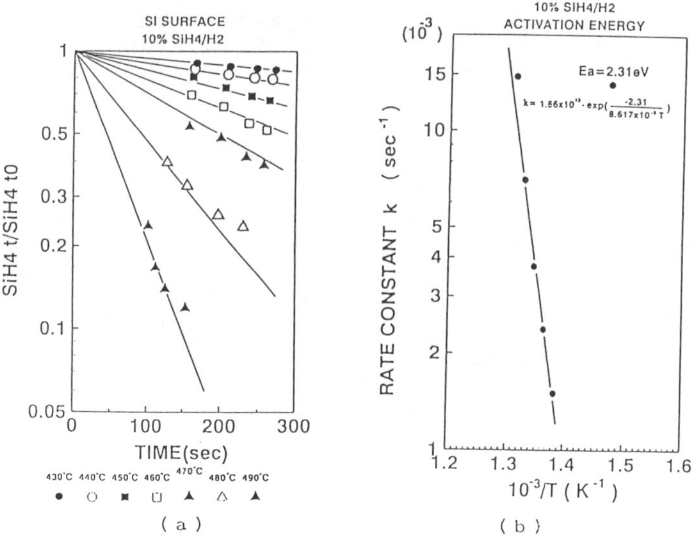

Fig–14 Thermal decomposition characteristic with H_2 dilution.

activation energy, their values are smaller than those in Ar and N_2 over the entire range of concentration of the experiment. Table 2 summarizes these values for three diluent gases. Figures 15,16 and 17 show the rate constants as a function of initial SiH_4 concentration for Ar, N_2 and H_2 diluent gases respectively. The vertical axis is the rate constant and the horizontal axis the concentration. We observed that for SiH_4 thermal decomposition in Ar and N_2, the rate constants were smaller as the initial concentration incresed. On the other hand, the rate constants for the case of H_2 appeared to peak at about 0.1% but overall speaking, the rate constants were much smaller than in Ar and N_2 and the dependence on the initial concentration was also not as significant. The reason for this is that first, the surface adsorption of H_2 would interfere with the SiH_4 adsorption and suppress its surface decomposition rate and second, H_2 is a SiH_4 decomposition product and therefore by considering the reaction equilibrium, it is more difficult to produce H_2 and to desorb in the excess H_2 atmosphere. In other words, the SiH_4 decomposition is suppressed.

3.3. Effects of different metal surface conditions on SiH4 decomposition characteristics

Specialty gases are usually delivered through a stainless steel gas piping system to the process chamber made of various metals. Let us review Figure 2 once more. We saw that SiH_4 decomposition rate varied with each Si deposition on the inside surface of the reaction tube. This result would imply that in the process chamber, the CVD film growth rate would vary with the surface condition which varies with deposited reaction by-products and consequently the process reproducibility can not be guaranteed even in the same process equipment. To achieve a complete reproducibility for film formation, it would seem best to deposit the entire surface adequately with Si but this approach would cause the particle contamination on the wafer surface as a result of detachment of adsorbed impurities from the chamber surface and deteriorate the device yield. Therefore, it is necessary to use metal surfaces in the gas delivery system and the process chamber that would not cause the decomposition or promote any catalytic activity on the specialty gases.

Figures 18 and 19 show the results of SiH_4 thermal decomposition characteristics on various surface modified stainless steel tubes. Figure 18 is for 100 ppm SiH_4 decomposition behaviors on the various surfaces. The vertical axis shows the unreacted SiH_4/Ar concentration measured at the reactor exit and the horizontal axis shows both the time after 100 ppm SiH_4 was flowed through the reactor and the temperature. Figure 19 shows the corresponding H_2 concentration of the SiH_4 decomposition product. These results indicated that SiH_4 began to decompose at about 360 °C for both Fe_2O_3 and Cr_2O_3 passivated surfaces and completed the decomposition at about 470 °C. In contrast, SiH_4 began to decompose at lower temperature of about 310 °C for electropolished surfaces, hesitated at around 360 °C

Table–2 The activation energy dependence on dilution
gas and dilution rate.

(eV)

		Diluted SiH$_4$ Concentration			
		0.01%	0.1%	1%	10%
Diluted Gas	Ar	2.0	2.0	2.6	2.8
	N$_2$	2.0	2.1	2.5	2.8
	H$_2$	1.6	1.4	1.7	2.3

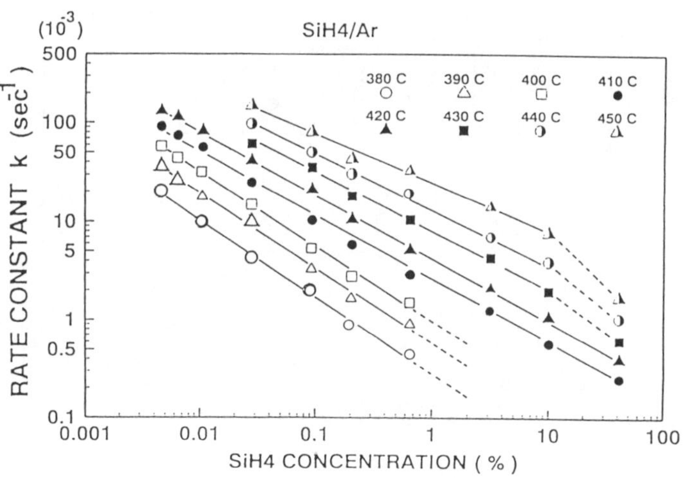

Fig–15 The relationship between diluted SiH$_4$ with Ar
and rate constant.

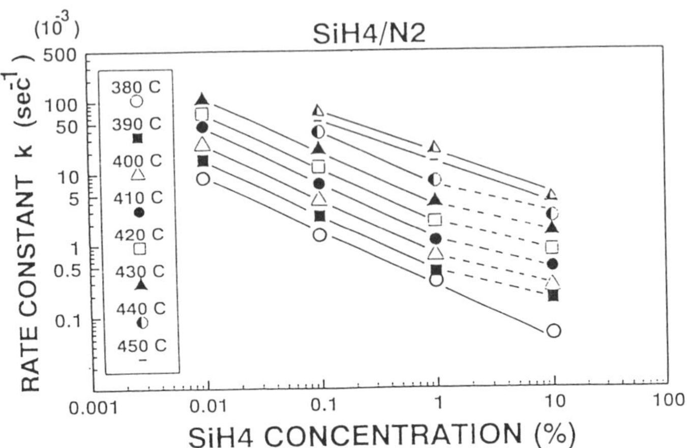

Fig-16 The relationship between diluted SiH_4 with N_2 and rate constant.

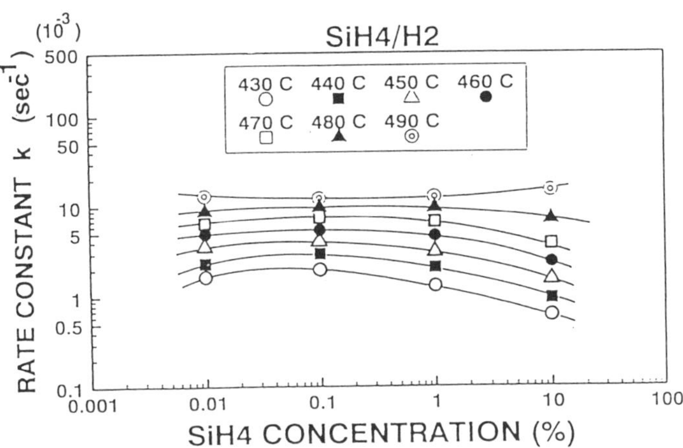

Fig-17 The relationship between diluted SiH_4 with H_2 and rate constant.

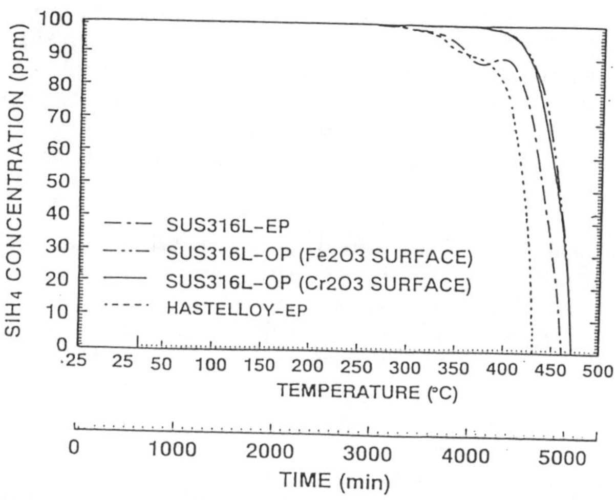

Fig-18 Thermal decomposition of 100% SiH_4 in Ar on various surface (SiH_4).

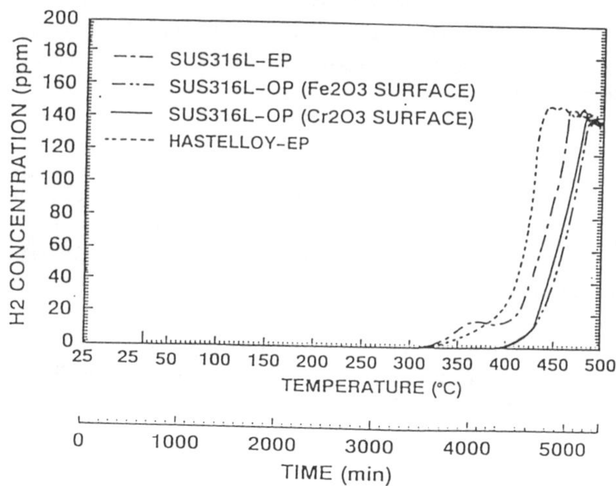

Fig-19 Thermal decomposition of 100% SiH_4 in Ar on various surface (H_2).

then accelerated after 400 °C, and completed the decomposition at about 460 °C in a stepwise manner.   On the electropolished Hastelloy tube surface, SiH_4 began to decompose at about 310 °C and complete the decomposition at a lowest temperature of about 430 °C in comparison with other surfaces. From these results, we understood that both electropolished SUS316L and Hastelloy surfaces have higher catalytic activity for SiH_4 decomposition and lower decomposition attemperatures than Fe_2O_3 and Cr_2O_3 passivated surfaces.

Next, we show the results of SiH_4 decomposition behaviors in Figures 20 and 21 on baked all-stainless filters which have relative large effective areas. Figure 20 shows the change of 500 ppm SiH_4 /Ar concentration as a function of contact time. The vertical axis is the measured SiH_4 concentration measured downstream of the filter element and the horizontal axis the total contact time.   Figure 21 shows the corresponding H_2 concentration from the decomposition reaction. For filter A with no Cr_2O_3 passivation, SiH_4 completely decomposed for 50 minutes with no SiH_4 detected downstream of the filter. In other words, during this 50 minute period, all SiH_4 fed to the filter was completely consumed by reacting with the filter. After this period, SiH_4 concentration gradually increased but did not quite return to the 500 ppm inlet level for 5 hours. On the other hand, for the Cr_2O_3 passivated filter, the period for no SiH_4 detection was shorter, about 30 minutes, and the amount of H_2 detected was also smaller.   For filter B, with and without Cr_2O_3 passivation, the time of no SiH_4 detection was only 10 minutes. In addition, we observed that Cr_2O_3 passivated filter B had only a very slight effect on retarding the SiH_4 decomposition, From these results, it became evident that baking alone can activate the stainless steel surface even at room temperature and the Cr_2O_3 passivation can suppress this catalytic effects.   Table 3 shows the order of catalytic activity for various oxides. Stainless steel is made of oxides Fe, Cr, Ni and Mn. Cr_2O_3 is the lowest among these oxide in terms of it catalytic activity. The present evaluation results appeared to be consistent with this information. We also concluded that Cr_2O_3 passivation appears to be the best approach for the metal surface treatment. In comparing filter A with B of the present evaluation, Filter B, with and without Cr_2O_3 passivation, appeared to better than filter A. The reason for this may be that their effective areas are different but more importantly the filter B as received may already been Cr_2O_3-passivated while filter A was not. Nevertheless, if that is the case, then the Cr_2O_3 passivation on the stainless filter B was still not quite complete. The technique for a more complete Cr_2O_3 passivation needs to be further advanced.

4. SUMMARY

We have evaluated in detail the thermal decomposition behaviors of SiH_4, which is most often used specialty gas for thin film formation in the LSI processes,

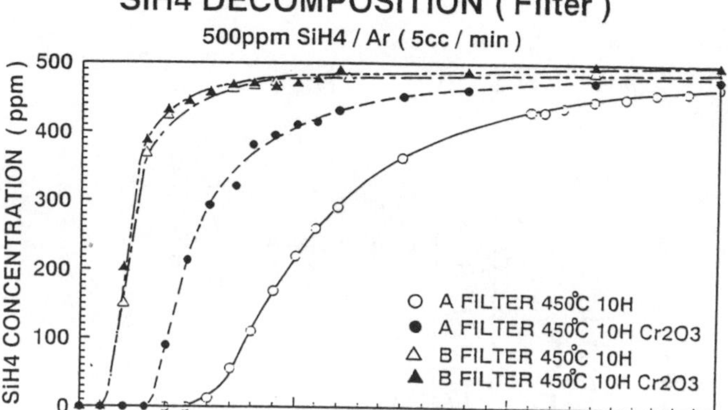

Fig-20   Thermal decomposition of 500ppm SiH_4 in Ar on various all-metal filter surfaces (SiH_4).

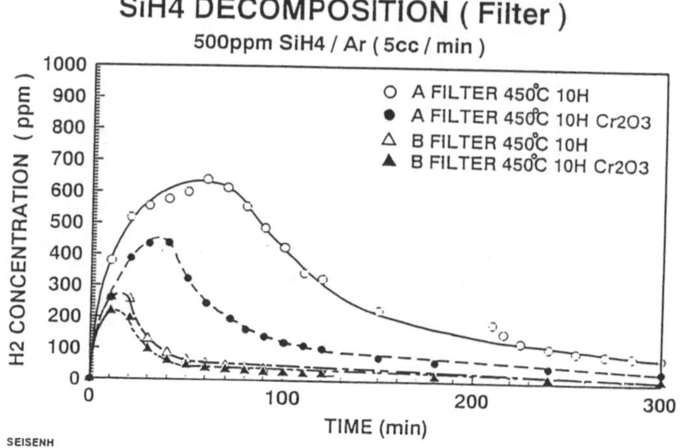

Fig-21   Thermal decomposition of 500ppm SiH_4 in Ar on various all-metal filter surfaces (H_2).

Table-3 Order of metal oxide catalytic activities.

Order of Metal Oxide Catalytic Activities
(Oxidation-related Reaction)

Reaction	Order of Catalytic Activities
O2 Isotope Exchange	$Co_3O_4 > MnO_2 > NiO > CuO > Fe_2O_3 > ZnO > Cr_2O_3 > V_2O_5 > TiO_2$ E 16 22 24 26 33 40 42 46 (39)
H2 Oxidation	$Co_3O_4 > CuO > MnO_2 > NiO > Fe_2O_3 > ZnO > Cr_2O_3 > V_2O_5 > TiO_2$ E 11 13 14 14 15 24 18 18
CO Oxidation	$MnO_2 > Co_3O_4 > NiO > CuO > ZnO > TiO_2 > Fe_2O_3 > V_2O_5 \sim Cr_2O_3$ Stone $CoO > NiO > MnO_2 > CuO > Fe_2O_3 > ZnO > TiO_2 > Cr_2O_3 > V_2O_5$

E:Activites Energy (Kcal/mol)

with respect to different diluent gases and different dilution rates. The results clearly indicated that even for the same gas, the starting decomposing temperatures, the decomposition behaviors and the activation energies are radically different depending upon diluent gases and dilution rates used. We observed that in the cases of Ar and N_2 diluent gases, the decomposition rate constants decreased with increasing SiH_4 concentration and that H_2 had a strong tendency to suppress the decomposition rates. Next, we observed the different catalytic effects on SiH_4 decomposition by various stainless surfaces. Electropolished SUS316L and Hastelloy surfaces appeared to decompose SiH_4 at lower temperatures than Fe_2O_3 and Cr_2O_3 passivated surfaces. We also demonstrated that baking would activate the surface and catalyze the SiH_4 decomposition even at room temperature. However, Cr_2O_3 passivation could minimize this catalytic activation effect. In typical volume production of ULSI, all related parameters must be controlled at desired levels regardless how and when these parameter would change. In the CVD process, it goes without saying that the diluent gas and dilution rate must be controlled perfectly. To ensure the reproducibility, the inside surface of the so-called Poly process chamber was traditionally allowed to be deposited of the reaction products. However, this technique can no longer be tolerated because the dust particles induced by these deposited matters is the most important reason for the wafer contamination. To suppress the surface deposition that causes the dust particle contamination problem and to maintain a constant condition on the inside stainless chamber surface in contact with SiH_4, we recommand the use of 100% Cr_2O_3 passivated stainless surface. Meanwhile, it is also essential to introduce the In-situ Cleaning technique [12] in order to restore the surface condition each time after the process chamber is used.

5. REFERENCES

1. Ohmi, Tadahiro, "Proposal for Advanced Semiconductor Manufacturing Equipment - An Approach to Automated IC Manufacturing," V. Akins, ed., Automated Integrated Circuits Manufacturing, PV 90-3, pp 3-18, The Electrochemical Society, Pennington, NJ, 1990.

2. Ohmi, Tadahiro, and Tadashi Shibata, "Closed Manufacturing System for Advanced Semiconductor Manufacturing," Vaughn E. Akins and Hiroyuki Harada, eds., Automated Integrated Circuits Manufacturing. PV91-5, pp. 3-63, The Electrochemical Society, Pennington, NJ, 1991.

3. Hogness, T. R., T. L. Wilson and W. C. Johnson, "The Thermal Decomposition of Silane," J. of Am. Chem. Soc., **58**, pp. 108-12, 1936.

4. Ohmi, Tadahiro, Masakazu Nakamura, Atsushi Ohki, Koji Kawada, Keiji Hirao and Tsuyoshi Watanabe, "SiH_4 and Si_2H_6 Thermal Decomposition Characteristics and The Influence of Residual Moisture and Oxygen," The 15th Workshop on ULSI Ultra Clean Technology, Physics and Chemistry of Specility Gases for Advanced Semiconductor Processings, Ultra Clean Society, pp 125-156. Tokyo, November 29-30, 1991.

5. Ishikawa, Kohichi, Hiroshi Mihira and Tadahiro Ohmi, "The High Sensitivity Gas Analyzing System," The 15th Workshop on ULSI Ultra Clean Technology, Physics and Chemistry of Specility Gases for Advanced Semiconductor Processings, Ultra Clean Society, pp 113-122, Tokyo, November 29-30, 1991.

6. Nakamura, Masakazu, Tadahiro Ohmi, Kazuhiko Sugiyama, Yasumitsu Mizuguchi, Koji Kawada and Atunobu Ohkura, "Total Ultra Clean Gas Delivery System", ULSI Ultra Clean Technology Workshop No. 7, All-metal O_2 Passivation Piping Technology- The target for Ultra Clean Gas Systems, Proc., pp 5-33 October, 1990.

7. Ohmi, Tadahiro, Atsushi Ohki, Masakazu Nakamura, Koji Kawada, "Moisture Measurements in HCl, SiH_4 and Si_2H_6 Gases," The 15th Workshop on ULSI Ultra Clean Technology, Physics and Chemistry of Specility Gases for Advanced Semiconductor Processings, Ultra Clean Society, pp 73-83, Tokyo, November 29-30, 1991.

8. Briesacher, Jeffrey L., Masakazu Nakamura and Tadahiro Ohmi, "Gas Purification and Measurement at the ppt Level," J. Electrochem. Soc., **138**, 12, pp 3717-23, December, 1991.

9. Abe, Mituso, Kazuhiko Sugiyama, Tadahiro Ohmi et al, "Palladium Membrane Unit for Hydrogen Purification," in LSI Advanced Manufacturing Technology I - 1. Gas Delivery System, ed. by Ohmi and Nitta, pp 189-213, March, 1989.

10. Ohmi, Tadahiro, Takeshi Okumura, Kazuhiko Sugiyama, Fumio Nakahara and Junichi Murota, "Outgas-free Corrosion-resistant Surface Passivation of Stainless Steel for Advanced ULSI Processing Equipment," Extended Abstracts, 174th Electrochemical Society Fall Meeting, Chicago, Abstract No. 396, pp. 579-580, October 1988.

11. Ohki, Atsushi, Masakazu Nakamura, Koji Kawada, Tsuyoshi Watanabe, Shinji Miyoushi, Shinji Takahashi and Tadahiro Ohmi, "Various Metal Surface treatment Technology," ULSI Ultra Clean Technology Workshop No. 18, Proceedings, pp. 177-201, June 1992.

12. Aoki, Yasuo, Shotaro Aoyama, Hedetoshi Wakamatsu, Jinzo Watanabe and Tadahiro Ohmi, "Formation of High Quality Refractory Metal Thin Films by Low Energy Ion Bombardment," Extended Abstracts, 180th Electrochemical Society Meeting, Phoenix, Abstract No. 427, pp. 625-626, 1991.

THERMAL DECOMPOSITION OF DISILANE
AND TRISILANE IN A TUBE REACTOR

Michael S. K. Chen, Koji Kawada, Shinji Miyoshi, Masakazu Nakamura,
Atsushi Ohki, Shinji Takahashi, Tsuyoshi Watanabe and Tadahiro Ohmi

Department of Electronics, Faculty of Engineering,
Tohoku University, Sendai, Japan 980

Thermal Decomposition Characteristics of ultra
high purity Si_2H_6 and Si_3H_8 gases were studied in a
polysilicon deposited tube reactor over 320 - 430 °C and
inlet concentrations from 0.01% to 10% in argon and
hydrogen diluent gases. The rates of Si_2H_6 and Si_3H_8
decomposition were found higher than SiH_4. Deviation
from the first-order kinetic law was significant at high
inlet concentrations and long residence times. Rates in
H_2 diluent gas are lower than in Ar. Activation energies
at low concentrations are in the range of 2.0-2.2 eV.
Si_3H_8 decomposition rate on SiO_2 was much lower than
polysilicon surface.

Keywords: Thermal decomposition. Silane. Disilane.
Trisilane. Activation Energy. Polysilicon surface. Tube
reactor. Argon. Hydrogen. Dilution. Ultra high purity.
Semiconductor.

1. INTRODUCTION

In the present semiconductor manufacturing for ULSI and TFT in LCD, various
thin-film formation technologies are used. The trend is to reduce the feature dimension
and to improve the device performance. This necessitates to develop techniques to
make higher quality thin-films at lower temperatures. In the Chemical Vapor
Deposition (CVD), which has traditionally been the key process for thin-film
formation, a variety of specialty gases such as SiH_4 are used.. To lower the film
forming temperature, higher silanes such as disilane (Si_2H_6) and trisilane (Si_3H_8) have
been suggested [1]. However, the properties of these very reactive specialty gases are

not well understood. Even in the current CVD process, important operating parameters such as gas concentration, flow rate, diluent gas, film forming time, temperature as well as chamber pressure etc. are by and large empirically determined [2]. To achieve higher film quality at lower temperature, it is essential not only to remove impurities completely from these specialty gases but also to gain a better control over the aforementioned process parameters. For these reasons, we undertook a study of the physical and chemical properties of these specialty gases.

Studies on silane decomposition went as far back as in the 1930s. For example, T. R. Hogness et al studied the extents of silane thermal decomposition by measuring the pressure changes in a static system [3]. Gas chromatography was used to measure the gas composition of the reactor at the end of reaction. Although their studies had revealed many valuable observations, the results are not well suited for the present semiconductor applications because the surface cleanliness in the gas delivery system, and the purity of the specialty gases used were not as high as the present day standards. In recent years, although there were many more studies on the reaction mechanisms of these specialty gases, they were mainly related to the prediction of film growth rates on silicon wafers under a variety of operating conditions. We therefore felt that it was necessary to conduct new experiments in an ultra clean environment to gain new fundamental understanding of the intricate interactions of these specialty gases with the surface of the gas delivery system from the source to the point of use as well as with the silicon and chamber surfaces.

We had previously reported the characteristics of SiH_4 thermal decomposition in a tube reactor by varying dilution rates, diluent gases, as well as on different surfaces [4]. These experiments were conducted in a completely closed system made of all-metal parts, free from the influence of the moisture-laden process chamber environment except for very minor effects of outgassing from the internal metal surface and external leaks. In this paper, we would like to report resutls of a similar study on disilane and trisilane in an attempt to elucidate their thermal decomposition characteristics and to compare them to monosilane.

2. EXPERIMENT

Figure 1. shows a schematic flow diagram of the experiment. The system is composed of a precision gas dilution system, a thermal decomposition tube reactor and a high sensitivity gas chromatography (GC) [5]. Ultra high purity Si_2H_6 or Si_3H_8 gas at a chosen dilution rate near atmospheric pressure was fed to the reactor and the extent of thermal decomposition was measured by GC.

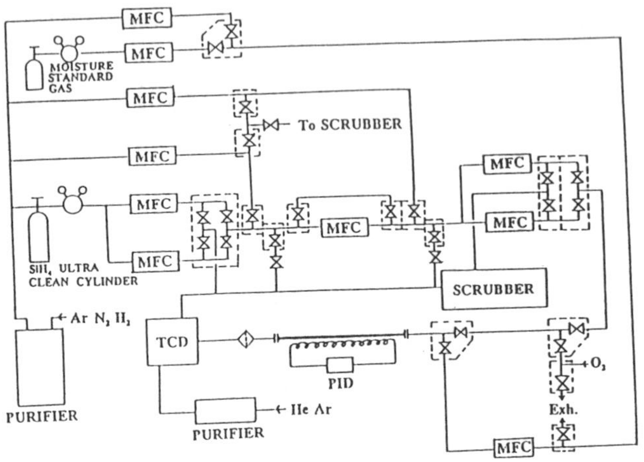

Fig-1   Schematic flow diagram of thermal
decomposition evaluation system.

table-1   The first-order rate equation.

$$-\frac{d[Si_mH_n]}{dt} = k \cdot [Si_mH_n]$$

$$t_0 \Rightarrow [Si_mH_n]_0 \quad \cdots \text{ Initial}$$

$$t \Rightarrow [Si_mH_n]_t$$

$$[Si_mH_n]_t = [Si_mH_n]_0 - x$$

$$x \cdots \text{Decomposed } Si_mH_n \text{ Concentration}$$

$$-\int_{[Si_mH_n]_0}^{[Si_mH_n]_t} \frac{1}{[Si_mH_n]} d[Si_mH_n] = \int_0^t k \cdot dt$$

$$\text{Decomposition Rate} = \frac{[Si_mH_n]_t - [x]}{[Si_mH_n]_0} = e^{-kt}$$

$$k = A\exp\left(-\frac{Ea}{RT}\right) \qquad k = \text{Rate Constant}$$

In the gas dilution line, all of plastic materials that tend to trap the moisture were replaced by all-metal, dead space-free, highly integrated parts [6]. All parts used in the system are made of SUS316L and their internal surfaces were electropolished and mirror-finished. All-metal C rings were used in the fittings with an external leak rate less than 2×10^{-11} Torr-l/sec. This ultra clean system made it possible to study the true thermal decomposition characteristics of specialty gases free from the effects of gas impurities, the outgassing from the metal surface and the external leaks. The mass flow controller (MFC) used in the precision dilution system can perform a precise dilution and flow control of the gas mixture fed to the reactor to achieve the desired residence time through the tube reactor.

The reactor itself was made of 1/2" OD x 40 cm long electropolished SUS316L tube with an 1.24 mm wall thickness and an effective reactor volume 32.8 cc. The reactor was wrapped with alumina foils, sheath heaters and insulation. Temperature was measured with a thermal couple inserted in center of reactor and controlled to within $2°C$ by using several PID controllers along the reactor tube length.

Downstream of the reactor, a high sensitivity GC equipped with TCD (Thermal Conductivity Detector) was used to measure the specialty gas concentration. This high sensitivity GC was made of all-metal parts; all internal gas contact area was electropolished and mirrored finished to avoid trapping any residual impurities. 1/16" tubings were used to minimize the sample gas contact area and the outgassing effect. Ultra high purity electronic-grade disilane and trisilane gases with a moisture of 6 ppb or less were used [7]. For Si_3H_8, 2% Si_3H_8/Ar gas was used.

Only the reactant disilane and trisilane concentrations were measured in these experiments. The reactor gas pressure was slightly about one bar, just enough to overcome the pressure drop through the system. For the most part of experiments, the polysilicon surface was used, which was prepared a priori by decomposing pure silane at 450 °C. Some experiments was performed on silicon oxide surface (SiO_2), which was prepared by oxidizing the polysilicon surface with 100% O_2 at 600 °C for 10 hours. The residence time calculations were based on the reaction volume and the gas flow rate at the reactor temperatures. The volume expansion due to hydrogen production was properly taken into account. The data were interpreted with a first-order kinetic rate equation shown in Table 1. The activation energy (Ea) was then determined from the slope of the Arrhenius plots of these first-order rate constants (k).

3. RESULT AND DISCUSSION

3.1 Diluent gas and dilution rate effects on Si_2H_6 decomposition

Figure 2 (a), (b), (c) and (d) show the Si_2H_6 concentration profiles on the semilog scale as a function of the residence time in Ar on the poly-Si surface for 0.01%, 0.1%, 1% and 10% respectively. Appreciable decomposition commenced at 330-340°C, which was lower than 370-380 °C of SiH_4, previously studied. At low inlet concentrations, 0.01%-0.1%, the rates appeared to follow the first-order kinetics reasonably well but at high concentrations, 1%-10%, the deviation from it was quite significant, especially at high temperature and long residence time regions. Qualitatively, this deviation may be explained by a surface reaction model of the reactant adsorption, surface reaction and product desorption, in additon to gas phase reaction. This surface reaction is judged to be significant especially for higher inlet concentration where the surface is saturated with the adsorbed disilane but the overall decomposition rate is limited by the surface reaction rate of some intermediate species such as SiH_3, SiH_2 and SiH, final product H_2 and possibly other polysilanes. High molecular weight polysilanes may act as inhibitors at long residence time and low disilane concentration.

Figure 3 (a) through (d) displace the similar resutls for 0.01%, 0.1%, 1% and 10% Si_2H_6 in the H_2 diluent gas. The starting temperature was at 350 °C, a litter higher than the argon case. Unlike the inert argon gas, hydrogen can adsorb to the polysilicon surface and compete with disilane for sites. The general behavior of Si_2H_6 decomposition in H_2 was similar to Si_2H_6 in Ar but the rates were lower at the same temperature and concentration. At high concentrations, the deviation from the first order kinetics was also significant.

The first order rate constants were deduced from the above data by the least-square method. Figures 4 shows the Arrhenius plot of these rate constants for Si_2H_6/Ar, from which the activation energies for each inlet concentration were calculated. Figure 5 shows a similar plot for Si_2H_6/H_2. At 0.01% and 0.1% levels, the activation energies for Ar and H_2 diluent gases are about the same, at 2 eV (within 10%) but the rate for Ar were about 2-3 times higher the H_2 case. At higher disilane concentrations, the differences between the two diluent cases began to narrow because disilane can compete with H_2 for adsorption sites more effectively and the effect of hydrogen was reduced.

3.2 Diluent gas and dilution rate effects on Si_3H_8 decomposition

Figure 6, (a), (b) and (c) show the concentration of Si_3H_8 in Ar as a function of residence time for 0.01%, 0.1% and 1% inlet concentrations respectively. The starting decomposition temperature was at 310 °C, about 20 degrees lower than disilane. While the first order kinetics was reasonable at low concentration, significant deviation was observed at 1% level.

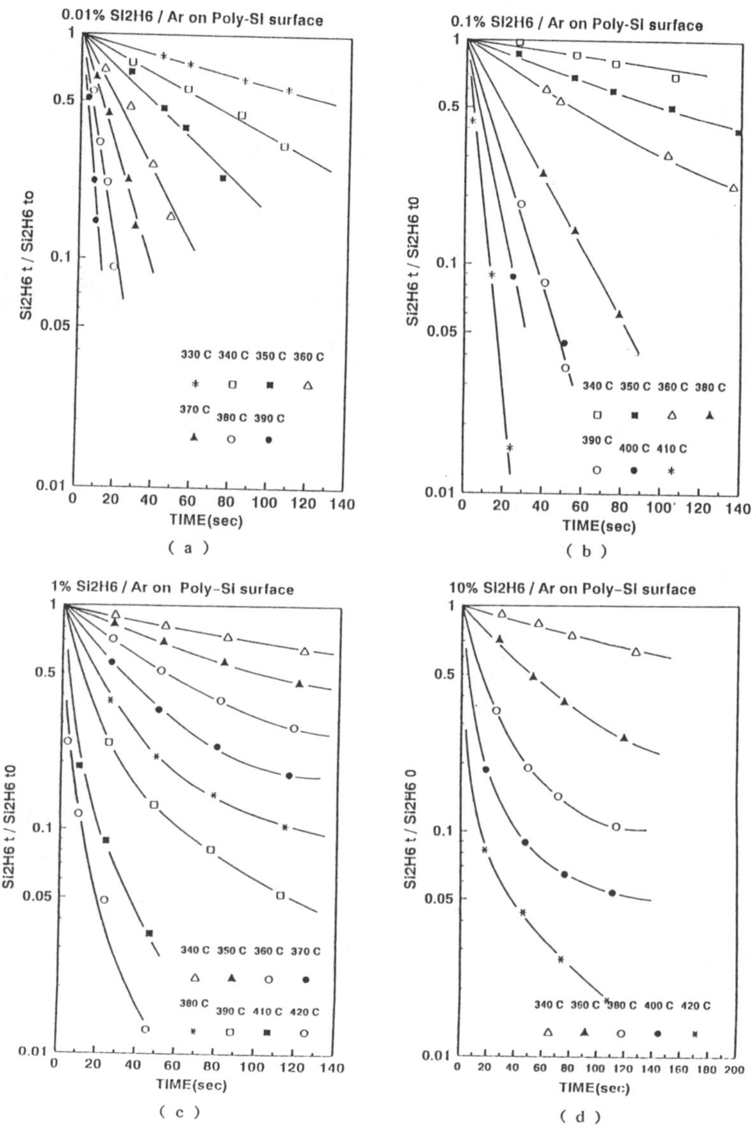

Fig-2 Thermal decompoition characteristics of
Si_2H_6 with Ar dilution.

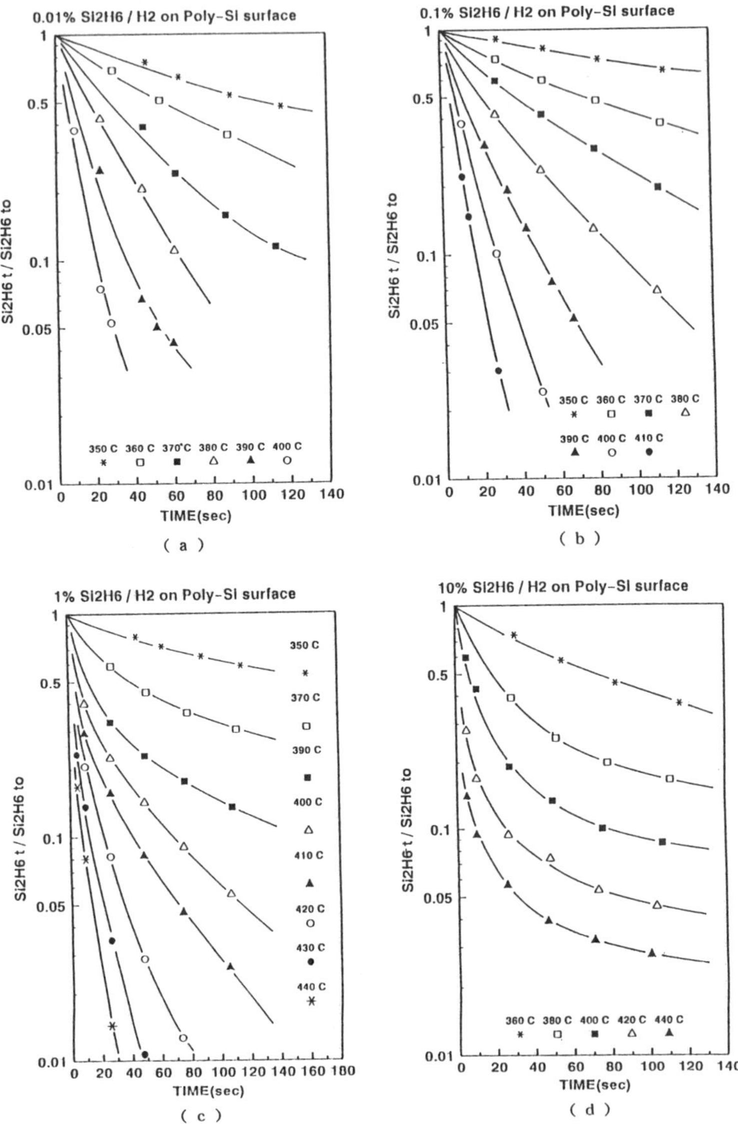

Fig-3 Thermal decompoition characteristics of Si_2H_6 with H_2 dilution.

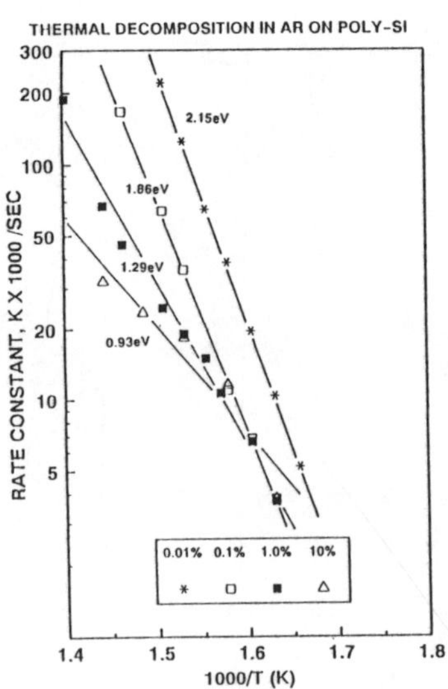

Fig-4   Arrhenius plots of Si_2H_6 thermal decompoition rate constants with Ar dilution.

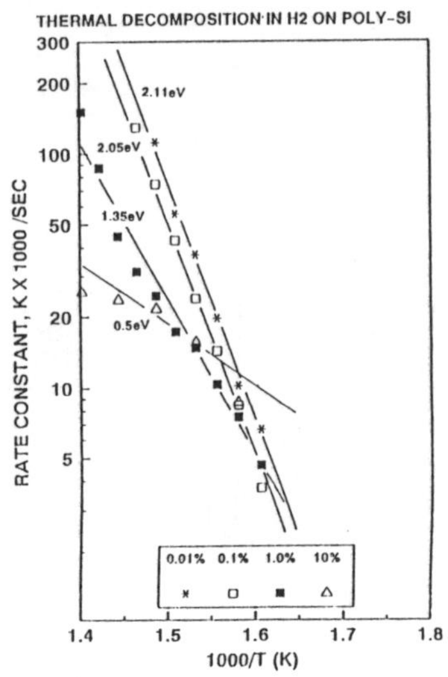

Fig-5   Arrhenius plots of Si_2H_6 thermal decompoition rate constants with H_2 dilution.

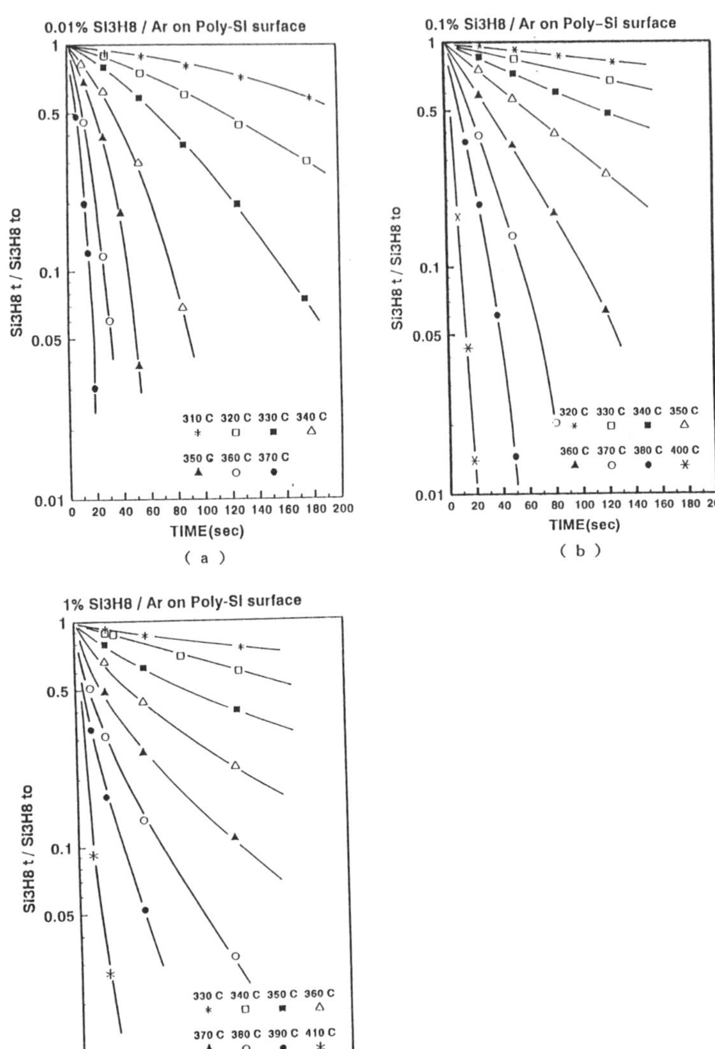

Fig-6   Thermal decompoition characteristics of
Si_3H_8 with Ar dilution.

486

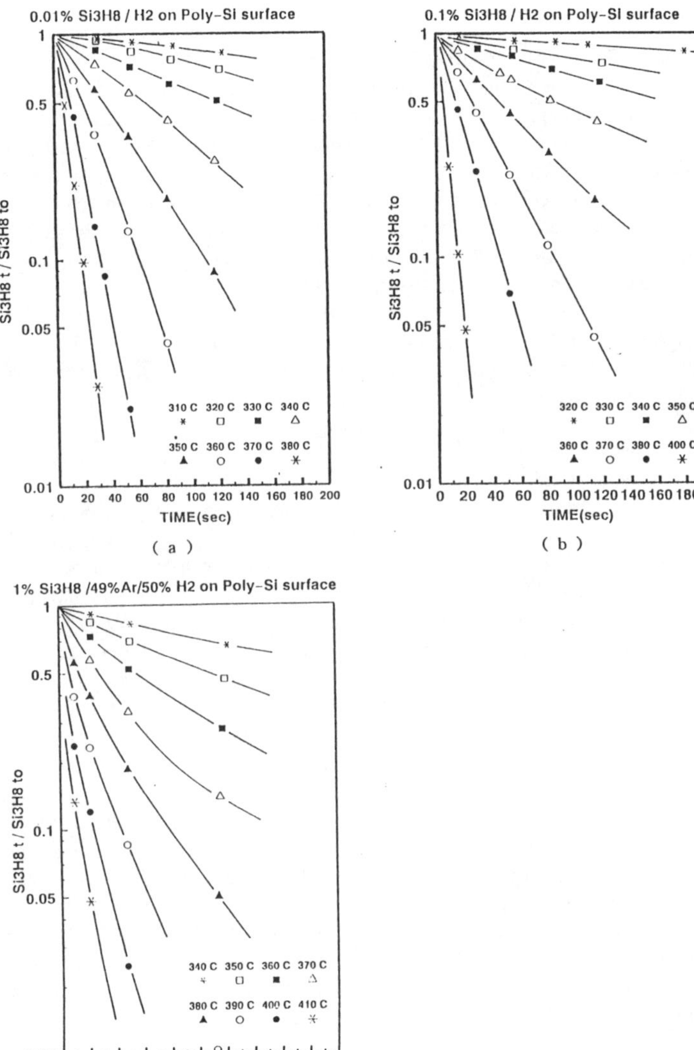

Fig-7   Thermal decompoition characteristics of
Si₃H₈ with H₂ dilution.

487

Figure 7 (a), (b) and (c) show the similar data of Si_3H_8 decomposition in H_2 diluent gas. Note that in the case of 1% Si_3H_8, the diluent gas was made up of 49% Ar and 50% H_2 as explained in the Experimental section. We observed from the data of 0.01% trisilane concentration that in general, for the same degree of decomposition, it was about 20 °C higher than the argon diluent gas case. At higher concentrations, however, the differences between the two diluent gas cases began to narrow. Also, similar to the disilane cases, at higher concentrations, the rates decreases and deviates from the first order kinetic law, due probably to the competition from hydrogen and other intermediates competing for adsorption sites.

Arrhenius plots of the rate constants are shown in Figures 8 and 9 for Ar and H2 diluent gases respectively. Activation energies were calculated from these Arrhenius plots. We observed that the differences between these cases were smaller than the disilane case. The activation energies for 0.01% and 0.1% are essentially identical. This may be explained by the fact that Si_3H_8 is a heavier compound than Si_2H_6; it is easier to adsorb onto the surface and compete with H_2 more effectively for adsorption sites. As the result, the effect of H_2 adsorption competition is much less.

Disilane and trisilane results of this study are compared with previously published silane results [4] in Figure 10 for argon as the diluent gas and in Figure 11 for hydrogen as the diluent gas. The corresponding activation energies are summarized in Table 2 (a), (b) and (c). The first-order rate constants as a function of temperature and the activation energies for 0.01% initial concentrations only are further contrasted in Figure 12 and Figure 13. From these comparisons, we noted that the activation energies of the argon diluent cases are about the same for all three silanes falls in the range of 2.0 to 2.15 eV. This interesting results would suggest a similar rate limiting step on the surface for all silanes. We also noted that at the same temperature, the rate of disilane thermal decomposition is about 10 times faster than silane and the rate of trisilane decomposition is about twice that of disilane. This may be explained by the fact that Si_2H_6 and Si_3H_8 more strongly adsorb on the surface than SiH_4 and that the Si-H bond in SiH_4 is more difficult to break than Si-Si bond because bond energy of Si-H in SiH_4 is 3.3 eV while the bond energy of Si-Si in Si_2H_6 or Si_3H_8 is about 2.0 eV. The reduced effect of H_2 diluent gas for Si_2H_6 and Si_3H_8 can be explained by the fact that because Si_2H_6 and Si_3H_8 adsorb more strongly than SiH_4, they can compete with hydrogen more effective for sites than silane.

3.3 The effect of different surfaces on thermal decomposition rates

Figure 14 shows the thermal decomposition characteristics of 0.01% Si_3H_8/Ar on the SiO_2 surface. Comparing this with Figure 2(a) on the poly-Si surface, we observed that at the same temperature, the rate on the SiO_2 surface is substantially lower. Figure 15 compares the two Arrhenius plots and their activation energies. The

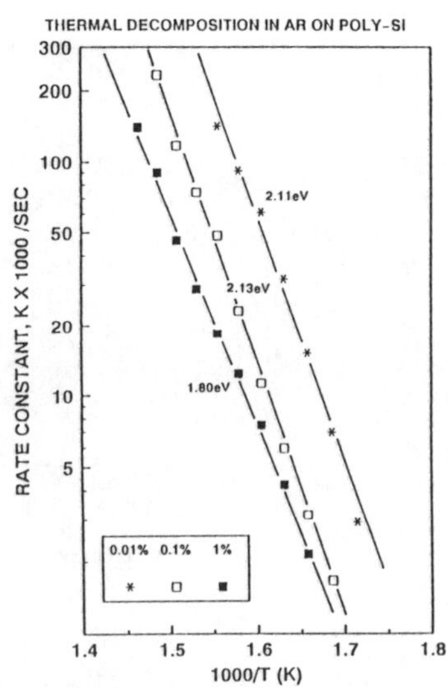

Fig-8   Arrhenius plots of Si₂H₆ thermal decompoition rate constants
with Ar dilution.

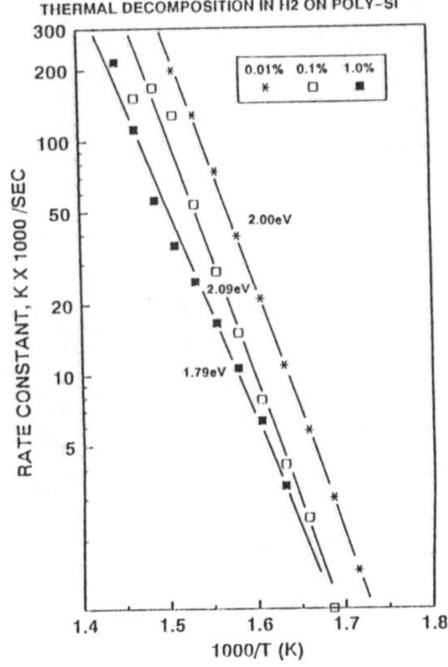

Fig-9   Arrhenius plots of Si₂H₆ thermal decompoition rate constants
with H₂ dilution.

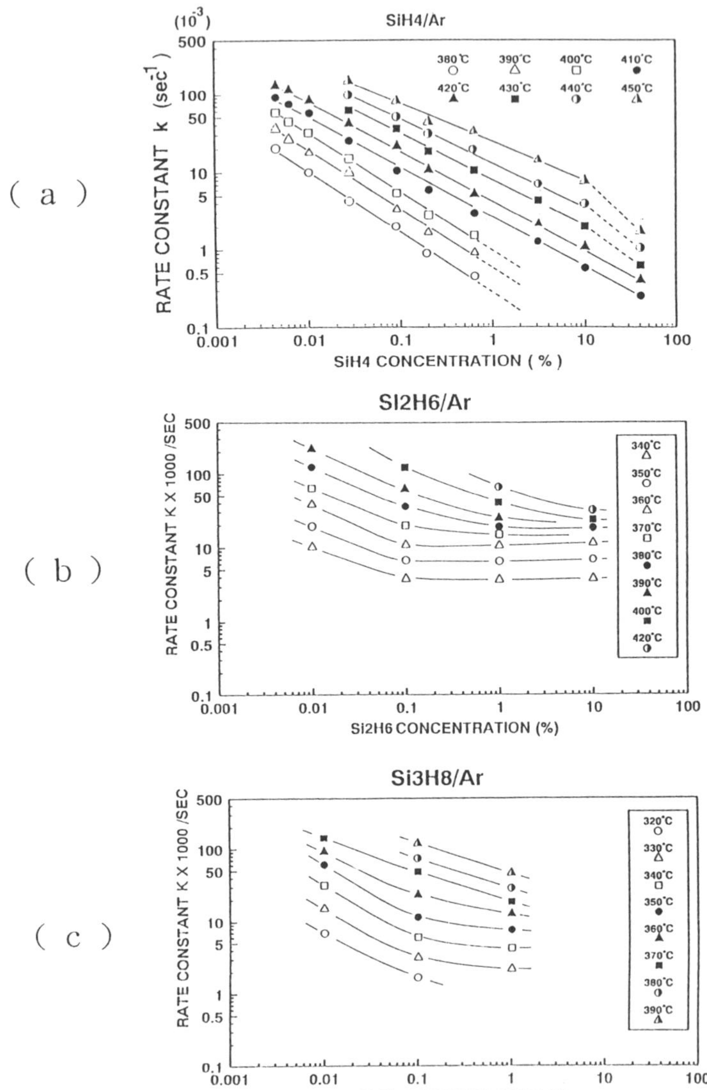

Fig-10   The relationships between the concentration and
rate constant of SiH₄, Si₂H₆, Si₃H₈ with Ar dilution.

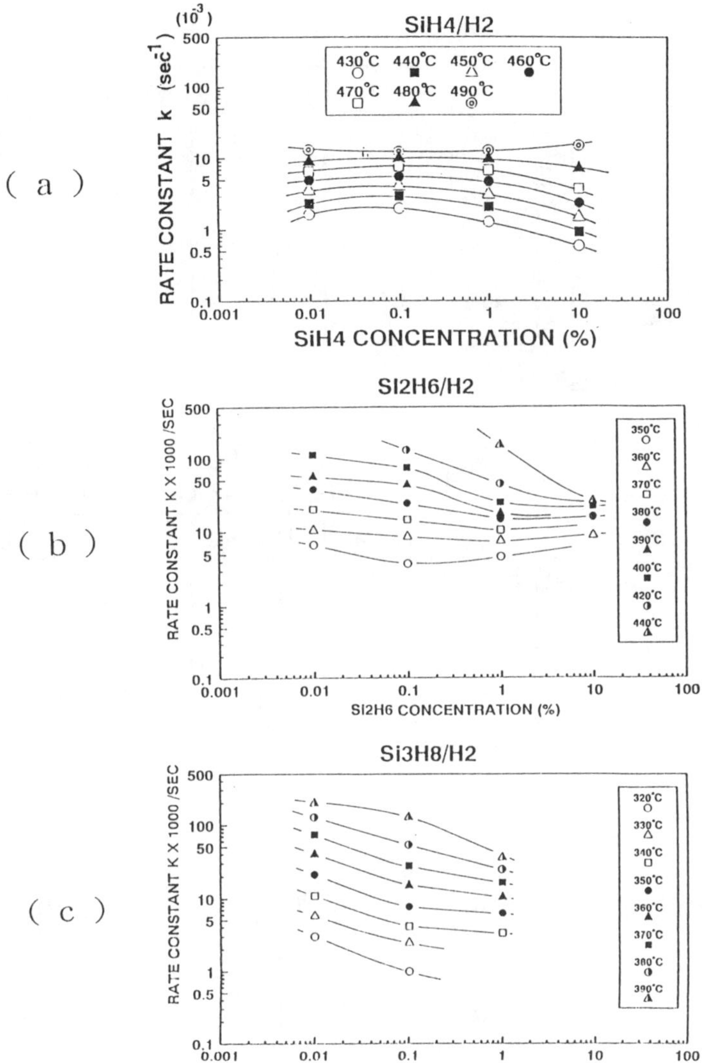

Fig-11   The relationships between the concentration and
rate constant of SiH_4, Si_2H_6, Si_3H_8 with H_2 dilution.

Table-2   Dependence of activation energies of SiH₄, Si₂H₆, Si₃H₈

(eV)

		Diluted SiH$_4$ Concentration			
		0.01%	0.1%	1%	10%
Diluted Gas	Ar	2.0	2.0	2.6	2.8
	H$_2$	1.6	1.4	1.7	2.3

(a)

(eV)

		Diluted Si$_2$H$_6$ Concentration			
		0.01%	0.1%	1%	10%
Diluted Gas	Ar	2.15	1.86	1.29	0.93
	H$_2$	2.11	2.05	1.35	0.5

(b)

(eV)

		Diluted Si$_3$H$_8$ Concentration		
		0.01%	0.1%	1%
Diluted Gas	Ar	2.11	2.13	1.80
	H$_2$	2.00	2.09	1.79 (49% Ar)

(c)

492

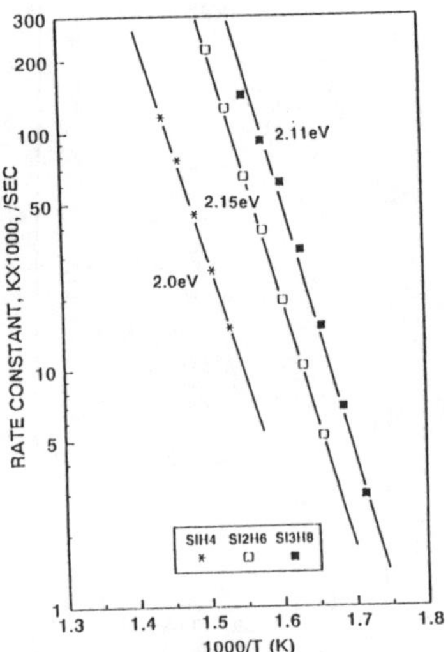

Fig-12   Arrhenius plots of 0.01% SiH_4, Si_2H_6, and Si_3H_8 / Ar thermal decompoition rate constants

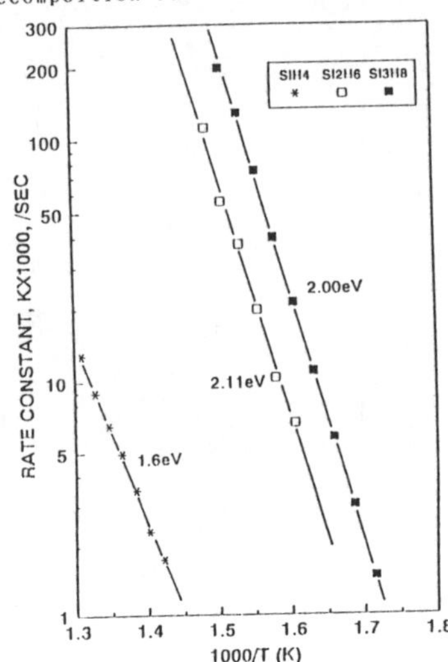

Fig-13   Arrhenius plots of 0.01% SiH_4, Si_2H_6, and Si_3H_8 / H_2 thermal decompoition rate constants

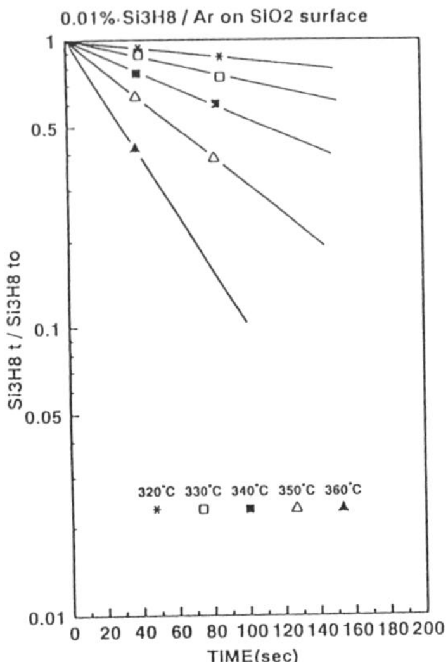

Fig-14   Thermal decompoition characteristics of·
0.01% Si₃H₈ / Ar on SiO₂ surface.

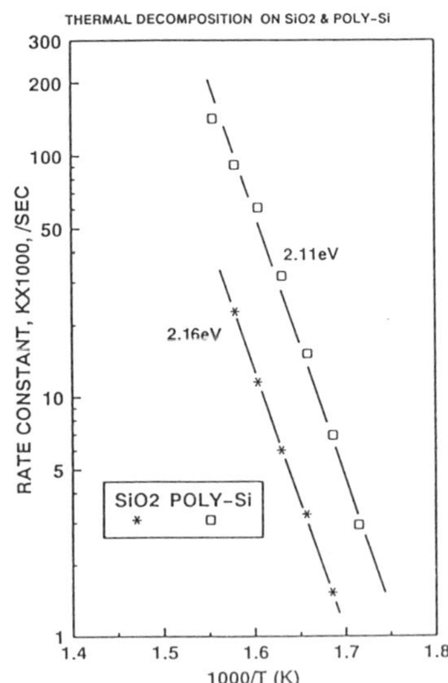

Fig-15   Activation energies of 0.01% Si₃H₈ / Ar thermal decompoition
on poly Si surface and SiO₂ surface.

494

rate on Poly-Si is about 5 times higher than on the SiO_2 but the activation energies are essentially the same, 2.16 and 2.11 eV. The lower rate may be explained by the fact that the oxygen is more electronegative as compared to silicon. The presence of Si-O on the SiO_2 surface to reduce its ability to adsorb trisilane molecules. These experiments were thought to be valid at the low concentrations and short residence time because silicon deposit on the surface change is small. However, at higher concentration and longer residence time, the SiO_2 surface would gradually turn into a poly-Si surface and the decomposition rate would increase accordingly. To restore the surface to the original SiO_2 surface, it is necessary to reoxidize the surface periodically.

4. SUMMARY

We have evaluated the thermal decomposition characteristics of disilane and trisilane systematically on the polysilicon surface in the heated tube reactor in an all-metal, plastic-free closed tube reactor system in order to minimize the effects of trace impurities, outgassing and the external leaks. The results revealed a varying degree of the diluent gas and dilution rate effects. Notably, as the silane reactants vary from SiH_4 to Si_2H_6 and Si_3H_8, the decomposition temperature decreased and rate increased as expected. The lower bond energy in Si-Si than Si-H is thought to be responsible for this. The rates in argon were higher than in hydrogen. This is due to the competition of hydrogen adsorption but the effect of diluent gas H_2 diminishes with increasing silane homologous series, due probably to stronger adsorption of higher silanes on the surface and thus competing more effectively with hydrogen for sites. The deviation from the first-order kinetic law also increases from silane to disilane to trisilane. Qualitatively, this is thought to be due to the complex interplay between the reactant adsorption, the surface reaction of intermediates and the product desorption in addition to the gas phase reaction. Some intermediates may act as inhibitors and hence to reduce the rate when the reactant concentration reached a low level at long residence time.

The decomposition rates of Si_3H_8 on SiO_2 at 0.01% was briefly studied. Its rate was shown to be lower than that on the Poly-Si surface. A implication of this result is that in the actual CVD process chamber, the presence of native oxide on the silicon wafer or the trace O_2 impurity would reduce the film growth rate and made the precision control of film thickness very difficult. Therefore, we must ensure an oxide-free process environment for precise film deposition.

5. REFERENCES

1. Breddels, P. A. et al., Japan J. of Appl. Phys. **30**, 233, 1991.

2. Ohmi, Tadahiro ,"Proposal for Advanced Semiconductor Manufacturing Equipment - An Approach to Automated IC Manufacturing," V. Akins, ed., Automated Integrated Circuits Manufacturing, PV 90-3, pp. 3-18, The Electrochemical Society, Pennington, NJ, 1990.

3. Hogness, T. R., T. L. Wilson and W. C. Johnson, "The Thermal Decomposition of Silane," J. of Am. Chem. Soc., **58**, pp.108-12, 1936.

4. Ohmi, Tadahiro, Masakazu Nakamura, Atsushi Ohki, Koji Kawada, Keiji Hirao and Tsuyoshi Watanabe, "SiH_4 and Si_2H_6 Thermal Decomposition Characteristics and The Influence of Residual Moisture and Oxygen," The 15th Workshop on ULSI Ultra Clean Technology, Physics and Chemistry of Specility Gases for Advanced Semiconductor Processings, Ultra Clean Society, pp. 125-156. Tokyo, November 29-30, 1991.

5. Ishikawa, Kohichi, Hiroshi Mihira and Tadahiro Ohmi, "The High Sensitivity Gas Analyzing System," The 15th Workshop on ULSI Ultra Clean Technology, Physics and Chemistry of Specility Gases for Advanced Semiconductor Processings, Ultra Clean Society, pp. 113-122, Tokyo, November 29-30, 1991.

6. Nakamura, Masakazu, Tadahiro Ohmi, Kazuhiko Sugiyama, Yasumitsu Mizuguchi, Koji Kawada and Atunobu Ohkura, "Total Ultra Clean Gas Delivery System", ULSI Ultra Clean Technology Workshop No. 7, All-metal O_2 Passivation Piping Technology- The target for Ultra Clean Gas Systems, Proc., pp. 5-33 October, 1990.

7. Ohmi, Tadahiro, Atsushi Ohki, Masakazu Nakamura, Koji Kawada, "Moisture Measurements in HCl, SiH_4 and Si_2H_6 Gases," The 15th Workshop on ULSI Ultra Clean Technology, Physics and Chemistry of Specility Gases for Advanced Semiconductor Processings, Ultra Clean Society, pp. 73-83, Tokyo, November 29-30, 1991.

8. Briesacher, Jeffrey L., Masakazu Nakamura and Tadahiro Ohmi, "Gas Purification and Measurement at the ppt Level," J. Electrochem. Soc., **138**, 12, pp. 3717-23, December, 1991.

9. Abe, Mituso, Kazuhiko Sugiyama, Tadahiro Ohmi et al, "Palladium Membrane Unit for Hydrogen Purification," in LSI Advanced Manufacturing Technology I - 1. Gas Delivery System, ed. by Ohmi and Nitta, pp. 189-213, March, 1989.

KEYWORD INDEX